Reclamation of contaminated land

Proceedings of a Society of Chemical Industry conference held at the Congress Theatre, Eastbourne, England, 22-25 October 1979

LONDON: SOCIETY OF CHEMICAL INDUSTRY 1980

First published in 1980 by
The Society of Chemical Industry,
14 Belgrave Square, London SW1X 8PS, England,

ISBN 090100 168 6

Printed in England by Premier-Clay Litho Limited, London

First published October 1980

The Technical Programme Committee was composed of
representatives from:-
British Industrial Biological Research Association
British Gas Corporation
The Chemical Society
    Analytical Division
Construction Industry Research and Information Association
Department of the Environment
    Central Unit on Environmental Pollution
Department of Health and Social Security
Health and Safety Executive
Institution of Chemical Engineers
Institution of Civil Engineers
Institution of Corrosion Science and Technology
Institution of Gas Engineers
Institution of Public Health Engineers
Institution of Structural Engineers
Institution of Water Engineers and Scientists
Ministry of Agriculture, Fisheries and Food
Safety in Mines Research Establishment
Society of Chemical Industry
    Agriculture Group
    Industrial Health and Safety Group
    Materials Preservation Group
    Road and Buildings Materials Group
    Water and Environment Group
Water Research Centre

This book has been reproduced directly from manuscripts supplied by the authors, who are solely responsible for the accuracy of their contents.

Contents

CORRIGENDA

Paper C3

Page 5, penultimate sentence should read:-

This material was throughly homogenised by a mechanical excavator before spreading to a depth of 500mm.

Paper C6

Page 2, penultimate line of top paragraph - "aversely" should read "adversely".

Page 2, last line of paragraph headed "Zinc" - insert "soil" before "extractable".

Page 4, 4th line "kg/mg" should read "mg/kg".

Page 8, in the first paragraph of "SUMMARY" in 8th line, delete the "and" after "pH" and substitute with a comma.

Page 8, 2nd line of last paragraph of "SUMMARY", "in" should read "is".

Page 8, end of 5th line of last paragraph of "SUMMARY" insert "molybdenum" after "mecury".

Page 8, penultimate line of last paragraph of "SUMMARY" delete "upon".

Page 11, in Table 8, the units below "Cd in plant dry matter" and below "Extractable Cd in soil" should read "ng/g" not "mg/kg".

In Table 8, last line, "1.75" should read "1.78".

INTRODUCTION

D. H. Sharp, PhD., FRIC., FIChemE., FRSH (General Secretary, Society of Chemical Industry).

In holding a major conference concerned with the reclamation of contaminated land, it was recognised that new ground was being broken. The Society of Chemical Industry was originally asked to sponsor and organise such a conference by the Department of the Environment as a follow up to the situation that had been revealed by its own committee concerned with this subject. (The Interdepartmental Committee on the Redevelopment of Contaminated Land set up in 1976).

It was recognised that the problems were many sided and that many different disciplines were involved. To prepare for the conference therefore meant assembling quite a large committee drawn from representatives of the disciplines that had been identified. As Chairman of this Committee and of the Conference that resulted, I must thank all who contributed to the undoubted success of the Conference.

To say that the Conference was a success, however, does not imply that it was a complete success. Reviewing the situation afterwards, the Committee found a number of matters for comment.

The first major matter for comment is that there is no 'hard' information on the real size of the problem that is faced by this country. Since the area of usable land is a country's prime natural resource, and since frighteningly large amounts of valuable land are utilised each year in the construction of roads, airports, and the like, the recovery of disused land would seem to need attack on a national scale. Many Local Authorities in the older industrial areas have a very praiseworthy record of recovery of disused land but much more needs to be done. The Committee feels that funding should be by no means a local matter since land is a national resource.

The Conference was however concerned with contaminated land, not simply disused or derelict land. And here, 'hard' information is still more difficult to come by. Sections of the Conference were indeed devoted to this particular problem - ascertaining historical usage of land from local records and the like but ultimately, in doubtful cases, the matter has to be resolved by sampling and analysis.

Methods of sampling and analysis received reasonable attention at the Conference but no real consensus emerged. This does seem to be an area on which further work is needed, possibly by the British Standards Institution to arrive at acceptable sampling and analysis methods.

Indeed, the actual reclamation of land once identified as contaminated can be reasonably straight forward. Again, research is needed to devise methods to deal with certain particularly persistent and difficult contaminants but, in a surprising number of cases, treatment by such readily available substances as lime proved most effective. But treatment is not cheap. Usually land recovery involves substantial use of labour and heavy earth moving machinery. Transport of soil from one site to another is inevitably very expensive. The Conference was at some pains to elicit estimates of cost from those presenting papers and some of the results are given in my own summary of the Conference prepared at the time. Costs of £40,000 per acre seemed to be about average. At a time when farmland is worth say £2,000 per acre and prime building land say £100,000 per acre, these costs seem high. This adds force to the necessity of producing a national land register of contaminated land, together with an indication of the nature of contamination and the estimated cost to rehabilitate, to set against the ultimate value of the land so brought back into use. One impression from a number of speakers at the Conference was that comparing costs of rehabilitation with the value of the land recovered, in market terms, was not a sufficient guide to the value of removing industrial blight and thereby improving 'the quality of life'. Nevertheless some assessment has to be made of the value of recovering the land and removal of blight. Again it is difficult to see how this can always be done on a purely local scale. Despite this a number of speakers, notably from South Wales, spoke of the community effort that is often given to worthwhile reclamation schemes and the use that can be made of voluntary enthusiasm and labour.

The Conference devoted a good deal of attention to industrial safety on site. This again is a subject which requires further study. Almost by definition, it may well not always be known just what contaminants will be uncovered once reclamation work begins - deposition of dangerous substances is usually extremely localised (making recommendations on adequate sampling and analysis very difficult). By definition, on site, few amenities are available in the way of showers and changing facilities for protective clothing and the like. There certainly seems to be a need to develop well thought out and constructed mobile ablution facilities.

The Conference therefore - and as a result the Conference Volume - cannot claim to be comprehensive but the papers taken together do give a very good 'state of the art' account of reclamation of contaminated land.

The various professional and learned societies that co-operated in the preparation of the Conference have each undertaken to consider particular aspects of the problem for further study in depth, perhaps leading to another general conference in three or four years time.

CONFERENCE ON RECLAMATION OF CONTAMINATED LAND

OPENING ADDRESS

A W KENNY, CBE

Department of the Environment

It was an honour and a pleasure to be invited by the Society of Chemical Industry to give the Opening Address of this timely and important Conference, whose objective is to bring together the experience of workers who have been faced with the problem of reclaiming and restoring to normal use land which has been contaminated by former use. In that former use the land (which in legal parlance generally includes water in, on or under it) has been rendered potentially dangerous by contact with some substance which has remained in it, the danger arising from that substance. Thus it is different from devastated or derelict land, which is not the concern of this Conference, though such land may be as dangerous, for example, by reason of abandoned mineshafts, and there are some common problems of restoration, for example landscaping and revegetation.

Land may be contaminated by atmospheric fall-out; by flooding or spraying or seepage of liquids; or by deposition of solids either directly as in waste disposal or by contaminated buildings. In order to set in perspective the part of this wide field with which this Conference is concerned it is worth listing some well-known and widely discussed examples in each category:-

a. Fall-Out

   i. Nuclear Explosions.

   ii. Acid Rain.

   iii. Fluoride Emissions.

   iv. Oxide Fume Emissions (Lead, Cadmium, Arsenic).

   v. Accidents eg Seveso.

b. Contamination by Liquids

   i. Salt from Sea Flooding.

   ii. Sewage Lagoons.

   iii. Industrial Waste Lagoons.

   iv. Sewage Sludge Applied to Land.

   v. Pesticide Applications.

   vi. Leakage of Waste from Pits, for example, Plating Wastes.

c. Solid Deposits

i. Mining Spoil.

ii. Industrial Slags.

iii. Landfills of Domestic and Industrial Waste.

iv. Buried Waste.

v. Abandoned Waste.

vi. Accidents.

This Conference will be concerned principally with b.ii., b.iii., b.iv., c.iii., and c.iv., if only because some limit has to be set to the discussion. There are noteworthy problems in some of the other categories, for example, accidental dispersion of plutonium from an aircraft accident, nitrate contamination of ground and surface water, reclamation of flooded land but the cases which are reported in the papers of this Conference bring out the main causes of contamination and illustrate how the reclamation may be tackled.

It may seem strange in retrospect that until quite recently, in fact until the last decade in this country, there was no specific legislation to control the contamination of land. For more than 200 years of our Industrial Revolution, even when the initial philosophy of laissez-faire gave way to Victorian reforming zeal, there was no formal control of disposal of waste on land. Though a succession of Alkali Acts, Clean Air Acts, River Acts, Water Acts and Public Health Acts developed a progressively stricter and ultimately sophisticated control over discharges to air and freshwater, deposition of waste on land remained uncontrolled. In consequence, of course, wastes were diverted from the atmosphere and rivers and sewers to land or sea, and usually to land.

Of course, there were general legal restraints, such as those provided by planning, nuisance, water protection and common law but they were weak and uncertain and for the most part a man could do much as he pleased on his own land. If he chose to use it for waste disposal, he need not keep a record of his activities or disposals. Only finally when the land came to market might enquiries be made, as often as not without much success. Indeed so long as the issue was one between private seller and private buyer and so long as there was no development of the land for housing, possible contamination of the land might not appear important and there would probably not be any public comment.

About the beginning of the present decade the possible dangers of this lack of control were high-lighted by public accusations of irresponsible dumping of chemical wastes on land. At the same time the feeling was growing that exposure to chemicals was in itself undesirable. No doubt the feeling derived partly from a gut belief that the human has evolved in natural conditions and is more or less suited to them, whereas changes may be, probably are, deleterious.

The scientist is naturally hesitant to subscribe to such sweeping generalisations without convincing proof but the proposition is not without its passionate supporters.

A more solid cause for concern is the increasing body of evidence accumulating from animal experiments that the effects of abnormal constituents of the diet, and, indeed of normal, may be detected at very low doses, so low as to suggest that there is no threshold or no-effect level. Indeed it has been suggested that the principal cause of cancer is not a genetic pre-disposition but environmental. If we could recognise these environmental carcinogens and eliminate them, it is urged, we might significantly reduce the incidence of these diseases; meanwhile, until we do let us be cautious and so far as practicable adopt a zero discharge policy for strange new chemicals. A few proven cases of harm to the public from such releases - mercury, cadmium, vinyl chloride, polybrominated biphenyls are examples - underline the point.

To these possible public health hazards may be added the possibility of hazards of safety, fire risk or explosion of methane, both to the public and to workers, and the possibility of interaction of the buried chemicals with building foundations or service pipes. In fact in this country the possibilities have not materialised despite the absence of controlling legislation. With the Control of Pollution Act 1974, whose provisions are still being brought into effect by the appropriate orders, the country at last has effective control. It is to be expected that few new problems, if any, will occur; what we have is the unrecorded legacy of the past and as its heirs we take the obligations with the benefits.

A comprehensive survey of the country to identify contaminated sites, such as the atlas of sediments compiled by the workers at Imperial College from which the cadmium contaminated land at Shipton was identified, is not practicable for most substances. The information comes in case by case as the land is developed. If in the past unwary developers have bought land only to find themselves with unexpected contamination and unexpected expense, it is likely that in future such cases will be rare for the problems have now been well publicised. But the old legal maxim, let the buyer beware is still good advice. The buyer should critically examine the past use of the site and if in the slightest doubt, arrange for expert survey.

Not that a survey will necessarily give the answer. For a small site or for a site contaminated with radioactive substances, many of which may be detected at the surface if not buried more than a foot or so, it may be possible by survey to obtain a reasonably clear answer and to give a fairly clear opinion. But if little is known about the nature of the contamination or where waste may have been deposited, a survey will be of only limited value. A random selection of soil samples based on some sort of grid of the site will give a result limited in certainty by the size of grid. It can be extremely expensive, perhaps prohibitively so, to obtain a result with a reasonably low error.

The more the proposed development of the land adds to its value, the more practicable it may be to spend the sum of money required. It is then important that the nature of the development should be linked to the nature of the survey, which can be concentrated on those areas which may be potential hazards. Land under paving, such as car parks, will not present hazards of ingestion or inhalation as the garden of a house may. The survey of land on which buildings are to be constructed may be linked to specific characteristics easily measured such as corrosivity and load bearing capacity. If the development and the survey are kept in line from the outset, the former can be modified to suit the results of the latter and the latter can be directed to the needs of the former.

Since the cost of an extensive survey may not be insignificant and since its results may affect the possible development of the site, it should preferably be done before the sale is negotiated. To complain later may earn sympathy but is unlikely to bring a cash refund.

The post Second World War years have seen mounting pressure for the use of land as the rising costs of imported food have encouraged greater home production and rising living standards have allowed more people to live in separate accommodation. All land is now a potential housing or industrial estate and there is an official policy to recover derelict land. At the same time there have been some industrial developments which have made land available: the closure of railway lines, the rationalisation of the water and power generation industries, the switch from town to natural gas, the move away from coal for domestic heating, the switch to newer chemical industries. And as always land is reclaimed by filling in or filling up with refuse. These general factors point to the main problems which have presented themselves on specific sites.

The use of domestic refuse for landfill has been practised for generations. Yet there has been no planned survey of the results and no study of the conditions in housing and industrial estates built on landfill. Moreover, since the nature of domestic refuse has changed considerably over the years, it is not clear that past experience would necessarily be a good guide for the future. It is known that the organic content ferments anaerobically, generating methane and carbon dioxide; in fact there have been proposals for collecting the gas for fuel by capping and piping. There have been reports of methane collecting in buildings and service ducts on sites previously used as landfills. There is no reliable rule of thumb - sometimes 20 years is quoted - and it is better to judge each case on the results of test sampling.

It has been the policy of some local authorities to allow local industry to deposit industrial solid waste or slurries on domestic landfills and sewage works, although they were under no legal obligation to do so. Sometimes special areas were set apart for these wastes and thereby an area of specialised contamination may have been created. In one case in my experience the local authority had been collecting in one place all the metallic sludges arising in their area in the hope that in the future, some constituent might be profitably recovered. Where this policy has been followed, extra care will be needed in developing the site.

Since the Drainage of Trade Premises Act 1937, though the practice is older, it has been deliberate policy to encourage the disposal of trade effluents to sewers instead of direct to rivers, since the sewerage system provides a buffer to mitigate the effects of accidental discharges and at the treatment works some removal of undesirables in the sewage effluent will be effected. Of course this is at the expense of collecting the undesirables in the sewage sludge. Thus we may find at a sewage works land contaminated by the industrial constituents of domestic sewage. The most significant have proved to be metals; and of these cadmium has given most concern.

That cadmium may be a highly toxic element has been recognised only in recent years. There is not very much information on its toxicity in the human but what there is, mainly the experience of Japanese who ingested cadmium-contaminated rice and the correlation at post-mortem of kidney dysfunction and cadmium content, combined with its long half-life in the body, suggests that we should be more than cautious in permitting human intake. Moreover, unlike the heavier metals, it is fairly mobile in soil and is taken up by crops, but it is not easy to quantify this relationship with soil conditions. In fact it is not unlikely that in sewage sludge the cadmium may be bound in a form unavailable to crops, though it may slowly mineralise and become available. In fact the whole matter is very uncertain and calls for more research.

A third common contaminated set of sites is redundant gas works. Often tar containing phenolics was disposed of on site in pits and in fact there have been cases of movement of organics from such disposal pits into ground water. This suggests what may become a bigger problem in the future from disposal of chemicals from the newer petrochemical industry which are often not so readily absorbed in soils or dispersed in ground water as the inorganic wastes of the older industries. Of the chemicals from the older industries chromium as chromate is perhaps the most mobile in ground water and several instances of movement of this anion from plating waste disposal sites have been reported. Some movement also occurs from chromium wastes arising in the extraction from the ore, especially in the modern process, where sodium has replaced calcium. The stabilisation of old chromium waste has been described and a report of a study of the movement of chromium in a recent disposal site has been published.

It is not usually possible to apply chemical methods to reclaim contaminated land. There are exceptions such as applying calcium salts to sea flooded land in order to displace sodium by ion-exchange or the application of fertiliser to oil-contaminated land to increase the rate of biodegradation but generally the methods are physical:-

a. Cover the contaminated parts.

b. Remove contaminated soil and/or contaminants.

Often a mix of both methods may be the proper solution. Until the Hazardous Materials Service was set up by AERE Harwell as part of its diversification from radioactive work, based on the expertise built up in handling one class of hazardous materials, there was no recognised service in this country for clearing sites formerly used for dumping chemical wastes.

Of course it is not usually possible to remove all the contaminated material from a site. It is then necessary to consider whether the residual levels of contamination are tolerable for the proposed use. For land proposed to be used for housing or growing crops, which probably require the strictest standards, consideration has to be given to the following:-

a. Inhalation of dust, particularly by those who may spend most of their lives on the site.

b. Ingestion of the crops, particularly by those who may take a substantial proportion of their diet from the site, for example, allotment holders.

c. Ingestion of soil by pica children.

d. Ingestion of water from shallow wells.

e. Fire hazard.

f. Hazard from contact with chemicals.

g. Damage to buildings.

Proper consideration beforehand may suggest ways of developing the site so that the hazards are minimised and made acceptable at minimum cost.

Some of these considerations can be quantified with sufficient accuracy but often the models which have to be used are crude. Combined with the uncertainty of the original toxicity assessments, they may lead to predictions which are well short of the precision needed for sound decision making. It would be nice to check the predictions, for example by analogy with other sites where the inhabitants could be shown, by direct measurement of suitable biological parameters, to be living healthy lives in a similar environment. Unfortunately suitable parameters are not always available and when they are suitable analogous sites are not often to be found.

With the dual purpose of collecting such data from previous experience as might be available and giving what advice one could to local authorities as housing authorities, the Department of the Environment in 1977 established under my chairmanship an Inter-Departmental Committe on Redevelopment of Contaminated Land to provide a central source of technical advice to local authorities wishing to redevelop contaminated sites. We have received inquiries in respect of nearly 200 sites. In this Conference, some of the more important will be discussed.

In fact, either by good luck or by sound common sense, not many cases of past building on sites contaminated to undesirable levels have been found in UK. But one of the problems is that reclamation of contaminated sites has not usually been written up and published. One sees a newspaper article and a brief record in local authority council minutes but rarely a well documented report. One valuable result of this Conference will be the publication of case histories properly documented, so that they are available for study and criticism by others.

The papers in this Conference illustrate the wide range of disciplines which may be involved in the reclamation of a contaminated site and emphasise how necessary it is that they should all be brought in at the beginning before the site is bought, since quite extensive investigations may be needed as a result of which expensive works or restrictions on site development may be indicated. As some of the case studies will show, it may frequently be desirable for the studies to continue as the site is developed because our understanding of many of the processes which affect hazard to workers and occupants of the site or which affect foundations and services is by no means complete. But to end on a more optimistic note: these sites can be reclaimed and present legislation goes far to ensure that in future they will be few and properly recorded.

In fact, if there is good look at present [illegible] many
areas of [illegible] documentation [illegible]
[illegible] of the problems [illegible] collection of
[illegible] has [illegible] been [illegible] published [illegible]
[illegible]
[illegible]
[illegible]
[illegible]
them.

The [illegible]
[illegible]
[illegible]
[illegible]
[illegible]
[illegible]
[illegible]
[illegible]
[illegible]
[illegible]
[illegible]
[illegible]
[illegible]

"STANDARDS" FOR THE REDEVELOPMENT OF CONTAMINATED LAND

M A Smith
Central Directorate on Environmental Pollution

SUMMARY

The principles which should govern guidelines, codes of practice and other forms of guidance, relating to the contamination with substances which may be hazardous to human health or the environment are discussed. The building and construction industries are already familiar with the problem of contamination with substances which may threaten the durability of materials and the integrity of buildings. Protective, remedial or ameliorative measures are generally required to be both durable and robust.

INTRODUCTION

'Contaminated Land' is land which is polluted with a sufficient quantity of toxic substances to present a threat to the health and safety of users or occupiers of land, or of workers engaged in its development or reclamation. It also includes land contaminated with an 'unacceptable' amount of substances which are harmful to animals or plant life, and land from which food and fodder crops may draw excessive quantities of substances harmful to animals or man. In addition to the potential hazards affecting land itself, there may be potential hazards to adjacent areas for example in the form of air pollution (eg toxic dusts), the migration of liquid or gaseous pollutants and the pollution of water courses and aquifers. Further hazards to human health, safety and well being may arise because 'contamination' may threaten the integrity of buildings or services to buildings. All these various hazards have associated economic and social costs. There are two ways of classifying the hazards and problems associated with contaminated land:

A. Based on who or what is at risk ('targets') for example:

i. the health of construction workers;

ii. eventual residents (housing) or users (of schools etc) including particularly sensitive groups in the population;

iii. a wider population due to pollution of aquifers and water courses, or aerial dispersal of dust;

iv. animals, plant life, aquatic life etc;

v. building materials and structures eg by corrosion;

vi. financial investment if site closure or evacuation is required.

B. Based on the nature of the hazards to the targets listed above

i.	physical	a. explosion
		b. subsidence
ii.	inhalation	a. dust
		b. toxic gases
iii.	direct ingestion	a. from fingers etc (particularly children)
		b. contaminated food
iv.	indirect ingestion	a. uptake of toxicants by edible plants
		b. contamination of water
v.	contact	a. skin irritation etc
		b. retardation or death of plants etc
		c. corrosion of structures

A classification of this type was first proposed by Chapman et al (1).

Whatever the nature of the contamination, the 'target' for that contamination (for example human or material), or of the nature of the hazard (for example of explosion or corrosion), two basic questions have to be answered:

i. what levels of contamination are 'unacceptable' and thus require some form of response?

ii. what are the appropriate remedial, protective, or ameliorative measures?

The present paper discusses these two problems mainly in relation to contamination with toxic metals, with an emphasis on establishing general principles, although some tentative numerical guidelines are suggested, which should be of wider application. Indeed, the need for "durability" and "robustness", which it is suggested should govern the selection of remedial measures are already well established within the building and construction industries. An attempt is made to identify areas and topics where some form of guidance may be required. Some of these, such as the evolution of methane and other gases, combustibility of fill materials and attack on building materials will be discussed in other papers to the Conference.

The chosen title for this paper is "Standards for the Development of Contaminated Land". The ambiguity in the choice of the term 'standard' is deliberate and it is intended to apply in the widest sense. It is apparent that when contaminated land is being considered judgements will be made ranging from an assessment of what is considered an acceptable environment or degree of risk to the precise specifications of methods of chemical analysis. Even within the relatively narrow field of British Standards a number of descriptors can be applied such as Codes of Practice, quantitative, materials, qualitative, performance and methods and there will be a need for all of these in dealing with contaminated land.

A definition of 'contaminated land' has been provided above. 'Development' also needs definition since it is taken in a wider context than is usually the case. The construction of buildings is obviously included, but so also is any activity involving disturbance of ground and likely to bring workers or other sections of the population into some form of temporary or continuing contact with the land. Thus the construction of roads and sewers is included; as also are the creation of recreational amenity or public open spaces areas; and the restoration of land to agricultural or horticultural purposes. Not included, although they may have important environmental effects, are schemes for reworking spoil heaps etc to extract residual materials of value. Such schemes may be an integral part of restoration or an agent in enhancing the environmental problems which must be dealt with.

Examples of land which has become contaminated as a result of man's activities include: old mining areas, scrap-yards, sewage works and farms, land-fill sites (refuse disposal) and waste dumps in general; areas where foundry wastes, slags and pyrotechnic wastes have been deposited; smelting works; agricultural or horticultural land overdosed with sewage sludge containing toxic metals; former iron and steel works; former railway land; coal carbonisation plants; and land on which silt dredged from docks and associated canals has been deposited. However not all contamination is man made: there are areas in the country where natural soil conditions are such as to present a threat of some kind to either buildings or health and the effects of 'contamination' in this broad sense are not new to the building and construction industries. They have always for example had to deal with aggressive soil conditions such as those associated with high sulphate levels or low pH's brought about by the presence of peat and on occasion methane emissions from buried layers of peat or underlying coal-bearing strata. Guidance has been provided in various forms for many years but BRE Digest 174 on 'Concrete inssulphate-bearing soils and groundwaters' (2) and CP on Cathodic Protection (3) are two good examples. The section in the former on sampling, analysis and interpretation of results is worthwhile reading for those unfamiliar with it.

Hazards to human health may be short-term, for example ffrom highly corrosive acid, toxic or flammable gases, or high concentrations of metals such as lead; or longer-term from low-level exposure as in the case of direct ingestion of contaminated soil or dust, the uptake of toxic metals such as cadmium by food crops, or from exposure to carcinogens. In some cases the harmful effects may not manifest themselves for thirty or more years. It should be remembered that a resident of a house built on contaminated land may be exposed to a potential hazard for twenty or more years and that groups of the population who may be more sensitive to toxic effects, such as small children and pregnant women, must be taken into account. There are parallel factors to be taken into account for other target/hazard combinations.

GUIDELINE LIMITS FOR TOXIC SUBSTANCES

General Considerations The establishment of guidelines or limits for 'acceptable' levels of a contaminating substance requires more than producing a list of numbers. They can not be set without a proper regard to sampling techniques and data interpretation, building processes, the response it is envisaged will be made when 'unacceptable' levels are found, practical ameliorative and remedial measures and the activities likely to take place on the site (including change of use) once development is completed. The analytical techniques to be employed must also be defined.

Guidelines for toxic substances will also have to be capable of being understood by people with a range of backgrounds including environmental health specialists and the professions involved in building and construction; they must be flexible and allow for the exercise of 'professional judgement' on the basis of personal experience and knowledge about a particular site; and they should pay proper attention to practical and economic implications and regard to the levels which are commonly encountered in rural and urban soils and in street dusts.

Guidelines which are set in relation to development of sites which are already contaminated should only be applied with caution in other situations. They should not be taken as indicating conditions which should be voluntarily created eg by importing contaminated material into a clean area or deposition of wastes within an industrial area. Nor should they be taken as indicating the standard which should be aimed at when ameliorative measures are undertaken - for example the degree of contamination may be such as to require the provision of not less than 1 metre of clean cover in order to provide a durable and lasting solution to the contamination, and allowance must also be made for the difficulty of carrying out a particular remedial treatment to an acceptable standard. They should also not be applied without careful consideration to an existing development which is found to be on contaminated ground: there is a proper distinction to be drawn between creating a new situation and assessing the hazards presented by an existing situation.

It is also import to distinguish between limits or upper guideline values which indicate the levels above which it is considered 'unacceptable' that a contaminant should be present and consequently require some form of response, and sets of values which are indicative of pollution being present and consequently indicate a need for thorough investigation of a site. I would put the values listed by Chapman et al (1) in this latter category. They have been widely misinterpreted as indicating the maximum permissible levels, whereas in general they are only the upper end of the range of levels to be found in 'normal' soils. It should be noted however, that higher values may occur naturally in certain places in the United Kingdom. It is also important to distinguish between what could be termed 'undesirable levels' and 'unacceptable' levels.

The discussion which follows is restricted to a selection of toxic elements; consideration must also be given in due course to: other toxic substances (eg fluorides), to the chemicals associated with gas works sites such as simple cyanides, complexed cyanides, phenols, coal tar and its derivates etc, and to a number of more complex chemicals such as PCB's. It is also largely in terms of acceptable levels for single elements or substances eg lead or cadmium, but it should be remembered that there may be interactions between substances which may either multiply the potential problem (synergistic effects) or sometimes ameliorate the problem. Other factors such as the pH or the amount of organic matter in the soil may also be important.

For each element or substance a matrix of guideline values is required based on the intended end-use of the land and the 'hazard' which is being considered. To set a single set of guidelines for all land would be likely to present practical difficulties and would have serious economic consequences since any general limit would have to be based on the "worst" situation. However as pointed out below, there may be times when a more stringent limit needs to be applied, than would be required by the immediate end-use which is planned.

The categories of land to be considered are: agricultural and horticultural, allotments and 'large gardens'; "small gardens", amenity and recreational areas, public open space, "inaccessible land". The discussion below concentrates on those categories of land likely to be involved in comprehensive physical development ie allotments and 'large gardens'; housing with 'small gardens'; amenity areas, recreational and educational land; public open space. Additional categories of dand which might be considered are 'inaccessible land' (eg wooded land or land remote from centres of population); industrial and commercial land; and agricultural and horticultural land. Clearly, however, not all land will fall conveniently into one of these categories and there will always be 'special circumstances' to be taken into account. To some extent standards have already been set for the major toxic elements in as much as guidelines have been issued for the disposal of sewage sludge on land by the Department of the Environment and the National Water Council (4). This document provides a useful summary of the phytoxic and human and animal health effects as represented by the deposition of sludge on land and the limits suggested (see Table 1) provide a useful starting point when other types of land are considered. It is anticipated that revised guidelines will be issued in the near future.

Primary concerns for public health in relation to the four categories of land listed above are: uptake by plants of elements toxic to man; direct ingestion (or in some circumstances inhalation); and dermal effects (eg contact with skin irritants such as chromium). In addition guidance is needed on acceptable levels in respect of phytotoxicity.

Which hazard will be most important will depend on the type of land and the element or substance concerned. For example the most acute situation as far as uptake by food crops is concerned would be presented by allotments or large gardens where substantial amounts of vegetables and other food crops might be grown, whereas for small gardens other hazards such as direct ingestion of soil by children are likely to be dominant. For some elements (eg zinc) the practical need to have viable plant growth will require limits to be set well below those at which a hazard to human health is likely to exist. Further hazards may occur from the pollution of water supplies and this might be of particular concern when certain organic substances, which can penetrate some plastics are considered, and such effects will clearly be important whenever a water supply has to be provided.

Site Investigation A judgement on the suitability of a site for redevelopment and on appropriate remedial measures usually has to be made sometime in advance and often with little idea about the final layout - indeed Local Authorities are advised against making firm decisions before information on the degree and extent of contamination is available. In this way the layout and type of development can be adjusted to reduce the impact of contamination.

The usual practice on a housing site is to strip and store the immediate surface layers (containing plant material etc). Concern is therefore for the metal content of this "top-soil" (if it can be described as such) which would usually be respread, or if it is clearly unsuitable as top-soil, taken from the site; for the metal content of the stripped surface; and for contamination at greater depths.

For playing fields, public open space, flower and shrub beds around industrial development, allotments and some garden areas, the immediate response (where contamination is not thought to be a problem), is to accept the existing "top-soil", with a covering of a suitable depth (often only about 100 mm) of imported top-soil, when considered necessary to sustain plant life. Indeed there have been considerable efforts made to reclaim land with minimal or nil addition of cover or amendment materials. Cost has often been a dominant consideration as pointed rut by Bridges et al (5).

The first steps in the assessment of a site which it is thought may be contaminated are a site visit and an examination of historic records, maps, plans, aerial photographs etc to draw up as comprehensive picture as possible of the location of structures and the way the site was operated. Former employees and nearby residents can also be valuable sources of information. This topic will be fully discussed by Dumbleton (6) in another paper to the Conference.

The scope of the sampling programme should be related to the stage any development has reached and its intended nature eg a pre-purchase survey is likely to be less comprehensive than the programme needed as a prelude to drawing up detailed plans for remedial treatment and subsequent development. It is sensible however to try to ensure that the results of a preliminary survey can be integrated into a subsequent fuller survey. It should always be remembered that no sampling programme can guarantee all 'hazards' will be found, but the effectiveness of the survey can be greatly improved by careful forethought. The ground and surface waters should also be examined for contamination.

The most straightforward approach to sampling, particularly for a large site, is to take samples on a regular grid (10, 25, 50 or 100 m) with individual samples being taken from the surface layers (say to 150 mm), at about 0.5 m depth and at greater depths as seems necessary. Other patterns of sampling - for example 'W' or 'X' shaped could also be appropriate. Any formal pattern of sampling may have to be varied to take account of physical obstructions such as concrete slabs and old foundations, topography and knowledge about the former use of the land and the layout of any pre-existing plant. On a small site, however, no formal pattern of sampling may be possible and often only a few sampling points will be practicable. It should be remembered that the normal activities on a building site are likely to cause a lot of sub-surface material to be distributed around the site.

The nature of the investigation, and in particular the frequency of sampling will clearly have an effect on the confidence which can be placed on results of the investigation and will consequently affect the way in which they must be viewed in relation to any guidelines.

When agricultural land or horticultural land is being sampled, either to obtain information relating to plant health or human health, the recommended practice, is to take numerous surface sub-samples on a 'W' pattern across the site or field in question, and to combine the sub-samples to give a single bulk sample. If the areas are particularly large then it is divided into sub-areas and a composite sample prepared for each sub-area.

This approach is appropriate for agricultural and horticultural land where the crop is dispersed to the population at large and the particular product will form only a small part of the total diet (employees on market gardens etc may be an exception to this general view). It is not appropriate for the assessment of a site destined for use for example for houses or allotments. In these circumstances, the concern is for individuals who may have intense and prolonged contact (30 years or more) - either direct or indirect, through plant uptake. Knowing the average level of cadmium for an area to be occupied by 100 houses is insufficient - attention has to be paid to likely levels in individual gardens.

Where public open space (which will vary greatly in the intensity of its use) or recreational land is concerned, the degree of exposure will usually be more limited, but an individual child may play repeatedly on the same spot on an area of amenity grass, or in an adventure playground, for example.

It is therefore suggested that 'spot' samples should be taken for analysis, rather than composite samples (although the latter may be of value in some circumstances) and that judgements and limits should be based on the results obtained on these spot samples. If sufficient 'spot' samples are taken from the surface, for example, it may be possible to estimate statistically, the proportion of soil with a metal content in excess of a particular value and the probability of a certain elevated value, not found in any of the samples, being present somewhere on the site. It would in practice be difficult to obtain a representative composite sample for other than the immediate surface. Composite surface samples may be appropriate for public open space of an informal nature where close contact by children is deemed unlikely.

Limits Relative to the Uptake of Metals by Plants The greatest concern is for the uptake of metals (or other elements) toxic to humans by food crops. However, the uptake of these metals by non-food plants is also relevant as plant growth and decay processes, may cause the toxic metals to be concentrated or brought back to the surface. When appropriate, toxicity to animal life, particularly farm animals must be considered. Toxicity to plants is also of practical concern since soil which will not support adequate plant life cannot be considered as suitable for domestic gardens, amenity areas etc. A distinction should perhaps be drawn between those areas where continuing health of plants on a long-term basis is essential, such as in domestic gardens, and those areas where, whilst loss of plant cover is undesirable, it is more easy to rectify any failures or deficiencies, such as landscaped areas around commercial or industrial development.

The literature on phytotoxicity is extensive and any set of guidelines would need to cover a wide range of plant types and of elements or combinations of elements. The longstanding ADAS guidelines for agricultural and horticultural land which have been accepted into the guidelines for the disposal of sewage sludge to land can form a basis for the domestic situations. These are based on the concept of 'plant available' metals - in particular zinc, copper and nickel. But of course other elements and substances can also be highly toxic. These ADAS guidelines, in which economic considerations such as effect on yield play an important part, may not be entirely appropriate, for example for amenity areas, public open space and inaccessible land. In these areas the requirements is to produce a qualitatively acceptable, durable, sward or other plant growth (indeed low yield may sometimes be advantageous) not necessarily optimum growth, and options available include the use of toxin-resistant strains of plants - thus we are looking to a performance standard of the final product - perhaps underpinned by some guidelines on soil levels - subject of course to the overriding considerations of minimal hazard to human health. In this connection the use of metal resistant strains of plants which in many cases have less tendency to accumulate metals in stems and leaves may be advantageous in reducing the amount of metals reaching the surface of the ground.

Although this discussion concerns direct toxicity to plants it must always be remembered that many other factors govern the viability of plant life - such as nutrition, physical soil conditions and selection of appropriate plant types (12).

A useful starting point for defining acceptable metal levels, in respect of both human health and plant health are the limits recommended in the Report on the Disposal of Sewage Sludge to Land (4) which are summarised and interpreted in Table 1. These limits are set in terms of the quantity of metals which may be added to an 'uncontaminated' soil over a period of 30 years. The upper limits are thus implied rather than explicit and are obtained by adding the permitted addition to the upper limit of the 'normal range'. The samples on which adherence to the guidelines is judged should be 'composite' samples obtained in accordance with ADAS advice. It is suggested that the limits for maximum permissible contents listed in Column 4 of Table 1 should be adopted for allotments and large domestic gardens and those in Column 5 for small domestic gardens.

Whatever guidelines are set, the interpretation of the limits is all important. Thus it is suggested that something along the lines as follows may be appropriate:

"No individual 'spot' sample taken from the top 450 mm should exceed the limits and there should only be an acceptably low probability (say 1 in 100 if adequate data are available) that any significant proportion of soil exceeds the limit". The depth of "acceptable" soil suggested as necessary is based on the need to prevent contamination being brought to the surface during normal cultivation and may have to be increased if the underlying areas are highly contaminated. If the limits are exceeded then ameliorative measures should be carried out. If a

TABLE 1

TENTATIVE GUIDELINES FOR METALS IN SOILS APPROPRIATE TO PLANT UPTAKE & PHYTOXICITY[e]
(DERIVED FROM GUIDELINES FOR DISPOSAL OF SEWAGE SLUDGE TO LAND)

1	2	3	4	5
			SUGGESTED ACCEPTABLE LEVELS[b]	
ELEMENT	"NORMAL RANGE IN SOILS[d]	IMPLIED MAXIMUM CONTENT FOR ARABLE SOILS[c]	Allotments & large gardens	Small gardens
	mg/kg ds	mg/kg ds	mg/kg ds	mg/kg ds
Zinc	10 - 300	) 225 zinc)	)	
Copper	2 - 100	) equivalent[a])	225 ZE	225 ZE
Nickel	5 - 500	) (450 for pasture))	)	
Chromium	5 - 500	955	960	1000
Cadmium	0.1 - 1.0	3.3	3	5
Lead	2 - 200	655	700	1000
Mercury	0.1 - 0.3	1.2	1.5	1.5
Molybdenum	2	4.3	4.5	4.5
Arsenic	0.1 - 40	44.6	45	45
Selenium	0.2 - 0.5	2.8	3	3

ds = dry soil

NOTES

a Zinc equivalent = ('Available' Zn) + 2 (Available copper) + 8 (Available Nickel)

b Levels measured on 'Spot' samples. The average values should consequently be below these values

c i. Based on additions of sewage sludge to land over 30 year period.

ii. Assumed contamination restricted to top 200 mm soil and that "Density" of soil is 2.2 tonnes dry solids/200 mm depth over 1 ha.

iii. Levels measured on composite samples representing field or part of field.

d Naturally occurring levels above those reported are found in some parts of the United Kingdom (13)

e These limits are not necessarily appropriate where direct ingestion or dermal effects are being considered.

judgement is required on an individual garden, allotment or well defined area of reasonable dimensions, then a further composite sampling programme may be appropriate.

Serious difficulties would be presented if an attempt were to be made to base limits on 'available' metals. Plant availability is dependent upon many factors including for example pH and the content of organic matter. Whilst these factors can be controlled in agriculture and horticulture, and need to be for good commercial reasons, there can be no certainty that this will be done where allotments, domestic gardens, public open space, amenity areas etc are concerned. Thus some allowances must be made in judging a particular situation for possible changes of availability with time. This will be important whether human or plant health is of concern. Thus some guidance on acceptable levels of total copper, zinc and nickel is also required. Indeed, total metals can vary with time: they may be reduced by leaching processes or increased due to various phenomena including the decay of organic matter for example on heavily sludged land or refuse sites.

Of course the most direct methods would be to measure the actual metal values in vegetables grown on the site in question. But this is not practicable, and because of the transient nature of some of the factors limiting availability, would not in any case provide a guarantee for the future. However, such measurements can be useful when it comes to monitoring sites which have been reclaimed or already developed.

Limits in relation to Direct Ingestion The primary concern is for small children who may accidentally or deliberately ingest small quantities of soil. The problem of the ingestion of toxic metals by children has already been tackled for toys and graphic materials in 'The Toys (Safety) Regulations 1974' and 'The Pencils and Graphic Instruments (Safety) Regulations (1974)' respectively and a consideration of these regulations could possibly be used as a basis for setting guideline values for soils for domestic gardens and similar situations, with the acceptable levels being reduced by a suitable factor to allow for the greater volume of soil likely to be ingested. It is debateable as to the extent that any limits should take account of the chilld with a tendency to 'pica' (the voluntary ingestion of non-food objects) as opposed to the 'average' child who merely has dirty hands or drops his sweets on the ground.

The limits are based on an assessment of acid (dilute hydrochloric acid) soluble metals - an implied assessment of availability of the metal to a child when the material is ingested. The same reservation must be made about 'child availability' as about plant availability ie that it may change as other soil parameters change. The methods may also not be appropriate for certain soils - particularly for example calcareous soils (containing calcium carbonate) which might react with and consequently neutralise the acid which might otherwise react with or dissolve the metals, although the Regulations do require the tests to be made at a fixed pH of 1.5. Total metal contents may therefore be more appropriate.

Samples should be taken on a similar basis as for an assessment in relation to plant uptake and plant toxicity and therefore, for garden areas (domestic and public): "No individual 'spot' sample taken from the top 450 mm should exceed the limits and there should only be an acceptably low probability (say 1 in 100 if adequate data are available) that any significant proportion of soil exceeds the limit". If the limits are exceeded then ameliorative measures as discussed below should be adopted. For grassed areas the depth of soil required to be beneath the limits could be reduced to say 250 mm.

The depth of 450 mm for garden areas is considered necessary to prevent contaminated material being brought to the surface during normal cultivation. No guarantee can be given that an area laid to lawn will not be converted for vegetable growing. However, the depth required must depend upon the level of contamination of the underlying soil and it may be necessary to have 1 metre of 'clean' cover when the pollution is particularly bad. Similarly the depth of 250 mm should often be sufficient for areas of amenity grass etc but may have to be increased if the underlying strata are particularly contaminated - otherwise plant growth and decay, earth-worm activity and moisture movement, may lead to the surface becoming recontaminated.

Appropriate limits in respect of direct contact Industrial hygiene standards, adjusted to allow for the different exposure patterns probably form suitable basis for deriving limits in respect of chromium and other substances which are of concern as far as direct contact is concerned. Sampling and the setting of limits would be as for direct ingestion.

REMEDIAL, PROTECTIVE AND AMELIORATIVE MEASURES

The measures which should be adopted to render an 'unacceptable' situation 'acceptable' will depend upon the degree and extent of contamination and the end use to which the land is to be put. Standards will be concerned with some concepts of performance requirements which will rarely be capable of interpretation in quantitative terms. It will not always be possible to eliminate all risk and the measures adopted will be designed to produce an 'acceptable' level of risk. A choice will often have to be made from options ranging from low cost/high risk to high cost/low risk. The consequences of failure, which may not occur for 15 to 20 years or more, and the steps which must be taken to remedy failure, should be consciously considered. The building and construction industries are only too aware that failures do occur - for a variety of reasons including lack of knowledge, a failure to allow sufficiently for bad workmanship, a lack of quality control and 'Acts of God'.

Any remedial, protective, or ameliorative measures which are adopted should be both 'durable' and 'robust'. Durable, in the sense that they continue to be effective for the required time period - which in the case of housing is at least 60 years; but might be only 20 years for warehousing. This means that they should not reduce in effectiveness too rapidly due to metal movement, however produced, or likely to be rendered ineffective for example by a chance flooding such as might be caused by a burst water pipe. They must be robust in the sense that they will not be easily undone either during the immediate construction process or subsequently should a service trench be re-excavated or someone decide to build an extension to his house. They should not rely too heavily on good workmanship or site supervision.

Two further concepts need to be accepted for many cases of contaminated land: proper monitoring of conditions over a prolonged period of time and where appropriate maintenance either on a planned basis or in response to problems revealed by the monitoring programme. It must be recognised that it will not always be possible to achieve a final and complete solution and subsequent action may sometimes be required. The first essential in any monitoring programme is to establish a base-line - ie it will usually be advisable to make a check once remedial measures have been taken that an 'acceptable' condition has been achieved. This monitoring will often produce data not only of relevance to the site in question, but also in our present limited state of knowledge, to wider problems. We must learn from one another's experience and the publication of proper accounts of problem sites is to be encouraged.

Remedial and ameliorative measures for toxic metals Consideration of what measures might be appropriate for land contaminated with toxic elements can provide a useful insight into measures which might be required to deal with other forms of contamination. Options available include removal of all or a proportion of the contaminated material. If the contamination is restricted to well defined areas (eg such as arise from the burning of car batteries or tyres on waste land) then selective removal of the offending material is appropriate. If the 'average' degree of contamination is acceptable but there are unacceptable 'hot-spots' it may be possible to produce an acceptable condition by stripping the soil, stockpiling it and then respreading. The soil from any particularly 'hot' areas can be selectively removed and disposed of during the stripping operation. This process should reduce the overall level of contamination. A similar result can be obtained if a proportion of relatively clean material is incorporated during the stripping operation. This may happen naturally if contamination is restricted to the immediate surface layers. The liming of the sub-soil or incorporation of lime in the top-soil itself before it is respread may also be useful in reducing 'availability'. A further ameliorative technique may be to 'deep plough' or 'cultivate' the contaminated lower levels, again with application of lime as appropriate.

The existing surface or a stripped surface can be covered with clean top-soil or a combination of top-soil and other materials which might be as little as 100 mm or as great as 2 metres in total thickness. Where the degree of contamination is relatively low and the applied layer of material is thin, it may be appropriate to mix the clean and contaminated layers together by ploughing, rotavating etc. Amendment materials such as lime and organic matter can be incorporated. Any imported fill or top-soil should be analysed. It should always be remembered in any filling or reclamation process that soil is a complex living material that requires handling with care and sensitivity.

If the underlying strata are contaminated, due allowance must be made, for any movement upwards that may occur. This means that the acceptable level of contamination of the surface layers may have to be reduced to allow for this enrichment process. Enrichment may also of course occur for other reasons, such as; the decay of organic matter; root absorption, translocation, death and decay of plants; and aerial deposition. Allowance should be made for these processes. There are examples where toxic effects have taken a considerable period of years to develop (5). Whilst some enrichment may be acceptable, say for public open spaces, it is unlikely to be acceptable for housing, - thus in selecting remedial measures on a scale of high risk/low cost to low risk/high cost, where risk is related mainly to the durability of the treatment, the tendency must be towards the low risk end. The information on which to judge the likely durability and robustness of remedial measures is at present limited and the Department of the Environment is financing research designed to provide relevant information. Reference to this research programme will be made in other papers to this Conference.

Another precaution often adopted is to specify that all service trenches etc are back-filled with clean material, this reduces immediate problems arising from the contamination whether due to possible attack on materials or health hazards, and should make re-excavation of the services in future years less hazardous and less reliant on fallible memories and checking procedures. I am not aware of any experimental evidence however on the long-term effectiveness of such solutions. In many instances they will be appropriate but on a typical housing site, they add considerably to the complexity of operations and thus add to costs and the demand for proper supervision, so that if they form a critical facet of the protective measures, they may be too susceptible to bad workmanship (ie insufficiently robust) to be acceptable. The provision of uniform cover across at least the relevant parts of the site - and if water pipes are to be properly buried this will mean about 1 metre of cover - may be a much more reliable solution. Possibly, initially more costly, but in the long-term far more satisfactory.

Although the emphasis in any housing or similar development the accent must be on long-term solutions there are occasions when purely temporary treatment is required, either for cosmetic reasons or to solve an immediate environmental problem and in such circumstances it will often be appropriate to have some relaxation of standards. This is alright provided these important criteria are met:

i. It is fully recognised that the works are temporary and preferably a time limit is set. Temporary works have a habit of becoming permanent.

ii. health and safety standards are maintained at a reasonable level.

iii. It is accepted by all concerned that further action may be required.

The role of temporary or semi-permanent cosmetic work on revitalising areas of dereliction has been recognised for many years.

The selection of remedial measures calls for the most thorough consultation between specialists from various disciplines to ensure that remedial measures adopted say to protect health are practical and do not for example have unacceptable adverse implications for the construction process or on horticultural viability. Similarly these professions need to understand the nature of the remedial process so that they do not unwittingly undo all the good work. It is essential that account of any problems from contamination are taken into account at the earliest stage of any development that is possible, which means when site use, zoning and layout are being decided. Where there are existing buildings or plant the demolition and reclamation of the site should be seen as a unified process: it is very easy to add to contamination during demolition and to remove from the site materials, such as brick rubble which could be of value in its reclamation. Indeed let us hope that the process will go further back and that we will be successful in avoiding the creation of so many polluted sites in the future.

If it is decided to cover up the problem, rather than to remove it, it should not be 'out of sight, out of mind'. Proper records must be kept of the nature of the contamination, the remedial measures adopted, and the reasons for their adoption and of any essential constraints which must be imposed on any further development or change of use of the land. This means that land owners, local authorities and statutory bodies may need to keep separate records, and to consider how such information should be made available to people who may have an interest in the site in future years.

"THE INTERDEPARTMENTAL COMMITTEE"

The origins, functions and progress of the Interdepartmental Committee on the Redevelopment of Contaminated Land, which was set up in 1976, have been described elsewhere (7). A number of documents describing qualitatively the problems which may be encounted on certain types of contaminated sites have been produced (8-11).

CONCLUSIONS

Standards and codes of practice are required which set guidelines of acceptable levels of contamination and which indicate the performance that is required of remedial, protective and ameliorative measures. They must be realistic in taking account of conditions usually encountered on building and construction sites and of economics. Remedial measures must be durable and robust, and in view of the limited knowledge on many topics and the long time scales involved, will often err on the side of low risk- high cost in order to provide the degree of confidence necessary.

REFERENCES

1. W B Chapman, P Baker and Burns, Journ Assoc Public Analysts, 1977, - ( ) 9 and Development & Materials Bulletin (GLC) 1976 (98) 1-8.

2. 'Concrete in Sulphate-Bearing Soils and Groundwater', Building Research Establishment Digest 174, (HMSO, 1975).

3. BS CP1021: 1973 'Cathodic Protection', (British Standards Institution, London, 1973).

4. 'Report of the Working Party on the Disposal of Sewage Sludge to Land' (Department of the Environment and National Water Council, London, 1979.

5. E M Bridges, D S Chase, S J Wainwright, Preprint Lower Swansea Valley Conference, Swansea 1979 (University of Swansea, to be published).

6. M J Dumbleton, Paper to be presented to RECLAN 79.

7. 'Progress Report of the Interdepartmental Committee on the Redevelopment of Contaminated Land Department of the Environment, London, 1977).

8. 'Redevelopment of Contaminated Land: Notes on the Development of Landfill Sites' (Department of the Environment, London 1979).

9. 'Redevelopment of Contaminated Land: Gas works sites' (Department of the Environment, London 1979).

10. 'Redevelopment of Contaminated Land' Development & Materials Bulletin (GLC), 1978 (118) (and Department of the Environment, London 1979).

11. M A Smith 'The Redevelopment of Contaminated Land', Proc Conf on Heavy Metals in the Environment, London 1979.

12. M S Johnson, Landscape Architect, 1978 (May) 33-36.

13. 'Wolfson Geochemical Atlas of England and Wales', (Clarendon Press, Oxford, 1978).

TABLE TENTATIVE GUIDELINES FOR 'ACCEPTABLE LEVELS' OF CERTAIN CONTAMINANTS IN SOILS WHERE NEW DEVELOPMENT OR REDEVELOPMENT IS TAKING PLACE

1	2	3	4	5
ELEMENT	'SMALL'[c] GARDENS	'LARGE' GARDENS & ALLOTMENTS[c]	AMENITY[ac] GRASS	PUBLIC OPEN SPACE[c]
	mg/kg dry soil	mg/kg dry soil	mg/kg dry soil	mg/kg dry soil
ZINC-) COPPER) NICKEL)	280 zinc equivalent[b]	280 zinc equivalent	-560 280 zinc equivalent	-560 280 zinc equivalent
CADMIUM	5	3	12	15
LEAD	1000 250 soluble[d]	550 250 soluble[d]	1500 250 soluble[d]	2000
MERCURY	1.5	1	50 25 soluble[d]	50 25 soluble[d]
TOTAL CHROMIUM	1000 60 soluble[d]	600 60 soluble[d]	1000 60 soluble[d]	1000
HEXAVALENT CHROMIUM	- 25 soluble[d]	- 25 soluble[d]	- 25 soluble[d]	- 25 soluble[d]
MOLYBDENUMf	4	4	5	5
ARSENIC	20 25 soluble	10 25 soluble	40 25 soluble	40
SELENIUM	3	3	3	6
BORON	3.25	3.25		
THALLIUM	-	-	-	-
SILVER	-	-	-	-
BARIUM	125 soluble	125 soluble	125 soluble	1000
ANTIMONY	60 soluble	60 soluble	60 soluble	500
FLUORINE	800	800	800	1000

a Includes schools, play areas, recreational areas, etc where small children likely to be regularly exposed.
b See Table 1 for definition of zinc equivalent
c All values totals on 'spot' samples unless specified otherwise.
d Soluble in 0.07NHCl or solution corrected to pH = 1.5 if alkaline substances present.
e Assumed attempt will be made to keep pH of soil at about 6.5.
f Probably too conservative?

from: DOE Consultation Paper on 'Acceptable levels' (ICRCL 16/78(6)) - October 1979. (Available from CDEP, Room N19/10, 2 Marsham Street, London SW1P 3EB).

NB. 1. The tentative nature of the guidelines is stressed
2. It is important that they are seen in the context described in the paper
3. Concentrations above the guidelines are not necessarily unacceptable - each case must be judged on its merits.
4. For advice on interpretation contact CDEP at address given above or telephone 01-212-4485 (M A Smith).
5. As the Table is derived from a Consultation Paper it follows that the numbers in the Table may be adjusted following the consultation procedure - it is consequently advisable to check that you have the latest version.

SITE INVESTIGATION AND MATERIALS PROBLEMS

R.T. Kelly

Scientific Adviser to the Greater London Council

In 1976 Chapman, Baker and Burns published a paper which was based on the experiences the Scientific Branch of the Greater London Council had had, up to that point, in the investigation of contaminated land. Such land had included sites formerly used for gas works and for sewage farming. Since then other types of sites such as railway marshalling yards, coal yards and the river-side wharves have had to be examined and it has become quite clear that more precise terminology is needed to describe to the non-specialist the former use of the land. For instance:

(i) the failure to differentiate between manufacturing industries and warehouse storage sites, and in particular the operational areas; and

(ii) the failure to differentiate between random industrial contamination associated with the former use of the site on the one hand and the planned depositing under closely specified conditions of refuse material on the other -

has tended to create confusion.

Chapman's paper classified the hazards which may arise in two ways, those at risk and the type of risk. He gave the following division of the groups at risk

(i) Construction workers

(ii) Eventual residents (of housing) or users (of schools etc) including particularly 'sensitive' groups in the population and neighbouring residents etc

(iii) Animals, plant life, including aquatic life etc

(iv) Building structures etc, eg by corrosion.

It has emerged that -

(i) another group needs to be added which includes 'testers', those carrying out site investigations and those analysing the samples taken. It has proved necessary to ensure that sample handling was undertaken in a manner which, for instance, covered the possibility that all samples were capable of creating dermatitis, since at the preparation stage there would be insufficient information to establish their 'safety'. Additionally, the safety of 'samplers' on site has to be covered in more detail, not merely from the point of view of the chemicals which may have been present in the samples as contaminants but also with respect to their physical safety. Some sites are badly over-grown and contain considerable underground structures badly marked, into which the unwary can easily fall. The site needs to be screened for the presence of underground services which may exist on the site ie, gas, electricity, sewers, before the sites are investigated. In spite of these precautions damage to services can occur when uncharted cables or pipes are struck;

(ii) Linked with the safety of the site investigation team is the question of the safety of those who come onto the site perhaps illegally during the investigation stage. No matter what attempts are made to make the site secure the public, particularly children, are likely to gain access perhaps with unfortunate consequences.

With regard to the type of risk to be guarded against which Chapman listed viz -

(v) Physical	(a) explosion (b) subsidence
(vi) Inhalation	(a) dust (b) toxic gases
(vii) Direct ingestion	(a) from fingers etc (particularly children) (b) contamination of food
(viii) Indirect ingestion	(a) uptake of toxicants by edible plants (b) contamination of environmental waters
(ix) Contact	(a) skin irritation etc (b) retardation or death of plants etc by contact with contaminated soil (c) corrosion of structures (metal, concrete etc),

he risk to building structures and services and people arising from the generation of methane and from the combustion of the material on the site (particularly in the form of underground fires), although originally regarded as a possibility, have now been shown to be real hazards.

The original paper went on to consider the mechanism of the study of the site and gave the following as methods of obtaining information -

(a) History of site	(i) from previous occupiers (ii) from library records, archives etc
(b) Visual inspection	(i) direct - eg obvious contamination (ii) indirect - appearance of plant life, water courses etc
(c) Tests on readily accessible material	(i) ambient air (ii) surface water - eg ditches, pools etc (iii) surface soil
(d) Tests of materials after borings, digging of pits etc	(i) gas analysis (ii) water (iii) soil

It also gave a decision tree indicating the stages which needed to be worked through in making an investigation. This decision tree is reproduced in Appendix I. In addition to greater attention now being paid to the possible combustion of the materials on site (see above) attempts have been made to ensure that as far as possible a simple mechanism for site appraisal has been developed to investigate all possible factors affecting the development of the site, in particular those associated with air, noise pollution (including vibration). This will be discussed in more detail later in the paper.

Whatever the information derived from an appraisal of the history of the site, site inspection and the examination of readily available materials, samples of the soil itself and of the ground water associated with it has continued to be necessary and the methods used for obtaining such samples have changed little. The following summarises the original approach made -

(i) In sampling agricultural land, the assumption is generally made that reasonably sized areas of land, say a field, do not vary greatly in composition from one part of the field to another. The sampling strategy is therefore designed to obtain a number (say 25) samples distributed reasonably evenly over the whole area and to mix these to give one composite sample which may be used as a basis for any advice on type of crops or treatment etc.

(ii) Because the objectives of sampling industrially contaminated land are different from (i) above, a different sampling strategy is therefore necessary. It is important to know, for example, what variation can occur in the concentration of contaminants from one part of the site to another and to know the depths at which the contamination may be present on the site. In addition because of the various purposes for which a site may have been used the variation from point to point cannot be assumed to be insignificant and small pockets of contaminants may well be distributed irregularly across the site.

To meet all these points, whilst at the same time attempting to keep the amount (and therefore the cost) of testing within reasonable limits, yet obtaining adequate information for the initial investigation of the site, we have found that for relatively clear sites (other than 'virgin land'), boreholes or pits etc on about a 100 metre grid across the site is an acceptable compromise, provided that any areas visually identified as being of importance are investigated in more detail. With land which is initially known to be more heavily contaminated, or which the above 100 metre grid testing indicates need for more extensive sampling, pit excavations etc, on the basis of a 50 metre grid are undertaken. Sampling points closer than 50 metres may be considered necessary in the case of small industrial sites, or where physical obstructions, concrete bases etc prevent the adoption of a grid pattern of sampling.

The depth to which such borings are taken will depend upon a number of factors such as the initial levels of the land, the eventual levels of the finished development, expected presence of trees, possibility of growing edible crops, use for 'grassing' only, expected 'hard core' areas such as roads, concrete structures etc. Generally, where relatively low rise housing is concerned, which usually involves the possibility of vegetable growing etc, samples down to at least one metre below either the initial or the final level of the ground is essential. Deeper pits are often advisable especially on heavily contaminated land, and are particularly important if trees etc are likely to be needed since the roots of some of these may reach 5 or even 10 metres. In relation to the actual sample, we favour spot samples at different levels rather than for example a composite sample representing say a 1 metre layer.

No sampling strategy short of an examination of the site so complete as to be inordinately expensive, can guarantee not to 'miss' a local pocket of contamination. For this reason, visual observations during boring, digging etc operations are a most valuable additional safeguard. If it is not possible to arrange for adequately qualified staff to be present at all times during such operations, it is most important to ensure that any unusual factors are noted by the contractor undertaking the sampling so that the regular sampling suggested above can be supplemented by any other 'spot' samples considered necessary in the light of the site condition discovered.

In the period since the publication of the original paper -

(i) the sampling of the underlying material has become more selective and

(ii) dug pits are preferred in many cases to boreholes, as pits give more information on the character and disposition of the fill material.

When dug pits are used it has been the practise to collect water samples by placing a short standpipe within the pit before refilling it. The collected water is examined at a later stage after pumping to waste any 'disturbed' ground water (see below).

Although the above sampling strategy is developed principally for soil samples, similar considerations apply to water sampling. For water sampling it is usual to obtain samples of water from each standpipe, as far as is practicable. These samples should also be supplemented by samples from any surface water whether ditches, pools, lakes etc, which are present on the site, or which develop during building operations. It should be noted that specialised equipment and techniques are necessary to obtain representative water samples and the involvement of specialist staff in the sampling is essential.

Where methane testing is to be carried out an additional requirement which has emerged is the need for careful positioning and preparation of the boreholes for methane testing. These may on occasions have to be taken much deeper than is usual for soil sample purposes.

At the time of the preparation of the Chapman paper, it was recognised that even on the smaller sites it was economically impractible to cover all possible levels and types of contamination and that a compromise had to be reached between what might be desirable to do and what it was not merely financially acceptable to do but what was possible given the availability of the resources. The need to reach such a compromise has been emphasised since then. Even by using outside agencies to augment the resources available within the Council, it has proved difficult to meet the timescale for development of land within the Council's programme. Certainly if much increased sampling rates had to be employed "to be on the safe side" there would have been little chance of avoiding a serious disruption to the planned development programme.

Thus although the number of contaminants for which soils may have to be tested has increased, the approach from site to site is not standardised in the sense that all samples are analysed for the full range of constituents. In some cases a staged approach is used so that, for instance, complex cyanide determinations are carried out only if the total cyanide exceeds 20ppm. Appendix II gives a list of the constituents which have had to be determined on the samples obtained. In addition, where the samples contain significant quantities of "coal", "coke", "ash", and "slag", (using these terms in some cases very loosely) calorific value and proximate analysis have been carried out. Also as mentioned above methane determinations have been made on some sites particularly those which occur over silt beds or on which to any marked extent dredged river silt has been deposited.

With regard to the analysis of samples it has proved necessary (when using outside agencies) to specify precisely the techniques to ensure consistency of approach. In many cases it has proved possible to use methods contained in reference works but in some cases techniques have had to be developed by the Scientific Branch.

An option which was explored in order to improve throughput and to reduce costs was that of using automatic analytical equipment, (of which the Branch had a considerable experience in a number of fields), to handle the main analytical load for soils and water analysis. Although it would have proved advantageous in certain respects to have done so the lead time for such a system to become fully operationsl was considered too long to meet the immediate pressing requirements and so it was decided not to proceed along these lines.

Interpretation of Data

Chapman considered the following in relation to drawing up guidelines -

(i) direct ingestion
(ii) inhalation
(iii) indirect ingestion
(iv) phytotoxicity.

Now to these perhaps could usefully be added -

(a) possible dermatitis problems, and
(b) effects on fauna.

A number of problems and possibilities have emerged in attempting to draw up guidelines for the various contaminants. The hazards from direct and indirect ingestion of soil and dust and toxicity to plants were discussed in the paper by Chapman et al. These principles have been used in drawing up our guidelines. We have also used our experience of the typical values obtained in "uncontaminated" London soils as a base for our considerations. The range of values for each contaminant is then considered against the risks to health and the environment. Thus the risks from cadmium are critical with regard to its uptake into vegetables, whilst for lead the risks would be from direct ingestion or inhalation. On the other hand boron and nickel are of more concern with regard to their phytotoxic effects.

Using these principles the present approach is shown in Appendix III. However, it must be emphasised that these are by no means fixed attitudes and the range of values have to be modified occasionally as new information is obtained.

Although originally drawn up as a result of the Council's investigation, some of the contamination levels given in Appendix III have been endorsed by other investigators such as the CUEP. What is now clearly needed are central government guidelines on the levels which can be regarded as acceptable. Additionally, some guidance needs to be given on sampling strategy and the reduction of the data obtained to parameters which can be compared with any guideline set. The values given in Appendix IV indicate the range of results which can be obtained from one site. It can be seen that whatever guideline is used a completely different opinion can be formed on the degree of contamination depending on whether, for instance, the median or the maximum values are used as indicating the level of contamination.

The results given in Appendix IV illustrate an additional problem which has been encountered - that of the difficulty in providing adequate protection to the structures including reinforced concrete from all the materials which may be present in the ground. The presence on some sites of such large concentrations of phenol, toluene, cyclohexane extractable material etc indicates that not merely will plastics piping be at risk of attack but also contaminants may transfuse through the plastics materials. Such transfusion may cause unpleasant tastes in, for instance, drinking water without causing the physical degradation of the pipes in question. Additionally, such organic contaminants may cause the degradation of materials currently available for use as protection to metallic components which need to be in contact with the contaminated soil.

A matter not discussed in the Chapman paper but important in the option given in the decision tree in Appendix II is that of disposal of contaminated soil. Both costs of removal and the limited availability of sites to which contaminated soil can be taken dictate that sites suitable for the most contaminated of materials are not used up unnecessarily with cleaner soils. To ensure that this is avoided, attempts have been made to draw up a scoring system to divide material coming from the sites into

(1) clean - can be moved to a normal "muck away chute";

(2) contaminated - not notifiable under Deposit of Poisonous Waste Act 1972 but should be treated as industrial waste - it is suitable for non-hazardous industrial or domestic waste disposal sites;

(3) highly contaminated - notifiable under the Deposit of Poisonous Waste Act 1972 - must go to an appropriately licensed waste disposal site.

The toxicity and hazard of any particular contaminant is determined by comparison to a series of working limits for that contaminant and scored accordingly -

eg for Cadmium - below 4ppm - clean - score 0
between 4ppm and 12ppm) contaminated (score 1
between 12ppm and 20ppm) (score 2

over 20ppm - highly contaminated - score $2x\left(\frac{Cdppm}{20}\right)^2$

High values of contamination are thus heavily weighted by the square factor (which it must be emphasised has at the moment no theoretical basis) which helps to counterbalance the smoothing effect of the rest of the scoring system.

The ten worst resultant scores from the contaminants analysed are averaged to give a score for the sample. This can be further combined with other results from the same trial pit, depth or part area of the site being surveyed. The designation of the material on the site is made against two fixed limits defined in relationship to the scoring system. Any material that averages a score greater than 1.5 is notifiable under the Deposite of Poisonous Waste Act and material that averages significantly below 0.25 is regarded as clean. For results in between, the volume domain of trial pits and vertical stratification of the contamination needs to be carefully considered.

Such a crude hazard assessment depends much on the setting of the individual working limits of each contaminant and at the moment is deficient in that it does not take into account any synergistic effects and the like. These were set by comparison with a series of data, eg consumer protection regulations, average soil concentrations, phytotoxicity, TLV's, LD_{50s} etc. In many cases they are no more than careful 'guesstimates' as data is extremely limited and some are difficult to apply to this use.

So far this paper has concentrated on an up-dating of the technical basis and the approach to the problems discussed by Chapman. However, perhaps the most significant developments which have taken place since Chapman published his paper have been organisational. It has been found that the task of adequately appraising sites under consideration for development by the Council or falling within the statutory control of the Council, has required considerable professional inputs from disciplines other than those normally associated with sampling, testing and drawing up guidelines. As far as possible the aim has been to ensure that a single all embracing procedure has been used to appraise sites in advance of design work beginning and to ensure that soil contamination, important as it has proved, was not seen as the only aspect of site appraisal which needed consideration. In particular, attempts have been made to ensure that air pollution and noise considerations to which the Council has paid considerable attention for some years, were taken into account as far as possible in a revised form of site appraisal rather than being dealt with separately. Such attempts have been aided by two early decisions -

(i) to devise methods of investigating the air and noise pollution which were less labour intensive than those generally used; and

(ii) to draw up guidelines for the most important air pollutants (see Appendix V) and the most significant aspects of noise pollution (see Item No.6 GLC Development and Materials Bulletin No.96 June 1976 "GLC Guidelines for Environmental Noise and Vibration).

Such an all embracing approach is obviously beneficial. For instance, if the noise climate can be established early in the development process, suitable preventative measures can be considered which do not involve more expensive construction techniques, for instance, by varying the layout. In some cases it is not possible to have all the site appraisal data available before design begins. Not only are some sites not immediately available for such investigation but also when deciding suitable anti-corrosion measures for underground piping, it would be uneconomical to do so. The cost of the investigation to decide such measures is high and it is more cost effective to wait until the layout has been decided and the position of the pipe runs finalised so that these small areas can be investigated in detail more cheaply than dealing with the whole site.

Nevertheless, the main aims organisationally have been -

(i) to coordinate as far as possible the different site investigations so that repeated visits by different specialists are avoided;

(ii) to ensure that the different specialisms undertaking the organisation of such appraisals and involved in the various associated activities make their input at the appropriate time so that the necessary timescales set by the Council can be adhered to.

To meet these aims a code of practice has been drawn up by officers of the Council and it has helped considerably in ensuring that sites have been adequately investigated at minimum cost and with the minimum of delay. However, this code of practice is based on the needs of the Council and its organisational method and there is little doubt that a similar more general document should be prepared by some national body for use by other organisations so that additional costs are not incurred by developers due to unnecessary delays etc.

Appendix I

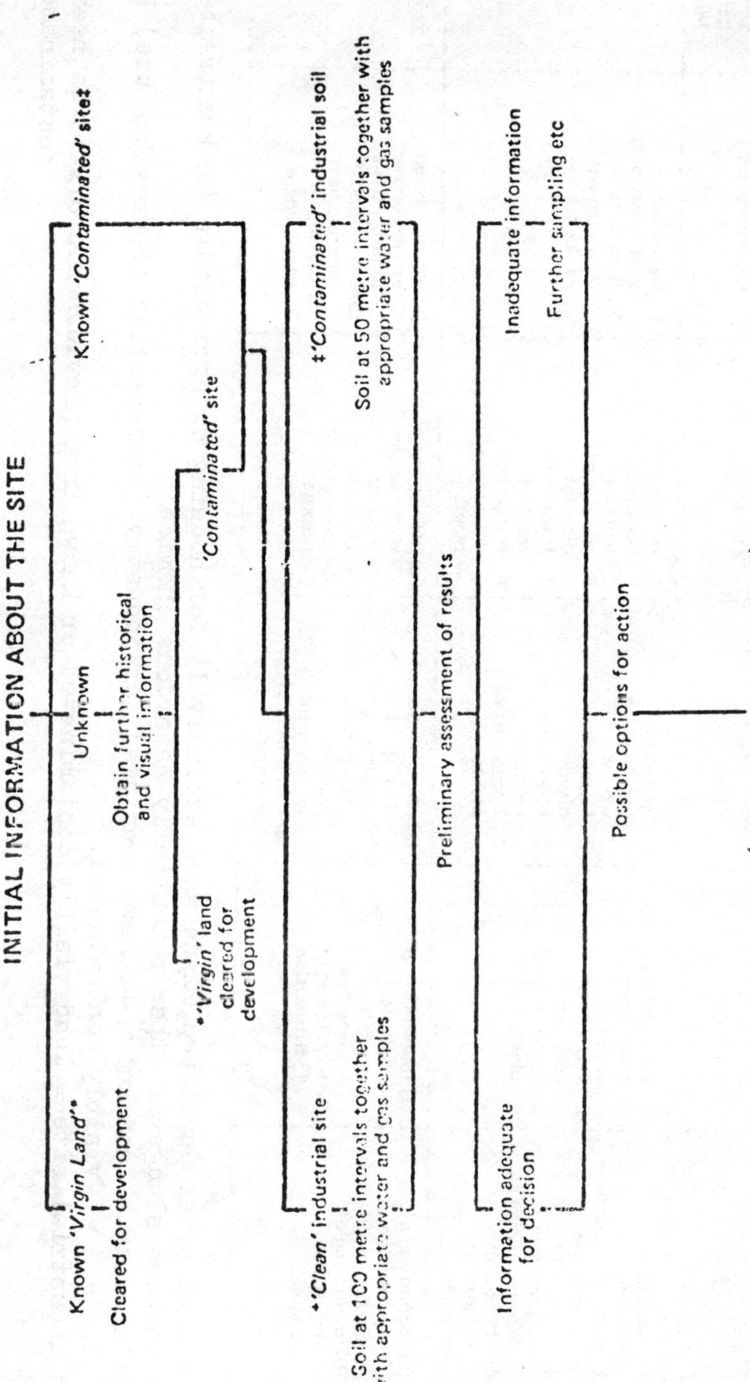

(continued overleaf)

Appendix I continued

Possible options for action

	A	B	C	D	E	F	G
	Hazards insufficient to need treatment → Cleared for development	Capping with X feet of cover	Redistribution on site	Removal over part of the site and replacement with X feet cover	Removal over whole of site to X feet and replace with cover	No development for X years → Decision needed on temporary use	No development possible for any foreseeable future → Decision needed on permanent use, eg, long term car park
Area of concern		1 Hazards to residents 2 Hazards to deep-rooted trees 3 Hazards to construction workers in digging 4 Hazards of disposal of diggings	1 Hazards to construction workers 2 Hazards to residents 3 Hazards to all plants	1 Hazards to construction workers 2 Hazards of disposal 3 Hazards to residents 4 Hazards to all plants	1 Hazards to construction workers 2 Hazards of disposal 3 Hazards to residents 4 Hazards to deep-rooted plants	1 Hazards to temporary users 2 Hazards to reinstatement workers	
Further testing		(a) monitoring of construction workers environment (b) monitoring of disposal of diggings	(a) monitoring of construction workers environment (b) monitoring of redistribution	(a) identification of 'hot' spots (b) monitoring of construction workers environment (c) monitoring of disposal	(a) monitoring of construction workers environment (b) monitoring of disposal	(a) monitor water and gas on say monthly basis (b) monitoring of reinstatement workers environment	

*Virgin land = land never developed, land used exclusively for housing or agriculture land untreated with sewage sludge or other potentially hazardous chemicals

† Clean industrial site = land known to have been used for industry (irrespective of previous use) but not visibly contaminated.

‡ Contaminated industrial site = land visibly contaminated or known from previous history to be heavily contaminated.

1. (a) pH (values outside the range 5-9, dependent options i/acidity ii/ alkalinity)
   (b) Acid Soluble Sulphate (5% HCL extract)
   (c) Magnesium (dependent option - when sulphate exceeds 1%)

2. (a) Elemental Sulphur
   (b) Sulphide
   (c) Total Cyanide, (dependent options FerroFerricyanide, Thiocyanate and Free Cyanide, when total cyanide exceeds 20ppm)

3. (a) Phenols
   (b) Toluene Extract
   (c) Cyclohexand Extract
   (d) Coal Tar (dependent option - when toluene or cyclohexane extract exceeds 5000ppm)
   (e) Organic Chlorine
   (f) Mineral Oil (dependent option as for Coal Tar)

4. Total Metals:- (a) Lead
   (b) Cadmium
   (c) Mercury
   (d) Antimony
   (e) Arsenic

5. Available Metals:- (a) Zinc
   (b) Nickel
   (c) Lead
   (d) Copper
   (e) Boron

6. Miscellaneous:- (a) Manganese (Total)
   (b) Chromium (Total)
   (c) Selenium (Total)
   (d) Barium (Total)
   (e) Vanadium (Total)
   (f) Beryllium (Total)

Appendix III

Guidelines for Contaminated Soils - Suggested Range of Values ($mg\ kg^{-1}$ on air dried soils, except for pH)

Parameter	Typical values for uncontamnt'd soils	Slight Contamination	Contaminated	Heavy Contamination	Unusually heavy contamination
pH (acid)	6-7	5-6	4-5	2-4	> 2
pH (alk)	7-8	8-9	9-10	10-12	> 12
Antimony	0-30	30-50	50-100	100-500	> 500
Arsenic	0-30	30-50	50-100	100-500	> 500
Cadmium	0-1	1-3	3-10	10-50	> 50
Chromium	0-100	100-200	200-500	500-2500	> 2500
Copper (avail)	0-100	100-200	200-500	500-2500	> 2500
Lead	0-500	500-1000	1000-2000	2000-1.0%	> 1.0%
Lead (avail)	0-200	200-500	500-1000	1000-5000	> 5000
Mercury	0-1	1-3	3-10	10-50	> 50
Nickel (avail)	0-20	20-50	50-200	200-1000	> 1000
Zinc (avail)	0-250	250-500	500-1000	1000-5000	> 5000
Zinc equiv.	0-250	250-500	500-2000	2000-1.0%	> 1.0%
Boron (avail)	0-2	2-5	5-50	50-250	> 250
Selenium	0-1	1-3	3-10	10-50	> 50
Barium	0-500	500-1000	1000-2000	2000-1.0%	> 1.0%
Beryllium	0-5	5-10	10-20	20-50	> 50
Manganese	0-500	500-1000	1000-2000	2000-1.0%	> 1.0%
Vanadium	0-100	100-200	200-500	500-2500	> 2500
Magnesium	0-500	500-1000	1000-2000	2000-1.0%	> 1.0%
Sulphate	0-2000	2000-5000	5000-1.0%	1.0-5.0%	> 5.0%
Sulphur (free)	0-100	100-500	500-1000	1000-5000	> 5000
Sulphide	0-10	10-20	20-100	100-500	> 500
Cyanide (free)	0-1	1-5	5-50	50-100	> 100
" total	0-5	5-25	25-250	250-500	> 500
Ferricyanide	0-100	100-500	500-1000	1000-5000	> 5000
Thiocyanate	0-10	10-50	50-100	100-500	> 2500
Coal Tar	0-500	500-1000	1000-2000	2000-1.0%	> 1.0%
Phenol	0-1	2-5	5-50	50-250	> 250
Toluene extract	0-5000	5000-1.0%	1.0-5.0%	5.0-25.0%	> 25.0%
Cyclohexane extract	0-2000	2000-5000	5000-2.0%	2.0-10.0%	> 10.0%

Appendix IV

Contaminant	Median (Parts per million)	Maximum (Parts per million)
Free cyanide (CN)	1	80
Free sulphide (S)	16	8,600
Ferri-ferroxyanide ($FeCN_6$)	20	10,000
Thiocyanate (SCN)	2	4,500
Arsenic (As)	16	463
Mercury (Hg)	1	41
Total lead (Pb)	430	6,840
Soluble lead	60	3,120
Cadmium (Cd)	0.5	62
Zinc equivalent	150	11,000
Phenols	0.5	195
Coal tars	0.1 per cent	25 per cent

Appendix V

GUIDELINE CONCENTRATIONS FOR AIR QUALITY ASSESSMENT

Pollutant	*Method(s) of measurement*	*Period of measurement*	*Guideline Concentrations averaged over the measurement period*
Sulphur dioxide	(a) British Standard procedure (b) Electrochemical	Over one year at daily intervals	60 μg/m³ with 98% of observations below 200 μg/m³
Suspended particulates	British Standard procedure	Over one year at daily intervals	40 μg/m³ with 98% of observations below 120 μg/m³
Carbon monoxide	(a) Electrochemical (b) Gas chromatography (c) Dispersive and non-dispersive infra-red	8-hourly intervals 1-hourly intervals†	10 mg/m³ 40 mg/m³
Ozone	(a) Ultraviolet absorption (b) Chemiluminescent	1-hourly intervals	160 μg/m³
Lead in suspension	Whatman GFA filter trap and dithizone analysis	3-monthly intervals	3 μg/m³
Elemental mercury	Ultraviolet absorption	8-hourly intervals	10 μg/m³

μg/m³—micrograms per cubic metre
mg/m³—milligrams per cubic metre

†The chairman of the Planning Committee, in submitting the report, altered it by the substitution in line 6 of the third column in the appendix of '1-hourly' for '4-hourly'.

HISTORICAL INVESTIGATION OF SITE USE

M J Dumbleton

Transport and Road Research Laboratory, DOE DTp

The past use of a site may have left chemical or biological contamination or physical intrusions, including fill, voids or foundations, which may affect its suitability for the intended future use, and which must be taken into account in the course of new construction or redevelopment. A preliminary appreciation of natural ground conditions and the remains and effects of previous use allows the suitability of the site for the proposed future use to be assessed at the planning stage, the cost of any remedial treatment or special construction measures to be estimated, and the ground investigation required before the site is redeveloped to be effectively and economically designed. This paper reviews the sources of information available for the preliminary appreciation of past site use, and discusses gasworks and brickworks as examples.

INTRODUCTION: INTERFERENCE WITH THE SITE BY PAST USE AND ITS SIGNIFICANCE

Previous use of the site, for example for industrial or mining purposes or for the disposal of waste material, may have left interference which can be conveniently divided into two types:

Chemical or biological contamination (including toxic or dangerous substances and pathogenetic organisms).

Physical intrusions (including foundations, fill, voids and disturbed ground).

Chemical and biological contamination can affect the health and well-being of plants, animals or people living on or frequenting the site, including site investigation and construction workers, and the wholesomeness of food plants and water supplies. It can also affect the serviceability of engineering materials, for example sulphates may attack concrete or organic chemicals may attack plastic pipes.

Physical intrusions will themselves have to be taken into account in construction work on the site. They can affect design as when fill or mining voids call for special foundations, or construction as when a tunnel runs into fill material or an excavation encounters massive buried masonry. Physical remains also have relevance to contamination. The physical remains of a contaminating activity may themselves be contaminated, and are a pointer to the possible presence of contamination in their vicinity. Masonry may be chemically or biologically contaminated, tanks, lagoons or pits may contain contaminated material or this material may have leaked from them into the surrounding ground, and underground cavities may have collected liquid contaminants or dangerous gases.

PROCEDURE FOR INVESTIGATING PAST USE OF THE SITE

Guidance on the site investigation as a whole, related particularly to road and tunnel projects, has been given elsewhere. (1,2) Sources of information for the preliminary appreciation of the site, particularly those relating to geology and natural conditions, are given in TRRL Report LR403, Preliminary Sources of Information for Site Investigations in Britain (3).

In this paper the preliminary appreciation of the past use of the site is developed and examples are given. The following paragraphs of this section discuss the aims and content of the Preliminary Appreciation, and the related observations made in the later stages of the site investigation.

Preliminary Appreciation of Past Use and Ground Conditions

At this stage information is searched for and examined to assess the suitability of the site for the proposed use, to assess the extent to which special treatment may be required, to make preliminary estimates of cost, and to plan more detailed investigations.

It is important that the effects of any previous use should be identified as early as possible in the planning process, so that they can be taken into account in the decision to acquire the site, and in the planning of its development so that use is compatible with any contamination remaining after treatment.

At the same time an early indication is also required on the suitability of the geological ground conditions for the proposed use, and their effect on design and costs. In addition, interference by man needs to be distinguished from natural conditions and the effects of natural processes. It is therefore appropriate that the preliminary appreciation of geological conditions and past site use should be made together, especially as they will often be related.

Ground Investigation Before and During Construction

This is the investigation to obtain the full range of information required for the treatment of the ground and the design of the works. To carry it out it is necessary to make disclosure of intentions, to obtain access to the site, to incur considerable expenditure, and in some cases to break into or risk damaging existing structures, with the additional possibility of producing leakage of contaminants. The work will be planned more effectively and economically if the Preliminary Appreciation has provided information on the nature of the undisturbed ground and the probable nature, location and depth of physical intrusions and chemical contaminants.

Observations should continue during construction, when previous predictions can be checked against reality and further observations can be made as the site is opened up.

INFORMATION ON PAST USE

The information required on past use includes:

- the general nature of the site, and the owner or occupier.

- the layout of the site: buildings, surface and underground installations, etc; the extent of persisting remains.
- the activities and processes carried out, the raw materials and products, the waste products and method of disposal.
- the nature and extent of any contamination of the ground or of remaining installations.
- the composition of fills and other deposited material.

If possible the information should be specific to the site under consideration, but information relating to similar sites may be of use to indicate the conditions that are likely to occur.

A greater or lesser amount of information on the past use of a particular site almost certainly exists, probably distributed in a considerable number of places. The problem is to bring it to light. The range of possible sources of information is large but only some of them will be relevant or available for a particular site.

WHERE TO GET INFORMATION

Site Owners and Occupiers

Previous owners or occupiers should ideally be able to give the fullest account of the site, and every effort should be made to get the maximum amount of information from them.

Local Collections

These should be explored before the national collections because they are likely to have local information that is relatively easily found. Local, city, county and university libraries often have particularly good collections of published texts relating to the industries of the district. Their local collections include directories, maps, Acts of Parliament, histories and trade catalogues of local firms, and unpublished material relating to the locality.

County Record Offices and City Archivists Departments contain large collections of maps, plans, estate records, deeds, leases, sale catalogues, rate books and many other classes of document, or this material may be held by a local library. Museums and university departments may have libraries and collections of documentary material and their staff may be able to offer advice and suggestions on local conditions.

Local Authorities, Regional Offices and Further Afield

District and County Council Offices have much information gained in the course of their planning and other functions, and in some cases will be the owner or occupier. Some members of their staffs have much personal knowledge of the affairs of the district. The Derelict Land Units set up by some local authorities should also be able to provide information and advice.

The local offices of the Factory and Alkali Inspectorates may be able to give information on the activities and processes carried out in particular premises. Records for coal mining, quarrying and opencast extraction are variously held by the offices of the National Coal Board and the Inspectorate of Mines and Quarries, and in the Mining Record Office in London. Firms of mining consultants and surveyors may have access to valuable plans and other information. Sources of information on mineral extraction are discussed further in TRRL Report LR403.

The Regional Water Authority may be able to give information on the pollution of streams and ground water, which may indicate the presence of a contaminated site, and will advise on any undesirable effects that the disturbance of a contaminated site may produce. It should also have information on underground water supply and sewerage installations. Information on wells and their pumping installations is available through the publications and records of the Institute of Geological Sciences Headquarters, and its regional offices (see LR403).

Large central libraries such as the National Libraries and the Science Museum Library have most of the published technical literature. Professional institutions and learned societies may have more easily accessible material relating to individual processes and sometimes to particular sites, both in their collections and published in their proceedings. Advice on technical processes and possible lines of investigation for local sites may be obtained from the Science Museum, South Kensington (Chemistry Department), and there are the National Museum of Wales, Cardiff (Department of Industry) and the Royal Scottish Museum, Edinburgh. University departments dealing with the history of technology include the Department for the History of Science and Technology, UMIST, Manchester, and the Centre for the Study of the History of Technology, University of Bath.

SOME TYPES OF SOURCE MATERIAL

Site Records of Owners, Developers and Occupiers

These are the documents produced in the course of the past development and use of the site, and should ideally be the most detailed source of information on it.

Classes of site record. These records include:

Site investigation records-

Reports, plans and sections, including information on the position, depth and method of backfilling of boreholes, trial pits and trenches, the results of observations and tests, and interpretations.

Site development and plant records-

Plans, sections, specifications and photographs of buildings, foundations, underground services and installations. Records of construction experience.

Technical process records-

Details of technical processes including composition of raw materials, products and waste materials. Disposal of waste materials. Production and control records.

Business records-

Deeds, leases and estate agents' or auctioneers' catalogues (for plans and description of the site and its installations). Trade catalogues and letterheads (for product and views of the works). Minute books, accounts, balance sheets and other non-technical records (the least useful).

Finding site records. The owner or occupier should be pressed to make available as much information as possible before the final decision to acquire the site has been made. The company's solicitor, architect or consultant civil engineer may have to be consulted. The records may have been deposited in repositories such as the County Record Office, or they may have passed to other companies in the course of acquisitions and reorganisations.

The National Register of Archives (part of the Royal Commission on Historical Manuscripts) is a central collecting point for information about manuscript sources and their location, including those relating to business and industrial undertakings. Its indexes are open to inspection and limited specific enquiries can be answered by post. The Business Records Advisory Service of the Business Archives Council locates and lists records of business organisations and sends these lists to the National Register of Archives and appropriate local record offices.

Preservation of site records. Unfortunately plans and records are often destroyed. The Business Archives Council provides advice on the preservation and management of business records, and on finding repositories for them such as the County Record Office, libraries or museums. The Archives Panel of the Institution of Civil Engineers has recently published a leaflet giving guidance on preserving and finding repositories for engineering records. The Panel for the History of the Industry, Institution of Gas Engineers, has produced model codes for the preservation of archives and artefacts, and provision has been made for the central recording of the whereabouts of the material on the Segas computer. All concerned should do their best to see that records are preserved and made available when required.

Maps and Plans

A series of maps and plans of different dates provides information on the nature and layout of past development of the site, and numerous sources are available. The Ordnance Survey will supply a complete listing of their maps relating to a named site. Good local series of OS maps are to be found in County Record Offices and libraries, and the Map Room of the British Library has a complete set and will supply photocopies. The 25 inch per mile map (actually 1:2500) was surveyed from 1853, and a typical series is: first edition surveyed 1870, revised 1890, 1910, 1930 and 1960. It shows development in detail and names features such as kilns, chimneys and retort houses. Nearly 400 town plans at a scale of 1:500 were published between 1843 and 1894, showing detail down to the level of lamp posts and drain gullies; the use of premises is frequently named, and street numbers are given allowing correlation with the rate books and directories.

From 1836 to 1860 a series of Tithe Survey maps was prepared for both rural and urban areas, usually at a scale of 1:4752 or 1:2376, with reference books showing owner, occupier and use of sites. Enclosure maps, at similar scales, are usually earlier. Plans have often been depositied for schemes such as gas works, sewerage, canals and railways, often accompanying a necessary Act of Parliament. For example in the nineteenth century many proposals both successful and abortive were made for railways, and any route passing close to the site may provide useful information. Plans for the proposed Staines and Woking and Wokingham Junction Railway, at a scale of 1:4752 and dated 1852, show a strip of land 600m wide following the route, and include plans of kilns and brick sheds at Bracknell, Berkshire. Plans submitted to the local authority in connection with planning and building regulations extend back to the middle of the last centry for some towns.

Private collections of estate records may contain plans, such as those for collieries among the manuscripts of the Duke of Norfolk (4), and for lead mining and smelting in the Bagshawe Collection, both in the Department of Local History and Archives of the Sheffied City Libraries, which also holds the Fairbank Collection of maps and plans (5). This collection includes 4600 draft plans dated 1739-1850, including 910 relating to engineering works.

For some towns insurance plans were prepared showing the structure of buildings including the number of floors and the internal layout and installations. Goad's Insurance Plan of Reading (...1895, 1901, 1907...) is at a scale of 1:480. It shows many industrial buildings including breweries, gas works, galvanising works, and a tannery with tan pits on the ground floor and various workshops above. Detailed plans for insurance purposes were also prepared for individual premises. That for Collier's Brick Works, Reading is reproduced by Major (6) at a scale of 1:1350.

Published Literature

Information on local industrial and mining activity may be found in the Victoria County History and other general local histories, industrial (7) and company histories (8), house magazines (9) (examples are given in the References), local and regional works on industrial archeology, local newspapers, and the proceedings and newsletters of local history, archeology and industrial archeology societies. Local directories can be used to identify the industrial activities carried out on a site, and trade directories cover individual industries such as quarries, pits and brickworks (Directory of Quarries and Pits, 21 edns 1925-72).

Details of individual processes are given in technical encyclopaedias and monographs, including both those treating the subject historically, such as Singer et al, A Short History of Technolgy (7 vols, OUP 1954-78) and modern and older texts such as King's Treatise on the Science and Manufacture of Gas (3 vols, 1876-82). The technical and professional journals of the relevant industries (such as Gas Journal, from 1849, and the Transactions of the Institution of Gas Engineers, from 1903) give details of technical processes and installations and descriptions of individual works.

Air Photographs

A good series of vertical air photographs will probably be obtainable from about 1945, and possibly earlier. They will show all features of the site visible at the time of photographing, and will often allow features seen on plans to be identified. The depth of pits and their history of filling can be examined and buried structures such as foundations and pipelines may be located during installation or from persisting ground markings. Oblique views of the site are shown on some RAF photographs, and Aerofilms Ltd have an extensive collection extending back to well before the war. Guidance on obtaining, examining and interpreting air photographs is given elsewhere (3,10).

Oral Information
Present or former employees, local government officers and local residents may be able to give much information on the former use of the site. The engineer should know as much as possible about the site before interviewing these informants, so that he can engage their interest and understand their comments. Large scale maps, air photographs, photographic views and a site visit are useful to stimulate memory and discussion and to relate information to individual parts of the site. Informants may also have photographs, press cuttings, brochures and catalogues.

Site Inspection
The site should be thoroughly inspected, before demolition or clearing if possible, and without delay if this is imminent. Further inspection should be carried out after a detailed study has been made of records of former site use. The investigator will by then have a better idea of what to look for, and be more able to recognise and interpret significant signs. He should if possible be accompanied by someone who knows the site or similar sites well. It may also be useful to inspect other sites of a similar character. Observation should continue during the clearing of the site, and during site investigation and construction work.

Structures and their uses should be checked against available plans or large scale air photographs, which should be modified or annotated where necessary. All unexplained features on the map or air photograph should be examined on site. Underground structures such as basements and tanks should be measured, and evidence for the position of former structures should be recorded. It may be possible to inspect or enter chambers, tunnels or mine workings. If this is considered, and throughout the work, great caution should be taken to avoid danger from unstable fabric, dangerous substances, or gases. Unauthorised use, for example as a scrap or car breaker's yard, should not be overlooked.

Besides structures and foundations, signs that should be looked for include:

- sewer, manhole and tank covers; old petrol pumps.
- pits, depressions, lagoons, ponds.
- signs of settlement, subsidence, filling and disturbed ground.
- materials such as slag, ashes, scrap and industrial or chemical waste at or near the ground surface.
- discoloured soil, blighted vegetation.
- polluted water and other fluids in seeps, puddles, ponds or streams.

Samples of solid and liquid substances should be taken where appropriate, for identification or analysis.

REPORT ON PRELIMINARY APPRECIATION AND RECOMMENDATIONS FOR GROUND INVESTIGATION

The final stage of the preliminary appreciation of the site is to prepare a report summarising the findings and drawing conclusions. The information brought to light may be sufficient to show that the site is substantially free from adverse features, it may allow an estimate to be made of the amount of remedial treatment needed to make it suitable for the intended development, or it may show that it is unsuitable for further consideration.

If there is still doubt about the status of the site it will be necessary to proceed to a preliminary ground investigation before the final decision is made. Recommendations should be made on the physical features of the site that are to be examined or searched for, the points at which pits or boreholes should be sited and the depths to which they will probably need to be taken. Recommendations should also be made for the sampling and testing programme, and chemical and biological materials that may be present should be named. Attention should be drawn to any possible hazards to site investigation workers.

EXAMPLES

Brickworks and gasworks sites will be considered, to illustrate the use of some of the sources that have been described. It is assumed that as much information as possible will have been obtained direct from the previous owners or occupiers.

Brickworks

Digging of clay will often have produced general lowering of the ground over a large area with loss of topsoil and interference with drainage, as well as deeper excavation with the possibility of fracture and deformation of the surrounding ground due to stressing. Ill-compacted, unstable or contaminated fill may have been placed in the excavations, with consequences for foundation design and possible effects on ground water and release of gases. Kilns have extensive undergound flues and tall chimneys have deep foundations that will probably be left when the site is cleared. The use of a kiln over a long period can produce drying of the clay soil beneath it to a great depth. The slow re-wetting of the soil after the kiln is demolished can produce ground movements over a period of many years, with damage to later buildings constructed on the site (11).

Sources of information. Information will be considered for a brickfield site west of Bracknell, Berkshire, on London Clay. The Binfield Brick and Tile Company was in occupation from 1899 as shown by local directories, and by the List of Quarries for later years (HMSO). A local historian photographed the works in 1964 and 1970, providing information on the kilns and on the structure of the heated floors of the drying sheds. In 1972 another local historian obtained information from a former employee, including a sketch plan identifying the kilns and buildings and showing their internal arrangement; a short account was published in the Berkshire Industrial Archeology Group Newsletter, September 1972, which gave the name of the occupiers after the works ceased operation in 1964 - who stated that they had destroyed the site records. The sketch plan and both sets of photographs were deposited with the County Industrial Archeology Representative of the Council for British Archeology Group 9, who made them available on request. The British Brick Society, which deals with the history of brick-making, made useful suggestions on sources of information and offered to circulate members through its newsletter. In 1978 a local society published an account of the works including information provided by another former employee who had known them since 1926 (12).

Ordnance Survey maps at scales of 6 inches per mile (1:10,560) and about 25 inches per mile (1:2500) were issued for the area in 1869-72 and revised in 1898, 1909, 1931-32 and 1964-65. A search of planning and building by-law applications made to the Local Authority brought to light a detailed plan of the site, dated 1946 and showing buildings constructed since the 1932 OS revision, including three beehive kilns which were demolished before the 1965 revision. The plan bore the name of the consulting engineers who were contacted, but they had destroyed their records relating to the site.

Air photographs (AP) were available from 1945, 1955, (1960), 1961, (1964, 1965), 1967, (1969, 1971, 1973 and 1976) - those in brackets were not seen in time to include in the analysis. They provided information on plan, elevation of buildings, and depth and filling of excavations.

Information on site use. At No 2 Works extensive buildings and a large Hoffmann kiln measuring 45 by 16m were constructed between 1898 and 1909 (OS). This kiln was replaced by nine beehive kilns in 1920 (information from employee but not mapped until OS 1932). Three more beehive kilns were added by 1945 (AP 1945 and building application plan 1946). All the kilns were in place in 1961 (AP) but demolished by 1965 (OS).

At No 1 Works major development also occurred after 1898. By 1909 (OS) there were extensive buildings and two Hoffmann kilns (identified from the sketch plan) - a 38 by 14m 'Brown's' patent kiln of 12 chambers with a 120 ft high chimney dated 1902 on a brick, and a 42 by 15m 'Osman' kiln of 14 chambers. Six beehive kilns and a Manchester type rectangular downdraft kiln had been added by 1932 (OS). The kilns are shown in elevation in the photograph of 1964, and a textbook gives plans and sections showing typical construction of the several types present (13). The Hoffmann kilns worked almost continuously until 1963, but one of the tall chimneys was demolished in 1964 (local newspaper report) and all kilns were demolished by the OS revision of the same year.

There were extensive areas of excavation, in parts deep and eventually flooded. These were studied on the air photographs, and measurements of depth of excavation and thickness of fill could have been made by photogrammetric methods. At No 2 Works little or no excavation took place after 1951 and the deepest part of the pit had been recently filled in 1967. At No 1 Works the east pit was flooded and no excavation took place from it after 1932, it appears. Between 1955 and 1961 three-quarters of its area was filled. The air photograph of 1961 shows an orderly and trim operation with a vehicle at work, and that of 1967 shows the same extent of fill, by then covered with vegetation. The north-east pit was deep and flooded in 1955 and vegetation had appeared around it by 1961. By 1967 it was being filled from the south and east sides. The Local Authority provided copies of more recent licenses to deposit fill stating permitted composition, and had measurements on fill composition and data on unauthorised depositions.

Gasworks

The gasmaking industry was widespread, with a gasworks at one time or another between 1806 and about 1970 in almost every town or village of any size. After an initial period of proliferation there were successive closures following amalgamations and concentration of production, so that some sites were succeeded by a later development while others were reconstructed and extended.

This development is illustrated by Stewart's (14) review of the 83 gas-works sites in the area served by the North Thames Gas Board. There were also at least 32 private works. The metropolis was supplied by the year 1835, the towns by 1840 and the villages by 1860, and the last new works was built in 1883. The maximum number of works producing at once, 53 between 1865 and 1905, was reduced to 18 by 1957. Since the introduction of natural gas, manufacture of gas has virtually ceased. Thus there are numerous gasworks sites available for redevelopment, some of them having been occupied by a succeeding phase of use. All of them must be assumed to be severely contaminated by disposal, spillage or leaching of products, waste and stored materials, and they may have underground structures including tanks which may contain fluids or contaminated fill material.

Gasworks in Reading. There were initially two companies operating in Reading and, in all, four gasworks sites and nine main gasholders (two of them small ones from early days). Their history illustrates the effect of pressure of space in confined urban sites, resulting from an eightfold growth in the local output of gas between 1872 and 1910 (15). A brief search in the Local Collection of the Central Public Library, Reading (CPL), and at the Berkshire Record Office (BRO) brought to light old Ordnance Survey and insurance maps, Parliamentary bills with their accompanying plans, and two short accounts issued by the Reading Gas Company, one titled 'Visit of the Southern District of the Association of Gas Engineers to the Reading Gas Works, June 30 1904' (reissued in 1905), and the other a 50-year retrospect issued in 1912 (15).

The Reading Gas Light Company was established in 1819 (copy of the trust deed, CPL) with works on a small central site in Bridge Street. Goad's Insurance Plan of 1901 shows a retort house and two 'gasholders', one of which was still in use in 1905, but the site had been redeveloped for other use by 1933 and yet again by 1957 (OS).

The rival Union Gas Company was established in 1836, and amalgamated with the earlier company in 1862 to form the Reading Gas Company. Its Kings Road Station was on a narrow site in the angle between the River Kennet and the Kennet and Avon Canal. Its gasholder is shown on the 1:450 scale map dated 1869 which accompanies the Parliamentary bill to enlarge the works, taking in properties including one described as workshop, tan-pits, land and buildings (BRO). The 1:500 scale OS map of 1877 shows the new developments, including retort house on the site of the earlier gasholder, purifying house, 'gasometer' over the old tannery, and tar tank. The 1879 map (BRO, see below) shows the internal arrangement of the purifying house.

In 1880 the company made a Parliamentary application for new works on open ground to the north of the River Kennet, alongside the railway, as shown on the accompanying plan dated 1879 (BRO). The account of 1905 (CPL) describes the works and identifies its plant on a 1:1200 scale map, and the account of 1912 (15) maps the position of gas mains and the internal layout of buildings. The retort house and coal store measured 94 by 36m and coal was delivered at first floor level on two railway tracks, each supported on massive brick arches, as shown by a photo-graph. There were three main gas holders, tower scrubbers, a purifying house with 8 purifiers, oxide of iron sheds, oil gas plant with oil tank and purifiers, four tar and liqour tanks, and sulphate of ammonia works with liqour tank and purifiers. One of the holders, started in 1882, was in a brick tank and there was much trouble with water flowing into the excavation, so the next holder, erected in 1901, was placed in a steel tank. The excavation met gravel at 4m, on which a bed of concrete 1m thick was laid to form a foundation for the 47m diameter and 7m deep tank. A fourth gasholder appears on the 1933 OS revision; the 1882 holder was gone by 1956 and the 1901 holder by 1971 (OS).

On the fourth site, to the north-east of the third and on the other side of the railway, a gasholder had been built by 1933 and a second by 1956 (OS).

This account illustrates the information that could be quickly obtained from immediately available sources. A more complete set of maps could have been examined, as well as air photographs. Technical journals could have been searched for descriptions of the works, as illustrated by the lists of references given by Stewart (14). Most important, any records of the Gas Board itself could have been called upon.

CONCLUSIONS

Some of the numerous possible sources of relevant information on the past use of a site have been described, but only a few of them are likely to prove relevant for any particular case. After the more accessible and potentially fruitful sources have been examined it will be a matter of judgment how much further effort should be put into obtaining information by these means.

The suggested procedure for investigating the past use of the site and its relevance to future development may be summarised as follows:

1. Get as much information as possible from the previous occupier, and supplement it with information from others who know the site including employees, local residents, local historians and industrial archeologists.

2. Make use of documents and published information on the site, including maps and plans, air photographs, local and company histories, and papers relating to the site in technical and professional journals.

3. Supplement this information with general information, from the technical literature, on the characteristics of the types of plant and processes used.

4. Inspect the site, before demolition or clearance if possible, to record evidence of previous use and details of physical remains including foundations, fill and contaminated ground.

5. Obtain as much of this information as possible in a Preliminary Appreciation stage of the site investigation. Use it at the planning stage to assess the suitability of the site for the proposed use, and later to design the ground investigation to obtain the more detailed information that will be required if development is to proceed.

The work described in this paper forms part of the programme of the Transport and Road Research Laboratory and the paper is published by permission of the Director.

REFERENCES

1. DUMBLETON M J and G WEST. Guidance on planning, directing and reporting site investigations. Department of the Environment, TRRL Laboratory Report LR625. Crowthorne, 1974 (Transport and Road Research Laboratory.)

2. DUMBLETON M J and G WEST. A guide to site investigation procedures for tunnels. Department of the Environment, TRRL Report LR740. Crowthorne, 1976 (Transport and Road Research Laboratory).

3. DUMBLETON M J and G WEST. Preliminary sources of information for site investigations in Britain. Department of the Environment, TRRL Report LR403 (revised edition). Crowthorne, 1976 (Transport and Road Research Laboratory).

4. Catalogue of the Arundel Castle manuscripts in the Sheffield City Libraries. 1965.

5. Guide to the Fairbank Collection of maps, plans and surveyor's books ... in the Reference Library. 1936. (Sheffield City Libraries).

6. MAJOR J K. Fieldwork in industrial archeology. London, 1975 (Batsford).

7. CHANDLER D and LACEY A D. Rise of the gas industry in Britain. 1949 (British Gas Council).

8. EVERARD S. The history of the Gas Light and Coke Company. 1949.

9. The Co-Partners Magazine of the Gas Light and Coke Company 1911-1950.

10. DUMBLETON M J and G WEST. Air photograph interpretation for road engineers. Ministry of Transport, RRL Report LR 369. Crowthorne, 1970 (Road Research Laboratory).

11. COOLING L F and W H WARD. Some examples of foundation movements due to causes other than structural loads. Proc. 2nd Int. Conf. Soil Mechanics and Fndn Engrg, Rotterdam. 1948. 2, 162-67.

12. DUMBLETON M J. Brickmaking: a local industry. 1978 (Bracknell and District Historical Society).

13. CLEWS F H. Heavy clay technology. London 1969. 2nd edn (Academic Press).

14. STEWART E G. Historical index of gasworks past and present in the area now served by the North Thames Gas Board, 1806-1957. (1957),

and Supplement (1970).

15. HELPS D H. The Reading Gas Company. A Retrospect. 1912.

THE SAMPLING AND ANALYSIS OF POLLUTED SITES.

P.S. WATERHOUSE

I.C.L.S. LABORATORIES LIMITED.

By way of introduction, I.C.L.S. Laboratories have gained considerable experience in recent years in the development of contaminated sites. This experience has led to the formulation of a site management programme, which may be summarised as follows:

1) Collation of available information including site history, plans, hydrogeological data etc.

2) Site visit for detailed inspection and sampling grid layout.

3) The sampling programme.

4) Analysis of the samples.

5) Reporting, with detailed comments and recommendations for remedial actions.

6) Monitoring on site during reclamation and development.

7) Long term monitoring where appropriate.

This paper deals with stages 2) to 4) of this programme.

It is convenient to divide the contaminated sites into three groups, to assist mainly in determining sampling frequency, but also in many instances the analytical programme. The three groups are as follows:-

Group 1 - Single pollutant, single usage site. e.g. Coal wharf, cyanide hardening shop.

Group 2 - Multiple pollutant single usage site. e.g. Gas works.

Group 3 - Multiple pollutant multiple usage site. e.g. Industrial waste tip.

A further division is possible in terms of end usage of the site:-

a) Hard cover with limited exposure time e.g. Car park.

b) Hard cover with medium exposure time e.g. Industry.

c) Soft cover with medium exposure time e.g. Parks, playgrounds.

d) Soft cover with high exposure time e.g. Residential housing.

SITE VISIT.

During the site inspection attention is paid to the general appearance of the site; in particular note is taken of discreet surface tipping, evidence of asbestos, surface colour, general vegetation, surface and subsurface structures general geography of the area and slope of the site. A photographic record of the site is kept for further reference. During sampling all excavations, boreholes and underground structures are monitored for toxic gas emission, a typical range being hydrogen sulphide, hydrogen cyanide, methane, and carbon monoxide. Whenever ground water is encountered, samples are taken for analysis, to assist in estimating the solubility, spread and mobility of pollutants.

SAMPLING.

1) Frequency.

The statistical significance of sampling frequency has been of considerable interest to our organisation. We have concluded that a 95% confidence level of location of major pollutants on a site is an acceptable standard, and therefore base the sampling frequency to achieve this end. Typical examples are:

0.5 Hectare site	10 metre grid	66 sample points.
5.0 Hectare site	20 metre grid	140 sample points.
16 Hectare site	30 metre grid	200 sample points.

Deviation from these sampling grids is acceptable in a limited number of cases, typically:

a) Group 1 sites such as coal wharfes, where calculation of site average pollution may be valid. On such a site of say 10 hectares, a 50 metre grid, generating 55 sample points, is considered satisfactory.

b) Group 3 sites may be very large and financial expediency dictates that larger grids, of say 50 metres, may be employed.

2) Depth.

The depth to which it is required to sample at each point shows greatest variation. In our experience the extent of contamination from a coal wharf rarely extends beyond 0.5 m; pollutants from industrial sites such as Plating Shops may be found at depths of 5 metres. Domestic or Industrial Tips should be examined to the depth of fill. Whilst deep sampling is 'very' costly in itself, sampling in conjunction with the drilling of bore-holes for structural purposes provides a viable alternative. Clearly the technique used in sampling shows a similar variation, from digging trial pits to depths of 2-5 metres with a JCB excavator, to drilling to obtain core samples for both chemical and physical analysis.

3) Analysis.

A comprehensive analytical programme allows for all major means of potential hazard to be covered. These hazards may be classified as follows:-

a) Direct Contact.

The work-force during development and construction should be advised of contact-hazards, in order that appropriate protective equipment is made available. In our experience this requires analysis for skin irritants, which may also include potential carcinogens. Typically phenols and thiocyanates are determined on a basic soil extract. On Gas Works sites a hot toluene extract is prepared, its value giving an indication of the extent of the hazard. The toluene extract is further analysed to determine the proportion of mineral oil and coal tar derivatives therein; coal tar derivatives include polycyclic aromatics, polyhydric phenols and ammonia liquors.

b) Ingestion.

The hazard of ingestion may be subdivided into two groups:

i) Direct Ingestion.

The potential hazard of soils containing significant levels of cyanides or toxic metals is dramatically illustrated by children with PICA. A basic extract is therefore examined for total and free cyanide, and a further extract prepared to analyse for total contents of lead, cadmium, arsenic, selenium, antimony, chromium, and mercury (should there be any indication from the site history of any other toxic metals being present, these clearly are also included).

ii) Indirect Ingestion.

The ability of some plants, including vegetables, to concentrate some toxic metals is also taken into consideration, wherever the possibility exists that in its end use the site may be used for the growing of vegetables. Accordingly the concentrations of "available metals" are determined as appropriate. Similarly the "zinc equivalent" of the sample is determined, to assess the phytotoxicity of the soil.

c) Inhalation.

In addition to regular on-site monitoring of gas samples, the soil samples are also examined for sulphide, and cyanide as indicated previously, to assess the potential hazard should the environment change to favour release of hydrogen sulphide or hydrogen cyanide during development. This can be relevant particularly where, for example, changing ground water levels might be anticipated. The hazard of inhalation of toxic metals in airborne dust is less significant, though also worthy of consideration. Specific airborne pollutants, typically asbestos, should also be taken into account.

d) Flammability.

Whilst less obvious than the previous hazards, the flammability of soil is a significant factor in certain circumstances. Coal Wharfes and railway sidings may contain large quantities of coal, and some gas works sites may contain up to 10% of elemental sulphur (determined via the toluene extract). The calorific value of the soil is used to assess this hazard; a figure of 3000 BTU/lb is accepted as a safe maximum for the contaminated site, compared with less than 1000 BTU/lb for normal arable soils.

e) Structural.

Concretes may be severely attacked by water-soluble sulphates, particularly if there is a steady ground water recharge from slow dissolution of the sulphate present. Both total and soluble sulphate levels are therefore established, as well as the pH value of the soil, which has a significant bearing in sulphate attack. Other aspects of structural attack include corrosion of ironwork by acidic soils, and attack of plastic's by phenols or coal tars.

This summarises our general approach to the sampling and analysis of soil samples from polluted sites. We are seeking to improve the analytical aspects of the programme. To this end we now include on-site gas analysis using a scanning Infra-Red analyser to detect and quantify unspecified gaseous pollutants, and Gas Chromatography-Mass Spectrometry to identify organic contaminants. The inclusion of a Quantometer in the next twelve months will enable the quantification of a far wider range of metallic contaminants at very low cost, and minimise the rist of potential hazard passing unnoticed.

SOME EXPERIENCES OF TESTING SOILS AND MATERIALS FROM CONTAMINATED SITES

J. W. Figg[1], K. W. Souch[2] and A. P. Tublin[3]
[1]Head, [2]Deputy Head of Chemical Department and [3]Deputy Head of Soils Section
Harry Stanger Limited, Elstree, Herts

INTRODUCTION

Changes in technology and designation of new zones for industry or residential areas have resulted in considerable changes in land use and in particular the redevelopment of older industrial areas. Examples are changes from Victorian small manufacturing complexes to light industrial use (such as the construction of the award-winning Sainsbury's warehouse at Charlton on the site of an old glassworks) and the redevelopment of redundant coal-gas manufacturing works (such as Beckton in East London which was once the largest gas-works in the world and part of which is now in process of redevelopment as an Industrial Park).

A Materials Consultancy can be called in to investigate ground conditions in a redevelopment project at several stages in the contract, by different organisations involved in the work and for a variety of reasons.

Thus the Developer (or his agent) shortly after or even in advance of acquisition of the site may ask for information on the contents of old tanks, pits etc as part of the general evaluation of the area. The Architect may need a preliminary soil survey before deciding on the layout of the new buildings to be erected. The Consulting Engineers will require a comprehensive survey to determine the presence of substances likely to be aggressive towards concrete foundations and buried services (including metal pipework, plastics, natural and synthetic rubber seals etc) in order to determine the precautions which must be taken to prevent premature failure of underground components.

Sometimes the onus is put on the Contractor to check the extent of contamination and hence decide the requirements for special cements or protective coatings for foundations or services. Once work commences Safety Officers and Union Representatives may demand further analyses to determine the nature and possible hazards as pockets of suspicious substances are uncovered during demolition and groundworks. During construction, smells, staining, abnormal setting of concrete and sometimes even actual health problems may all need the services of the Materials Consultant. Finally, the ultimate Client, Housing Authority, Tenant or Estate Agent may require long term reassurance regarding the presence and effects of contaminants in the ground.

It is seldom, if ever, that a single Materials Consultant will be asked to deal with all aspects of the redevelopment of an old industrial site. More usually it will be but a passing involvement in the project and often only a request for assistance when a sudden emergency has arisen during the contract.

EVALUATION OF CONTAMINATED SITES

An ideal testing and evaluation strategy for contaminated sites has been described by Chapman, Baker and Burns (1) and this paper includes some sensible criteria for assessing health risks to both construction workers and residents. Some of the problems associated with the redevelopment of old industrial sites are outlined by Eglinton (2) and similar information is included in the GLC information item (3) which gives part of the background to the Deposit of Poisonous Waste Act, 1972.

The Department of the Environment is well aware of the paucity of information for many redevelopment sites and some of the considerations involved have been outlined by M. A. Smith (reprinted in Reference 4). However, the disparate interests involved in redevelopment projects unfortunately militate against comprehensive studies and, as Bromley (5) has pointed out "analysis is expensive" so there is every incentive to skimp on site evaluation - a short-sighted attitude strongly to be deprecated.

It is easy for the principals of a redevelopment to neglect contamination problems which seem remote hazards but as Jarrett (6) warns, past practices have resulted in actual loss of life and costly remedial works. Old industrial sites are very heterogeneous and inevitably contain pockets of wastes that are easily missed by even the most comprehensive sampling programme. Wherever possible historical records should be checked to supplement the laboratory evaluation programme.

It is against this background that the following Case Studies should be considered.

CASE STUDIES

Case A

This study was a comprehensive investigation for a proposed residential development consisting of both high and low-rise dwellings with associated car-parking and communal facilities. The site had been a gas-works until about 1900 when a chemicals manufacturing company took over the premises. Over the years a wide variety of chemicals had been manufactured or stored on site including acids, esters, pigments, essential oils and gums.

Site investigation commenced as the works were being run down at which time it was found that the site surface was grossly contaminated with coloured crystalline deposits, oily liquids, unpleasant scums on the many pools of water present, lumps of stained and corroded concrete, and there was an all-pervading pungent 'aromatic' smell.

Partial results of the chemical investigations are presented in TABLE 1.

It was concluded that the site was dangerously contaminated from both the environmental and civil engineering points of view. Because of the depth of contamination it was recommended that the site surface should be stripped off, that a water and vapour-proof barrier should be laid consisting of compacted cement-stabilised pulverised fuel ash 1m thick, followed by restoration of levels with selected fill, and covering by 250mm of top soil. After further discussion, it was decided to excavate additional contaminated material, to dispense with the vapour barrier and to use "hoggin" instead of p.f.a.

Case B

Here again the redevelopment of an old gas-works was involved but in this case for an industrial site consisting of warehouse units (up to $7000m^2$ in plan) and offices. The ground was generally level and the soil a mixture of clay, loam, sands and gravel and occasional cemented pebble-beds.

A preliminary soil survey was commissioned by the Developer which included some chemical analyses. (TABLE 2). Tests on similar samples found lead present up to 600m, copper up to 0.2ppm, zinc to 0.1ppm with mercury and chromium present in trace quantities. Barium was not detected in any sample.

From these results it was concluded that the depth of contamination would make it impracticable to remove all the affected soil and it was recommended that partial removal should be carried out followed by blanketing with a substantially water and vapour-proof barrier (cement-stabilised pfa or polyethylene sheet) and covering with selected fill and topsoil. It was also recommended that below-ground concrete works should be designed for sulphate-resistance.

Again, that was the end of the story - until we received a panic telephone call from the Contractor. Trenching works had broken into an old brick sewer which contained an unpleasant, chemically-smelling liquid. Chemical analysis (TABLE 3) showed that the material was clearly residues from gas-works spillage and a further six samples from drains in the area found water with tarry matter up to 2330ppm, phenols up to 250ppm and sulphate up to 1970ppm. The pH was found to be in the range 6.5 to 8.1.

It was recommended that specialist contractors should be called in to dispose of the accumulated liquid and that operatives should be warned of the hazard of skin contact with the liquids. It was reiterated that concrete foundations, pipes etc would require precautions to be taken against sulphate attack and that the setting of fresh concrete could be adversely affected by the heavy metals and phenols present.

Case C

This involved a development located on a site, part of which had included a photographic chemicals factory. In this case the history of the site was known including the fact that considerable quantities of chrome alum had soaked into the ground (ammonium chromium sulphate was a by-product of the manufacture of hydroquinone). It was also known that an explosion in 1927 had caused a major spillage of dimethyl ether and sulphuric acid to produce dimethyl sulphate. The latter was found to be still present whilst excavating for an unexploded bomb in 1940. General contamination with tarry matter, sulphates and chromates was expected.

A preliminary survey for the Consultants (TABLE 4) showed that contamination was patchy. No dimethyl sulphate was detected.

Upon appointment, the Contractor arranged for further samples to be examined where concrete foundations were to be situated (TABLE 5A) and for analysis of soils in the general area of old plant (TABLE 5B). Again, no dimethyl sulphate was detected. From these results recommendations were made for the manufacturer of sulphate-resistant concrete and for the protection of workmen on site.

Two years later the telephone call'. An uncharted tank had been uncovered which contained a murky liquid. After analysis (TABLE 6), it was recommended that specialist chemical disposal contractors should be engaged to remove the material. Shortly thereafter another telephone call reported that groundworks had encountered a suspiciously yellow-coloured soil. Again this was analysed and found to contain a high proportion of chromium (TABLE 7) and warnings were reinforced regarding hazards for site personnel.

The site is now largely covered with a multi-storey car-park.

Case D

This case is one where we had no early involvement with the project which was a small housing development. Subsequently it transpired that part of the site had been previously occupied by a paint works.

The Contractor telephoned us when an excavator contacted a buried tank which contained a strongly-smelling straw-coloured liquid. Chemical analysis (TABLE 8) showed that significant proportions of heavy metals were present and recommendations were made for disposal.

Once the site staff were alerted to the situation it was not long before the uncovering of suspicious looking soil led to a fairly comprehensive survey of conditions prevailing on site (TABLE 9).

In this case our recommendations for action were not sought, but later due to the concern of the Housing Authority about the high lead and zinc values obtained, we were commissioned to survey the surface soil contamination over the site. The results of these analyses (TABLE 10) revealed some disturbingly high concentrations of zinc in an adjoining area previously designated for domestic horticulture but not forming part of the site.

Case E

The site in question was intended for a town centre redevelopment and a large supermarket was planned for an area which had previously been occupied by an electroplating works. At a late stage, and shortly before the Contractor took over the site, the Architects became concerned about possible contamination of the site by toxic substances and commissioned a small survey to determine the depth of any contamination. Samples were taken of surface material and at 300, 600 and 900mm depths. Gross contamination, including cyanides, was detected at all levels and an additional survey at depths of 1, 2 and 3 metres was undertaken. The results of the chemical analyses of the samples are presented in TABLE 11.

In view of the deep penetration of contaminants (even extending laterally beyond the development boundaries) it was thought prudent to remove all soil to a depth of 3m. Contaminated soil was diluted with clean material and transported to a specially designated disused clay pit some 50 miles away. Covered lorries were used to transport the soil and protective clothing was worn by all personnel. Strict precautions were instituted to ensure that no contaminated clothing was worn in areas where food was supplied, prepared or eaten. The removal of affected material involved costs of about £20.00 per m^3.

IN CONCLUSION

The Case Studies illustrate very well how seldom a single organisation will be aware of all the problems and the resultant actions that occur during the redevelopment of contaminated sites. A seemingly well organised initial site investigation has no conclusions (as far as we are concerned) and a minor enquiry leads to extensive and unbudgetted remedial operations. A piecemeal approach appears to be the rule rather than the exception.

There is a need for a nationally-organised data-bank for contaminated sites in which the results of soil surveys, especially chemical analyses can be collected, together with details of remedial action adopted. Hopefully this information could be correlated with the long-term performance of the redevelopments so that the effectiveness of the remedial actions could be monitored. There must, at present, be a great deal of unnecessary work and much replication of investigation because of the lack of co-ordination in site studies. National collation of information could save time, money and expensive mistakes and also indicate shortcomings in existing knowledge and methods of investigation. Chemical hazards can and should be avoidable.

REFERENCES

1. W.B. Chapman, P. Baker and D. Burns - Some guidelines on the re-use of industrially contaminated land. J. Association Public Analysts, 1977, Part 1, pp. 9-20. (Also in GLC Materials and Information Bulletin, No. 98, August/September 1976, item No. 1, pp 1/1 - 1/8).

2. M.S. Eglinton - Chemical considerations in construction on filled ground. Symposium on Engineering Behaviour of Urban and Industrial Fill, Midland Geotechnical Society, University of Birmingham, 23-25 April 1979.

3. The Deposit of Poisonous Waste Act, 1972 - Site problems. GLC Materials and Information Bulletin, No. 82, February 1975, item No. 3, pp 3/1 - 3/4, 3A1.

4. Development of contaminated land. GLC Materials and Information Bulletin, No. 118, August/September 1978, item No. 3, pp 3/1 - 3/4.

5. J. Bromley - Monitoring of landfill sites and derelict land for pollution. CAMLAB Seminar on Water Pollution, Cambridge, 5-6 September 1978.

6. P.J.D. Jarrett - Control and treatment of leachate from industrial and domestic wastes. Gas emissions from waste landfill sites - its regulation and control. Symposium on Engineering Behaviour of Urban and Industrial Fill, Midland Geotechnical Society, University of Birmingham, 23-25 April 1979.

Test Pit No.	Sample Depth (m)	Chemical Analysis								Other Information
		tarry matter (per cent)	pH of aqueous extract	total water-soluble matter (per cent)	total acid-soluble sulphate (as SO_3) per cent	Chloride (as Cl) per cent	Semi-quantitative analysis: Lead (as Pb) per cent	Barium (as Ba) per cent	Copper (as Cu) per cent	
1	0.3-1.0 1.5-2.9	0.26 -	6.8 7.0	1.30 -	0.53 0.04	0.02 -	0.1 -	< 0.1 -	< 0.05 -	Brown scum, aromatic odour
2	0.3-1.2 2.1-2.9	0.44 -	7.5 7.3	0.71 -	0.51 0.16	0.03 -	0.05 -	< 0.05 -	0.1 -	Yellow, blue, green crystals present. Oily sludge
3	2.3-4.0	0.01	8.5	-	0.02	trace	-	-	-	Smell of ammonia
4	1.0 2.6-3.3	0.55 0.47	7.5 7.7	- -	< 0.01 0.15	- 0.06	- -	- -	- -	Tarry odour and naphalene smell in places
5	1.0-2.1 1.4-2.0 1.5	0.15 0.10 -	5.2 4.9 3.6	1.62 - -	0.90 0.80 0.11	0.01 - -	- - -	0.5 - -	< 0.01 - -	Green crystalline material present. Sulphurous smell.
6	0.2-2.1 0.5-2.8	0.89 0.31	7.8 8.0	0.95 0.73	0.27 0.21	0.03 0.03	0.1 < 0.02	< 0.1 < 0.1	0.1 < 0.05	Oily contamination, strong benzaldehyde smell. 'Soapy' smell.
7	0.3-3.3	0.34	7.4	-	0.01	trace	-	-	-	Green oily liquid present, strong terpene odour.
8	1.0-2.5 3.0-4.0	0.02 -	6.4 5.4	0.59 -	0.02 0.01	trace -	< 0.02 -	< 0.1 -	< 0.05 -	Sample adjacent to corroded concrete, blue crystals present.
9	0.6-1.8 2.4-3.7	0.02 -	6.7 6.3	0.64 -	0.08 < 0.01	- trace	< 0.05 -	< 0.05 -	0.1 -	
10	1.0-1.8 2.4-3.1	0.07 0.11	5.0 5.5	0.47 -	0.01 0.01	0.02 -	< 0.05 -	< 0.05 -	< 0.01 -	White crystals present, rubbery smell.
11	0.2-1.8 2.4-3.1	0.02 0.01	7.9 5.2	1.41 -	0.85 0.05	0.02 -	< 0.05 -	< 0.05 -	< 0.01 -	Grey clay-like material present. Pungent odour.
12	0.5 0.6-2.5	0.03 1.98	6.4 5.9	- 0.62	0.28 0.08	- 0.01	< 0.02 < 0.02	< 0.05 < 0.1	0.1 < 0.01	Oily contamination, strong aromatic odour.
13	0.5-1.8 2.3-3.1	0.03 -	6.1 -	0.45 -	0.17 0.07	0.01 -	- -	- -	- -	White/green/blue crystals, brown oily liquid present.

Other elements sought, but not detected were As, Sb, Cd, Cr, Co, Hg, Mo, Ni, Sn, Zn

TABLE 1 PARTIAL DATA FROM EVALUATION OF OLD GAS-WORKS/CHEMICALS MANUFACTURING SITE (CASE A)

Soil Sample No.	Sample Depth (m)	Chemical Analysis: pH of aqueous extract	Total acid-soluble sulphate (or SO_3) per cent	Chloride (as Cl) per cent	Phenols p.p.m
1	1.5 13.5	8.6 8.1	0.85 0.40	- -	- -
2	1.5 12.0	7.7 7.3	7.78 0.74	- -	- -
3	1.5 9.0	7.0 7.0	7.55 0.23	- -	- -
4	4.0	7.5	0.10	0.02	1
5	2.0 4.0	7.0 7.9	0.15 0.09	0.18 0.12	5 3
6	2.0 6.0	7.6 7.5	0.06 0.06	0.03 0.15	1 1
7	2.0 6.0	7.8 7.8	0.04 0.08	0.02 0.08	0.5 3

Water Samples	Chemical Analysis: pH	Sulphate (as SO_3) p.p.m	Tarry matter p.p.m	Phenols p.p.m
A	8.1	410	310	200
B	6.5	910	1330	250
C	7.1	1360	150	50
D	7.3	990	280	4
E	7.4	480	not detected	2
F	7.2	1630	310	50
G	6.9	1430	85	16
H	7.3	1410	66	314

TABLE 2 RESULTS OF PRELIMINARY SOIL AND WATER ANALYSES (CASE B)

pH	5.3
Total suspended matter	430 ppm
Total dissolved solids	1650 ppm
Sulphate (as SO_4)	802 ppm
Chloride (as Cl)	149 ppm
Total phenols	250 ppm
Tarry matter	9590 ppm

TABLE 3 RESULTS OF ANALYSIS OF LIQUID FROM OLD SEWER (CASE B)

Soil Sample	Chemical Analysis					
	water soluble			acid soluble		loss-on-ignition per cent by wt.
	pH	organic matter	chlorides	sulphate (SO_3) per cent by wt.	chromium (Cr_2O_3) per cent by wt.	
1	7.6	present	trace	0.12	trace	7.95
2	6.9	present	trace	0.11	trace	8.31
3	7.4	present	trace	0.17	trace	10.92
4	7.4	present	trace	0.31	trace	11.03
5	6.6	present	trace	0.22	trace	9.07
6	6.2	major amount	high concentration	0.37	trace	15.02
7	5.7	moderate amount	present	0.33	not detected	21.91
8	4.9	present	trace	1.71	1.27	9.34
9	7.1	present	high concentration	0.15	0.38	10.79
10	8.0	present	trace	0.15	trace	6.25
11	4.9	present	trace	1.56	0.91	7.37

Heavy metals (Pb, Cu, Zn), cyanide and chromate were not detected in any of the samples.

TABLE 4 PRELIMINARY SITE SURVEY FOR CONSULTANTS (CASE C)

Soil Sample No.	Chemical Analysis			
	water-soluble		acid-soluble	
	pH	sulphate (as SO_3) g/litre	sulphate (SO_3) per cent by wt.	chromium (Cr_2O_3) per cent by wt.
1	7.6	-	0.78	0.02
2	4.7	1.8	2.44	0.03
3	7.6	-	0.11	0.03
4	6.8	1.8	2.16	0.02
5	7.9	-	0.09	0.03
6	6.8	-	0.28	0.06
7	7.8	-	0.23	trace
8	7.8	-	0.14	trace
9	7.7	-	0.13	0.03
10	7.2	-	0.07	trace

TABLE 5A RESULTS OF ANALYSES OF SOIL SAMPLES FOR CONTRACTOR (CASE C)

Trial Pit No.	Soil Sample	Chemical Analysis			other results
		pH of aqueous extract	acetic acid-soluble material		
			lead (as Pb) p.p.m	chromium (as Cr) p.p.m	
1	A	7.1	133	133	Soil sample had yellow colour, tar and phenols present.
	B	11.5	80	present	
	C	7.0	not detected	not detected	Calcium carbonate a major component.
2	A	6.8	trace	70	
3	A	6.6	present	115	Tarry matter present.
	B	6.1	38	present	Sample had purplish colour
4	A	6.4	trace	55	Tar present.
	B	6.0	trace	900	Crystals present in sample.
5	A	6.4	trace	21	
	B	4.5	not detected	not detected	Yellowish-brown colour.
6	A	7.0	15	12	
7	A	6.8	10	58	Tarry matter present.
	B	7.1	13	22	

Barium, cyanide and sulphide were not detected in any of the samples.

TABLE 5B RESULTS OF ANALYSES OF SOIL SAMPLES (CASE C)

pH	7.0
Analysis of Solution	p.p.m.
Calcium (as Ca)	not detected
Magnesium (as Mg)	19
Sodium (as Na)	928
Potassium (as K)	56
Chromium (as Cr)	not detected
Chloride (as Cl)	440
Sulphate (as SO_4)	306
Total dissolved solids	2600
Total suspended matter	3950
Analysis of Suspended Solids	(dried at 105°C)
	p.p.m.
Arsenic (as As)	25
Barium (as Ba)	300
Cadmium (as Cd)	60
Chromium (as Cr)	3520
Copper (as Cu)	920
Mercury (as Hg)	20
Lead (as Pb)	775
Antimony (As Sb)	not detected
Sulphate (as SO_3)	3600
Cyanide (as CN)	not detected
Loss at 450°C	23.4 per cent by wt.

TABLE 6 ANALYSIS OF RESIDUAL LIQUID IN TANK (CASE C)

Total chromium (as Cr)	= 0.18 per cent
Water-soluble Chromium (as Cr)	= 0.11 per cent
Water-soluble sulphate (as SO_3)	= 0.61 per cent

Hydroquinone and hypo (sodium thiosulphate) were not detected

TABLE 7 ANALYSIS OF SUSPICIOUS YELLOW-COLOURED SOIL (CASE C)

pH	= 5.7
Total dissolved solids	= 1030 p.p.m.
Organic matter	= 0.47 per cent
Heavy metals	present
Lead	= approx. 1000 p.p.m.
Barium	= approx. 100 p.p.m.

TABLE 8 ANALYSIS OF STRAW-COLOURED LIQUID FROM TANK (CASE D)

Soil Sample No.	Chemical Analysis											Other Observations
	pH of aqueous extract	organic matter (per cent by wt.)	lead (as Pb) p.p.m.	barium (as Ba) p.p.m.	arsenic (as As) p.p.m.	cadmium (as Cd) p.p.m.	zinc (as Zn) p.p.m.	copper (as Cu) p.p.m.	nickel (as Ni) p.p.m.	mercury (as Hg) p.p.m.	organic liquid content per cent	
1	8.0	1.5	18	62	8	not detected	42	11	14	0.07	0.1	Brown clay with black areas, burnt charcoal smell
2	7.0	0.2	14	61	1	0.4	45	13	24	0.01	-	Brown clay, musty smell
3	6.9	12.8	8	51	1	not detected	30	11	17	0.01	1.6	Black soil with areas of yellow and red material. Waxy odour.
4	8.6	0.5	20	59	8	2.0	67	13	13	0.10	-	Pale clay. Musty odour.
5	6.7	0.3	13	73	3	1.8	55	12	47	0.06	-	Sandy-coloured clay. Odourless.
6	5.8	0.7	19	73	12	not detected	66	17	23	0.01		Brown clay soil containing black areas, musty odour.
7	6.1	0.2	99	172	5	not detected	78	11	39	0.01		Brown clay soil containing black areas, musty odour.
8	6.5	0.1	1140	412	6	1.7	2310	64	30	0.89		Brown clay soil containing black areas, musty odour.
9	6.3	0.8	41	104	10	0.9	109	27	20	0.01		Brown clay with black areas, slight 'organic' odour.

Cyanide, sulphide and phenols were not detected in any of the samples.

TABLE 9 RESULTS OF ANALYSES OF SOIL SAMPLES (CASE D)

Soil Sample No.	Chemical Analysis	
	Zinc (as Zn) p.p.m.	Lead (as Pb) p.p.m.
1	1400	60
2	420	80
3	430	70
4	220	70
5	390	80
6	190	70
7	260	80
8	300	80
9	280	60

TABLE 10 RESULTS OF ANALYSES OF SURFACE SOIL SAMPLES (CASE D)

Sample No.	Depth of Sample (m)	Chemical Analysis						Comments
		pH of aqueous extract	sulphate (as SO_3) per cent	cyanide (as CN) p.p.m.	chromium (as Cr) p.p.m.	nickel (as Ni) p.p.m.	cadmium (as Cd) p.p.m.	
1	surface	11.0	-	35	3640	41	38	Trial pit.
	0.3	8.1	-	6	396	111	14	
	0.6	7.0	-	4	200	41	14	
	0.9	8.2	-	2	103	16	5	
1A	1	7.3	0.01	28	110	218	14	Deeper excavation adjacent to original trial pit.
	2	7.2	-	0.5	31	40	0.4	
	3	7.2	-	0.5	11	22	4	
2	surface	10.6	-	203	48	290	79	Trial pit.
	0.3	9.4	-	454	53	66	14	
	0.6	7.5	-	14	27	26	3	
	0.9	6.9	-	1	16	21	3	
3	surface	11.4	-	187	52	312	84	Trial pit.
	0.3	7.6	-	32	300	324	144	
	0.6	7.4	-	385	556	516	180	
	0.9	8.1	-	104	340	212	148	
3A	1	6.3	0.18	1500	180	146	210	Deeper excavation adjacent to original trial pit.
	2	4.4	0.08	230	62	51	51	
	3	5.5	0.08	105	57	26	22	
4	surface	9.4	-	7	34	57	13	Trial pit.
	0.3	8.9	-	6	10	9	4	
	0.6	9.1	-	755	176	122	188	
	0.9	8.5	-	28	56	63	52	
4A	1	5.7	0.09	0.5	34	27	80	Deeper excavation adjacent to trial pit.
	2	7.7	-	trace	4	22	6	
	3	7.2	-	trace	6	28	6	
5	surface	10.0	-	154	98	312	125	Trial pit.
	0.3	8.9	-	600	291	752	12	
	0.6	8.0	-	149	238	888	484	
	0.9	7.6	-	320	200	82	52	
6	1	6.3	0.01	trace	16	7	0.4	Excavations in ground approximately 4m outside the boundary of the development.
	2	6.5	-	trace	7	5	0.2	
	3	6.8	-	trace	6	8	0.2	
7	1	6.6	0.01	trace	17	8	4	
	2	7.0	-	trace	19	5	0.2	
	3	7.0	-	trace	4	11	0.2	

TABLE II RESULTS OF CHEMICAL ANALYSES OF SHALLOW AND DEEP EXCAVATION SOIL SURVEYS (CASE E)

RECLAMATION OF A DISUSED SEWAGE FARM

.. D. HEEPS. E. R. PIKE

City Engineer's Department, Leicester/Public Analyst, Leic.County Council

General Background

The Beaumont Leys Estate is a site of around 800 hectares situated at the northern boundary of the City of Leicester. In 1890 it was commissioned as a sewage treatment works and farm and continued in use until 1964, when a new 'activated sludge system' works was opened at Wanlip. The original treatment consisted of settling out the majority of the solids as sludge and distributing the supernatant liquor by field irrigation. The sludge was then spread over an area of about 80 hectares, allowed to dry and ploughed into the soil.

Over the years refinements such as percolating filters were added, but the basic method of treatment remained the same.

The effluent distribution area was extensively and productively farmed, raising cereals, root and green vegetables, sheep, cattle and large flocks of turkeys. On the closure of the sewage works in 1964, the land was let to tenant farmers and the produce was from then onwards almost entirely potatoes and cereals.

In the later sixties there was an overall shortage of building land and an acute demand for houses. Obviously, Beaumont Leys was therefore considered ripe for development and plans were proposed for the erection of a satellite town on this land, incorporating residential and industrial areas. This proposal sparked off representations doubting the fitness of the land from the bacteriological aspect. Any aesthetic objections from this viewpoint were quickly dispelled by the Public Health Laboratory's assurance that the land was devoid of pathogenic bacteria, but meanwhile, the Agricultural Advisory Service had found the land, especially the sludge-spreading area, to be highly contaminated with zinc and copper such that the so-called zinc equivalent content of the land exceeded the desirable maximum of 250 parts per million, figures up to 6,000 p.p.m. being recorded.

At this juncture, about 1970, the Leicester City Analyst's services were enlisted to examine the toxic metal contents of the soil from a plot of the Beaumont Leys Estate near the Anstey Lane and adjoining the existing residential developments on the City outskirts. This area became known as the 'Pilot Development Plot'. From the results of these analyses, concern developed regarding the lead content of the soils and ultimately it was decided to monitor the heavy metal contents over the whole 800 hectares of the estate. Thus, in 1970-71, samples were taken at 200 feet intervals on a rectangular grid basis and submitted for analysis for total contents of Lead, Cadmium, Copper, Zinc, Chromium, Nickel, Arsenic and Mercury. The 'Available' metals (as defined by the metal soluble in 0.5 M Acetic Acid under prescribed conditions) were also determined for Lead, Cadmium, Copper,

BEAUMONT LEYS SEWAGE FARM

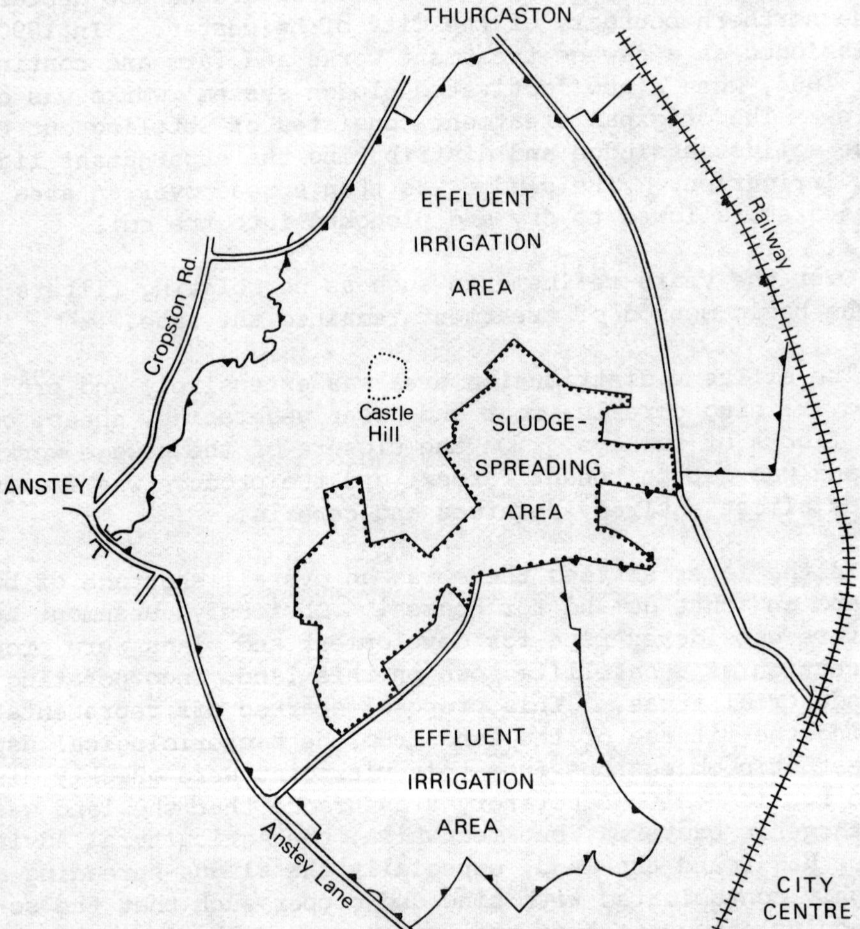

Scale 1 : 35000

KEY : Boundary of Sewage Farm
Boundary of sludge-spreading area

Zinc, Chromium and Nickel. These results were subsequently published in the Journal of the Association of Public Analysts (Pike, Graham & Fogden, J.A.P.A., 1975, 13, 19-48).

The Original Survey 1971-72

The above-mentioned survey was conceived and carried out in little more than a year and the work was subject to considerable political pressure to meet deadlines of committee meetings and the resultant public meetings once the possibility of toxic metallic pollution of the land had become common knowledge. The methods of analysis adopted had, therefore, to combine the advantages of simplicity and speed, coupled with acceptable accuracy. The samples were taken by members of the City Engineer's Department at the rate of approximately one per acre, this being a sample at the intersections of the 200ft. grid mentioned above. The results obtained for each metal (total and 'available') were then plotted and the contours of the metal deposits obtained.

A map of the area is provided showing the effluent irrigation area and the sludge spreading area. Obviously a proliferation of maps would be required to show the deposition of all the individual metals over the estate - both total and available. However, the results showed that the heaviest deposits of all metals coincided in extent with that of the sludge spreading area. The following summary therefore, gives a good indication of levels at which the metals were found in the effluent irrigated and sludged lands.

Distribution of Metals in Surface Soil (In mg/Kg on the dried soil)

	Sludged Area				Effluent-irrigated Area			
	Range	Quartiles			Range	Quartiles		
		1st	2nd	3rd		1st	2nd	3rd
Zinc								
Total	69-3050	740	1320	1850	42-1070	140	205	302
Available	3-1980	215	465	785	3-635	25	56	104
Copper								
Total	24-1415	205	370	755	13-365	41	63	96
Available	3-165	22	35	48	0.3-40	3.9	6.6	9.9
Nickel								
Total	42-385	110	145	210	20-245	70	86	105
Available	5-200	21	38	66	2.2-56	9.0	14.5	21.0
Lead								
Total	47-2470	465	780	1040	7-870	75	120	240
Available	1.0-42	5.3	9.4	13.3	0.5-50	3.0	3.6	6.1
Cadmium								
Total	1.2-61	7.2	14.5	27	0.4-25	1.8	3.0	4.5
Available	0.2-25	2.3	5.6	9.2	0.1-6.9	0.55	1.0	1.7
Arsenic								
Total	2-59	17.4	25.0	33.5	0.1-51	7.4	12.1	17.8
Chromium								
Total	23-2020	370	800	1060	18-1600	83	165	310
Available	0.2-39	3.3	4.6	6.3	0.2-20	1.7	3.3	5.1

No. of samples analysed on the Sludged Area = 1235
No. of samples analysed on the Effluent-irrigated Area = 1713

Accepting that the Sludged Area has toxic metallic contamination heavier than natural soil, it is informative to assess the excess contamination in the soils of the Effluent-irrigated Area. For this purpose soils were obtained from private gardens and allotments; the comparative contents of the toxic metals are shown in the table below:-

	Total		Available	
	Range	Median	Range	Median
Zinc				
Private Gardens (26)	70-575	136	5-166	28
New North End Allotments (10)	125-390	159	21-104	31
Washbrook Allotments (10)	65-222	109	8-39	18
Effluent Irrigation Area	42-1070	205	3-635	56
Copper				
Private Gardens	19-100	37	2-22	5
New North End Allotments	40-87	45	2-5	4
Washbrook Allotments	20-80	53	4-7	4
Effluent Irrigation Area	13-365	63	0.3-40	7
Nickel				
Private Gardens	42-105	70	5-22	7
New North End Allotments	75-90	80	5-10	7
Washbrook Allotments	65-110	75	4-8	4
Effluent Irrigation Area	20-245	86	2.2-56	14.5
Lead				
Private Gardens	27-707	93	2-80	6
New North End Allotments	110-560	162	1-33	8
Washbrook Allotments	20-200	80	3-10	5
Effluent Irrigation Area	7-870	120	0.5-50	6
Cadmium				
Private Gardens	1.1-3.3	1.6	0.2-1.1	0.4
New North End Allotments	1.4-2.4	1.6	0.2-0.5	0.4
Washbrook Allotments	1.0-4.0	1.6	0.2-0.5	0.35
Effluent Irrigation Area	0.4-25	3.0	0.1-6.9	1.0
Arsenic				
Private Gardens	3-24	9	-	-
New North End Allotments	7-12	9	-	-
Washbrook Allotments	7-13	8	-	-
Effluent Irrigation Area	0.1-51	12.1	-	-
Chromium				
Private Gardens	19-82	23	1-8	2
New North End Allotments	27-38	32	2-5	3
Washbrook Allotments	23-55	36	4-7	5
Effluent Irrigation Area	18-1600	165	0.2-20	3.3

Examination of this table shows that median values for most metals to be in general only marginally higher than values obtained from soils of the allotment plots, sampled in this investigation. Surprisingly, values for lead are no higher but those obtained for cadmium are about double for median contents and the range is much wider, reaching 25 p.p.m. for total cadmium and up to 6.9 p.p.m. for available cadmium. Values for arsenic and chromium were also of a greater order.

Distribution of Metals in the Depth of the Soil

In an attempt to monitor the penetration of the metals in the depth of the soil, twelve core samples were taken and samples of the cores examined, corresponding to the depths of 0, 2, 4 and 6 feet. The results for each metal are set down in the following table (the values representing parts per million (mg/Kg) of the metal in the dry soil).

Sample Depth (ft)	Range	Mean	Median
Zinc			
0	87-1105	244	221
2	57-125	79	70
4	31-122	67	64
6	53-174	77	68
Copper			
0	20-292	54	70
2	27-48	36	37
4	28-87	42	37
6	27-77	39	38
Nickel			
0	45-140	93	93
2	65-130	103	108
4	45-150	115	120
6	110-155	127	123
Lead			
0	40-820	225	160
2	25-75	45	40
4	20-65	40	40
6	25-70	45	40
Cadmium			
0	2-11	4	3
2	1-5	2	2
4	2-8	4	4
6	2-7	4	4
Arsenic			
0	3-23	11	11
2	1-18	6	5
4	1-19	6	7
6	1-12	4	4
Chromium			
0	40-980	245	159
2	38-493	87	48
4	32-56	44	46
6	32-95	50	46

This table indicates that, in general, the higher values occur in the first two feet of soil. This opinion has been strengthened in later investigations and we subscribe to the opinion that values obtained for the lower depths have been exaggerated to some extent, owing to contamination by some of the upper soils due to the samples having been obtained by means of a hand auger.

Advice from the 'Physical Environment' Sub-Committee of the Committee on Medical Aspects of Chemicals in Food and the Environment (DHSS)

The above-mentioned Committee, having considered the data summerised above, advised the Leicester City Council that development could proceed on the Beaumont Leys site, provided that the top two feet of soil was removed from the sludged area before building was undertaken.

Some development for residential purposes has since taken place but has been restricted to the effluent distribution area south of the sludged area.

Subsequent evaluation of the costs of removing the top two feet of the soil of the sludged land indicated that this would be uneconomic. Also, there was the physical problem of where to deposit the contaminated soil. It was therefore planned to develop this area only for concreted areas or non-agricultural turfed areas such as industrial, shopping areas or playing fields.

The building that has already taken place consists in the main of three and four-storey residential property with relatively small garden space. The problem of contaminated soil is therefore limited, in that it is not possible for a householder to grow enough fruit or vegetable such that any toxic metallic uptake could prove a hazard. This aspect of toxic hazard is dealt with more fully later in this paper.

Further Consideration of the Toxic Hazard

During 1975-76 concern was expressed regarding the cadmium levels found in certain 'hot spots' in the effluent irrigation area. Meetings were therefore convened between representatives of the D.O.E., C.U.E.P., D.H.S.S., M.A.F.F, The City Engineer's Department and the County Analyst to review the matter. Hitherto, emphasis was mainly levelled at the lead pollution, but now more concern was placed on the cadmium content of the soil. After consideration of the maximum likely size of garden attached to proposed housing development and the amount of fruit and vegetables it would be possible to grow, limits of 600 mg/Kg of lead and 5 mg/Kg of cadmium were decided upon for replaced top soil in the development areas. To achieve these limits the whole of the topsoil for an area was to be removed by mechanical scrapers dumped into stockpiles and then re-distributed to the gardens of the subsequently built houses. In this process it was assumed that a reasonable degree of mixing and equalisation of lead and cadmium contents would be achieved.

As an early check samples were taken from stockpiles already in existence and compared with the original results of the area from which they came. The original samples showed lead well below 600 mg/Kg, but cadmium varied between 1.8 to 7.0 mg/Kg, with an average of 4.07 mg/Kg. Samples from the stockpiles averaged 2.17 mg/Kg, with a maximum of 3.0 mg/Kg, indicating that mixing had occurred. Other areas were checked with similar results. However, it was realised that some areas, due for development in 1979 onwards, had areas of higher levels of cadmium in the top soil, and greater variations within the area. Obviously, these areas would require a greater degree of control over soil movements than had previously been necessary and so 'Heatherbrook', the first of the areas to be developed, was chosen as the site of study.

The Heatherbrook Survey 1979

This survey relates to a much smaller area than did the 1971-72 Beaumont Leys survey. Heatherbrook stretches north from the borders of the sludged area, towards the village of Anstey and is situated to the east side of Anstey Lane. The area investigated covers approximately 10 acres (4 ha. approx), and for the purpose of sampling was surveyed on a 200ft. and 50ft. grid.

The programme included sampling of topsoil at the above-mentioned points, then after removing the topsoil, the subsoil was sampled. It is also proposed to assess the toxic metal contents of the stockpiled topsoil to obtain an indication of the expected averaging of metallic contamination.

Comparison of Lead and Cadmium in topsoil and sub-soil specimens taken from Heatherbrook site (results in mg/Kg of the dried soil) were -

Sample	pH Value		Lead		Cadmium	
Point	Topsoil	Sub-soil	Topsoil	Sub-soil	Topsoil	Sub-soil
Di 31	7.3	7.2	31	17	0.55	0.05
Di 32	7.2	9.2	93	25	2.03	0.20
Di 33	6.8	8.3	120	19	3.46	0.20
Di 34	6.8	7.8	149	53	3.49	0.90
Di 35	5.6	7.6	265	37	4.35	0.40
Di 36	5.7	7.4	322	38	5.63	0.40
E 31	6.7	8.7	84	35	1.82	0.40
E 32	6.8	7.5	129	70	3.53	1.20
E 33	6.5	7.9	136	35	3.88	0.40
E 34	6.6	7.2	143	73	2.03	1.10
E 35	6.6	6.6	191	23	3.40	0.40
E 36	5.9	7.6	398	36	6.15	0.40
Ea 31	7.2	7.4	94	39	1.44	0.40
Ea 32	7.1	7.8	317	39	1.73	0.50
Ea 33	6.6	7.1	164	45	4.05	0.70
Eb 31	6.2	6.6	110	40	2.54	0.40
Eb 32	6.6	6.2	118	81	1.59	0.70
Eb 33	6.0	6.8	170	25	2.24	0.30

	Lead		Cadmium	
	Topsoil	Sub-soil	Topsoil	Sub-soil
Range	31-398	17-81	0.55-6.15	0.05-1.2
Average	168.5	40.5	3.00	0.50
Std. Deviation	96.2	18.2	1.49	0.306
Upper 99% Tolerance Limit	447.3	93.2	7.3	1.39

This Table shows relationship between lead and cadmium contents in the top and subsoils taken from specific sites. It is interesting to note that in general, the pH values of the top soils were slightly acid whilst the subsoils were, in the main, alkaline. More important though is the marked drop in lead and cadmium contents of the subsoils, suggesting that little translocation of these metals take place from the topsoil.

Uptake of Toxic Metals by Vegetable Crops

The presence of toxic metals in soil would be of no consequence if these metals were not able to enter the metabolic processes of plants or animals. It has long been established that excess of copper, zinc, nickel (and possibly cadmium) constitute a toxic hazard to plants and for this reason A.D.A.S. have established a maximum of 250 p.p.m. for the so-called zinc equivalent, a parameter used to measure possible phytotoxicity of the soil. The zinc equivalent is calculated from the available zinc, copper and nickel contents of a soil according to the formula, zinc equivalent = zinc + 2 copper + 8 nickel all measured in terms of available metal in mg/Kg of the dried soil. The available metal being that amount of metal rendered soluble by treating a definite weight of soil with a specific extractant under specified conditions.

Obviously the health of a plant is dependent upon the presence of possible toxic elements in the soil. If these elements are absorbed into the crops intended for animal or human consumption, then they also form a possible health hazard to man and cattle etc. It was in order to assess this risk that a number of allotment plots were set up at Beaumont Leys and crops raised thereon, analysed for toxic metals including zinc, copper, nickel, lead, cadmium and arsenic.

The following table attempts to summarise the results obtained from the various crops grown on soils highly polluted with the metals concerned:-

Range of Metal Content of Soil	Zinc	Copper	Nickel	Lead	Cadmium
(mg/Kg) Total	350-1740	89-470	105-355	170-982	2.8-18.0
Available	92- 805	8-48	10- 60	2.8-15.0	0.9-6.9
Lettuce	2- 16	0.1-0.8	0.4-2.7	0.1-0.4	0.1-0.5
Radish	5- 35	0.3-0.7	0.1-2.9	0.2-0.35	0.03-0.07
Cabbage	2- 16	0.3-1.4	0.3-4.2	0.05-0.39	0.02-0.14
Savoys	-	0.4-1.7	0.9-3.0	0.10-0.40	
Cauliflowers	5- 11	0.2-1.9	0.2-3.1	0.1-0.24	0.03-0.08
Potatoes	0- 21	1.0-4.4	1.1-1.8	0.11-0.25	
Carrots	2- 13	0.1-0.7	0.3-6.5	0.03-0.35	0.01-0.06
Sprouts	0- 8	0.9-1.5	1.7-9.8	0.2-0.4	
Tomatoes	2- 5	0.2-0.9	0.7-2.4	0.10-0.38	0.03-0.04
Beans	3- 7	0.3-0.9	1.1-3.3	0.10-0.29	0.02-0.04

This table relates to data obtained from eleven samples of each vegetable grown during 1972. This year was a particularly dry growing season and the figures obtained must be judged with this information in mind. The statistics show that with this limited investigation no definite evidence was found to substantiate the view that abnormal amounts of toxic metals, other than zinc, are taken into the tissues of vegetables grown on Beaumont Leys soils. Little or no arsenic was found in any of the samples submitted. Values for lead, cadmium, copper, nickel and chromium were normal for vegetables and in no instance were statutory or recommended limits exceeded. It was noted that lettuce and radish grown on high zinciferous soils did absorb zinc into the tissues, but in no case was the value higher than the recommended figure of 50 p.p.m.

The above investigation lasted for only one season and therefore factors such as climatic conditions, absorption by deep rooted perennial plants such as fruit trees etc. cannot be assessed and obviously a long-term investigation is called for to evaluate such matters. For these reasons, an allotment plot has been set up at Beaumont Leys under the auspices of the Department of the Environment and managed by the Department of Botany of the Liverpool Polytechnic (Dr. N. W. Lepp). The main objective of the work is to provide information on the effectiveness of the treatments now being recommended in providing long-term amelioration of potential health hazards:-

i. When the recommended treatment is carried out and maintained to an average standard.

ii. When the treatment is carried out and maintained to a 'low standard'.

The basis of the study is the measurement of mobility of the contaminating ions (especially cadmium) in terms of uptake by plants and movement through the soil profile.

The basic experimental technique will be to place various depths of cover over a contaminated site, and to measure both movement of metals through the soil, and metal uptake by a variety of vegetables and soft fruit, over a period of not less than six growing seasons. Such a medium to long term approach is essential, as the mobility of metal ions will be effected by time dependent changes in the soil such as organic content.

This experiment is described in detail in the paper to be presented by Drs. Lepp and Harris later in the Conference.

Practical Development

Having set limits for lead and cadmium in housing areas, means of achieving these and also for the disposal of all soil not meeting these requirements were necessary.

Investigation has shown that where a "green field" site is developed, only 60% of the original topsoil is required for the house gardens. Further calculations showed that space must be found for the disposal of 1.75 million cubic metres of soil from Beaumont Leys, this being made up of the remaining 40% of topsoil and the surplus sub-soil from sewerage, roads and building construction.

Two sites situated in the areas of highest contamination and on what is, in winter, a rather bleak and windswept plateau, were designated as public open spaces in the "Development Plan". Schemes have been designed for both sites which will not only provide tipping space for the surplus material, but eventually result in pleasant and attractive parks. Soil with the highest contamination levels is being tipped first, covered with several metres of the uncontaminated sub-soil and the final layer of topsoil is to a standard of not more than 12 mg/Kg Cadmium for the grassed play areas, but any flower beds will be to the same standard as housing areas.

Control of soil movement is maintained by a colour coded and numbered ticket system and the origin and final destination of each load is monitored.

Conclusion

The problems have been identified and limits set for maximum metal content of soil. By careful planning and co-ordination, these limits can and are being achieved and the end result will be a safe and acceptable environment.

Acknowledgements

We are indebted to our respective Authorities for permission to present this paper and also to D.E. Sharpe, City Engineer Leicester, for his advice and unfailing encouragement.

GLC DEVELOPMENT AT THAMESMEAD

INVESTIGATION AND RECLAMATION OF CONTAMINATED LAND

G. W. LOWE

Greater London Council, Housing Department

The contents of this paper which is presented by permission of the General Manager, Thamesmead have been prepared to given an account of experience in dealing with investigation and reclamation of contaminated land. The paper does not necessarily reflect the policy of the Greater London Council which cannot accept responsibility for the authors opinions or conclusions.

1 The Thamesmead Site

The development area of Thamesmead which covers approximately 650 Hectares (1600 acres) is situated on the South Bank of the Thames between Woolwich and Erith with a frontage to the river of 5.5 km ($3\frac{3}{4}$ miles). The land is composed almost entirely of drained marsh at the foot of rising ground with open space and woodland forming a backdrop to the initial stages of new development.

Most of the land which has been developed and occupied to date was previously used for farming and rough pasture, with a large proportion given over to allotment gardens. This area has no known history of industrial usage and has not been included in the investigation for industrial contamination. The remainder has an area of approximately 247 Hectares (1000 acres) and lies to the North of a raised embankment containing the Southern Outfall Sewer which terminates at Crossness sewage works to the east.

Ground Conditions and Drainage

The ground is composed of silty clay, up to $1\frac{1}{2}$m thick overlying peat and alluvium 5 to 7 metres in depth. At the base of the alluvium is a stratum of water bearing gravel on Thanet sand and chalk.

The site forms a natural drainage basin between the Thames which is its Northern boundary, and the Wickham Valley which lies at the base of the rising ground to the South. Historically, surface water run-off from the hills drained into the area by way of a series of culverts, and then after flowing naturally across the site through a network of dykes and ditches, was discharged into the Thames via tidal sluices. Typically of marshland, the site is flat and naturally featureless except for trees. All prominent features are man-made.

The high water table and poor load bearing capacity of the ground has demanded that all construction work be piled, or preloaded in the case of major road works. Action of a similar kind was undertaken in earlier times including the filling of undrained marsh, and this led to many of the present pollution problems due to the nature of the materials used for reclamation and consolidation purposes.

2 Industrial History of the Site

The original river bank was probably built up by the Romans and there is evidence of a settlement at nearby Plumstead. In the 12th Century an Augustinian Abbey was established on the rising land to the South and from that time until 1524 the monks from Lesnes played a part in reclamation and drainage of the land.

Following a period of neglect during which severe flooding was experienced, the responsibility for maintaining the river wall and draining the marshes passed to Commissioners for Sewers who raised local taxes to do so from about 1600 onwards. In the meantime during the reign of Henry VIII the industrial development of the area was forshadowed by the establishment of the Woolwich Naval Dockyard. Downstream of the dockyard which included the site of Thamesmead the land was known as 'The Warren' and in the 17th century the area became a naval and military centre. Ranges were established on The Warren and storehouses for naval guns were erected.

The earliest manufacturing industry in this area started with construction of a brass foundry by Sir John Vanbrugh in 1716 where the first cannon were cast. As the output of the foundry grew, the Royal Artillery moved its barracks to Woolwich in 1719 to be near its main source of supply. The Royal Artillery depot remains in Woolwich to this day.

Land was reclaimed from the marshes as the establishment grew and land filling became an occupation of the convicts normally incarcerated in prison hulks anchored in the river adjacent to The Warren. Mud and debris from the construction of St Catherines Dock near the Tower of London was taken to the site for filling purposes and it is almost certain that this material was heavily augmented by waste products from the gun foundry and other industrial installations. By the beginning of the 19th Century, The Warren had been renamed 'The Royal Arsenal' and a period of major expansion began, no doubt to provide essential equipment for the concurrent expansion of the Empire. By 1890 the Arsenal covered 324 Hectares which included practice and experimental ranges. A further 160H. of marsh was added at this time and eleven moated magazines were constructed, some of which are being preserved for their open space and amenity value.

Manufacture of heavy weapons, ammunition and explosive devices reached its height in the war of 1914-18 when some 56,000 people were employed in armaments work. Coal and raw materials entered the area via a spur from the adjacent railway, by barges into the Arsenal dock or by ships using piers to deep water in the river.

The Arsenal was self supporting in that it produced its own town gas (from about 1860) and later a steam raising plant and generating station were added to the facilities. The prodigious quantities of ash and waste material produced by these plants alone over their long life have remained and make their contribution to the topography of the site and to the contamination problem now being dealt with.

The Arsenal continued with production and testing of modern armaments through the war of 1939-45 when it proved to be vulnerable to aim attack as had long been suspected. A very large number of bombs fell on the area and over 1000 people lost their lives as a result. Following the cessation of hostilities, vast quantities of surplus explosives were destroyed on the site by burning, the products of which added to the growing quantities of waste material from industrial/experimental processes and debris from destruction of obsolete installations.

The area finally sold to the GLC was handed over in stages during the 1960s, and with the demolition of remaining buildings, the stage was set for the development of Thamesmead, which at that time was proposed to be a town for 60,000 people.

3 The Discovery of Industrial Contamination

Development of Thamesmead started in the area to the south of the Arsenal boundary and it was some time before work of any magnitude began on land where there had been any concentrated industrial activity. In consequence it was not until 1975 that the extent of industrial contamination became fully manifest, although its presence had been known for some time. The initial impact on site development was spectacular in that the discovery of heavily contaminated material during excavation work was accompanied by a chemical fire and the penetration of buried tanks containing liquid residues (phenols) from the former gas works.

The situation was considered to be sufficiently serious to suspend contractual operations on the site in question to avoid risk to personnel and until a full investigation had been made. The financial penalties arising from stoppage of a major civil engineering contract was enough in itself to prompt a project-wide investigation, but it was clear by this time that all the land known to have been used by the Arsenal must be thoroughly examined. Long and short term hazards to health had to be assessed and means of providing permanent protection had to be determined.

4 Background to the Investigation

The working party of officers which was established initially by the Thamesmead Architect to investigate and deal with the problem, were faced with a number of difficulties at the outset, not least of which was the diverse nature of former industrial activity. Detailed information on much of this was not available but in any event, buildings housing industrial processes had been constructed on marshland reclaimed with industrial waste and filling materials of unknown origin. Furthermore, a long established practice of using industrial waste for foundations to the Arsenal network of railway lines, vehicle tracks, filling under floor slabs and in massive mounds as protective cover to sensitive installations led to a land make-up of considerable complexity.

The placing of rubble from damage by bombing in East London as backing to the river wall was not necessarily a cause for concern, but waste tipping from the Arsenal was carried at the same time leading to random localised filling patterns.

Destruction by burning of war surplus explosives, incendiary devices and other dangerous material was carried out either on areas of unused marsh or on tips of industrial waste which again tended to complicate investigation procedures.

In contrast to the mixed and apparently serious contamination which was sometimes evidenced by severe distress to vegetation and by unnatural mounds, the Eastern sector of the site was typical of pastureland on drained marsh which supported a healthy plant growth and habitat for wild life.

To add to the already complex situation it soon became clear that isolated areas of severe contamination existed in the generally clean, Eastern part of the Arsenal, while in the heavily industrialised West and Central parts there remained zones of thickly wooded land which have subsequently been earmarked for public open space and amenity use.

5 Scope of the Investigation

It is clear from the foregoing, that the complexity of land use and abuse, necessitated a total examination of the site previously occupied by the Royal Arsenal. Apparently clean areas would be subject to investigation as there was evidence to suggest that healthy plant coverage on a seemingly undisturbed ground surface was not necessarily a sound indication of contamination free conditions beneath.

The determination of sampling strategy from the beginning of the investigation was, and still is undertaken by representatives of the Council's Scientific and Medical Advisers, in conjunction with the Thamesmead Derelict Land Unit which now forms part of the Thamesmead General Manager's Operations Division. They rely principally on:-

(a) Such history of the site to be tested as is possible to collect.

(b) A visual inspection of the ground.

(c) Details of the intended land use.

Item (c) determines the scale of long term risk; houses with gardens being at the top of the scale, while items (a) and (b) provide an indication of the condition and previous usage of the ground. A judgement of these items is then made to determine the intervals of sampling which has ranged in practice between trial pits at 100m centres to as close as 10.0m where conditions are considered to be particularly bad.

As stated previously the original use of long demolished buildings has been difficult to determine but it has been possible to identify the following activities:-

Heavy Machine Shop and Forging work.
Non-Ferrous metal foundries.
Cadmium and other metal plating.
Town gas manufacture.
Laboratory testing of paints.
Manufacture of Acetylene.
Testing of Weapons and Explosives.
Destruction of Explosives by burning.
Destruction of Incendiary devices.
Storage of Coal stocks on the surface.
Dumping of waste products from Gas Works.
Dumping of Industrial Waste.

Such activities prompted a broad spectrum of analysis and as the investigation progressed across the site to suit the programme of development it was necessary to modify the list of determinations in the light of experience. From the outset the bulk of soil analysis has been carried out by external consultants to meet the fluctuating and often heavy load of essential laboratory work. Determinations needed and methods of test were specified by the Scientific Adviser not only to satisfy the needs of the Medical Adviser in the assessment of health hazards but to provide data on possible deterioration of sub-structures and, any special requirements for health and safety of site personnel.

The determinations most commonly called for are indicated below, but variations are made to suit particular circumstances:-

1 (a) pH.
  (b) Acid Soluble Sulphate.
  (c) Magnesium.

2 (a) Elemental Sulphur.
  (b) Sulphide.
  (c) Total Cyanide (dependant options (i) Ferro-Ferricyanide, (ii) Thiocyanate (ii) Free Cyanide).

3 (a) Phenols.
  (b) Toluene Extract.
  (c) Cyclohexane Extract.
  (d) Coal T[illegible] (depend[illegible]ption) and mineral oil.
  [illegible]orine.

4 Total Metals: (a) Lead
  (b) Cadmium
  (c) Mercury
  (d) Antimony
  (e) Arsenic

5 Available Metals: (a) Zinc
  (b) Nickel
  (c) Lead
  (d) Copper
  (e) Boron

6 <u>Site Sampling Procedures</u>

Before action is taken on any section of the site, a careful examination is made by the Site Inspection Team in order to agree the sampling strategy. The team is composed of a Scientist, an Environmental Health Officer, a Structural Engineer and an officer from the Derelict Land Unit. When the sampling strategy is agreed, the Derelict Land Unit undertakes or arranges the following tasks in the order shown:-

(a) Preparation of a sampling drawing based on a 1:500 scale survey. The drawing indicates the operational grid for Thamesmead and trial pits are related to this for ease of setting out. Each pit is given a code reference to enable its location to be easily determined on site.

(b) Surveyors from the Valuer and Estate Surveyor's Department set out the trial pits locating each with a driven stake marked with the code shown on the drawing.

(c) The Structural Engineer responsible for site investigation employs a contractor for excavation by JCB and provides an officer for his supervision. An officer from the Derelict Land Unit directs the operation and ensures samples are taken to a set routine. The success of this routine depends initially on a great deal of tedious office work which includes the writing of identification labels for every sample and the drafting of log sheets.

Soil samples are taken at set intervals in the trial holes down to 3.0m. If deeper holes are required a track driven excavator is employed. Bore holes are not favoured unless deep samples are required, e.g., in the centre of tips.

Samples are placed in 1 litre plastic buckets with swing handles and snap-on lids, and previously coded self-adhesive labels are simply applied to the side of the containers.

Pre-coded log sheets are filled in as sampling proceeds, care being taken to use a standard nomenclature given the most accurate possible description of the sample including features which may be transient, such as colour and smell.

(d) Samples tests for pH and temperature are carried out in appropriate circumstances and these are recorded on the log sheets.

(e) Sample containers are batched and sent to consultant analysts together with log sheets, and instructions from the Scientist as appropriate.

During more recent investigations colour photographs of trial pits have been taken to reinforce data on log sheets and supplement general site photographs taken at site inspection stage.

It is vital that sampling procedures are carried out with care and precision as simple mistakes are very costly to rectify, particularly on the larger sites. Needless to say, the possibility of error is very significant when teams are working in bad weather on overgrown sites or areas where fly tipping is a feature. It is reduced considerably, however, by the initial preparation in the office, which shortens the site time to the minimum.

Twelve trial pits producing 60 samples is the target for each day's operation, but when conditions are particularly good, up to 20 pits producing over 100 samples, have been excavated.

<u>Water and Methane tests</u>

Tests on water samples have not figured greatly in the Thamesmead

Investigation to date and little testing for methane has been undertaken. It is intended, however, to carry out a comprehensive test on the water system over the whole development area in order to check the relationship between conditions in ground water, canals and lakes and the contamination of the site generally. The main value of the tests will be threefold in that:-

(i) The movement of contaminated material through the sub stratum can be evaluated.

(ii) Assistance can be rendered to the Council's Parks Department in their provision of planting at the margins of lakes and other bodies of water.

(iii) Loss of amenity and the creation of health hazards can be avoided by remedial action at an early stage, through later selective monitoring.

Methane bores are to be sunk in test sites behind the existing river wall where earlier soil sampling has uncovered organic refuse at depth mixed with industrial waste and rubble fill. Together with the possibility that the river wall may have enclosed bodies of silt, a situation exists which is similar to other riverside sites in London where methane emanation has proved to be a serious problem. As all buildings on the site are piled without exception, hazards created by the release of pockets of methane, should they exist, would be very real.

Fibrous Asbestos Deposits

A search for deposits of fibrous asbestos is being carried out simultaneously with the investigation for industrial contamination. This has been necessitated by the discovery of asbestos lagging on the surface, along the lines of a long removed network of overhead steam and hot water pipes which extended for several kilometers across the site from each of two boiler plants. Such mains distribution was presumably prompted by the need for heat without the risk of explosions or fires.

Asbestos lagging so far subjected to test contains both crocidolite and chrysotile and must be removed before site clearance and construction work can take place. A 10m x 10m search pattern is carried out based on setting out from the 100m operational grid. It is a tedious and exhausting process for the officers concerned as no other satisfactory way of locating asbestos deposits in densely overgrown conditions has been established.

Clearance of areas is currently undertaken by licensed firms who remove the asbestos off site, but cheaper and more certain methods of disposal on site are to be put into operation in the near future when asbestos will be buried in the Thamesmead licensed disposal facility.

7 Assessment of Site Conditions and Remedial Measures to deal with Contamination

Responsibility for assessment of each site and the framing of recommendations for remedial works rests with an Inter Departmental Assessment Panel composed of officers with widely differing professional skills. They provide specialist knowledge in dealing with the various aspects of contamination and represent the particular interests of their own departments.

The panel is composed of the officers forming the Site Inspection Team noted in para. 6 together with representatives of the Council's Poisonous Waste Unit, the Licensing Unit for Disposal Sites, the Parks Department Horticultural Advisor and an Officer from the Valuer's Department which formally 'holds' the land until handed over for development. A Structural Engineer with a special interest in explosion hazards and structural deterioration also attends.

An officer from the Thamesmead Derelict Land Unit co-ordinates the work of the Panel, distributes all documentation and provides the input on intended construction work for the sites in question.

The main item for consideration is a detailed report prepared by the Scientific Adviser which includes and is based upon, the soil analysis data, trial pit logs, and drawings indicating the site sampling pattern. Reports on methane emission combustability and ground waters are also included where relevant.

The panel considers the following items in respect of their possible affect on health and the general environment in framing recommendations for remedial works:-

(a) Levels of contaminants in the soil.
(b) Calorific value where appropriate.
(c) Condition of ground water.
(d) Problems of phytotoxicity.
(e) Methane Hazards.
(f) Corrosion of below ground structures and buried services.
(g) Disposal of excavated contaminated material.

Where possible, the condition of the land is considered against the layout of intended development in order that recommendations can be framed to suit the proposed land form and constraints imposed by fixed structural elements. The long term hazard to health of land users is the primary consideration and this tends to set the pattern for remedial works. Occupants of houses with gardens where produce may be grown and consumed over many seasons are regarded to be most greatly at risk, and as the bulk of development is taken up by housing, it is to this which most attention is directed.

Schools are given particular consideration especially where there are possibilities of horticultural activity by pupils, but the buildings (being large) and the provision of hard play surfaces are seen to be satisfactory forms of protection in themselves. The bulk of remedial measures are therefore included in soft landscaped areas.

There are many forms of development at Thamesmead ranging from housing to industry and public open space and it suffices to say that each case is dealt with on its merits.

Short Term and Occasional Risks

The panel through the medium of the Medical Advisor's Chief Environmental Health Officer determines if there are any special requirements in respect of the health and safety of site construction personnel. Fortunately, with one exception to date, the condition of sites has not prompted special needs beyond the use of ori-nasal masks and/or water spraying at most. The provision of washing facilities and accommodation for consumption of food under the Health and Safety at Work Act are satisfactory for the majority of sites examined so far. The exception to this is the site of the old Arsenal gas works which is referred to below.

'Occasional' risks will occur when repairs or alterations are made to buried services. Low voltage electricity, telephone and TV cables are usually buried about 0.5m below the surface but HV cables, water and gas mains are laid with not less than 0.9m cover and drainage is usually deeper still.

It would be unrealistic to assume that statutory or local authority operatives will be kept informed of possible health risks in the long term, so where necessary, recommendations are made for backfill of trenches to take a form which will reduce the risk of contact and the spreading of foul material over clean surfaces during re-excavation.

Reclamation

The methods of reclamation recommended by the assessment panel have been derived to resolve the complex nature of the contamination by the simplest possible means and this has usually involved:-

(i) Overfilling with clean imported material.

(ii) Excavation and removal of contaminated soil and replacement with clean imported fill where the original ground level has to be maintained.

(iii) The covering of sloping surfaces when the contaminated inner cores of raised mounds are exposed, or deep cuts are formed for canal construction.

(iv) Special arrangements for prevention of direct contact and erosion when excavations for lakes bring sub surface contamination into conflict with open water.

(v) Change of land use.

Oversite filling on large areas of land poses many difficulties in locating supplies of suitable material and transporting this to the site. If for instance a site of say 16 Hectares (40 acres) is filled to a depth of one metre, the import of 160,000 cubic metres of material involves possibly 20-25,000 vehicle movements between site and source of fill. This is clearly undesirable.

Ideally, the material to best deal with heavy metals and indeed most other contaminants would be clay but this is unsatisfactory as a working base for construction and would lead to many problems later, in respect of surface drainage.

Estuarine sand containing say 10% silt or pit dug sand with the same silt content would provide the desired rate of cationic exchange to deal with metals but experience with such material has not been satisfactory. Dredging for estuarine sand of the correct constituency may not be permitted at the time it is needed for development and in any event, the unloading of vessels by conventional means which would be essential to preserve the silt content, would be inordinately expensive.

Pit dug sand providing a minimum silt content of 10% is obtainable, but a workable and consistent upper limit of silt in local sources is virtually impossible to achieve. Such material has been used already at Thamesmead and found wanting in that excessive silt in the sand causes the site to become a quagmire during bad weather.

Hydraulic fill using sea sand

Large tracks of the Western area of Thamesmead have been filled with hydraulically placed sea dredged sand for civil engineering purposes. The sand is taken from the North Sea by dredger and pumped ashore using river water as a transport medium in the proportion of approximately 10 to 1 by volume. Water is drained back to the river leaving the sand which is sharp and though containing shell, has little or no silt content.

This method, previously regarded as unsatisfactory due to the lack of silt is now likely to be adopted for capping contaminated areas by taking into account the additional quantities of surface soil necessary to create a suitable environment for plant growth on sand. The next major site due for reclamation at Thamesmead will be capped with 800mm of hydraulically placed sea dredged sand and 300mm of soil for garden areas. Extra top/sub-soil is necessary for shrub planting and large tree pits are essential, but the advantages of the fill method are considerable, particularly in respect of its lack of environmental impact on surrounding areas.

The slightly increased depth of filling overall, the beneficial shell inclusions and 300mm of topsoil are together considered to be satisfactory for the site in question. The site will be self draining and provide a clean working platform for construction work, and is likely to be free of the majority of its salinity in approximately 18 months. The topsoil is placed late in the construction period.

The Arsenal Gas Works

The highly dangerous condition of the site of the old Arsenal gas works prompted the Assessment Panel to strongly recommend a change of land use whereby building work within the area would be prohibited. The ground had been saturated in phenolic liquids during 3 generations of gas works going back to approximately 1860, and this in company with high levels of other contaminants created a below ground environment of considerable risk to site operatives.

It is fortunate that an area planned for open space of a similar size was able to be relocated, leaving a more suitable area available to make up the loss of housing land. The gas works site will be covered with at least 1.0m of broken concrete and rubble from demolition works which will be blinded with shingle and then capped with a 300mm layer of clay graded to falls, and taken down at least 1.0m below the existing ground at the perimeter. Landscaped open space and playing fields will be established on the areas but trees will only be planted outside the capping.

Corrosion Problems and Deterioration of Concrete

As a matter of course, all sites to be developed by the Council are subject to investigation to establish their geophysical properties and levels of acid sulphates. Such action as is necessary, is taken by the Architect in conjunction with the Scientific Advisor's Building Sciences Group, and in a number of cases of Thamesmead, excessive levels of sulphate and severe acid conditions have demanded special protection to piles, especially where other concrete damaging materials are in close association. In some cases, pile shells have been coated with 2-3 applications of an epoxide resin based coating.

8 Disposal of Excavated Contaminated Material

The requirements of the Disposal of Poisonous Waste Act and Control of Pollution Act are such as to have placed the development of a large part of Thamesmead in jeopardy had it not been possible to provide on site disposal facilities.

Construction of a new river wall as part of the London flood defences created a space between it and the old river wall of some 350,000 cubic metres in volume which has been licensed as a disposal facility suitable to receive the kind of material likely to be excavated at Thamesmead. The facility, which is the only one in London, has proved to be a major success in the rehabilitation of contaminated land while providing an additional bonus in the reclamation of an area of river bed without the need for importing material. The reclaimed land suitably capped with clean material will eventually be used for development and has been designated as areas 4J and 4K on the attached map. An illustration of the facility follows.

A similar space will be created with the construction of the next section of river wall and it is hoped that a licence will be granted for the use of this for disposal of contaminated material.

It is perhaps worthy of note that the licence has recently been modified to enable the 4J/4K facility to receive bagged asbestos cleared from development areas. The asbestos will be buried in deep trenches under the same strict controls governing the use of the facility for disposal of contaminated soil.

9 Research

The most difficult aspect of reclamation at Thamesmead has been the lack of any authoritative guidelines on such matters as the upward migration of metals through covering layers and the effectiveness of different forms of fill as barriers to contamination. It must be said that under such circumstances a 'fail-safe' attitude has had to be adopted in deciding on remedial measures in the knowledge that this may involve additional and heavy costs both in advanced civil engineering work and the construction of buildings.

It is fortunate that research into these matters is now under way and gratifying that Thamesmead has been chosen as one of the sites where long term research can be conducted. Liverpool University are currently operating an experimental site which was set up in the Spring on behalf of the Department of the Environment Central Unit on Environmental Pollution (CUEP).

It is to CUEP that the Council's Inter Departmental Assessment Panel submits its report and recommendations for comment before passing them on to the Thamesmead Architect for action. It is hoped that the research being conducted will eventually enable CUEP to publish the firm guidelines so urgently needed by developing authorities in order that the appropriate level of protection to the public can be provided at a cost which does not endanger the economic viability of development.

10 The Thamesmead Experience

Much of Thamesmead's 1000 acres still needs to be examined, not least of which is the conservatively estimated 1,000,000 cubic metre elevated area at Tripcock Point. The mound which is covered in dense thicket is composed almost entirely of ash, industrial residues and gas works waste. It has been shaped to enclose a network of narrow road and railways between steep embankments into which have been built scores of explosives magazines, test facilities and other buildings, all contained in massive masonry enclosures.

The Derelict Land Unit is also being taxed with problems of two areas of below ground spontaneous combustion, but it is probable that difficulties of such varied kinds on one site are peculiar to Thamesmead. If any value is to be found in the experience of dealing with the rest of the site it may stem from the commitment to tackle the many and varied problems of land under severe environmental stress on a comprehensive basis. Like many sites in the older run down riverside industrial areas of London and other cities, the land at Thamesmead is not only contaminated with the waste products of manufacture but contains the remains of several generations of substantial buildings. The process of reclamation on sites elsewhere demands the import of clean fill and the disposal of excavated spoil and masonry as at Thamesmead but it is apparently impossible to draw development advantage from these activities without the special interest of a central controlling authority.

The controlled use of contaminated material to reclaim other land and the re-use of clean excavated spoil to form sound barriers to majorroads or as base fill to landscaped forms are examples of the comprehensive approach at Thamesmead. Another is the concrete and masonry in old foundations being removed before landfilling and crushed to provide a source of hardcore for new roadworks, etc., at substantial savings in cost - a demonstration of profitable recycling. In other places such material is often regarded as a nuisance and arrives on someone else's land through the clandestine activity of fly tippers.

If the advocates of revitalising the inner cities are to see any progress at all, they will take note of the value in comprehensive management at Thamesmead and encourage adequate provision for environmental legislation to be made effective and similarly turned to advantage.

HG/T/OD/GWL
July 1979

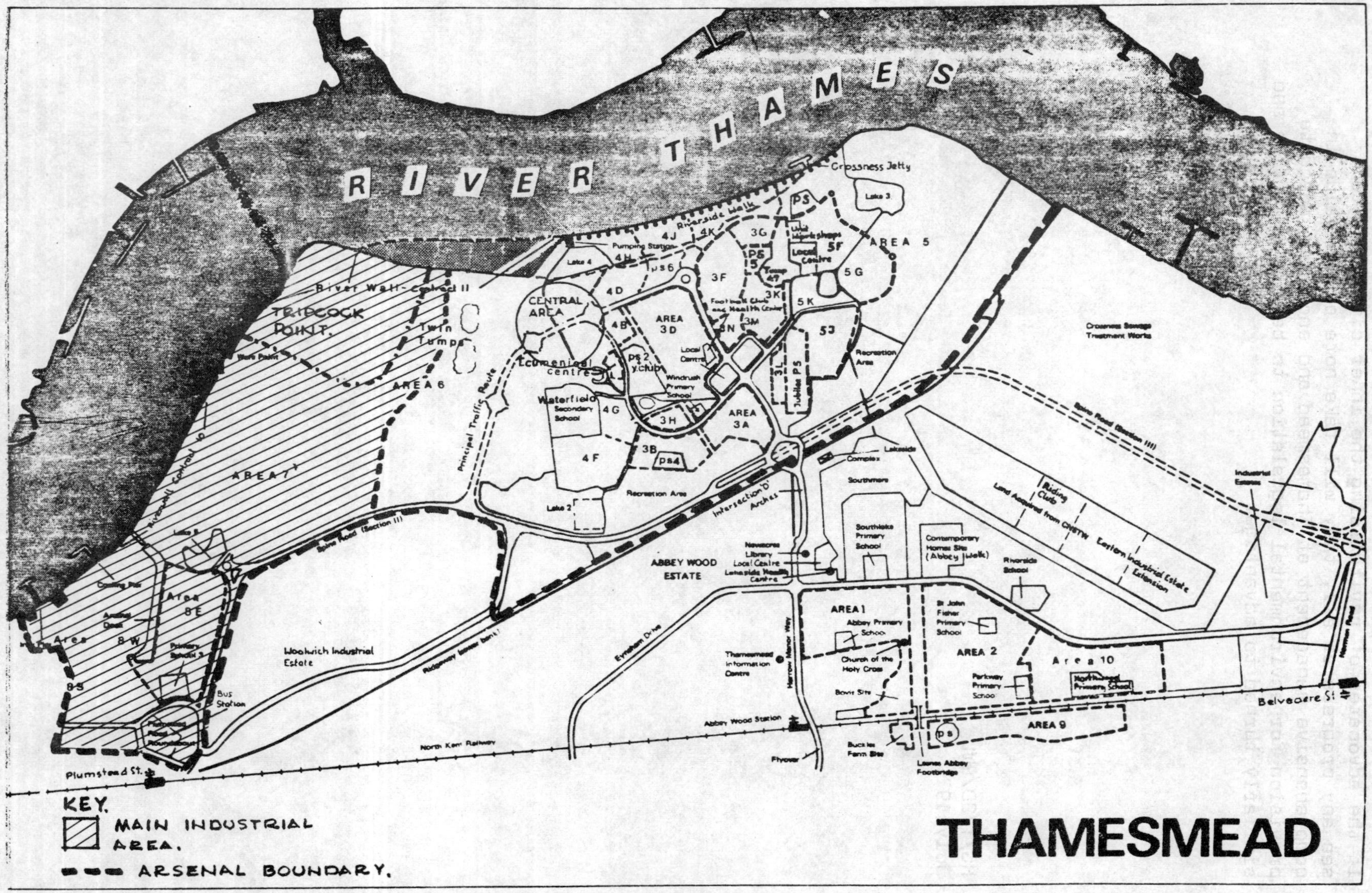
RIVER THAMES
Crossness Jetty
Lake 3
AREA 5
5F
5G
5K
5J
Local Centre
3G
3F
3K
3M
3N
4J
4K
4H
4D
4B
4G
4F
3D
3H
3B
3A
PS
PS4
Riverside Walk
Pumping Station
Lake 4
CENTRAL AREA
AREA 3D
Ecumenical centre
Windrush Primary School
Waterfield Secondary School
Recreation Area
Lake 2
Jubilee PS
Principal Traffic Route
River Wall
TRIPCOCK POINT
Twin Tumps
AREA 6
AREA 7
Area 8E
8W
8S
Bus Station
Woolwich Industrial Estate
North Kent Railway
Plumstead St.
Crossness Sewage Treatment Works
Spine Road (Section III)
Riding Club
Land Acquired from CNSTW
Eastern Industrial Estate Extension
Industrial Estates
Lakeside
Complex
Southmere
Intersection 'D' Arches
ABBEY WOOD ESTATE
Southlands Primary School
Contemporary Homes Site (Abbey Walk)
Riverside School
AREA 1
Abbey Primary School
Church of the Holy Cross
AREA 2
St John Fisher Primary School
Parkway Primary School
Area 10
AREA 9
Thamesmead Information Centre
Abbey Wood Station
Flyover
Lesnes Abbey Footbridge
Belvedere St.
Norman Road
Eynsham Drive
Harrow Manor Way
KEY.
MAIN INDUSTRIAL AREA.
ARSENAL BOUNDARY.
THAMESMEAD

Above: The licenced tip between the old and new river walls will take 350 000 m³ of fill. The inner slope on the river side is stabilized with interlocking concrete blocks (below, left) which will be removed as the tip is filled to avoid interference with future piling

Right: The remainder of the mud flat which will eventually be reclaimed

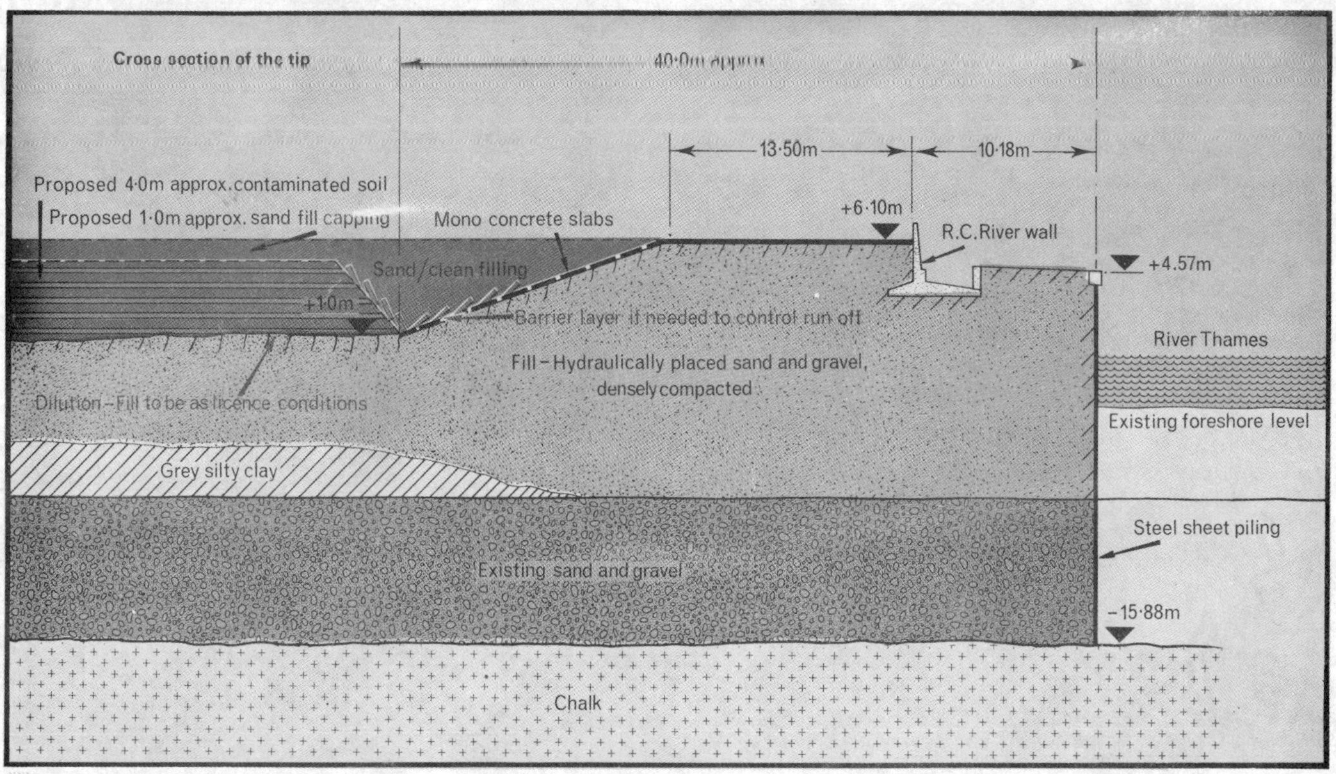

RECLAMATION OF A REFUSE TIP FOR OPEN SPACE AND HOUSING DEVELOPMENT

M.J. McCarthy Dip Arch RIBA AMBIM (London Borough of Hillingdon)

THE WILLOW TREE LANE JOINT DEVELOPMENT

The Site

The Willow Tree Lane Joint Development is in the Yeading area of the London Boroughs of Hillingdon and Ealing in West London. The site is 65 hectares in area. Willow Tree Lane bisected the site and was built to provide access from the Ruislip Road (B455) in the north to a heavy goods vehicle testing station and Post Office depot in the south. The western part of the site was rough grassland which sloped gradually towards the south. The eastern part of the site adjoining the Paddington branch of the Grand Union Canal was an old domestic refuse dump and had been tipped to heights of 11m above the general ground level. In this area a considerable amount of broken glass, tyres, and other debris remained from the activities of car breaking merchants. There was a Civil Defence training building in the centre of the site and several pillboxes and similar structures at various positions. The only pleasant feature of the site, apart from the canal frontage, was lost when Dutch Elm disease compelled the felling of mature elm trees along the northern and southern boundaries.

Planning Background and Land Acquisition

Development of the site had been under consideration for many years. Having had several planning applications for housing development rejected, the private landowners formed a Consortium in 1971 to work in partnership with the London Borough of Hillingdon, the London Borough of Ealing, and the Greater London Council to determine how the site could be satisfactorily developed. A Planning Brief was published in January 1973 demonstrating that the development could create a new community by providing 1600 homes together with schools, shops, community buildings, and a substantial area of public open space with recreational facilities. The Consortium immediately submitted outline planning applications. Outline planning permissions were granted in April 1974. Changes in the economy during the intervening period had prompted the Consortium to negotiate the sale of its land to the Authorities so that, having purchased other land from the Property Services Agency, they owned the whole site. The acquisition of 16 hectares of the tipped area to the east of Willow Tree Lane was at nil value and sums totalling the estimated cost of laying out this portion as public open space were deducted from the price of the remainder of the land.

Structure Plan[1]

As soon as outline planning permissions had been granted, the Authorities established Coordination and Design Teams under the chairmanship of Hillingdon officers to undertake detailed studies and prepare a Structure Plan. The political objective was to complete the Development within seven years.

The form of the Structure Plan emerged from detailed consideration of the planning objectives, the constraints on access to the site, the difficulty of providing adequate drainage facilities, and the existence of the tip in the east of the site. The Planning Brief stipulated that there should be 16 hectares of public open space and that it should be distributed to form a link between larger areas of open land to the north and south of the site. This meant that it would be necessary to build housing on some parts of the tipped area although most of it was to be converted to a canal-side hill for walking and sporting activities. Site investigations were progressed during the preparation of the Structure Plan which was substantially complete by late 1975.

SITE INVESTIGATIONS

Preliminary Investigations

The first site investigations were carried out in 1972 by the Consortium who drilled 12 boreholes to depths of between 7m and 15m. Representative samples of the different strata, undisturbed cohesive soils, granual soils, and ground water were taken. The penetration of granular soils was measured with the standard penetration tool, and laboratory testing undertaken for index property, bulk density, moisture content, undrained triaxial compression and sulphates. The results showed that to the west of Willow Tree Lane there was some made ground between 0.5m and 2m thick consisting of sands, gravels, and clays with brick fragments, whereas to the east there was made ground between 0.5m and 9m thick consisting brick fragments, ash, and organic material. The Geological Survey of England and Wales showed brick earth overlaying London clay over the site, and the boreholes proved stiff fissured brown silty clay to a depth of 7m below the made ground. The purpose of the investigation was to assess the characteristics of the site as they would affect the design of foundations and the results gave no cause for alarm.

In April 1973 the London Borough of Hillingdon dug 6 trial pits using a mechanical excavator and established that the firm brown clay was capable of sustaining a safe increase in pressure of at least $160 kN/m^2$.

Detailed Investigations

Analysis of samples of the tip material to assess the problems of obtaining successful tree, shrub and grass growth indicated salt concentrations sufficiently high to pose problems of growth for most plant species.

A borehole survey on a 100m grid across the whole site was commenced in November 1974. In addition to the normal range of tests, it included the taking and storing of soil samples from various depths. While the results were awaited, the Greater London Council made 3 trial digs into the southern escarpment of the tip. The digs showed topsoil 0.3m thick over a layer between 0.5m and 2m thick of white/grey substance, which was thought could be gaswork waste, and then mixed fill including bricks, bottles, and fond slurry over silty grey clay from canal excavations.

A number of borehole core samples were sent for testing as they became available. All samples were found to contain high proportions of calcium carbonate and the tests confirmed the earlier results of very high total soluable salt content. Consideration of possible methods of ensuring the leaching of salts from the tip material prior to topsoiling and planting led to discussion with the Thames Water Authority regarding acceptance of drainage liquor into the foul sewer system. The Authority stated that it would require additional analytical information before considering acceptance of effluent from the tip. It required the analyses to be performed on at least 20 water samples taken from different parts of the tip and to include heavy metals, boron, oxidised and organic nitrogen on filtered and unfiltered samples, phosphorous, chlorides and total organic chlorine, cyanides, sulphides and phenols. The Authority also wanted to know the probable discharge volume in terms of average daily and maximum flows. This point was reached in September 1975.

First Tests for Contamination

The borehole analysis became available in April 1975 for the Greater London Council area, June 1975 for the Ealing area, and July 1975 for the Hillingdon area. The first volume showed that more of the site had been tipped than had previously been thought. In the light of its experience with other contaminated sites, the Greater London Council undertook chemical analysis of 60 samples taken at 0.5m and 1.5m depths from the boreholes on its land. The total concentrations and the plant available concentrations of lead, cadmium, mercury, zinc, copper and nickel were determined. In the 0.5m samples, 71% had total lead concentrations greater than 1000ppm, 10% had total cadmium concentrations greater than 10ppm, and 83% had total mercury concentrations greater than 1ppm. These results indicated abnormal concentrations of the more toxic metals and although the concentrations decreased with the depth below the surface, they were still considered unduly high. The fact that these tests had been undertaken was revealed to the other Authorities in October 1975 but the results and the implications were not made known until December 1975.

IDENTIFICATION OF CONTAMINANTS

Contamination Working Party

Immediately the implications of the contamination were known, a Working Party was formed to consider what action was necessary. The objectives of the Working Party were to establish the extent and the type of contamination on site, to conclude whether the contamination presented short term or long term health hazards, if necessary to formulate proposals for remedial action, and to progress their investigations with the utmost urgency. The desire for speed stemmed from the fact that no financial approvals from central government would be forthcoming for any items of infrastructure or individual projects until it had been demonstrated that the problems of contamination could be overcome without destroying the economic viability of the Development. The Working Party included architects, landscape architects, quantity surveyors, civil and structural engineers, environmental health officers, medical officers and scientists specialising in soil and water chemistry.

History of Tipping

It was established that the area of the site was largely agricultural until 1895. The sequence of changes in site uses over the period 1895 to 1953 were recorded from Ordnance Survey information for 1895, 1914, 1935 and 1940, together with aerial photographs taken in 1953.

The site had been used for brick-making which involved clay extraction and the use of refuse as fuel. The tipping of refuse continued after the brick-making ceased. The site was operated by Thomas Clayton (Paddington) Limited and approximately 37000 tons of refuse from Paddington and 6000 tons from Kensington was sent to the tip annually. The refuse was transported to the site by canal barge, unloaded by grab-crane and distributed around the site by narrow gauge railway. Critical reference was made to the tip in 1929 in the Dawes Report[2]. which named it as the Yeading

"Clayton Dump". The report stated that the refuse was deposited indiscriminately and the methods used were primitive and unsatisfactory. A refuse incinerator was built and operated from 1939 as a result of the various criticisms. Most of the putrescible materials in the refuse from then onwards would have been burnt and only clinker and ash deposited. The tipping of refuse ceased in 1950 allowing 26 years for uninterrupted decomposition of the most recent refuse, but most of the deposits of organic material would have had a period of between 37 and 81 years to decompose.

The three canal docks serving the tip had been backfilled during the period of tipping. In later years the main tip had been used for car breaking activities, and in the early 1960's clinker and ash was excavated for motorway construction. There was no record of any industrial activity on other areas of the site and no evidence of industrial tipping.

Methane

It was considered advisable to check whether the tip was still yielding methane in spite of its age and evidence from the boreholes indicating that there were no barriers, as on a controlled tipping site, to prevent methane rising to the surface. Analysis of a series of samples taken from 36 locations showed that a large majority contained less than 10vpm. There was one exception which on investigation was found to be adjacent to old farm buildings where there was a substantial chicken manure pit. It was agreed that this could be rendered innocuous by removal and that the rest of the site had become innocuous for the purposes of the Public Health Act 1936, Section 54.

In later discussion of the results, it emerged that there was no standard method of testing for the rate of evolution of methane from landfills, and divergence of view on whether there was a correlation with the rate of generation of carbon dioxide.

Heavy Metal Contamination

The testing of samples for heavy metals was the largest and potentially the most time-consuming of the surveys undertaken. However, by dividing the samples and sending them to 3 laboratories for analysis the majority of the results were obtained within a month. Certain of the samples were divided between the laboratories as a means of verifying the consistency of the results. Determinations were made of the total and plant available amounts of lead, mercury, cadmium, zinc, copper, nickel and arsenic. The results showed similar concentrations of the metals as the earlier tests.

The principal hazards to health were considered to be inhallation of dust, ingestion of insufficiently washed vegetables grown in contaminated soil, and children eating soil. In later discussion, the possibility of absorbed heavy metals being transmitted to the leaves and fruit of domestic trees was considered but it was decided that such processes would not be hazardous.

Non-Metallic Contamination

Samples from various levels were examined for non-metallic contaminants. Tests were carried out for coal-tars, cyanides, sulphides, ferri-ferrocyanides, and thiocyanates. The results gave no indication that the previous site uses had resulted in non-metallic contamination.

Asbestos

At the suggestion of the Department of the Environment, a survey was carried out to determine whether asbestos had been dumped at the site. Examination by X-ray diffraction on 32 samples and additional examination by electron-microscope on 7 samples revealed minimal amounts of chrysotile. This was considered consistent with airborne deposition.

Ground Water Contamination

A list of potential contaminants of ground water was drawn up in consultation with the Thames Water Authority and the British Waterways Board. Analysis of ground water samples indicated that decomposition of the organic materials in the tip had virtually ceased, that there would be no hazards to site workers from contact with the water, that the ground water could be discharged into the foul sewers during works to the tipped areas, but that high concentrations of phosphate, mercury, iron, boron, ammonia, and high BOD could be a hazard if discharged to surface water sewers.

This conclusion allowed the design of the drainage system to be finalised. Periodic monitoring of its performance has been carried out, and high ammonia levels are still being recorded.

Conclusions on Health and Safety

It was recognised that the tip material was heterogeneous and pockets of severe contamination or items such as drums of industrial waste could have been missed in the sampling, but it was considered reasonable to assume that the samples were representative of the whole tip. Examination of the results of the investigations led to the conclusion that the greatest health hazards to persons living adjacent to the site and to site operatives during construction work would be the inhallation of dust. It was proposed that this problem could be overcome by using water sprays whenever necessary, and by providing operatives with suitable protective clothing, facilities for washing, and face masks for those working in trenches. Thus, the reclamation measures could be devised solely to ensure public safety in the completed Development.

REMEDIAL MEASURES

Economic Considerations

Removal of contaminated material from the site was dependant upon finding a suitable toxic waste tip which could accommodate the large volumes involved. Even if such a tip existed, the cost of transporting the material and importing clean soil to maintain the levels required for drainage would have been prohibitative. Numerous ideas for remedial measures were examined by the Working Party and methods of economising on cost were given paramount consideration.

A solution was devised whereby the designed contours of the new land formation allowed all earth moving activities to be confined within the already contaminated areas of the site. Flexibility was allowed by designing areas where the contours could be varied to accept increased or reduced volumes, and by creating a 'toxic tip' to accept surplus soil from building operations in the later stages of the Development.

Definition of Areas

Of the total area of 65 hectares approximately 43 hectares was found to be contaminated. The layer of contaminated material was completely removed from 3 hectares on the west of the site in order to create building sites which were either wholly contaminated or wholly uncomtaminated. This material was used in the recontouring of the eastern contaminated areas. The remaining 40 hectares of contaminated land had no effect on the content or layout of the Structure Plan and approximately 21 hectares was to be subject to building development.

Capping

It was decided that the most economical way of overcoming the contamination was to provide capping over the whole of the affected area. The purpose of the capping was to prevent contact between people and contaminated soil. Different compositions and thicknesses of the capping layer were evolved to suit the different end uses of the site. This also involved devising means of ensuring that sideways leaching of any contaminant was prevented.

The capping proposed for public open space comprised a 150mm base course layer of hardcore consisting of particles between 25mm and 50mm, a 25mm layer of fine porous material, and a layer of topsoil. The course layer was to deter any upward movement of salts by capillary action, to prevent upward movement of glass fragments, to provide a drainage blanket, and to minimise the danger of a presoiling contaminated dustblow. The second layer of fine porous material was to act as a blinding agent to prevent the loss of topsoil into the course layer. The final 125mm or 150mm layer of topsoil was essential to provide a growing medium for vegetation. It was proposed to increase the depth of soil to 550mm by the inclusion of a layer of subsoil where trees were to be planted. The proposed depths of the layers assumed compacted material. The use of a layered capping enabled the total depth to be substantially reduced over that provided previously at other sites.

The capping proposed for gardens was a minimum of 500mm of clean soil of which 300mm would be topsoil. Areas to be covered by roads, car parks, footpaths, or buildings did not require additional capping.

Foundations

The amount of excavated contaminated material to be deposited in the toxic tip was kept to a minimum by evolving a foundation design whereby either driven piles would be used, or excavated material from trenches would be used as filling under the ground floor slabs. The level of the ground floor slabs was raised to minimum of 650mm above the existing ground level to accommodate the fill material, consistent with the adjacent garden areas which were to be provided with a 500mm capping.

Initial Reclamation Contract

The scope of the initial reclamation contract was to fence the entire site, carry out the major earth moving operations, thoroughly compact the fill over the building sites and regrade as necessary to achieve the levels required for drainage, provide temporary capping by means of hydroseeding to the building sites to be developed in the latter phases of the Development, and to contour, cap, topsoil and plant the whole of the eastern area of public open space. The capping required on the building sites was to be provided as part of each building contract.

Approvals

A Contamination Report[3.] was prepared detailing the sample locations, analysis techniques and equipment used, results of the analyses, proposed remedial measures, and financial implications. It was submitted in October 1976 to the Department of the Environment, the Department of Health and Social Security, the Ministry of Agriculture Fisheries and Food, the Health and Safety Executive, the Thames Water Authority, and the British Waterways Board, and later made available to the press and public.

The Department of the Environment agreed to consider the Contamination Report on an ad-hoc basis in advance of establishing the Central Unit for Environmental Pollution and formal consultative machinery. Discussion culminated in a meeting attended by all parties in January 1977 at which the Department requested further information on the risks of corrosion and combustion. However, the Department confirmed that whilst the responsibility for ensuring health and safety rested with the Authorities, it saw no reason to dissent from the major conclusions in the Report as to the likely efficacy of the proposed remedial measures.

This approval enabled detailed financial negotiations to be concluded, and the Initial Reclamation Contract commenced in August 1977. The cost at May 1977 prices of the actual reclamation was £15,500 per hectare, and the development of the public open space £9,100 per hectare.

ADDITIONAL INVESTIGATIONS

Corrosion

A further 22 soil samples from 14 boreholes were analysed for calcium content, pH values, electrical conductivity factors and sulphates. The results showed slightly alkaline values of pH, due to the calcareous materials deposited and the presence of chloride-bearing materials, and high conductivity factors. The earlier surveys had shown the presence of soluable sulphates and this was confirmed. It was concluded that the soil would be corrosive and would have a specific conductance but that usual techniques adopted by the Statutory Undertakings would be sufficient safeguard. Sulphate resisting materials would be required in all substructures but this was a general requirement of the locality.

Combustion

In the period between submission of the Contamination Report in October 1976 and the meeting of all parties in January 1977, smouldering was noticed in an exposed position on the elevated southern edge of the main tip. The area affected was approximately 25m^2 in area and was facing the direction of the prevailing wind. An investigation using temperature probes showed that the highest temperature reached was 125°C, and the fire eventually burnt itself out. It was the occurance of this fire that raised the question of combustion.

The assessment of the risk of combustion in the tip materials after recontouring and compaction proved difficult and not without controversy. Detailed discussions were held with the Department of the Environment, the Building Research Establishment, the Fire Research Station, the Coal Board, the London Fire Brigade and experts in solid wastes. However, the general state of knowledge in this subject was remarkably limited, and there was no known method of quantifying the risk of combustion in soils.

Notwithstanding the site history of tyre burning and the sizeable logs of felled elm that had been burnt on the site in large fires kept alight for long periods, it was suggested that a full scale fire experiment be carried out to simulate a house burning to the ground or a large community fire on Guy Fawkes night. Before the significance of any results that might be obtained were established, two major fires occurred on the tip material. The larger of the fires was in an old timber stable and was partially extinguished by use of a bulldozer after 40 mintues. The other was a surface fire, and was extinguished by the local Fire Brigade within a few hours. Soil temperatures were recorded using a pyrometer with a range between 0°C and 600°C over a period of 5 days. Temperatures around the stable 9 hours after the fire exceeded 600°C and probably exceeded 900°C. Neither fire had an adverse effect on the tip material.

Samples of tip material were tested for loss of weight on ignition, and for calorific value on a dry ash-free basis, dry basis, and "as found". Fresh excavations were made and the interistitial gases analysed in the undisturbed profiles at various depths. However, difficulty was experienced in relating the results of laboratory tests to the actual site conditions. Factors that gave particular difficulty were the natural variation in the moisture content of the ground over the seasons, long term reduction in moisture content under buildings, and the significance of the soil compaction in relation to the availability of oxygen. The relevance of the calorific values, particularly on a dry ash-free basis, was questioned.

After considerable discussion, it was generally agreed that there was no possibility of spontaneous combustion caused by either microbiological action or exothermic reaction, but there was a remote risk of combustion triggered by an external heat source. The distinction made between the risk of spontaneous combustion and the risk of triggered combustion was important. If spontaneous combustion was possible no economical remedial measures could be taken to lessen the risk of an underground fire occurring. However, if spontaneous combustion was not possible, the residual risk of triggered combustion from an external heat source could be rendered even more remote by undertaking precautionary measures. Thus there was nothing to indicate that there were any reasons to stop or modify the operations being undertaken on the site as part of the initial reclamation contract, or to indicate that any of the land designated for building purposes would prove to be unsuitable for the proposed Development. Underground fires are slow burning and in the limited depth of compacted fill on the building sites would not present a danger to life, but would be a threat to buildings. As owners of municipal housing, the London Borough of Hillingdon could, and did, accept the remote risk involved.

Methane

Towards the end of the earth moving operations in January 1978, the reclamation contractor encountered material different to that found elsewhere on the site. It appeared that the material was domestic refuse dating from 1936 and that its decomposition had been retarded by an impervious capping layer of clay. Testing showed that methane was being generated. Sampling tubes were inserted around the area and the results indicated that the methane producing material was in an isolated pocket approximately 1600m^2.

It was decided that the material should be removed and spread in a shallow layer along the new escarpment adjoining the canal remote from any building. The material was later covered by capping and no additional precautions were taken.

Monitoring

Hillingdon officers decided that it would be prudent to carry out soil investigations after initial reclamation of its building sites had been completed. Soil samples were taken from the surface and a depth of 300mm on a 25m grid. The moisture content and concentration of lead were determined on 5% of the samples taken at random. The results showed no significant differences from the earlier surveys. All samples were dried and placed in jars and are being kept as a permanent record.

PROBLEMS OF STATUTORY UNDERTAKINGS

General Problems

On commencement of the initial reclamation contract in August 1977, firm programme dates could be established with the Statutory Undertakings in respect of the installation of distribution mains. Discussions were held with each Undertaking regarding the precautions to be taken against both soil contamination and corrosion. A meeting of all the Undertakings and the Factory Inspectorate of the Health and Safety Executive was held at the end of November 1977 but unfortunately it served only to confuse.

The Undertakings had joint meetings and sought further guidance from the Health and Safety Executive in March 1979. They were concerned about the health hazards to operatives during construction and the possible contamination of the water supply. A meeting was held on site in April 1979. The Authorities reiterated that any costs to the Undertakings necessarily incurred as a direct result of contamination would be reimbursed, and that there was no intention to avoid or minimise any problems. Nevertheless, any proposed precautions would be given detailed scrutiny to ensure that they were not excessive. It had previously been agreed that all service trenches through contaminated fill would be back filled with clean material so that emergency repairs or maintenance to the mains could be undertaken without special precautions for the health and safety of site operatives. Eventually agreement was reached on suitable protective clothing, facilities to be provided for eating on site, the storing of materials, and the cleaning of equipment and vehicle wheels.

Problems Particular to the Water Company

Protracted discussions were held with the Rickmansworth & Uxbridge Valley Water Company regarding the possibility of contamination of the water supply either from material entering the pipes during main laying or in the event of a fractured main. The Company delayed completion of formal Agreements required before main laying could commence until requested to explain its action with reference to its obligations under the Water Acts. At this point the Company sought an all embracing indemnity from the London Borough of Hillingdon and stated that the Agreement to lay mains in contaminated ground would be subject to the satisfactory outcome of discussions with its insurers. Where a water supply is requisitioned under Sections 27 and 37 of the Water Act 1945, there is a duty to provide water and no basis for seeking an indemnity. The Company claimed that the terms of the requisition made by Hillingdon on behalf of the Authorities was outside the scope of these Sections and thus it was entitled to supply water on reasonable terms including the provision by Hillingdon of an indemnity. The scope of the indemnity sought was modified to include loss, damage, or injury arising from the presence of contamination not then known to be present, or present in greater quantities than currently thought. The Authorities demonstrated they could bring legitimate requisitions within the scope of Sections 27 and 37. The Company completed appropriate Agreements and commenced work but sought to have restrictions regarding interference with the supply pipe from the mains to individual dwellings made the subject of land charges so that they were brought to the attention of individual householders. The Authorities could see no proper basis for this and it was left to the Company to initiate suitable notification procedures.

PROBLEMS INTRODUCED BY POLICY CHANGES

The Policy Changes

The political control of the Greater London Council changed in May 1977 and it immediately resolved to offer its sites for sale on the open market. The political control of the London Boroughs of Hillingdon and Ealing changed in May 1978 whereupon Hillingdon made a similar decision in respect of its sites.

Toxic Tip

The changes in policy radically altered the original phasing of the development resulting in a re-appraisal of the mode and period of operation of the toxic tip. It is unlikely that the private developers will be afforded the same facility as would have been available to Council development.

Control of Development

The sale of land to private developers raised the question of how the London Borough of Hillingdon could ensure that appropriate measures were adopted on contaminated sites. The Public Health Act 1936, Section 54 deals only with animal or vegetable waste, and there are no relevant provisions in the Building Regulations. Conditions relating to contamination have been imposed in outline planning permissions granted to developers, but there are grounds for doubting whether they are valid. In any event, the method of enforcement through Enforcement and Stop Notices is far from satisfactory.

SUMMARY AND COMMENTS

Summary

The Authorities acquired the site without knowing of the presence of contamination which was discovered more or less by serendipity. However, The Authorities have been at pains to take all necessary steps to ensure the health and safety of operatives on the site, persons adjacent to the site, and future residents.

Comments

The initial site investigations were carried out in accordance with the relevant British Standard Codes of Practice. Until the publication of the Building Research Establishment Digest No. 222 "Fill and Hardcore" in February 1979, there was no readily available warning of health or environmental hazards that might arise from fill materials or contaminated land. Appropriate revisions to the relevant Codes are urgently required.

Powers enabling local authorities to effectively control development of contaminated land are urgently required.

Guidance on the best sampling and testing techniques, and up-to-date threshold limits directly applicable to soils in the context of reclamation would be helpful.

Society, individuals, and property are all at risk all the time. Society readily accepts risks that are sufficiently small and in some instances tolerates greater risk. Individuals can accept very high risks voluntarily, but are likely to protest against risks imposed upon them unless they are very small indeed. Determining the severity of a risk is largely a scientific process, but the final judgement as to whether the risk should be incurred is not. The decision-maker has to combine objective analysis of data with subjective judgement of values, often working within a broad range of uncertainty. Despite research on the application of probability theory to decision-making under the conditions of uncertainty, it is evident that subjective judgement predominates. Experts can come to conclusions on a statistical assessment of a hazardous event occurring and a partially subjective evaluation of the risk against a partially subjective evaluation of the benefits. However, experts have no special qualification that sanctifies their decision to declare a level of risk justifiable. It is proper to take account of public perception of a potential hazard notwithstanding that it is distorted and inconsistent with a rational assessment. All technical information leading to proposals for land reclamation projects should be freely available and open to challenge.

Further information

Further information about the Willow Tree Lane Joint Development can be obtained from the Borough Librarian, London Borough of Hillingdon, Uxbridge.

1. Willow Tree Lane Joint Development Structure Plan, April 1978
2. Public Cleansing Report of an Investigation into the Public Cleansing Service in the Administrative County of London, 1929 (HMSO)
3. Contamination Report, October 1976.

New Industrial Site from Steel Works Tip Land

A R Gilchrist C.Eng. M.I.C.E. F.I.H.E.

Scottish Development Agency

The Scottish Development Agency, set up in December 1975, is now well known for its main role in providing finance and factories for industry. Closely following these main activities there is its environmental role where it has been charged with reversing the trends of dereliction in cities, towns and villages throughout Scotland and thus have the dual role of revitalising industry and improving the environment both where people work and live.

It is sometimes asked if there is not a conflict of interests in an organisation charged jointly with the tasks of industrial expansion and environmental improvement. Certainly the industrial revolution created physical devastation, the scars of which are still with us and during this century too, there has been much industrial and housing development which paid little regard to its surroundings and its impact upon them. The Agency sees no conflict in its dual role. Indeed it is one of great potential strengths that, perhaps, for the first time, these have been considered in one single organisation. The subject of this paper, the reclamation of the derelict steel works tip site to provide a modern industrial estate is a prime example of this dual function.

In the early 1970's it became apparent that, as a result of British Steel Corporation's reorganisation and modernisation programme, the Clyde Iron Works would be run down and eventually close. To offset the fairly major redundancies expected it was agreed that new industry, located on unused steel works land, should be set up on a new industrial estate. In July 1975 the Secretary of State for Scotland announced that the Scottish Industrial Estates Corporation, forerunners to the Scottish Development Agency, would carry out development of this site.

The site selected extended to some 32 hectares and consisted of a variety of uses (i) a derelict brickworks, (ii) the Orion steel works tip, (iii) the Orion colliery tip, (iv) the Bogleshole No. 4 pit colliery tip leaving about 5 hectares of low lying wet land adjacent to the river Clyde. The site had an appearance of total dereliction but was strategically located close to the steel works, to the village of Cambuslang, on the eastern perimeter of Glasgow and very near to the end of the M74 motorway.

From an earlier study of the whole steel works area certain problems had been identified in the potential development of this site due mainly to the silty nature of the natural deposits and to the need to substantially raise the level of the southern end of the site above the river Clyde flood level. A further hazard was in the form of air borne pollution arising from the operations of the Colvilles Clugston Shanks asphalt plant located just to the west of the site.

Extensive mineral and subsoil investigations were initiated.

The mineral investigation confirmed the earlier findings that a considerable area of the site had been underworked in a number of coal seams. From the various depths it was considered that only the upper coal seam was likely to cause concern. Loosely packed waste was encountered at depths between 42 and 50 metres below ground level. Over certain areas of the site it was considered that the rock cover was insufficient to guarantee the stability of structures. In total this area amounted to about 8 hectares and it was decided to design this area for use as lorry and car parking rather than proceed with the possible consolidation of the workings which, for a factory size of 10,000 m^2, was estimated to cost £1 m.

During the test boring an artesian condition was also established in the south eastern corner of the site giving a water level of 2.9 metres above ground level. It was known that effluents had been pumped into the old workings by the steel works and samples from the artesian borehole were checked. It was found however that the mineral constituents were consistent in composition with those of a deep borehole water.

Extensive tests were carried out on the Orion and Bogleshole No. 4 tips to check the suitability of these materials as infill for the low southern end of the site.

The assessment of material for fill purposes fell into three main categories.

1. Visual inspection.
2. Physical tests.
3. Chemical tests.

Orion Tip

Visual inspection revealed an overlay of domestic and industrial rubbish i.e. old cars, paint pots, plaster etc. Material from trial pits and boreholes yielded a mixture of ash, iron ore, slag, wire, timber, plastic, sand gravel cobbles, pottery and colliery discard. Such a mixture would have been difficult to compact onto a uniform mix without extensive control.

The specific gravity and compaction tests on the finer material showed a poor fill material in terms of low compacted density.

Tests for sulphate showed that normal concrete would suffer chemical attack when the results were compared with B.R.S. Digest No. 174.

Five combustibility tests revealed the presence of a significant amount of combustible matter of up to 28% in the colliery tip materials.

Bogleshole No. 4 Tip

The material in the colliery discard was of a flat and flaky nature which would be difficult to compact to a dense state. The constituents ranged from colliery discard, sand, gravel, ash, silty sand and cobbles. Fragments of timber were also found.

Sulphate tests were similar to those for the Orion tip.

Combustibility test also followed the same pattern.

Subsequent experience confirmed the judgement to reject this material.

The first site clearance contract dealt with the removal of the Orion waste tip down to a predetermined level. Much of this tip lay within the minerally unstable area of the site. About 120,000 m^3 of material was removed and compacted in an area adjacent to the river Clyde designed for a river walk. The compacted volume was 83,000 m^3 giving a compaction factor of 1.46.

Approximately 100,000 m^3 of unsuitable material was taken from the Bogleshole No. 4 tip and spread and compacted in the low areas of the unstable part of the site outwith the Orion waste tip area.

The major earthworks contract was carried out during the period June/December 1976 when some 263,000 m^3 of burnt colliery shale was imported from West Lothian by rail and spread and compacted to raise the site to the final required levels. The maximum depth of fill was 3 metres.

Ground Stabilisation

The existence of loose material in both the northern and south western areas of the site posed a foundation problem for the new factories to be built there. Consideration was given to piling but this was ruled out because of excessive cost due to depth to bedrock being of the order of 27 metres and the deep deposits of silt.

The factories were built using the vibro replacement technique and this has proved to be entirely satisfactory.

Air Pollution

The pollution was in the form of air borne grit, smut and dust from the steel works complex and further aggravated by the slag recovery and production of road making materials operations of Colvilles Clugston Shanks.

The factories now built have taken account of this problem and has meant the introduction of what is basically an air conditioned environment. The cost of incorporating these measures has meant an increase of about £7.5 per m^2 of factory floor area.

The preliminary site clearance contract started in December 1975 and the major infill and site servicing contract continued through into 1977 by which time work had started on the first four factories. Some of these are now let and the final contracts for the structure landscaping of the estate will be complete by the autumn of this year.

Final costs have not yet been established but best estimates of these are given below.

Surveys and site investigations	£ 14,000
Site clearance and preliminary landscaping	103,000
Major earthworks and infill	660,000
Sealing of mineshafts	49,000
Ground stabilisation	296,000
Service diversions	137,000
Provision of main water, gas and sewer services	318,000
Final landscaping	45,000
	£1,622,000

These costs represent a development cost per hectare of £50,700 based on the total site area of 32 hectares.

The effective area of the site for factory development after deduction of the various unstable and unusable areas is 23.3 hectares and this area related to the total development cost of £1.622 m gives a cost per hectare of £69,613.

These costs are about 2½ to 3 times the normal site development cost and reflect the problems of dealing with these difficult sites.

Unless these sites are tackled, however, the scars of dereliction will persist and further take-up of good agricultural land will continue to the detriment of the country as a whole.

Cambuslang's problems were similar to many that exist throughout the country and it is my view that engineers and others dealing with them should know as much about each others problems and solutions as possible through a frank exchange of information.

RECLAN 79 - PREPARATIONS FOR THE RECLAMATION OF A LAGOON CONTAINING CANAL DREDGINGS

P R EDWARDS AND N W J PUMPHREY

ICI MOND DIVISION

INTRODUCTION

Over the last ten years, due to increasing international awareness, the emission of toxic heavy metals from industry into the environment has been greatly reduced. There are still areas, however, where heavy metal contamination resulting from historical operations can be detected. As part of a continuing programme of work at all Mond sites where chlorine and caustic soda have been made by the mercury cell process we have carried out detailed studies of the distribution and fate of mercury in the environment arising from the adventitious release of trace quantities of this element in process effluents. This paper describes the studies carried out to ensure that when a lagoon full of canal dredgings contaminated with heavy metals is finally reclaimed it will be entirely safe for agricultural purposes.

The Weston Canal is a navigable waterway which runs alongside a chemical manufacturing complex in Runcorn, Cheshire. This complex includes the Mond Division Chlor-Alkali site at Castner-Kellner Works, where the mercury cell process has been operated for over 80 years. Many of the effluents discharged to the Weston Canal from this complex are sufficiently alkaline in nature to cause precipitation of calcium and magnesium compounds, thereby necessitating continuous dredging. The canal dredgings are deposited in Frodsham Marsh Lagoon which is a 60 acre site adjacent to the canal.

The dredgings contain a whole range of trace metals coming from a wide variety of industrial processes. The metals are either present as sulphides or hydroxides or are bound tightly onto the hardness compounds by adsorption or coprecipitation. A typical analysis of mud from the lagoon is shown in Table 1. Other important parameters are the alkalinity and salinity of the mud. A mixture of mud and water equilibrates to a pH in the range 9.2-9.5. The salinity of the interstitial water is about 3% NaCl. Experiments have shown that the trace metals are present in highly insoluble form. For example, in the interstitial liquor in the lagoon the mercury concentration is ∿0.0001 ppm, which is an order of magnitude lower than the WHO recommended level for mercury in drinking water. The available mercury in the mud, which is an indication of the fraction of mercury accessible to plants, was very low - less than 0.1 ppm as measured by an acetic acid extraction method. It has also been shown that mercury in the methylated form is only found in trace quantities, invariably less than 1% of the total mercury figure. Coupled with the fact that no mercury has been detected in the air above the lagoon, or in ground water beneath the lagoon, we believe that the lagoon has complete integrity as a deposit ground.

The reclamation of any large area of land contaminated with heavy metals raises a number of questions:

(a) Is the metal taken up by and perhaps concentrated in growing plants?

(b) What depth of topsoil cover is required to reduce any potential uptake to a negligible level? and

(c) If trace metals are taken up, does the extent of pick-up represent a threat to grazing animals, and thereafter to human beings?

We have carried out extensive trials to investigate these various aspects over a number of years. Many trace elements are present in Frodsham Marsh Lagoon sediments and although our work has been concentrated on mercury because of our specific interest in this element, we believe the results to be illustrative of trace metal behaviour generally in this particular location.

UPTAKE OF MERCURY FROM SOIL

Field Trial Experiments

The trials were carried out on 15 ft square plots. These were excavated so as to be filled with:

(a) 1 ft dredgings + 1 ft topsoil over

(b) 1 ft dredgings + 1 ft subsoil + 1 ft topsoil over

(c) 1 ft dredgings + 2 ft subsoil + 1 ft topsoil over

(d) 2 ft subsoil + 1 ft topsoil over

Each plot was replicated four times giving 16 plots in all, arranged in random fashion. Each plot was sown half with grass and half with clover. The trace metal content of each crop on each plot was measured each year for five years. The plants were analysed for Hg, Cr, Pb, Cu, Ni, Cd, Mn, Mo and V. A very large quantity of data was generated by this programme which was given full statistical treatment. Table 2 gives an example of results generated. The concentration of mercury in the crops was found to lie in the range 0.1-0.3 ppm dry weight, with no significant variation year by year or plot by plot. There is therefore no evidence of pick-up of mercury by growing plants in these trials. The concentrations of the other trace metals in crops grown on plots (a), (b) and (c) were also indistinguishable from those grown on plot (d). However, other experiments on the yield of sugar beet, a very deep rooted crop, show that it is necessary to cover the sediments with at least 1 ft of topsoil because plant roots are unable to penetrate the unweathered sediments due to their salinity, alkalinity and texture.

Pot Trial Experiments

When this work was started in the early 1970s, literature searches did not reveal sufficient information for us to be confident of the sort of results we might have expected from these experiments. Therefore, at the same time plant growth trials were done in plant pots at the Macauley Institute.

Rye grass and cocksfoot grass were grown in a series of mixtures of soil and dredgings. These mixtures had mercury contents of 90, 30, 10 and less than 1 ppm Hg. All grass samples grown in these mixtures contained less than 0.3 ppm mercury on a dry weight basis. Here of course the dredgings were intimately mixed with the soil, thereby giving the plant roots a chance to contact the trace metals directly. No difference was found between the mercury contents of the grass growing on the controls and the most heavily contaminated of the mixtures, confirming the results of the field trials.

ANIMAL GRAZING TRIALS

These trials were started in parallel with the field trials because (a) the conclusions from the field trials were not expected to be available for some years due to the slow mobilisation of heavy metals in soil, and (b) in response to growing international pressure on chlor-alkali manufacturers for information on the effects of mercury emissions on crops and animals. A flock of sheep (initially 12 ewes and their lambs) was kept from May 75 to Apr 77 on a field owned by ICI and known to be contaminated with inorganic mercury via aerial deposition from an adjacent atmospheric source. The mercury concentration in the grass growing in this field varied according to the season from 2 to 20 ppm Hg (dry basis). At intervals, sample sheep were slaughtered and their flesh and organs were analysed for mercury. Despite the presence of mercury in their diet, the mercury concentrations in flesh remained very low, <0.1 ppm on a fresh weight basis (Fig 1). In the kidney, the organ with the highest concentration, the level averaged only 1.4 ppm and after the initial rise in mercury concentration there was no steady increase with time (Fig 2). Despite the presence of methanogenic bacteria in the stomach of ruminants no methyl mercury was found in sheep flesh, and only trace quantities, typically <5% of the total mercury present, were found in the organs. Sheep therefore do not accumulate inorganic mercury, nor does methylation take place to any significant extent. The level of mercury in sheep flesh found in the trials was within the range found in meat generally in the UK, namely not detected up to 0.2 ppm (Survey of Mercury in Food, MAFF, 1971). It is interesting to compare these figures with the WHO recommended maximum concentration for mercury in fish tissue of 0.5 ppm which was derived making due allowance for the fact that (a) mercury in fish is present almost entirely in the methyl form, and (b) in some areas fish forms a very high proportion of the total human diet. In the USA the FDA has recently raised this recommended level to 1.0 ppm.

ACKNOWLEDGEMENTS

The authors wish to thank all those in Mond Division involved in these investigations over a long period of time, too numerous to mention individually. The staff of the Macauley Institute, Aberdeen, are thanked for carrying out the pot trials. Dr J Finney of Plant Protection Division, ICI, and Dr J Ishmael of Central Toxicology Laboratory, ICI, are also thanked for their guidance and advice on the conduct of the field trials and grazing trials respectively.

SUMMARY

Frodsham Marsh Lagoon contains canal dredgings contaminated with mercury and other trace metals, The integrity of the lagoon as far as the containment of heavy metals is concerned has been fully established. At the end of its lifetime it will be covered with a minimum of one foot of topsoil and reclaimed for agriculture. As a result of very extensive field, grazing and pot trials over a number of years it has been demonstrated that the presence of inorganic mercury and other metals will cause no problems, either with regard to pick-up by plants or accumulation by grazing animals, and therefore that future agricultural activities can be conducted entirely safely.

TABLE 1 - COMPOSITION OF A TYPICAL SAMPLE OF FRODSHAM MARSH LAGOON MUD

Metal	Concentration ppm, w/w, dry basis	Acid Radical	Concentration %	
Hg	400	Sulphate	1.5	(SO_2)
Cd	3	Carbonate	27.5	(CO_2)
Cr	75	Phosphate	0.5	(PO_4)
Ni	50	Chloride	0.1	(Cl)
Zn	325	Total Nitrogen	0.1	(N)
Cu	125	Insoluble Matter	18.0	(SiO_2)
Pb	250			
Mn	300			

TABLE 2 - FRODSHAM MARSH/WALLERSCOTE FIELD EXPERIMENT
CLOVER-TRACE ELEMENTS, 4 SEPT 75, ppm

	Hg	Cr	Pb	Cu	Ni	Cd
1 ft Topsoil, 1 ft Mud	0.18	12.3	7.0	11.4	2.1	0.99
1 ft Topsoil, 1 ft Subsoil, 1 ft Mud	0.24	13.3	6.3	11.9	2.2	0.82
1 ft Topsoil, 2 ft Subsoil, 1 ft Mud	0.20	11.5	5.5	11.8	1.7	0.58
1 ft Topsoil, 2 ft Subsoil	0.20	15.3	3.6	9.7	2.1	0.69
Standard Error (Single Plot)	0.06	2.0	3.7	2.1	1.1	0.43
Coefficient of Variation	31.2	15.5	65.6	18.8	53.2	55.4
Diff Between Treatment Means	N/S P=63.6	N/S P=12.0	N/S P=61.8	N/S P=44.9	N/S P=91.4	N/S P=58.5

FIG 1 - MERCURY CONTENT OF SHEEP MUSCLE

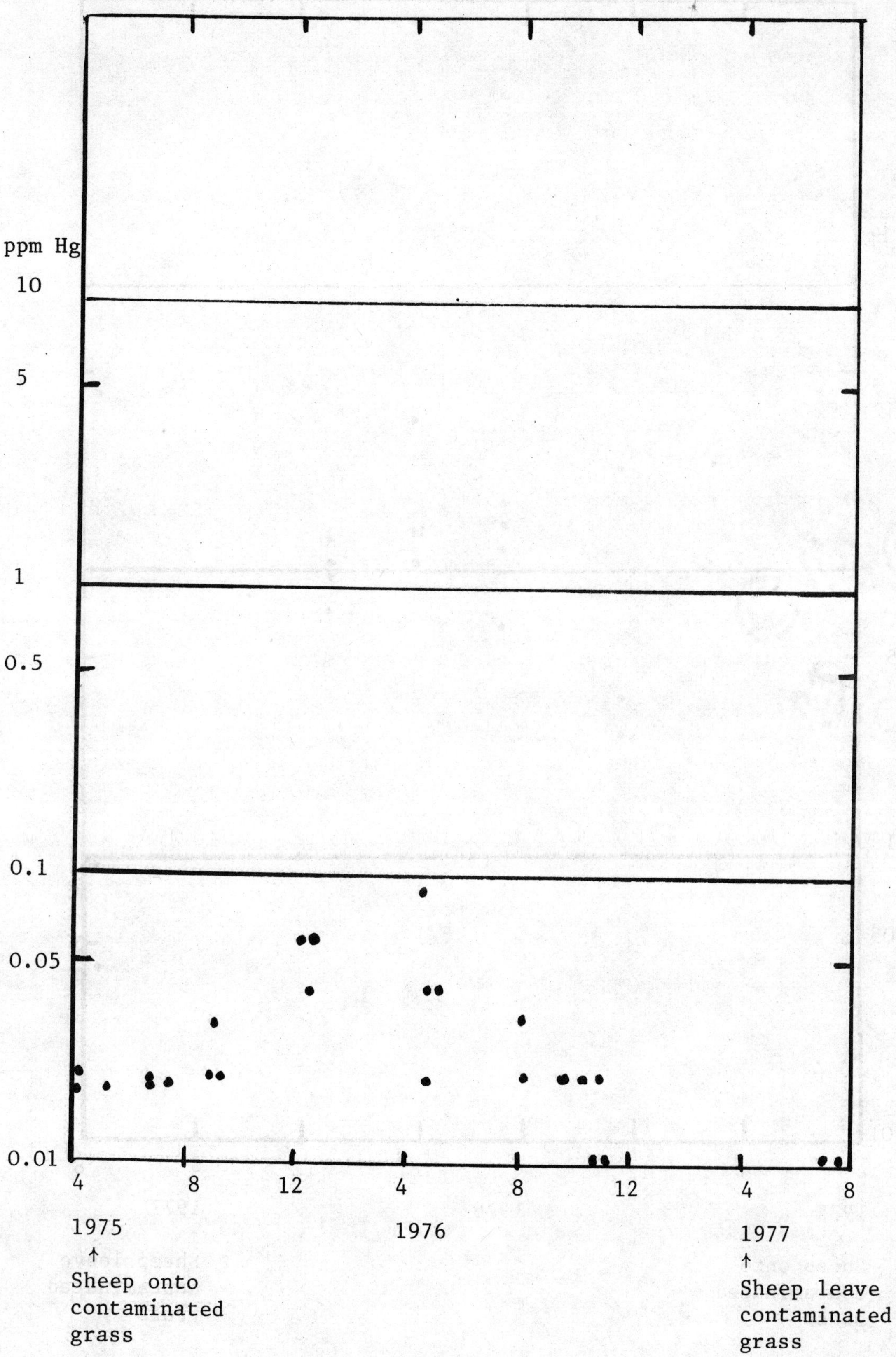

FIG 2 - MERCURY CONTENT OF SHEEP KIDNEY

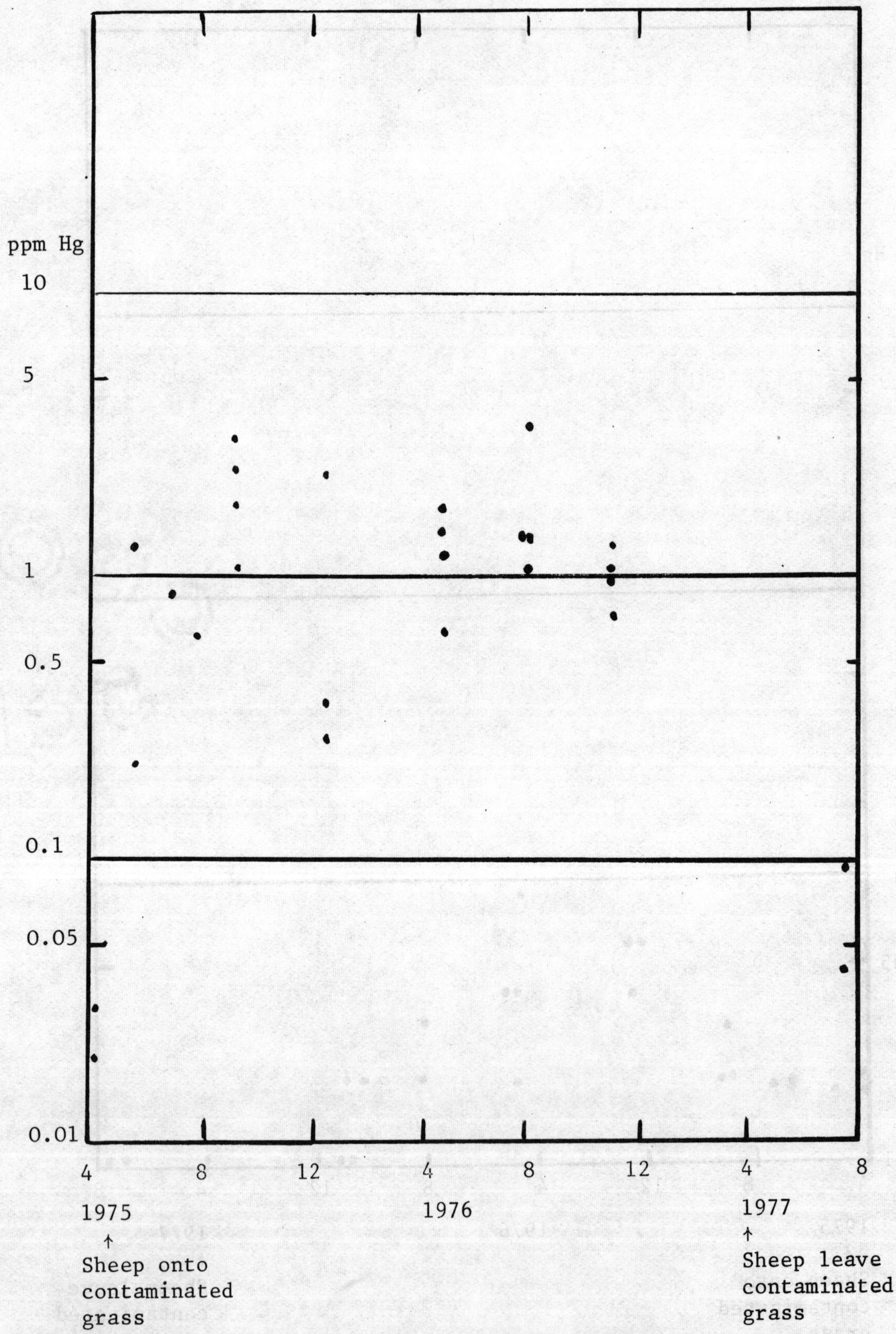

TRAMMEL FILTER TECHNIQUES FOR THE CONTROL OF LEACHATES

T.A. TAYLOR
MA (Cantab) AMBIM

The disposal of untreated waste by landfill is here to stay for many years. Whilst there is the potentially beneficial role of this simple means of disposal for filling and reclaiming worked-out gravel pits and other mineral excavations that would otherwise remain derelict land, adverse environmental effects such as water pollution, wind blown litter and smell need to be taken into consideration. By far the most serious environmental disturbance is the pollution of water sources, groundwater, streams and rivers by leachate generated from the waste.

Some Current Research

The growing public and government concern about the danger of pollution to health, welfare and the environment has led to a worldwide increase in fundamental research into the formation and control of leachates from landfill sites. This encouraging research indicates that both domestic waste and the soil bounding the landfill may play a significant part in attenuating the leachates of hazardous waste. Natural attenuation is due to numerous biological and physio-chemical reactions occurring during the passage of the pollutant through these various materials. It is believed that the existence of an unsaturated zone of soil beneath the landfill is particularly beneficial since the presence of both a liquid and gas phase delays the movement of leachate towards the water table and promotes reactions involving gaseous exchange, oxidation, adsorption on to clay minerals and biodegradation. (1)

Where due to shortage of space sites need to be close to water supply boreholes, or where the sites receive hazardous wastes not easily broken down or attenuated by natural processes, containment sites have been sought. Containment is achieved by locating the site on naturally occurring impermeable strata or by sealing less impermeable sites with flexible polymeric linings.

Present Limitations

In Utopia, the exercise would be that of matching wastes classified on the basis of their pollution potential and sites according to their ability to contain or attenuate the leachate produced. In practice, neither sites nor wastes can be sufficiently well categorised nor do they define all aspects of the problem. The evidence collected concerning the significance of biochemical processes within attenuating media is far from definitive. Different soils attenuate the movement of particular chemicals to varying degrees, so that, for example, gravel allows rapid biodegradation of T.O.C. but has little effect on contaminant ions which require soils possessing high ion exchange capacity. (1) Sites frequently comprise mixed mineralogies such as clay with sand lenses. Many quarries, especially sand and gravel are worked to the water table and below thus causing landfill to be in hydraulic continuity with the underlying formation. Hydraulic continuity also occurs when the site contains such large volumes of leachate that the local water table is raised. Once in the saturated zone the movement of leachate is governed by local flow patterns and pollution plumes spread. Some contaminants have an immediate pollution effect upon coming into contact with the groundwater. Others build up slowly in concentration and can be even more serious in the long term, persisting for many years.

Containment sites are prone to overspilling on to adjacent land or possibly into rivers. High infiltration will cause a build up of significant head of leachate which will increase seepage where the site is located on 'impermeable' strata. Impermeable liners are difficult to install even before tipping commences and when rupture occurs during the life of the site, large quantities of concentrated leachate can escape under circumstances where repair is impossible. Subsequent reclamation of the site is difficult if the landfill is too liquid due to excessive liquid waste or heavy and/or continuous rainfall.

It is hoped that the new legislation providing for hazardous waste control will go a long way to improve landfill practice but even so, parliament cannot legislate for the many associated unknown factors, nor for human failings and irresponsible behaviour such as depositing unspecified material which will act as a biocide and render the landfill system biologically inert. Hence the threat of pollution and its possible far reaching consequences is restricting the use of many potential sites that have few methods of control available other than those mentioned in the preceding paragraphs. This in turn is leading to high transport costs in order to dispose of waste at more suitable sites further away from sources. (2) Where natural attenuation or containment has been inadequate in existing landfill sites, and serious problems have arisen, legal actions have ensued.

The Trammel Drainage Cut-Off

With all the complex variables involved in the selection of site and type of waste permitted for disposal, there exists at present no universal solution to the problem of leachate control. The Trammel Drainage Cut-off is an inexpensive addition to the means at hand for combating the spread of pollution from landfill and has proven effective within its sphere of application.

By definition the Trammel Cut-off comprises a liquid channelling fin in the vertical or near vertical plane that encapsulates a perforated pipe at its base. (Fig.1) (3)

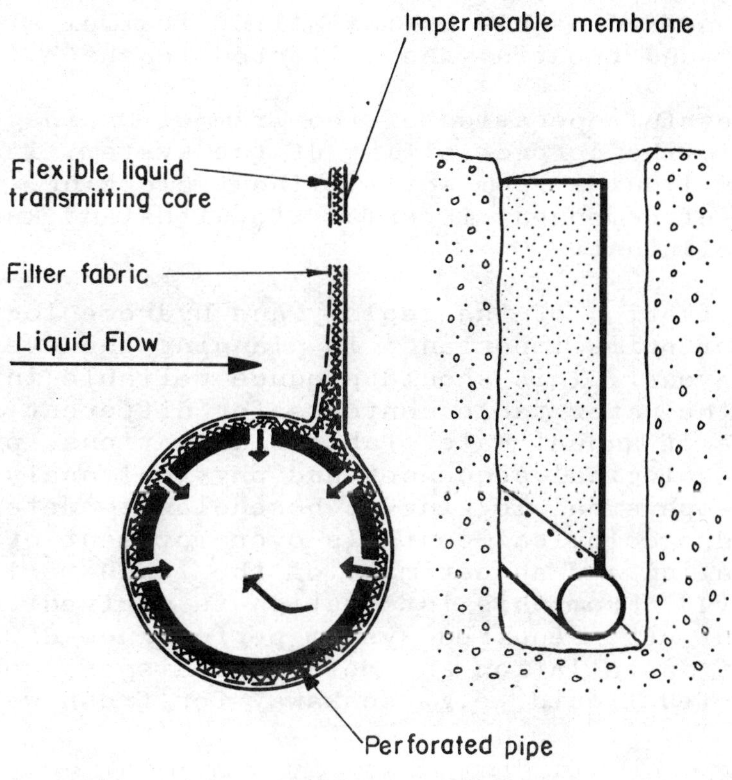

Detailed Section

FIG.1

The fin is a 'sandwich' of filter fabric, liquid conducting core and impermeable membrane. The filter fabric face of the fin is positioned upstream with respect to flow. Liquid therefore passes through the filter and flows into the pipe via the core system, and is prevented from passing straight through the fin by the impermeable sheet.

Ingress of Fresh Water or Egress of Leachate

At the heart of the concept of pollution control by drainage is the separation of clean and polluted water since water is the major leachate solvent. The Trammel Cut-off drain is capable of preventing surface water run-off, groundwater flow and spring issues from entering the landfill, thus reducing the quantity of leachate produced. It may be installed around existing sites to drain leachate that would otherwise flow horizontally from the landfill into the surrounding environment, or within the fill material itself, acting as a french drain with filter fabric on both sides of the core.

Sites that give rise to rapid leachate migration and possessing insignificant attenuation such as hard calcareous rock where flow is through discontinuities (4) can be rendered usable by installing Trammel along the perimeter and treating the collected leachate.

The successful operation of the Trammel Drainage Cut-off lies in i) the correct siting of the system, ii) the ability to filter liquids from soils without clogging, iii) the matching of component materials to withstand known chemicals in the pollutant.

An investigation of the geology and hydrogeology of the site is of prime importance in planning the overall drainage system layout. This should produce reliable information such as the water table contours for different seasons, magnitude of annual water table fluctuations, presence of complex geological sequences and physical analyses of rock cores, geophysical logging of boreholes to determine the major hydrogeological controls over movement of any contamination and an estimate of the leachate likely to be formed. (4) From this information is derived the positioning of the cut-off, required system performance data, pipe sizes, depths of installation and method of disposal or treatment of collected liquid, e.g. soakaway for fresh water, recycling of leachate.

Inducing Soil to Filter

Where drainage is the first means of controlling the production of leachate or its movement, it is essential that the drainage system should remain functional and not block or clog. Traditional drainage methods such as graded aggregates are not considered suitable because of their unreliability and propensity to block within relatively short periods. It is believed that blocking is a function of the random spread of size of the interstices in the filter and its thickness. Fine soil particles from the soil/filter interface migrate into the interstices causing a drop in effective permeability which in turn induces further fine material to settle out of suspension due to the reduction in permeability gradient caused by the initial migration. Trammel filters are based on the findings of extensive research into filter criteria using regular distributions of known hole sizes in thin meshes. It has been found that the mesh opening size can be related to the bridging coefficient of the soil in use. Hence a filter comprising a regular distribution of a discrete uniform mesh opening size can be designed for almost any soil.

The actual operation of the filter in non-cohesive soils is akin to the development of gravel filters around well screens. (Fig. 2).

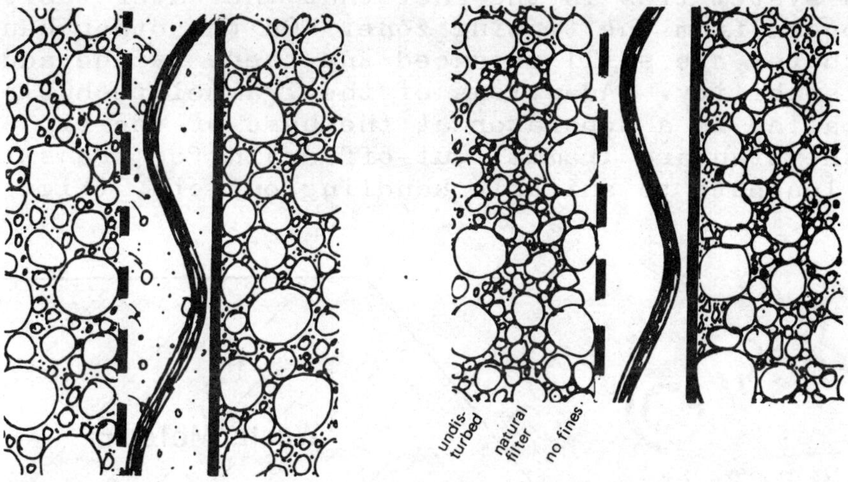

Initial flow of fine particles into the trammel mesh

After a period a natural filter forms

FIG. 2

The area adjacent to the filter is depleted of a specific, controllable quantity of fines, thus creating a zone of higher permeability and beneficial permeability gradient towards the Trammel fin.

Chemical Resistance of Trammel System Components

The selection of impermeable membrane will depend upon the character of the waste, soil conditions, performance required, design life, cost and availability. The potential for interaction between many substances within a landfill which could contribute to leachate composition is large. The concentration of particular substances in leachate will depend upon the waste from which it was derived, amount contained in that waste, the form in which it was deposited and other waste in the landfill with which it may interact. Wastes containing both aqueous and oily phases require a special membrane to resist simultaneously two fluids which are inherently different in their compatability with other materials. (5) It is therefore important that chemical analysis of the leachate be carried out in order to identify any species that has known deleterious effects on certain polymeric membranes in order to be able to select the appropriate material. So far as knowledge extends today, crystalline polymers have good all round resistance to chemical attack particularly emulsions containing aromatic compounds which can cause amorphous polymers to swell. (6) The Trammel filter fabrics and core are made from polypropylene and polyethylene respectively. The weakness of the polyethylene due to the degradation of the plasticiser with certain chemicals is compensated for by the appropriate bulk of the core structure.

European Disposal

Mention should also be made of above ground tip construction so favoured in some European countries. The advantage of this system lies in the fact that the water table is divorced from the tipping zone. On the other hand toxic leachates are still produced and there is the addition of tip stability. Adaptions of the Trammel technology can be installed as a separator at the base of the tip and connected to the standard Trammel Cut-off drain for transmission of the leachate to suitable handling outlets. (Fig. 3).

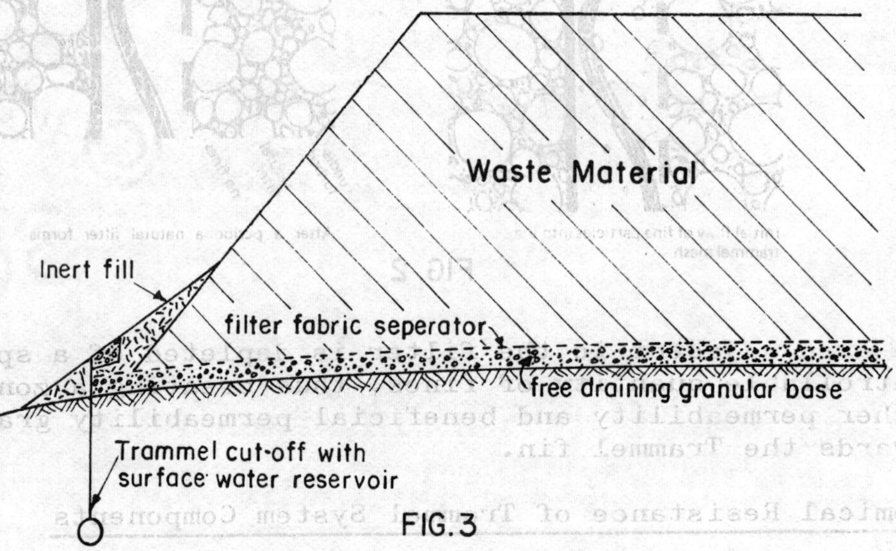

FIG.3

Supporting Facilities

The Trammel activity has in house all the facilities for the site investigations, testing of soil/filter/core combinations and chemical analysis of leachate so that a comprehensive service is available with ease of co-ordination of the various disciplines and greater efficiency towards reaching a solution.

On site, the Trammel system is assembled to suit the specific requirements of the contract. The following are a selection of completed and operational installations.

1. Leachate Cut-Off in made-up ground

   The made-up ground was part of a land reclamation scheme. The area had been previously used for tipping coke oven waste - mainly coal dust and tar. This had been capped with clay. Hydrocarbon liquid leached out seriously affecting adjacent arable land. (Fig. 4 overleaf)
   The Trammel cut-off collected the leachate and directed it to a petrol interceptor. Clean water from the interceptor was led to a soakaway comprising 5" x 3" aggregate.

   The system has been functioning for 18 months.

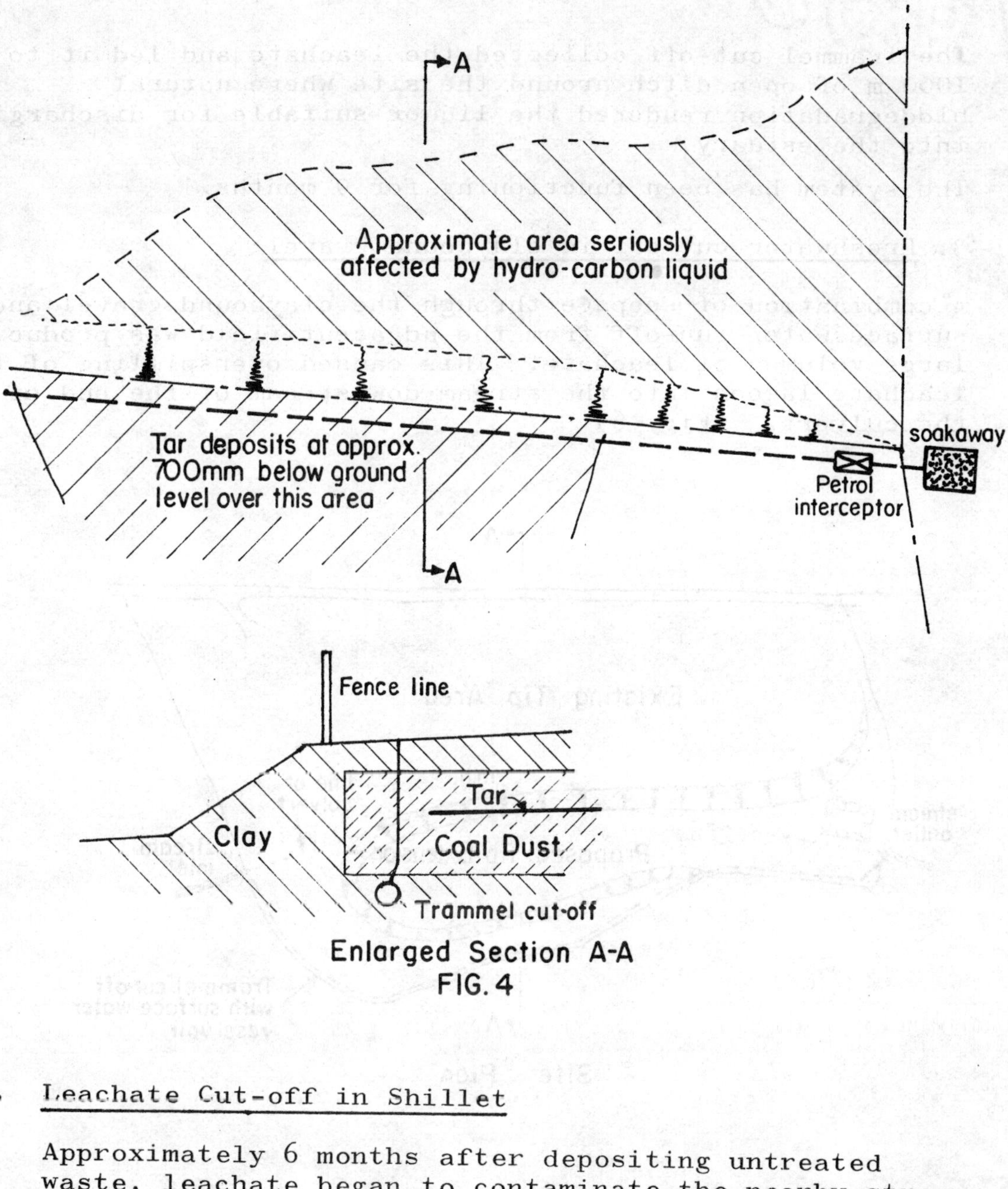

Enlarged Section A-A

FIG. 4

2. Leachate Cut-off in Shillet

Approximately 6 months after depositing untreated waste, leachate began to contaminate the nearby stream which had an outfall on to a local river beach. (Fig. 5)

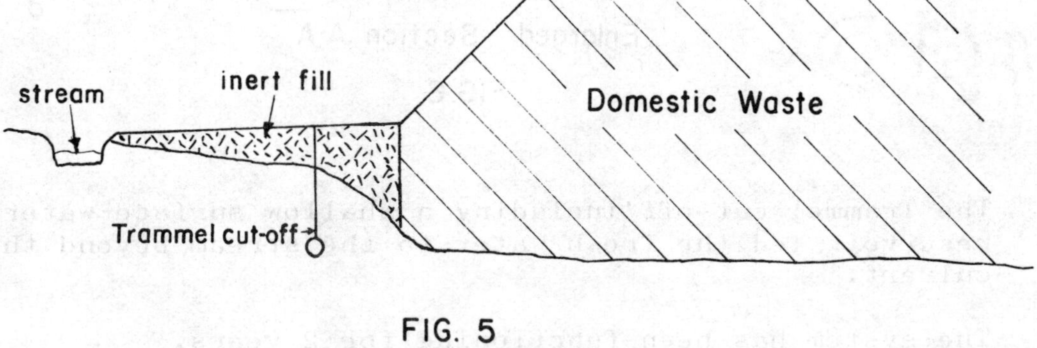

FIG. 5

The Trammel cut-off collected the leachate and led it to 1000 m of open ditch around the site where natural biodegradation rendered the liquor suitable for discharging into the estuary.

The system has been functioning for 9 months.

3. Freshwater cut-off in claybound gravel

A combination of seepage through the claybound gravel and surface water run-off from the adjacent field was producing large volumes of leachate. This caused overspilling of the leachate lagoon into the stream downstream of the end of the culvert. (Fig. 6)

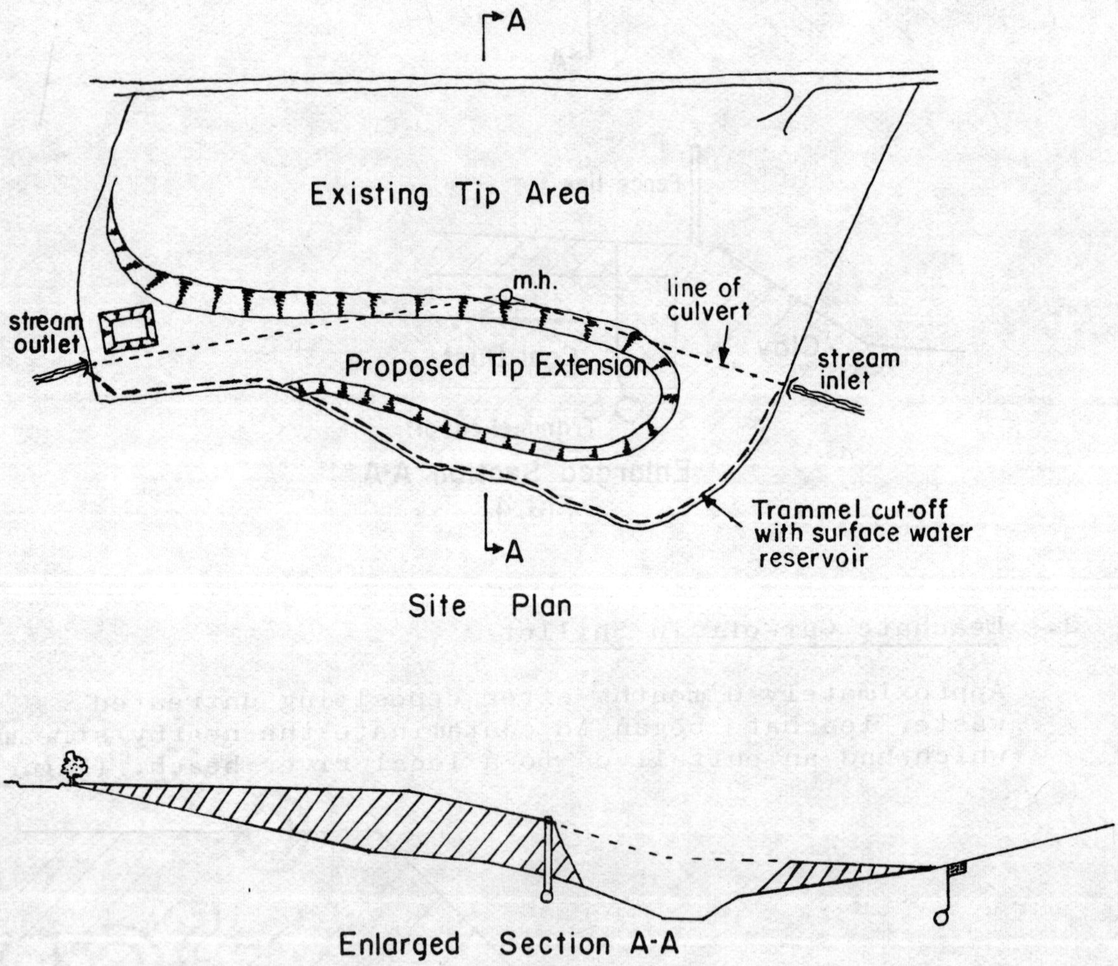

FIG.6

The Trammel cut-off including a shallow surface water reservoir led the fresh water to the stream beyond the culvert.

The system has been functioning for 2 years.

Today sites have to be found for the disposal of large amounts of waste materials of many different kinds and land shortage provides an inducement for their subsequent development. Whilst the disposal of hazardous liquid and solid wastes as landfill is increasing, the potential for pollution of surface and groundwaters by their leachates, intercepting and controlling seepage by means of Trammel drainage offers a means of reducing or eliminating such pollution. By control of leachate production and subsequent movement, the Trammel Drainage Cut-off opens up and creates new opportunities for Local Authorities to release many hectares of old quarries for refuse disposal and ultimate reclamation.

Acknowledgment: The author wishes to acknowledge the permission of H.G. Clapham, Managing Director of Ground Engineering Limited to publish this paper.

References:

1) Co-operative Programme of Research on the Behaviour of Hazardous Wastes in Landfill Sites. D.O.E. 1977.

2) Waste Disposal in S.E. England. Standing Conference on London and South East Regional Planning. (March 1978).

3) Liquid Channelling Method and Assembly. Pat. Spec. 1536551. C.O. Forestier-Walker.

4) Waste Management Paper No. 4. Licensing of Waste Disposal Sites. D.O.E.

5) Haxo H.E., Haxo R.S., White R.N. Liner Materials Exposed to Hazardous and Toxic Sludges. First Interim Report. EPA/600/2-77-081. (June 1977).

6) Haxo H.E. Compatability of Liners with Leachate. 3rd. Annual Municipal Solid Waste Research Symposium Proceedings EPA/600-9-77-026. (September 1977).

Today sites have to be found for the disposal of large amounts of waste materials of many different kinds and land shortage provides an inducement for their subsequent development, whilst the disposal of hazardous liquid and solid wastes at landfill is increasing the potential for pollution of surface and groundwaters by their leachates. Interception and control of seepage by means of Trammel drainage offers a means of reducing or eliminating such pollution. By control of leachate production and subsequent movement, the Trammel Drainage Cut-off opens up and creates new opportunities for Local Authorities to release many hectares of pit quarries for refuse disposal and ultimate reclamation.

Acknowledgements. The author wishes to acknowledge the permission of H.G. Clapham, Managing director of Ground Engineering Limited to publish this paper.

References:

1) Co-operative Programme of Research on the Behaviour of Hazardous Wastes in Landfill Sites. D.O.E. 1978.

2) Waste Disposal in S.E. England. Standing Conference on London and South East Regional Planning. (March 1978).

3) Liquid channelling method and assembly. Pat. Spec. 1,3651. C.O. Forsstrom Malkki.

4) Waste Management Paper No. 4. Licensing of Waste Disposal Sites. D.O.E.

5) Haxo H.E., Haxo R.S., White R.M. Liner Materials Exposed to Hazardous and Toxic Sludges, First Interim Report, EPA/600/2-77-081, (June 1977).

6) Haxo H.E. Compatibility of Liners with Leachate. 3rd. Annual Municipal Solid Waste Research Symposium Proceedings. EPA/600/9-77-026, (September 1977).

DISCUSSION

Dr. D.H. Sharp (Chairman) inviting Mr. A.W. Kenny (Department of the Environment) to give his opening address to the Conference, referred to the long and much valued association between the Society of Chemical Industry and the Department. This association and co-operation was much admired in other countries and he had no hesitation in saying that in very large measure its success was due to Mr. Kenny.

Discussion on Paper B2

Discussion on Mr. Kelly's paper was opened by Dr. Sharp who enquired how was it possible to ensure the adequacy and efficiency of ground sampling. "The speaker replied that no simple investigation mechanism could guarantee complete adequacy. However, by studying all the background information, arranging a sampling mechanism based on inspection of the site and careful watch being kept by supervisory staff during development, the possibility of missing anything of importance could be minimised.

With reference to a question on monitoring, Mr. Kelly felt that whilst the technical aspects were fairly easily dealt with, an important element of the success of a monitoring exercise was the extent to which those using a development could be informed of the objectives of monitoring and being involved in it. Users' concern must be dealt with openly and seriously whatever the technical opinion of the time indicated was the significance of some of the questions raised.

Mr. Burrows (EHO Sheffield) asked whether responsibility for investigation lay with the LA or the owner/developer. Mr. Kelly felt that a local authority was likely to be involved in some way in the development of a site even if it did not undertake development itself. It would seem prudent, therefore, for a local authority to take an interest in the level of investigation undertaken by the owner. He felt that there was a case for a code of practice to help decide on the level of investigation necessary to meet the different requirements on particular types of sites.

Mr. Khan (S. Yorks CC) asked if compounding a number of samples of agricultural land might not create problems and could it also give the impression that if satisfactory for agricultural purposes it could in future be used for other purposes without further investigation. Mr. Kelly outlined the difficulties associated with the sampling, sub-sampling and averaging of soil sampling and emphasised that the sampling plan used must be considered when deciding the significance of any analytical information obtained.

Paper B1

Mr. Smith presenting his paper showed an (unreadable) slide giving some tentative guidelines for small gardens, large gardens, amenity areas and public open space for a selected range of contaminants. He emphasised the tentative nature of the proposals and said that a consultative document was being circulated and would be available in the near future. The main table from this Consultative document is included at the end of this discussion but the reader is reminded that the table must be viewed in the context of Mr. Smith's paper and the consultative document (ref: ICRCL 16/79) 'Redevelopment of Contaminated Land - Accepted Levels of Contaminants in Soils' (Consultation Paper, Central Directorate on Environmental Pollution, Department of the Environment, London, 1979) and, in due course, any definitive document which is produced.

In the discussion Dr. I. Thornton remarked that he had distinct reservations on putting tentative guidelines for acceptable levels of heavy metals in soils for future development down in print, as while these levels may be for discussion at the present time and various caveats for their use may be applied, such figures are readily absorbed into the scientific literature and could in some years time appear in an EEC directive.

What scientific evidence do we have for these figures? Why for instance 4 - 5 ppm for molybdenum? The figures listed were for total concentrations and metals in soils occur in different forms which may vary in availability to plants and the food chain. He thought it dangerous to apply figures derived from the application of sewage sludge to land to, for example land contaminated by base-metal mining and smelting. The natural values shown in Table I must be of some concern as we already have large areas of Britain where soils exceed these values either due to natural geological sources of metal or in old mining areas. For example (a) research at Imperial College has shown that soils extending to some 720 sq. km. in South West England contain 100 ppm or more Arsenic due to mineralisation and the mining history of the area; (b) high lead concentrations occur widely in agricultural and garden soils, over much of the lead-zinc mining area in Derbyshire and other mining areas in England and Wales; (c) the molybdenum value quoted will be exceeded in soils derived from extensive areas underlain by the Lower Lias formations and other marine black shale parent materials.

Will the publication of guidelines not be a cause for concern for people in these areas or those farming the land? Mr. Smith replied by pointing out that there had been some revision of the figures since publication of the pre-prints but that there had to be some extrapolation from existing information. The level must be one which fell between a "low" that was virtually unachievable and a "high" of obvious risk. Environmental lead levels in soil could be greatly in excess of 100 ppm. He reiterated that use must be made of actual research results, e.g. Shipham. The object of the exercise was simply to offer a level below which there need be no cause for concern. Figures above these levels called for examination, sampling and testing of the soil. He had inserted caveats about the levels and hoped that readers would keep these firmly in mind when considering the figures given. To which Mr. Kelly added that guidelines were like traffic lights. Someone had to make a decision on the facts as known, that the present "openess" of local government meant that one must be prepared to change one's mind and consequently it was essential to use the guidelines correctly. Mr. Khan then suggested that smaller concentrations should be allowed on small gardens and allotments than on agricultural land (the reverse of the Table) since produce of agricultural land was widely distributed among the population as opposed to the family consumption of small gardens. He also queried the suggested difference in standards of treatment for temporary and permanent use. Mr. Smith emphasised that the allotment was the most acute situation and that small gardens do not furnish substantial crops. Thus the ingestion level is reduced. He explained that the temporary measures were for dealing immediately with a problem, e.g. temporary cover to prevent high metallic dust content being blown over occupied areas. He accepted that there could be a risk that temporary cover might be forgotten or regarded as permanent. Hence the need for accurate records and planning.

Paper B3

The discussion on this paper began with a reference to the heating up of ground beneath brick kilns in long-term use, with consequent drying out and shrinking of clay soils. The length of time over which the subsequent re-wetting and swelling of the ground took place was queried.

Dr. Dumbleton said that Cooling and Ward had examined the ground beneath a kiln that had worked continuously for 30 years on an Oxford Clay site. Seven years after demolition of the kiln, a building that had been constructed on the site three years earlier had settled 330 mm at one point and risen 180 mm elsewhere. The ground temperature at this time was 70°C at 5 m depth and 390°C at 11 m, probably due in part to slow combustion of the organic matter present in this type of clay. Under a kiln that had been fired for 6 years the soil was dried out to a depth of 4.3 m, and highly cracked. That re-wetting and swelling is likely to continue for a period of years was suggested by observations on ground after moisture removal by transpiring trees. Samuels and Cheney (1) measured the heave of a building constructed on London Clay shortly after the felling of large elms. The maximum upward movement in the first 7 years after construction was estimated at 100 mm and a further 50 mm was measured over the following 13 years. The former positions and character of the trees were studied from old air photographs.

In reply to a question on the cost of historical investigation of past site use, Dr. Dumbleton said that this depended on how far it was taken. An initial study was always worthwhile and could be quickly and cheaply carried out with searches in local record offices and libraries, examination of old air photographs, enquiries in appropriate quarters, and inspection of the site. The adequacy of the picture thus obtained depended on circumstances, and the amount of further effort that should be expended to fill it out would have to be judged for the particular case. A guiding consideration was that the cost would be relatively modest, and the penalties for overlooking trouble could be very high. The cost could be regarded in part as an insurance premium; finding nothing adverse should not be regarded as a loss.

REFERENCE

SAMUELS, S G and CHENEY, J E. Long-term heave of a building on clay due to tree removal. Proc.Conf. on the Settlement of Structures. Geotechnical Society, Cambridge 1974. Paper III/8, 212-220.

Paper B4

This paper was presented by Mr. Dugdale and there was some difference of opinion as to the basis of the claim that the sampling system suggested would yield a 95% degree of certainty. Mr. Dugdale maintained that a higher degree of certainty was neither feasible nor necessary and in fact, 95% was the maximum one could expect. Further questions related to the monitoring of TOC-COD ratios as a means of fingerprinting tip leachates. Dr. Ellis, who replied said he felt unable to comment as he had no experience of this. Later Dr. Ellis said that in his opinion, fatty-acid profiles could have a significant bearing on the analytical aspects of landfill site management. (Ref: Anal. Chem.

51, 1953 (1979))

Paper B5

Dr. Sharp opened the discussion with a request for clarification of the cost of removal of contaminated material quoted in Case E of the paper. He thought it might be about £240,000.

The authors thought that the figure might be incorrect as the site was small - an area of 600 sq. m. In fact 3000 m^3 were removed at a cost of about £20.00 per m^3. The total cost came to £58,000.

Further discussion centred on whether or not the levels of heavy metals shown in Case D were sufficient to justify removal of the soil. The number of samples (9) on which removal decision was taken was also queried. The authors agreed that perhaps in the light of current thinking, over caution had been exercised, but his was the action taken.

The sampling rate and the decision to remove soil was not made by Stangers.

Mr. Smith (DoE) said that as regards the registration of data, the Department had the beginnings of such a system and already possessed details of some 200 sites.

As to the question of availability of data generally, Mr. Smith intervened to say that as regards the registration of data, the Department had the beginings of such a scheme and already possesses details of some 200 sites.

Papers B6 - B11

When the area was let to tenant farmers, there was no concern for metal contamination. The presence of zinc in lettuce (although the amount was quoted on the dried material) was found to be 20 times above the acceptable level. Total zinc in the sludge was up to 3%. When the sludge caught fire, it took some months to put it out.

No particular hazard was discovered below 1 metre.

Mr Knan (South Yorkshire) asked if householders who live on this site are having their health monitored? The answer to this was no.

The future of parks and public open space built on disused sewage farms, where children might dig, brought the reply that at Leicester it would need the removal of $1\frac{3}{4}$ cubic metres of soil to get rid of all the contamination. And to a further question, the use of material for levelling and landscapes requires discipline in the use of areas.

To a question to Mr. Gilchrist as to where does all the liquid from the site go, it was stated that at first they went into the foul sewer and then into the surface water system. The Thames Water Authority did the monitoring.

The wearing of protective clothing when operators were sampling on contaminated land was raised. Whatever was required was supplied.

Geophysical techniques in the site investigations in Birmingham were discussed but do not seem to be the answer.

The trees which caught fire at the roots after twenty years was due to the dry summer of 1977.

Several questions were asked about mercury contamination, this gets into the food chains from eating fish which have inhabited seas containing mercury. If the pH of mercury contaminated land soil falls in a few years time, what happens to the mercury? There has been an occasion when a plot has grown a crop of grass but there was little contamination in the grass.

There is no mechanism to stop change in use of reclaimed land.

Mr. T. A. Taylor introducing the topic of Trammel filters said that the Trammel system is complementary to existing systems. 70% of all products to be used in the next decade have yet to be discovered. The Trammel system is to take out some of the unknowns. On the question of the longevity of materials, polythene changes somewhat over a number of years; it becomes a bit tacky. The design of the drainage pipes had changed since the commencement.

Work had been done on the growth of biomass on Trammel material and the work is proceeding.

TOXIC ELEMENTS IN FOOD

BY D G LINDSAY, FOOD SCIENCE DIVISION, MINISTRY OF AGRICULTURE, FISHERIES AND FOOD, GREAT WESTMINSTER HOUSE, HORSEFERRY ROAD, LONDON SW1P 2AE

Any assessment of the health risks associated with toxic elements in food has to take into account numerous variables which contribute to an assessment of the risk. Toxicity, that is the inherent ability of a pure chemical substance to cause adverse affects in isolation is clearly one of the variables but only one. Other factors which determine the severity of an effect on the presence of such a chemical in the diet include:

(a) the chemical form of the element present in food;

(b) the extent of absorption through the gastro-intestinal tract;

(c) the distribution of the trace element in the organism;

and (d) the half life of excretion.

Apart from a few elements such as mercury little is known about the chemical form of the elements which are present in the diet. Invariably it is only possible to know that the elements tend to be present in the diet in a loosely or tightly bound complex. Examples are known where the element can exist in a number of differing valency states which may have a profound influence on the inherent toxicity and overall effect. A good example is arsenic which is extremely toxic as inorganic arsenite but its detection in large quantities in certain fish and shellfish is primarily in a relatively non-toxic organically bound state. Although intermediate forms between inorganic arsenite and organically bound arsenic may occur these forms have not yet been characterised (1).

The extent of absorption is often difficult to ascertain. Individuals vary enormously in their ability to absorb lead and presumably trace elements (2) although the variations observed may be in part a reflection of the difficulty of carrying out long-term balance studies in humans. The effect will also depend on whether the element is principally concentrated in a sensitive organ, for example, methyl mercury in brain cells or cadmium in the kidney cortex, or whether it is widely dispersed in all cells and is an element essential for cellular metabolism in the form of a metalloenzyme complex, for example zinc and copper.

Finally excretion of an element may be rapid, for example manganese, indicative of a role of the element in cells with a high turnover rate or in transport. On the other hand elements such as cadmium are slowly excreted and are firmly bound to cells with a slow turnover.

A consideration of all these variables enables the trace elements to be divided into roughly 3 groups; A, B and C according to the nature of their effects which are characterised by the properties shown in Table I. Elements in class A are primarily those which have no known biological function and are slowly excreted from their sites of accumulation. They include, lead, mercury, cadmium and hexavalent forms of chromium. Elements in classes B and C are those which are nutritionally essential. Those in class B show toxic effects at exposure levels which are only at an order of magnitude or so higher than those which are required nutritionally and include selenium (as selenite), copper, cobalt and tin. Elements in class C are those in which cellular homoeostasis and regulation of intestinal absorption are well controlled and toxic effects from high doses are not observed. Such elements include zinc, molybdenum, nickel and trivalent chromium.

Table 1 PROPERTIES OF VARIOUS CLASSES OF TRACE ELEMENTS PRESENT IN THE DIET

	Biological function	Inherent toxicity	G.I. absorbtion	Organ specificity	Excretion
A	–	+++	+	+	+
B	+	++	++	±	++
C	+	+	+)[a]	–	+++)[a]

)[a] Cellular homeostasis regulates levels of absorbtion and excretion.

Class A elements include Pb^{2+}, HgR^{+}, Cd^{2+} and Cr^{6+}; class B elements include Cu^{2+}, SeO^{-}, Co^{2+}; class C include Zn^{2+}, Cr^{3+}, Mn^{2+}

Individual susceptibility to toxic elements can affect the level of risks such that the above generalisations do not apply. Wilsons disease is an inborn error of copper metabolism in which copper accumulates in the liver in contrast with normal individuals, and toxic effects can occur. Such deviations from normal responses must be subject to a separate risk assessment from the population as a whole since the elimination of certain foods from the diet, perfectly wholesome for normal individuals, may be necessary in such circumstances.

The toxicity of a trace element like any other toxic substance may be antagonised or agonised by other trace elements present in the diet. These complex interrelationships are only just now being studied in animals and any interpolation to the assessment of risk is premature. Nonetheless it has been shown in animal experiments that the toxicity of cadmium is antagonised by the presence of selenium (3) and copper (4) in the diet and more complex relationships have been demonstrated involving other trace elements.

The other source of exposure to trace elements present in uncontaminated land is through vegetable consumption. The trace elements are derived from either the uptake from soil or from the direct contamination of the food with soil. These factors will be discussed in depth in the contributions which are to follow but some assessment of the human risks involved is necessary.

Individuals with the highest consumption of vegetables, and therefore with the highest degree of risk, are likely to be those people who grow vegetables on their own land either in allotments or gardens. Examination of data in the South East and South West regional returns of the 1976 National Food Survey has shown that the top 10% of consumers growing their own vegetables in allotments or gardens have a consumption rate of fresh vegetables which is approximately three times the national average (5).

Of the other elements in group A, the available evidence suggests that the only element of concern is cadmium. Soil pollution by lead, arsenic and chromium do not in general result in a significant uptake by vegetables in terms of total dietary exposure and are presumed to be tightly bound to soil and un-available. However factors affecting uptake include soil pH and type as well as the type of crop grown and much more information is necessary on the behaviour of these elements in all the soil types and with a range of commonly consumed crops before one can be certain that the risks are small in all circumstances. The direct contamination of crops no doubt leads to higher exposure to these elements but normal food preparation is usually effective in making such a route much less significant than the direct uptake route.

There is no evidence of exposure to mercury through vegetable consumption apart from the known potential of fungi to accumulate a number of toxic elements including mercury (6). It is likely that soil micro-organisms can function to recycle mercury additions to soil. Thus both mercury, the organo-mercurials, and elemental mercury appear inter-convertible and the volatility of some of these forms may be responsible for the lack of the potential for accumulation in soils (7). Quite apart from these constraints on human exposure the uptake of mercury in plants would have to occur at similar levels to those which are found in fish before exposure from vegetable consumption became an important source in the diet. In the United Kingdom the hazards associated with the consumption of methyl mercury in fish have been judged to be negligible even amongst extreme consumers (8) and it is unlikely that any situation will arise where the uptake of mercury into vegetables would pose a comparable degree of risk.

In contrast dietary levels of cadmium do appear to derive almost exclusively from the uptake of cadmium from soil. The levels of cadmium in air in both urban and rural environments are low and little of the cadmium found in the diet can be attributed to sources other than soil with the exception of fish. It is well established that certain vegetables, particularly fast growing leafy vegetables, can accumulate quite high levels of cadmium in comparison to the levels generally found in the diet and for lettuce linear relationships have been observed between cadmium plant dry matter concentrations and cadmium concentrations in soil (9).

Much of the information required to make an adequate risk assessment is unavailable at the present time. Even if methods of analysis were capable of detecting the different chemical forms of an element present in the diet, which at present they are not, so much more information is necessary and I hope this brief introduction highlights many areas where fuller research is necessary. Because of these complex interrelationships the role of trace elements in the diet in the induction of chronic disease can often only be confirmed through the application of the most painstaking epidemiological methods.

Since it is not possible to assess the potential hazard of a given level of a trace element present in the diet in all circumstances, it is nonetheless prudent to keep the levels of elements which fall into class A and possibly B, and are present in the diet, as low as possible until such time as the necessary research indicates such a cautious approach is unnecessary.

Quite apart from the factors which affect the inherent toxicity of trace elements, consideration must also be given to other factors which determine the overall exposure. Information about the frequency and amount of food consumed is extremely important particularly for elements in class A. This is because the degree of risk can be determined as much, if not more, by the amount of any food consumed as from the concentration of element which is present in the diet. The degree of risk from doubling the concentration of a trace element may be judged to present an acceptable risk but if it is considered that consumption rates for certain foods can vary over a much wider range, the risk may become unacceptable to those people with the highest consumption rates.

Figure 1, illustrates the routes of transfer of trace elements in soil to the diet. Reclaimed land which is used for pasture can result in the contamination of meat and dairy products through the uptake of trace elements or directly by soil ingestion. In general the contribution these sources make to human exposure to trace elements is small for elements in class A. Ruminant metabolism ensures that the elements are concentrated in organs such as liver and kidney which are not consumed in large amounts. In addition the transfer to milk is generally limited to a fraction of the levels which are found in the animals blood.

Figure 1 : TRANSFER OF TRACE ELEMENTS FROM SOILS TO HUMAN DIETS

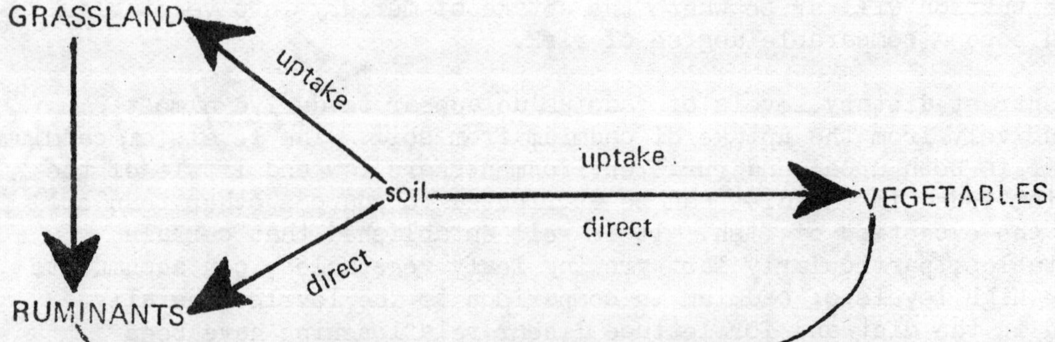

From Table (2) it can be calculated that average weekly intakes of cadmium in the UK are about $\frac{1}{3}$ of the tolerable weekly intake of 400-500 µg per week derived by the World Health Organisation (10). This tolerable intake is currently not thought to offer more than a 2-4 fold protection from toxic effects. Any enhancement of the current levels through the addition of cadmium to soil in excess of the average levels found in the UK (that is 1 mg per kg or less) must be viewed with concern particularly when it is realised that there are consumers with consumption levels three times the national average.

Table 2 DIETARY INTAKE OF CADMIUM IN THE UK IN 1977 ACCORDING TO FOOD CLASS

Food group	Estimated weight of food eaten (kg/day)	Number of samples	Range (mg/kg)	Mean value (mg/kg)	Estimated mean daily intake (mg/person)
1. Cereals	0.23	25	<0.01 - 0.04	<0.022	<0.005
2. Meat	0.15	25	<0.01 - 0.07	<0.020	<0.003
3. Fish	0.02	25	<0.01 - 0.05	<0.015	<0.0003
4. Fats	0.08	25	<0.01 - 0.03	<0.012	<0.001
5. Fruits and sugars	0.17	25	<0.01 - 0.03	<0.012	<0.002
6. Root vegetables	0.18	25	<0.01 - 0.04	<0.018	<0.003
7. Other vegetables	0.11	25	<0.01 - 0.02	<0.011	<0.001
8. Beverages	0.12	25	<0.002-0.004	<0.002	<0.0003
9. Milk	0.40	25	<0.005-0.005	<0.005	<0.002
Total	1.46				<0.018

Means have been calculated assuming that all figures quoted as less than the limit of detection are equal to the limit.

As far as elements in group B are concerned it is unlikely, on the basis of present knowledge, that there are any hazards associated with the uptake of many of these elements into vegetables grown on reclaimed land. The elements can be found in comparable levels in other important items of the diet, and if it is assumed that the forms of the element present in vegetables are no more toxic or more available than those present in other foods, overall exposure is unlikely to be increased to an unacceptable extent or the nutritive balance of trace elements in the diet disturbed. However there are trace elements whose presence in vegetables can be detected at low levels and about which little or nothing is known. The incidence of such elements in the diet and in vegetables generally must be determined through monitoring and the use of contaminated sites containing unusual trace elements profiles in the soil will provide very useful situations for further study.

Recently a range of commonly consumed vegetables covering different cultivars have been analysed at Harwell using neutron activation techniques for those trace elements which can be measured with an accuracy of 10% or less. The elements which were analysed and fall into class B include, silver, bromine, cobalt, chromium, copper, manganese, nickel, rubidium and antimony, scandium, selenium and tungsten. Table 3 summarises those samples where the range of trace element concentration exceeded 2 orders of magnitude. These results indicate the elements which through vegetable consumption are potentially more able to contribute to the total of dietary exposure and are liable to uptake from the soil. The results suggest that lettuce are a particularly valuable species for indicating such a potential since they are able to accumulate certain trace elements more readily than any of the other crops investigated. This ability may apply to all of the fast growing leafy crops, like the situation with cadmium, and this theory is worth further investigation.

Table 3 RANGE AND MEAN VALUES OF SOME ELEMENTS FOUND IN VEGETABLES BY NEUTRON ACTIVATION ANALYSIS

Vegetable	No. of samples examined	Range and mean levels of element in sample (mg/kg, fresh weight)						
		Silver	Cobalt	Chromium	Antimony	Selenium	Scandium	Nickel
Lettuce	5	<0.001-0.22 (0.07)	0.002-0.12 (0.05)	0.02-4.2 (1.06)	0.0002-0.026 (0.006)	<0.001-0.1 (0.02)	0.0006-0.09 (0.03)	–
Brussels tops	5	<0.005-0.10 (0.03)	–	–	–	–	0.0008-0.03 (0.01)	–
Parsnips	5	–	–	<0.01-10.7 (2.3)	–	–	0.0006-0.04 (0.016)	–
Swede	5	–	–	<0.008-6.9 (1.4)	–	–	–	–
Carrot	5	–	–	–	–	–	–	<0.001-0.63 (0.11)
Wheat	5	–	<0.01-0.83 (0.18)	–	–	–	–	–

(–) range of values observed less than two orders of magnitude.

Since other toxic trace elements other than those mentioned above do not show more than an order of magnitude variation in levels of trace elements, it is possible that foods other than vegetables dominate the overall intake. Further total diet studies should delineate the position.

In conclusion although there are many problems in making an adequate risk assessment of trace element uptake at the present time, further monitoring and research adopting the policy outlined in this paper, should help to improve the assessment procedure.

REFERENCES

1. J. Wrench., S. W. Fowler and M. Y. Unlu. Marine Pollution Bull. "Arsenic metabolism in a marine food chain." (1979), 10(1), 18-20.

2. D. Baltrop and C. D. Strehlow (1978) 3rd Intn. Symp. Trace Element Metabolism in Man and Animals 1978 pp 332-334 Technische Universitat, Munchen.

3. T. W. Clarkson Fedn. Proc. "Factors involved in heavy metal poisoning". (1977), 36(5), 1634-1639.

4. C. F. Mills and A. C. Dalgarno. Proc. Nutr. Soc. "The influence of dietary cadmium content on liver copper levels in ewes and their lambs" (1972) 31, 73-74A.

5. Ministry of Agriculture, Fisheries and Food. Unpublished data.

6. A. Hasuk. Müll und Abfall. "Investigations into the uptake of copper, zinc,lead, cadmium and mercury into mushrooms". (1977), 7(6), 172-6.

7. R. D. Rogers. J. Environ Qual. "Methylation of mercury in agricultural soils". (1976), 5(4), 454-458.

8. J. Haxton, D. G. Lindsay, E. J. Dixon et al. Environ. Res. "Duplicate Diet Study on Fishing Communities in the UK". (1979), 18 (2) 351-368

9. L. B. Wood, R. P. King and P. E. E. Norris, EEC Conference on the Utilisation of Sewage Sludge on Land Oxford. "Some investigations into sludge amended soils and associated crops and the implications for trade effluent control". (1978) (in press).

10. Joint FAO/WHO Expert Group on Food Additives. WHO Food Additive Series No. 4. FAO Nutrition Meetings Report Series No. 51A. "Evaluation of certain food additives and the contaminants mercury, lead and cadmium (1972).

INDICATOR CROPS

R. D. Davis

Water Research Centre, Stevenage Laboratory

ABSTRACT

The use of indicator crops for assessing hazard associated with accumulation of elements in soil is discussed and illustrated with results from a pot trial in which 33 crop varieties were grown on a contaminated soil. The results confirm that crops vary widely in their sensitivity to metals in soil, as considerable differences in the concentration of cadmium and zinc were observed amongst various plants and their parts all grown on the same soil. Interpretation of results depends on comparisons with critical concentrations in plant tissue known to be indicative of background levels (no contamination), deficiency levels (for the essential elements), and toxicity levels (to plants and the animals which eat them). It is concluded that tissue analysis of indicator crops can provide useful information for planning the future use of contaminated soils.

INTRODUCTION

Reclaimed soils often contain concentrations of various elements which are higher than background levels associated with uncontaminated agricultural soils. The significance or hazard of these raised concentrations depends largely on their availability for plant uptake. Thus it is necessary to measure availability as a basis for managerial decisions concerning future land use, and as a means of assessing the effectiveness of soil treatments (e.g. liming) designed to reduce uptake by plants and hence hazard.

MEASURING AVAILABLE METALS IN SOIL

Soil analysis

Determinations of total metals (extractable by strong acids) in soil indicate the extent of contamination, but tell us little about its significance. The use of milder extractants (weak acids, chelating agents) can be informative in indicating how metals are bound in soil. Unfortunately, conventional extractants tend to be insensitive to the soil properties which control the availability of metals to plant roots, for example pH value, cation exchange capacity, and contents of organic matter and phosphate, so it is often as difficult to relate extractable concentrations of metals in soil to probable plant uptake as it is total concentrations. For instance, the widely used extractant acetic acid (0.5 M) extracts the same proportion of metal from two soils of different pH value but containing the same total concentration of metals, although, in reality, plant uptake would be greater from the more acid soil. This problem is recognised, and the use of alternative extractants has been proposed for some purposes. Symeonides and McRae[1] have recommended ammonium nitrate (1 M) for assessing available cadmium (Cd) in soil and Lindsay and Norvell[2] found that DTPA was useful for identifying soils deficient in zinc (Zn). Concentrations of metals in water displaced from soil have also been compared with concentrations in plant tissue[3]. Soil analyses are convenient and examination of novel extractants would bear further investigation, although the comparatively low concentrations of metals in the solutions they extract can present analytical problems.

Plant analysis

Plant analysis offers a direct method of measuring available metals in soil. Concentrations of metals in plant tissue are the integrated result of all the factors which have affected uptake up to the time of sampling. No soil analysis can predict this, particularly since the genotype of the plant may be as important as soil properties in controlling uptake[4].

Plant analysis involves sampling plants either from the field or grown in controlled conditions. Field samples may be contaminated either with soil or as a result of aerial deposition. Washing of samples to remove contamination is subject to the criticism that elements may be leached from within the plant together with external contamination. Soil contamination is avoided by careful harvesting of tissues, or is estimated by determining the concentration of titanium (Ti) in the sample, since this element is abundant in soil but is not taken up by plants. More than 2 mg Ti kg^{-1} in dried plant material indicates significant soil contamination[5]. The impact of aerial deposition is likely to be minimal in samples taken when the plants are growing actively. Concentrations of metals in actively growing green tissue are more indicative of the physiological state of the plant than concentrations in moribund tissue, which may be anomalous. These considerations suggest that field sampling of plants is best undertaken in spring or early summer.

Sampling of plants grown in controlled conditions has the advantage that problems from contamination are minimised, and growing conditions are reasonably defined and can be adjusted to promote active plant growth. There is some evidence[6] that uptake of elements is greater in plants grown under glass than in field conditions, although this was from a study comparing plants of different age. If correct, it might be considered an advantage since the plant test conducted under glass takes account of the worst possible field conditions. Preliminary findings from a field trial using some of the crops and the same soil as the pot trial reported below suggest that results in the two environments are similar.

USE OF INDICATOR PLANTS

In general, named cultivars of crops, especially if inbred, are likely to be the most appropriate indicator plants because of their consistent genotype. Graded seed can be obtained which will also contribute towards a uniform response, especially if yield as well as the concentration of elements in tissue is to be measured. Genotype may vary widely amongst species of wild plants, leading to erroneous results if these are used as indicators of the availability of elements in soils at different locations.

Selection of crop

Selection of an appropriate indicator crop depends on the purpose of the analysis. Standardised tests, using plants to assess available metals in soil (almost as though they were chemical extractants), have made use of barley harvested at the five-leaf stage or crops which grow quickly to economic maturity, such as ryegrass and radish[3]. Alternatively, a range of crops thought suitable for possible cultivation on a particular site may be used as indicators to determine which are, in fact, the most appropriate crops. Where a soil is contaminated with one element, tests should use those plants most likely to be associated with possible hazard. For instance, molybdenum is non-toxic to plants and unlikely to affect humans, but upsets the copper metabolism of grazing animals. Therefore, evaluation of a soil contaminated with molybdenum should involve forage crops (e.g. grass, clover, kale, mangold) in order to assess potential hazard realistically.

Testing a contaminated soil

The histograms in Figures 1 and 2 illustrate respectively Cd and Zn concentrations in the leaves and edible parts of a series of crops grown in a glasshouse pot trial in a soil containing 67.8 mg Cd kg^{-1} and 1613 mg Zn kg^{-1}. Background concentrations of these elements found in uncontaminated soils are approximately 1 mg Cd kg^{-1} and 50 mg Zn kg^{-1}. Edible parts and leaves of crops were sampled at economic maturity except for those where the leaves would be moribund at this stage. Leaves of cereals, for instance, were sampled at the five-leaf stage.

Figure 1 presents results for Cd arranged in order of increasing Cd concentrations in the edible parts of the crops. Cd is principally zootoxic, being more toxic to animals than plants, so concentrations in edible parts are of most relevance for assessing hazard. Considerable variation in tissue concentration of Cd between the crops was evident; it ranged from <0.1 mg kg^{-1} dry matter in the tap root of carrot to 11.0 mg kg^{-1} in the leaves of cos lettuce. Leaf concentrations of Cd ranged from 0.2 mg kg^{-1} in peas to 15.5 mg kg^{-1} in one variety of beetroot. Differences between varieties of the same species were seen but were not as noticeable as in some previous studies[4,7]. There was a marked difference in concentration between the leaves and edible parts of some plants, exemplified by the beets and radish, where much higher concentrations occurred in the leaves compared with the storage roots. In all cultivars tested, leaf concentrations of Cd exceeded concentrations in non-foliar edible parts. Previous studies have suggested that Cd concentrations in plant parts decrease in the order fibrous roots > leaves > seeds = storage organs[8,9].

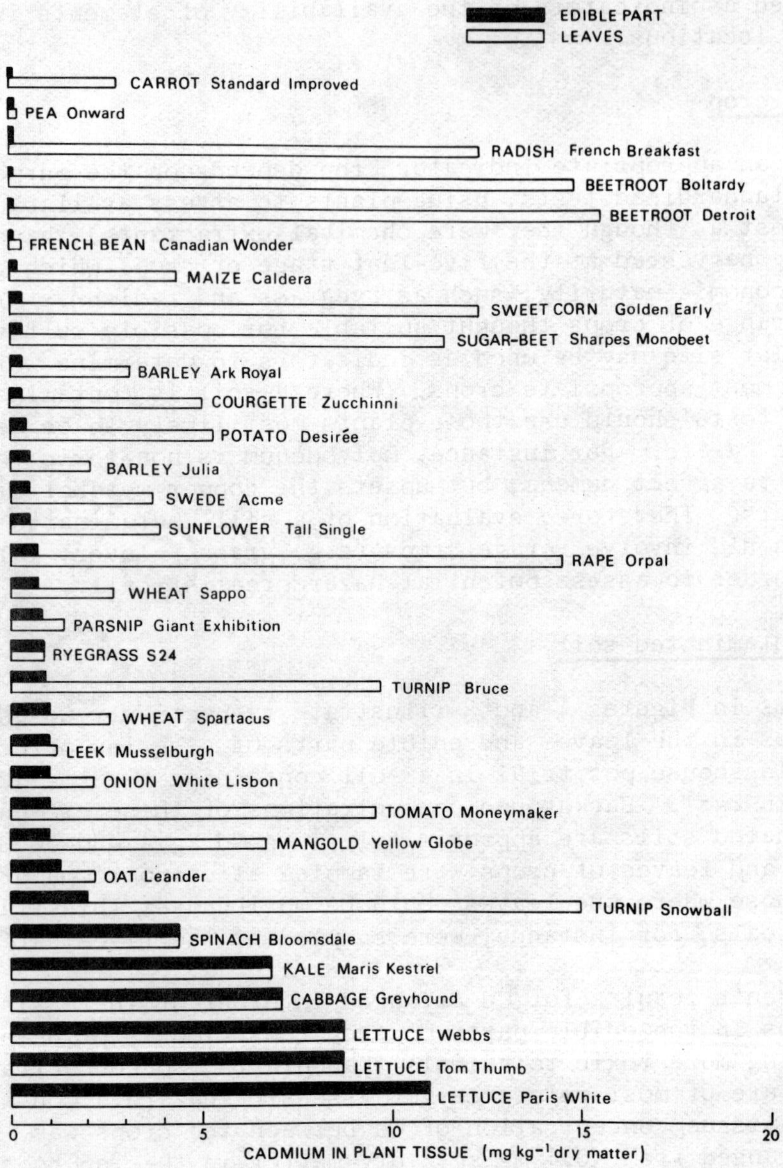

Fig. 1. Concentrations of cadmium in tissues of plants grown on a contaminated soil.

Figure 2 presents results for Zn arranged in order of increasing concentration in the leaves. Zinc is phytotoxic, and leaf tissue is most indicative of the physiological state of the plant. Again considerable variation of concentration in different crops was observed. Sugar-beet leaaves contained the most Zn (487.1 mg kg^{-1}) and onion leaves the least (27.6 mg kg^{-1}). Concentrations in edible parts ranged from 20.8 mg kg^{-1} (potato tubers) to 346 mg kg^{-1} (spinach leaves). As with Cd, differences in Zn concentrations

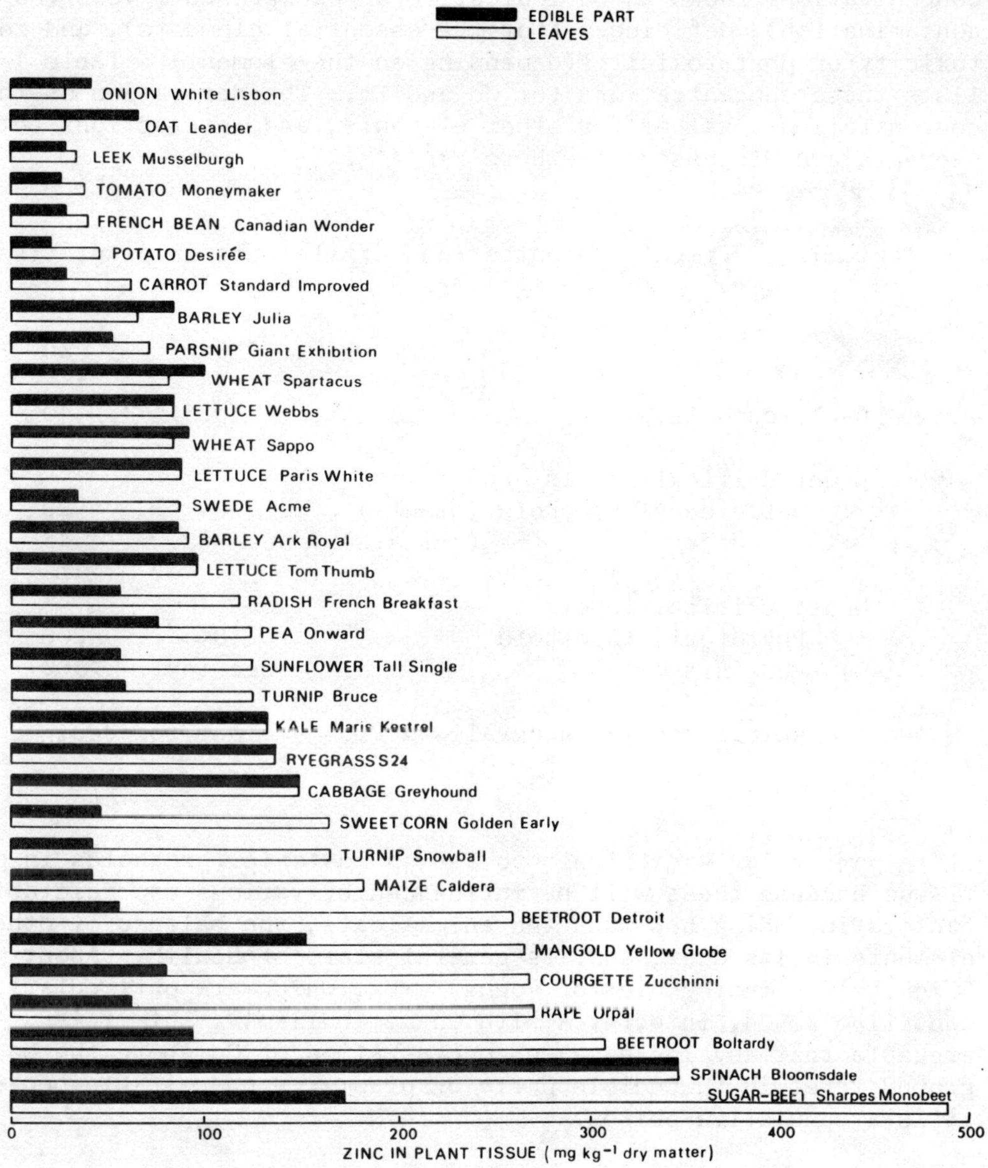

Fig. 2. Concentrations of zinc in tissues of plants grown on a contaminated soil.

between different cultivars of the same species were evident but were not as wide as inter-species differences. The trend of higher concentrations of elements in leaves compared with non-foliar edible parts, which was so marked in the Cd results, was shared by the Zn results but with some exceptions e.g. higher concentrations of Zn occurred in oat grains than in oat leaves. Concentrations of both elements were high in leaves of members of the beet family but other similarities between the crop concentration profiles for the two elements are less obvious.

Interpretation of results

This depends on comparisons of observed results with plant-tissue concentrations known to be indicative of background levels (no contamination), deficiency (for the essential elements), and zoo-toxicity or phytotoxicity (depending on the element). Table 1 lists these concentrations for Cd and Zn. The derivation of the concentrations, values for other elements, and interactions between elements are discussed elsewhere[10,11,12].

Table 1. Critical concentrations of elements in plant tissue ($mg\ kg^{-1}$ dry matter)*

	Cd	Zn
Background level	<0.5	40
Lower critical levels		
Deficiency threshold (plants)	-	20
" " (animals)	-	45
Upper critical levels		
Phytotoxic threshold	8	200
Zootoxic "	see text	>500

* Tentative data, generalised from various sources.

It is particularly difficult to assign zootoxic thresholds in plant tissue because these will be influenced by various unpredictable factors including how much the animal eats, the balance of other elements in its diet, and its general state of health. About 3 $mg\ kg^{-1}$ Cd in the diet of young lambs, which are particularly sensitive to Cd, interferes with Cu metabolism[13], but it is arguable that any increase in concentration of Cd above the background level in the edible parts of plants is undesirable because it is a cumulative poison.

Inspection of the histograms in Figures 1 and 2 together with the critical values in Table 1 permits an appraisal of the hazard associated with growing crops on a contaminated soil, supposing that agricultural use was contemplated for this soil. The results suggest that this soil would be unsuitable for growing any of the crops tested below sunflower in the list, on the grounds that edible parts of these crops exceeded the background concentration of 0.5 mg Cd kg^{-1}. Of the acceptable crops in terms of edible parts, radish, sweet corn, and the beets would be unsuitable because the concentration of Cd in their leaves exceeded the phytotoxic threshold of 8 $mg\ kg^{-1}$. Figure 2 suggests that zinc deficiency would not limit crop growth on this soil, but that phytotoxicity would reduce yields of crops lying below maize in the histogram. Cadmium is the limiting element of the two examined in this soil. In general the results suggest that contaminated soils are likely to be unsuitable for the cultivation of leafy vegetables, such as spinach and lettuce, but may be appropriate for potatoes, legumes, and some cereals and grasses subject to careful management.

CONCLUSION

Plant analysis has been subject to the criticisms that it is more laborious than soil analysis and is retrospective so far as fertilizer recommendations are concerned, because the results may come too late to correct deficiencies diagnosed in a growing crop. However, standardised plant tests are not difficult to set up, and although they take longer to complete than soil tests the time is often justified by the direct information they supply for the assessment of hazard and the management of contaminated soils.

ACKNOWLEDGEMENT

This paper is published by permission of the Director of the Stevenage Laboratory, Water Research Centre.

REFERENCES

1. Symeonides, C., and McRae, S. Journal of Environmental Quality, 1977, 6, 120.

2. Lindsay, W., and Norvell, W. Soil Science Society of America Journal, 1978, 42, 421.

3. Davis, R. D. Uptake of copper, nickel and zinc by crops growing in contaminated soils. Journal of the Science of Food and Agriculture (in press).

4. Hinesly, T. D., Alexander, D. E., Ziegler, E. L., and Barrett, G. L. Agronomy Journal 1978, 70, 425.

5. Scott, R. O., Mitchell, R. L., and Purves, D. Consultative Committee for Development of Spectrochemical Work, Bulletin,2. Macauley Institute, Aberdeen 1971.

6. DeVries, M. P., and Tiller, K. G. Environmental Pollution, 1978, 16, 231.

7. John, M. K., and Van Laerhoven, L. Environmental Pollution, 1976, 10, 163.

8. Jarvis, S. C., Jones, L. H., and Hopper, M. J. Plant and Soil, 1976, 44, 179.

9. John, M. K. Environmental Pollution, 1973, 4, 7.

10. Beckett, P. H. T., and Davis, R. D. New Phytologist, 1977, 79, 95.

11. Davis, R. D., Beckett, P. H. T., and Wollan, E. Plant and Soil, 1978, 49, 395.

12. Beckett, P. H. T., and Davis, R. D. New Phytologist, 1978, 81, 155.

13. Mills, C. F., and Dalgarno, A. C. Nature, 1972, 239, 171.

BIOLOGICAL ASPECTS OF THE TREATMENT OF HEAVY METAL CONTAMINATED LAND FOR HOUSING DEVELOPMENT SCHEMES

A.K Jones, B.Sc., M.S. Johnson, B.Sc., Ph.D., R.M. Bell, B.Sc and Professor A.D. Bradshaw
Department of Botany, University of Liverpool, Liverpool L69 3BX

SYNOPSIS

The origins of metal contaminated land in Britain are summarised. The inability of normal restoration strategies, based on after-use for amenity and recreation, to provide long-term solutions to the problems of heavy metal contamination sufficient to permit housing development is examined. A field trial designed to test the success of a variety of cover materials in providing the type of long-term solution necessary is described.

ORIGINS OF METAL CONTAMINATED LAND IN BRITAIN

The origins of heavy metal contamination of land in Britain are varied. They span the whole range of metal utilisation from the mining of metalliferous ores to the disposal of domestic and industrial wastes containing metals *via* sewage disposal systems.

The most severe contamination is usually associated with metal mining and smelting activities. Contamination of land by metal mining has resulted from the disposal of solid wastes from ore crushing and separating operations, as spoil heaps of coarse material or impoundments of water-deposited fine tailings. The problems are usually more severe from old mining operations where metal recovery techniques and environmental controls were poorly developed. Contamination from such sources can affect considerable areas around metal mines. Wind transportation of fine tailings can result in accumulations of high levels of metals in surrounding soils. Drainage waters from mine workings, waste heaps and tailings dams also serve to transport heavy metals away from the site of extraction and processing, producing high levels of metals in stream waters and sediments, and producing contamination of flood plains due to deposition of sediment at times of flooding (Harmsen, 1977). Several investigators have examined the extent of such contamination in areas of metalliferous mining in Great Britain (eg. Davies, 1971; Davies & Roberts, 1978; Johnson *et al*., 1978).

Metal smelters have, in the past, been a serious source of soil contamination through emissions of metal oxides from furnace stacks, and dust-blow from stock piles of metalliferous ore concentrates (Harmsen, 1977). Such contamination can be severe and cover considerable areas especially where a number of metal smelters have operated in close proximity, as for example in the Lower Swansea Valley, South Wales (Weston *et al*., 1965). Even modern smelters, with advanced emission control facilities, can cause elevated metal levels in the environment over considerable distances downwind of the plant (Little & Martin, 1972). More local contamination has occurred through dumping of metal contaminated smelter wastes on land (Gemmell, 1973; Weston et al., 1965).

Primary smelting of metal ores and concentrates generates relatively impure products so that secondary refining is necessary to remove the final contaminants. Electrolytic refining has become the preferred method of purification as it produces a higher grade final product and facilitates recovery of valuable metal impurities (Firth, 1978). Metal refineries are frequently located in or near urban areas, and although contamination from these sources is not as severe as that produced by mining and smelting, the proximity of the source to large centres of population, where garden crops may be grown, has led to concern (Muskett et al., 1979). Other industrial activities producing metal contaminated wastes contributing to the contamination of land include foundries, plating works and the land disposal of dockyard sediments (Patterson, 1971).

Sewage sludge, especially when derived from urban-industrial areas. frequently contains high concentrations of the heavy metals zinc, copper, tin and, in some sludges, nickel. If the sewage is deposited on to land this can produce problems of contamination. Other potentially hazardous heavy metals which frequently occur at elevated levels in sewage sludge include cadmium, lead and mercury (Department of the Environment, 1977).

RESTORATION OF METAL CONTAMINATED LAND

The restoration of land contaminated by heavy metals, such as that associated with mining and smelting activities, has been the subject of considerable research. Hitherto this research has concentrated on methods of overcoming the toxicity, low nutrient status, and extremes of pH of the substrates, in order to establish a permanent vegetation cover. This can achieve a significant degree of environmental quality control simultaneous with a marked aesthetic improvement. A wide range of surface treatments has been developed to enable grass swards to be established on materials severely contaminated by heavy metals. The usual technique is to alter the waste by surface addition, or incorporation of amendments such as topsoil, sub-soil, domestic refuse, sewage sludge and peat mixed with sand (Johnson & Bradshaw, 1979). Some of the most successful treatments involve the application of organic matter (as sewage sludge or domestic refuse) and lime which, under certain conditions, serves to reduce the availability of metal cations (Gemmell & Goodman, 1978).

However, such revegetation strategies frequently produce a good initial vegetation cover, which subsequently regresses. There are several possible causes of this phenomenon depending on the waste material and treatment. Breakdown of organic matter leads to release of bound metal ions, which become available for plant uptake (Gemmell & Goodman, 1978). Migration upwards of metal ions with capillary water through topsoil during periods of drought leads to accumulation of metals in surface layers of soil (Gemmell, 1974). The initial high fertility of many amendments leads to rapid growth of roots towards the amendment/toxic waste interface (Johnson, 1979). Other causes of dieback are recurrence of acidity due to breakdown of pyrites, and reduction in fertility through leaching of nitrogen and complexing of phosphorus (Johnson & Bradshaw, 1979).

In view of the lack of persistence and necessity for expensive maintenance of vegetation established as described, alternative strategies have been developed. One of the most successful is the direct seeding of metalliferous wastes with metal tolerant varieties of grass species (Smith & Bradshaw, 1972). In addition to their adaptation to high levels of metals, these varieties are also adapted to the low nutrient and low moisture conditions frequently associated with metal contaminated land. Moreover, because they are slow growing, sward maintenance costs are minimal (Gemmell & Goodman, 1978). Nevertheless the specificity of the tolerance mechanism means that this approach is not always viable (Johnson, 1979).

The above restoration strategies have all been developed with a view to revegetating metal contaminated land for pollution control, amenity and recreational uses only. However, in view of the increasing pressures on land in urban/industrial areas, it is desirable that reinstatement of derelict land with a view to the construction of housing should be a practical objective.

CONSTRAINTS ON DEVELOPMENT OF METAL-CONTAMINATED LAND FOR HOUSING

The requirements for development of metal contaminated land for housing are much more exacting than those for amenity use. Increasing concern over environmental quality in urban areas leads to demands for domestic gardens and allotments, and recreation areas. In this context it is not sufficient that plants can be grown on contaminated material. It is also necessary to prevent the movement of metals into the vegetation and the uppermost soil layers, so that metal levels of consumable produce are maintained within safe limits and the risks of direct ingestion of contaminated soil by children are minimised.

The traditional reclamation strategies are of little use in providing solutions to these problems. Down (1974) states that "it is wise to avoid the use of revegetated mine wastes for food production for man", since most restoration procedures do not prevent uptake of metals by plants. Root uptake can lead to accumulation of metals in edible parts of vegetable crops, making them unfit for human consumption. Moreover plant mortality leads to metal accumulation within organic residues of the upper soil horizons. The rates of uptake vary greatly between metals and plant species. For example, cadmium is readily taken up by plant roots and translocated, while lead tends to be bound in root cell walls (Pilegaard, 1978). Lettuce has been found to be an efficient heavy metal accumulator (Haghiri, 1973; Giordano et al., 1979).

A corollary of the more severe constraints imposed on housing development of metal contaminated land is that soil metal concentrations which are too low to be of importance in terms of phytotoxicity can be of serious concern if plants grown on such materials are intended for human consumption, since many species are able to tolerate tissue metal concentrations far in excess of levels of significance to health without showing visible damage symptoms. Several studies have shown that vegetables grown in garden soils affected by mine wastes contain excessive levels of heavy metals. For example, Davies and Roberts (1975) grew radish in gardens in the Halkyn Mountain area of Clwyd, North Wales, and found lead levels varying from 25 - 777 μg/g ash weight (5 - 155 μg/g dry weight). Radish grown in 14 out of 17 gardens had lead contents above that considered normal (12 μg/g dry weight in radish), and radish from 3 gardens exceeded the statutory limit for lead in food of 2 μg/g fresh weight (H.M. Govt. 1961) which in the case of radish is equivalent to 30 μg/g dry weight. These radish were grown in soils with total lead concentrations of 920 - 5,897 μg/g dry weight which, although elevated, are not high enough to produce serious plant growth problems.

An alternative system for reinstatement of metal contaminated land which has been developed in recent years is that of physically isolating the surface plant rooting zone from the toxic material by using deep layers of amendments (Johnson & Bradshaw, 1979). Several materials have been advocated for the formation of such barriers, eg. slate and limestone quarry spoils and pulverised fuel ash. A particularly promising example is that of non-pyritic colliery shale for the treatment of lead/zinc mine spoil (Johnson et al., 1977). Gemmell (1974) has described a similar system using a porous sub-soil to provide a barrier over chromate smelter waste.

The development of these techniques is likely to be of use in the context of housing development on contaminated land since they are specifically intended to prevent the movement of metal contaminants into surface soil and vegetation.

PROJECT OBJECTIVES

The aims of the research project, a part of which is described in this paper, are to examine the success of past reclamation techniques in preventing the reappearance of metal contamination problems in reclaimed areas, and to assess the efficiency of a range of cover materials in preventing the upward movement of metals from contaminated materials into surface soil and plants. A parallel project is being carried out by Liverpool Polytechnic to investigate other aspects of this problem (Lepp & Harris, 1979).

FIELD TRIALS

A major part of the research is centred on the area of London south of the Thames between Woolwich and Erith, now known as Thamesmead, where the Greater London Council have embarked on a large housing development (Lowe, 1979).

This area was occupied by the Royal Arsenal from at least the eighteenth century, and was used for a variety of military/industrial purposes such as the manufacturing and testing of weapons, explosives, gases and paints (Gellion, 1978). The area has been contaminated with a range of materials, including heavy metals, as a result of these operations, and also through the disposal of foundry and gas works wastes.

In view of its proximity to Central London the Thamesmead area represents a very valuable land resource for housing development, but it is important that the problems of contamination are controlled as an integral part of the development. With this objective in mind a series of trial plots has been constructed within the Thamesmead area to establish the ability of a range of cover materials to provide the long term effectiveness required.

A site was selected with the co-operation of the Thamesmead section of the Housing Department of the Greater London Council. In order to ensure the capability for a long term study the location was determined partly by scheduled development proposals and site security.

24 surface treatments, including a control, have been established in separate plots in each of three randomised blocks. Table 1 shows the amendments used. The underlying material was compounded from a

TABLE 1

THAMESMEAD PLOT TREATMENTS

Base	Barrier	Topsoil
	None	
	Ground limestone incorporation	
	Hoggin	Sandy loam
Compounded	Crushed demolition rubble	0, 150, 300mm
contaminated	Pulverized fuel ash	depth above
material	Clay	each barrier
	Sea-won sand	
	Land-won sand	

24 treatments replicated in three randomised blocks.

number of waste materials in the vicinity containing from 750 - 2000μg Pb/g and 1 - 10μg Cd/g dry weight. Lead and cadmium concentrations were chosen as the criteria for waste selection in view of their toxicity and widespread occurrence in contaminated soils. This material was thoroughly homogenised by a mechanical excavator before spreading to a depth of 1100mm. A preliminary analysis of the material is shown in Table 2.

TABLE 2

PRELIMINARY ANALYSIS OF METAL CONCENTRATIONS IN THAMESMEAD WASTE MATERIAL

	Total metal μg/g dry weight			
	Pb	Cd	Zn	Cu
Geometric mean	1114	2.0	884	654
Range	668-2171	1.5-5.6	233-8002	205-3382
Normal soil	2-200	0.1-1.0	10-300	2-100

The walls separating individual plots were constructed of corrugated asbestos covered with 500-gauge black polythene, the heights of the walls being 600mm. The plots were then filled with the appropriate combination of contaminated material, break layer and topsoil.

The area surrounding the blocks was then covered with a layer of sea-won sand to minimise contamination of plot surfaces, and the blocks surrounded by a 450mm high polythene wind break to eliminate cross contamination by wind blow.

SELECTION OF COVER MATERIALS

Cover materials were selected based on practical and economic considerations, with particular regard for availability of materials, both at the Thamesmead site and urban areas nationally. The materials under examination cover a wide range of physical and chemical properties liable to be of use in preventing upward movement and accumulation of metal contaminants in surface soil. For example, the coarse building rubble acts as a capillary break allowing percolation of water but preventing upward capillary movement. The clay layer acts as a barrier to water movement into overlying soils where these consist of coarser materials, since capillary movement is reduced where a fine clay underlies a soil of coarser texture (Ward, 1975). The converse applies where fine material overlies a coarser matrix. Thus, the topsoil over sea-won sand treatment will tend to increase capillary movement. However, a major influence on the dynamics of metal movement through soils is that of binding on to the soil cation exchange sites following upward movement under drought conditions, so that secondary mobilisation does not occur when a change in the net water balance favours downward movement. The presence of a layer of coarse sand with low cation exchange capacity can be expected to keep the metal ions in a state of flux in the system (Johnson, 1979), and, provided continuous upward movement over long periods of time does not occur, the metals should be prevented from reaching the topsoil. The other materials represent intermediates between these extremes, all of which are readily available and, in some cases, already in use at Thamesmead. The depth of 300mm was chosen as representing the minimum depth of break layer which could be applied reliably. Any reduction below this value is likely to result in zones with an unacceptably shallow depth unless extreme care is taken during large scale application.

The topsoil selected is already in use at Thamesmead for amenity areas, and comprises of a stony, sandy-loam developed on glacial outwash deposits. The two topsoil depths were 150mm for reasons similar to those stated previously, and 300mm. The latter was taken to represent the maximum depth which it is economic to apply on a large scale.

Prior to planting a general fertilizer was applied (NPK 12:12:17 at 200kg/ha) to the plots, and they were divided into two halves with plastic lawn edging.

Half of each plot was sown with an amenity grass mixture (30% *Lolium perenne* var. Stadion, 30% *Festuca rubra rubra* var. Novorubra, 30% *F. rubra commutata* var. Kobet, 10% *Agrostis tenuis* var. Highland and 5% *Trifolium repens* var. S184) at a rate of 100kg/ha. The other half of the plot was planted with crop species as follows:-

Strawberry var. Royal Sovereign
Lettuce var. Webbs Wonderful
Leek var. Musselburgh
Radish var. French Breakfast.

Lettuce is known to accumulate heavy metals from soil (Haghiri, 1973), and is thus useful in identifying soils from which metal uptake is likely to be a problem. Radish has been widely used in other research into plant metal uptake (eg. John, 1972; Davies & Roberts, 1975) enabling comparisons to be made between this and previous work. In addition both lettuce and radish are widely grown under domestic cultivation in the British Isles. Leeks were selected as a vegetable which is winter hardy, and frequently left in the ground until January or February. It is known that in some plants metal ions normally strongly bound in root tissues are mobilised during the winter months, (Mitchell & Reith, 1966) and it is possible that this mechanism might lead to an accumulation of metals in the edible portions of leeks. Strawberries were selected as the most widely grown soft fruit likely to be cultivated in gardens in an area such as Thamesmead.

Since varietal differences in plant metal uptake are known to occur (eg. Petterson, 1977) the varieties were selected to align with the parallel studies being undertaken by Liverpool Polytechnic.

Monitoring of the performance of the trial plots is by harvest and analysis of the vegetables, fruit and amenity grass sward. Grass establishment is being recorded photographically.

COLUMN EXPERIMENTS

In order to investigate further the relative effectiveness of the amendments, series of soil columns are being constructed. These will enable the amendment materials to be tested under conditions of sustained low rainfall and high evapotranspiration conditions when upward movement of water will be at a maximum. The column trials cover a wider range of trace metal contaminated materials. These include a lead/zinc/cadmium contaminated mine waste from Minera, nr. Wrexham (Clwyd), a zinc smelter waste from the Lower Swansea Valley, zinc/cadmium contaminated soil from agricultural land (Shipham, Somerset) within the former non-ferrous metal ore field of the Mendips, and material from an ammunition disposal area at Thamesmead.

CONCLUSIONS

If derelict land is to be usefully and safely developed to provide much needed housing in urban areas, then it is clear that major modifications of the traditional techniques for restoration of such areas for amenity purposes are required. It is essential that a detailed knowledge of the possible strategies is acquired before such development takes place on a large scale.

It is intended that detailed monitoring of the dynamics of heavy metal behaviour in the soil/plant ecosystem, based on the experimental system outlined, should provide valuable practical information of use in the planning of large scale redevelopment works on contaminated land. Allied with other studies investigating engineering and further biological aspects of this problem, the investigations described should assist planning authorities in the preparation of specifications to provide long term security of derelict sites being redeveloped for housing purposes.

ACKNOWLEDGEMENTS

The authors wish to acknowledge the assistance of the Greater London Council in providing the experimental area and assisting with its construction, and the Department of the Environment for funding the research.

REFERENCES

DAVIES, B.E. (1971). Trace metal content of soils affected by base metal mining in the West of England. Oikos, 22, 336-372.

DAVIES, B.E. and ROBERTS, L.J. (1975). Heavy metals in soils and radish in a mineralized limestone area of Wales, Great Britain. Sci. Total Environ., 4, 249-261.

DAVIES, B.E. and ROBERTS, L.J. (1978). The distribution of heavy metal contaminated soils in North-east Clwyd, Wales. Water, Air and Soil Pollut., 9, 507-18.

DEPT. OF THE ENVIRONMENT (1977). Report of the working party on the disposal of sewage sludge to land. Standing Technical Committee Reports No.5. Dept. of the Environment/National Water Council.

DOWN, C.G. (1974). Problems in vegetating metal toxic mining wastes. In Minerals and the Environment (ed. Jones, M.J.) Institute of Mining and Metallurgy, London.

FIRTH, J.N.M. (1978). The origins and exploitation of non-ferrous metals. In Environmental Management of Mineral Wastes. (ed. Goodman, G.T. and Chadwick, M.J.) NATO Advanced Study Institute Series No.7, pp 259-72.

GELLION, K. (1978). Pathways of toxic substances into biological systems at a proposed nature reserve at Thamesmead. Unpublished MSc. thesis, Imperial College of Science and Technology, London.

GEMMELL, R.P. (1973). Revegetation of derelict land polluted by a chromate smelter. Part 1: Chemical factors causing substrate toxicity in chromate waste. Environ. Pollut., 5, 181-197.

GEMMELL, R.P. (1974). Revegetation of derelict land polluted by a chromate smelter. Part 2: Techniques of revegetation of chromate smelter waste. Environ. Pollut., 6, 31-37.

GEMMELL, R.P. and GOODMAN, G.T. (1978). Problems of grassland maintenance on metalliferous smelter wastes. In Environmental Management of Mineral Wastes. (ed. Goodman, G.T. and Chadwick, M.J.) NATO Advanced Study Institute Series No.7.

GIORDANO, P.M., MAYS, D.A. and BEHEL, A.D. (1979).Soil temperature effects on uptake of cadmium and zinc by vegetables grown on sludge amended soils. J. Environ. Qual., 8, 233-36.

HAGHIRI, F. (1973). Cadmium uptake by plants. J. Environ. Qual., 2, 93-6.

HARMSEN, K. (1977). Behaviour of Heavy Metals in Soils. Agricultural Research Reports No.866, Pudoc Publ., Wageningen, Netherlands.

H.M.GOVERNMENT, (1961). The lead in food regulations. S.I. 1961 No. 1931 H.M.S.O.

JOHN, M.K. (1972).Uptake of soil applied cadmium and its distribution in radishes. Can. J. Plant Sci., 52, 715-19.

JOHNSON, M.S. (1979). Options for treatment of metal contaminated mine workings. In Reclamation of Contaminated Land. Proc. Society of Chemical Industry Conference, Eastbourne, Sussex. Oct. 1979.

JOHNSON, M.S. and BRADSHAW, A.D. (1979). Ecological principles for the restoration of disturbed and degraded land. Applied Biology, 4, 141-200.

JOHNSON, M.S., MCNEILLY, T. and PUTWAIN, P.D. (1977). Revegetation of metalliferous mine spoil contaminated by lead and zinc. Environ. Pollut., 12, 261-277.

JOHNSON, M.S., ROBERTS, R.D. and FIRTH, J.N.M. (1978). Lead and zinc in the terrestrial environment around derelict metalliferous mines in Wales. Sci. Total Environ., 10, 61-79.

LEPP, N.W. and HARRIS, M.R. (1979). A strategy for evaluation of soil covering techniques to reduce trace metal uptake by soft fruits and vegetables. In Reclamation of Contaminated Land. Proc. Society of Chemical Industry Conference, Eastbourne, Sussex. Oct. 1979.

LITTLE, P. and MARTIN, M.H. (1972). A survey of zinc, lead and cadmium in soil and natural vegetation around a smelting complex. Environ. Pollut., 3, 241-254.

LOWE, G.W. (1979). Thamesmead Development. In Reclamation of Contaminated Land. Proc. Society of Chemical Industry Conference, Eastbourne, Sussex Oct. 1979.

MITCHELL, R.L. and REITH, J.W.S. (1966). The lead content of pasture herbage. J.Sci. Fd. Agric., 17, 437-439.

MUSKETT, C.J., ROBERTS, L.J. and PAGE, B.J. (1979). Cadmium and lead pollution from secondary metal refinery operations. Sci. Total Environ. 11, 73-87.

PATTERSON, J.B.E. (1971). Metal toxicities arising from industry. In Trace Elements in Soils and Crops. Min. Agric. Fish. Fd., Tech. Bull. 21, 193-207.

PETTERSON, O. (1977). Differences in cadmium uptake between plant species and cultivars. Swed. J. Agric. Res., 7, 21-4.

PILEGAARD, K. (1978). Heavy metal uptake from the soil in four seed plants. Botan. Tids., 73, 167-177.

SMITH, R.A.H. and BRADSHAW, A.D. (1972). Stabilization of toxic mine wastes by the use of tolerant plant populations. Trans. Inst. Min. Met., 81a, 230-237.

WARD, R.C. (1975). Principles of Hydrology. 2nd edition, McGraw-Hill, Maidenhead.

WESTON, R.L., GADGIL, P.D., SALTER, B.R. and GOODMAN, G.T. (1965) Problems of revegetation in the Lower Swansea Valley, an area of extensive industrial dereliction. In Ecology and the Industrial Society (ed. Goodman, Edwards & Lambert) British Ecological Society Symposium 5, 297-325.

The Use of Selective Cation Exchange Resins for the Fixation of Phytotoxic Heavy Metals in Cultivated Soils

Dr. H. Heller
Bayer AG, FB-F/Lewatit Division, D-5090 Leverkusen

Introduction

Following the environmental damage that has been caused by heavy metals over the last few years, a large amount of work has been carried out on the heavy metal content of natural soils. It was found that, at particularly exposed places (close to traffic routes, near to certain industrial plants, in areas of high population density), the heavy metal content may be up to 100 times the natural content (cf. KLOKE (3) and the references listed there).

As we know, all heavy metals - including the essential ones - have a phytotoxic action when they occur in sufficiently high concentrations in the soil (SORAUER,6). The consequences are poor germination, dwarfism and necrosis on roots and leaves. The growth-inhibiting effects depend both on the tolerance of the plant genus and on the nature of the soil, i.e. on the amounts of heavy metals that may be taken up by the plants.

If we assume that heavy metals are taken up by the plants via the soil solution, then the actual amounts that can be taken up will be influenced by a variety of equilibria. To be taken into account are the equilibria between the soil solution on one side and less soluble heavy metal compounds (e.g. oxides, carbonates, phosphates), clay minerals containing heavy metals, heavy metal complexes of humus substances and the root system of the plants on the other side (SANTILLAN-MEDRANO and JURINAK, 5).

A large number of tests have already been carried out in an attempt to decrease the accessibility of heavy metals for the plants in contaminated soils (cf. ROHDE (4) and the references listed there). The most promising so far seemed to be the method of mixing lime to the substrate. More recent tests in this direction have not been entirely satisfactory yet (VETTER et al., 10), but they are being continued (VETTER,11). It must also be considered in this respect that the method of adding lime cannot be used in every case.

Tests and results with selective cation exchange resins

If we add cation exchange resins to substrates contaminated with heavy metals, then we are adding a further adsorptive component to the equilibrium system described above. As a result, there is a change in the position of equilibria and, at the same time, in the amounts of heavy metals that may be taken up by the plants.

Tests on sandy soils and coniferous substrates

In several studies carried out at the Laboratorium voor Fytopathologie en Plantenbeschirming, Katholieke Universiteit Leuven, headed by Prof. C. VAN ASSCHE, it has already been shown that cation exchange resins with selective fixing properties for heavy metals are capable of neutralizing or reducing the growth-inhibiting action of high heavy metal concentrations in sandy soils and coniferous substrates contaminated with heavy metals (VAN ASSCHE et al., 1, 2, 8, 9). For these tests, use was made of the cation exchange resins Lewatit OC 1029, OC 1030 and OC 1034*, which are noted for their high selectivity for heavy metals compared with alkaline earth and alkali ions. Lewatit OC 1029, OC 1030 and OC 1034 desorb calcium, magnesium and potassium ions respectively into the surrounding medium in exchange for heavy metal ions.

The vegetation trials were carried out in the greenhouse with various types of plants on naturally contaminated sandy soil and coniferous substrates as well as outdoors on a sandy soil contaminated with heavy metals in the immediate vicinity of two metallurgical works - and thus under constant exposure to emissions - with vegetables, ornamental plants and nursery plants. The results of the test prove that, by mixing selectively acting cation exchange resins to the substrate, the negative phytotoxic and growth-inhibiting effects of the heavy metals are virtually eliminated or greatly reduced. The crop yields increase considerably and the amount of heavy metals taken up by the plants is decreased. The use of ion exchange resins produced better results than adding lime. Particularly favourable was the combined use of lime and ion exchange resins. Garden cress proved to be very good as an indicator plant for determining the quantities required.

Figures 1 to 4 show the large differences in growth observed in vegetables and ornamental plants on a contaminated sandy soil with a heavy metal content of 600 ppm Cu, 250 ppm Pb and 150 ppm Zn under constant exposure to emissions outdoors. Figure 5 shows a germination trial with beans on a naturally contaminated sandy soil with a content of 1900 ppm Cu; 1500 ppm Pb and 530 ppm Zn (WESTERHOFF method) in the greenhouse. The many other test results can be found in the original papers (l.c.).

Tests on other soils and substrates

Similar results to those on sandy soils were obtained with leaf vegetables (lettuce) and root vegetables (carrots) on marshy soil in pots (Landwirtschaftliche Untersuchungs- und Forschungsanstalt Oldenburg; headed by Prof. Dr. H. VETTER). Here, too, the joint application of cation exchange resins and lime (pH 6,7) proved to be particularly effective. The increases in crop yield were nevertheless much smaller than on sandy soil. The tests are at present being continued outdoors. Of particular interest was also the effectiveness of the cation exchange resins in substrates containing composts contaminated with heavy metals, since the usability of such composts still appears problematical (cf. TIETJEN, 7). For this, pot trials with sandy and loamy soil as well as three different types of compost were carried out with a leaf vegetable (lettuce) and a root vegetable (carrots) at the "Lehrstuhl für Gemüsebau" at the Technical University in Munich (Chairman: Prof. Dr. D. FRITZ).

* Trial products of BAYER AG, OC Division, D-5090 Leverkusen.

Tables 1 and 2 show the crop weights and the analysis data of lettuce. It can be seen that the addition of ion exchange resins can also appreciably increase the crop in these substrates and can reduce the amount of heavy metals taken up by the plants. It must be taken into account in this connection that the heavy metals contained in the composts from municipal wastes are predominantly in a chemical bond which is still unknown. The test results with carrots were less pronounced - they reveal marked differences between the two types of vegetable.

For the above mentioned tests, use was made of a mixture of the ion exchange resins Lewatit OC 1029 and OC 1030 in equal parts. The amount was 100 equivalent %, calculated on the content of toxic heavy metals in the substrates. The mixing ratio of soil to compost was 1:1. The pH in the marshy soil treated with lime was 6,7 and the mixtures of compost and soil were weakly alkaline. The heavy metal contents of the composts produced of municipal wastes are shown in Table 3.

Studies of the adsorption rate and equilibrium exhaustion

Whether or not the fixing of heavy metals using selectively acting cation exchange resins in contaminated substrates is sucessful depends not least on the rate of adsorption. In several vegetation trials, finely ground ion exchange resins (grain diameter below 0,1 mm) proved, as expected to be fast acting.

More accurate investigations on the rate of adsorption were carried out with the various Lewatit resins in bead form (grain diameter: 0,3-1 mm), which can be separated again from the sandy soil when the test has been completed.
For this purpose, soil specimens of maximum moisture were thoroughly mixed with the cation exchange resins in a quantity of 100 equivalent %, calculated on the heavy metal content of the soil, and stored in plastic containers. During the first three weeks of the test, the plastic containers were shaken for one full day a week. The amount of heavy metals adsorbed by the ion exchange resins was determined at the end of the trial by separating and eluting with hydrochloric acid (15 %). Sandy soils were used with contents of 600 ppm Pb, 1220 ppm Cu and 540 ppm Zn (WESTERHOFF method), the pH was 5,7.

The results are shown in Figures 6 and 7. It can be seen from Figure 6 that of the three heavy metals Cu, Pb and Zn, the zinc is taken up relatively quickly by the cation exchange resin. This is evidence of the high mobility of the zinc in the soil samples. In comparison, the adsorption of the copper takes place more slowly and that of the lead even more slowly. It can also been seen from the results that the three heavy metals are adsorbed more quickly and - in the set test time - more completely by Lewatit OC 1034 than by the other two cation exchange resins, between which there was little difference in their effectiveness. Equilibrium exhaustion after 18 weeks can only be assumed for Lewatit OC 1034. As Figure 6 shows, around 90 % of the zinc and copper was adsorbed under the test conditions from this soil by the Lewatit OC 1034, but only about 30 % of the lead.

Figure 7 depicts the degree of exhaustion of Lewatit OC 1034. The cation exchange resin was originally charged 45% with potassium ions (the rest:

H-ions). In the equilibrium state (after around 13 weeks) under the selected test conditions, it is charged about 43 % with copper, 30 % with zinc and 2 % with lead. The remaining functional groups are occupied with potassium -(10 %), calcium - (10 %), magnesium - (2 %) and H-ions (3 %).

Long-term effectiveness and resin stability

The extent to which the proposed method can be applied is basically dependent on the long-term effect of fixing the heavy metals and on the stability of the cation exchange resins under environmental conditions. Investigations on this subject were carried out on resin specimens in a four-year field trial, in which Lewatit OC 1029 in bead form was used, the resin was charged with calcium - (77 %) and H-ions.

Table 4 shows the amounts of heavy metals adsorbed by the ion exchanger and the degree of exhaustion of the resin after 3, 11, 26 and 47 months. It is clear that the adsorption of heavy metals in this test was still continuing after 11 months. It must also be taken into account that the test field was constantly exposed to emissions. A re-exchange of heavy metal ions - apart from manganese - did not occur. It is also worth noting that only little iron was adsorbed. These findings can be explained by the selectivity properties of Lewatit OC 1029 and the low solubility of the iron-(III) compounds in the soil.

From the characteristic data of Lewatit OC 1029 given in Table 5 it is evident that the composition of the resin (ratio: C/N and O_2/N) has remained virtually unchanged during the course of the test. (If some of the cross-linking points had broken up, the water content would have increased considerably.) As far as the cation exchange capacity is concerned, there was slight decline of 8 % altogether. However, in view of the difficulty of regenerating the resin (some heavy metals are fixed very firmly), this slight decrease in capacity could hardly be taken as sign of a continuous loss of capacity.

Outlook

It emerges from these studies that mixing the described selective cation exchange resins to soil contaminated with heavy metals has a very positive effect on the development of the plants and thus on the crop yields. At the same time, there is a reduction in the amount of heavy metals taken up by the plants. Particularly beneficial was the combined use of ion exchanged resins and lime. At the present stage it cannot be predicted whether or not this will make it possible to lower the contents of heavy metals in plants intended for human or animal consumption to the levels of the proposed guide data. Independently of this, the process may nevertheless gain significance in certain sectors of horticulture and landscape management.

For the economy of the process, the long-term action of the cation exchange resins will be of vital importance. Four-year vegetation trials have so far been carried out and the results give good reason to assume that the cation resins will remain effective for a long time.

Summary

The use of selective cation exchange resins for the fixation of phytotoxic heavy metals in cultivated soils.

The growth-inhibiting effects of high concentrations of heavy metals in contaminated cultivated soils and substrates can be eliminated or greatly reduced with the selectively acting cation exchange resins ®Lewatit OC 1029, OC 1030 and OC 1034. Plant trials have shown that these cation exchange resins are active in sandy and marshy soils, in substrates containing composts contaminated with heavy metals and in contaminated coniferous substrates. The combined use of a cation exchange resin and lime has proved particularly beneficial. The use of the cation exchange resins also reduces the extent to which the heavy metals are taken up by the plants.

The adsorption of the heavy metals as a function of the time and of the type of cation the cation exchange resins are charged with was determined in the laboratory for a natural sandy soil with a low humus content which had been contaminated with copper, zinc and lead by environmental factors. Field trials performed on the same soil have shown that the heavy metals adsorbed by the cation exchange resins are fixed irreversibly for at least four vegetation periods and that within this time the cation exchange resins employed do not decompose under environmental conditions.

The use of the described method for the fixation of phytotoxic heavy metals would appear appropriate and beneficial in cases where corresponding damage has been suffered by the soil in certain fields of horticulture and landscape management.

Bibliography

1. JANSEN, G.; VAN ASSCHE, C.: The phytiatric influence of treatments with cation exchangers in a soil, contaminated with heavy metals, on the growth and the mineral composition of Zea Mays and Phaseolus vulgaris L. Plant and Soil 49, 229-249, 1978

2. JANSEN, G.; VAN ASSCHE, C.: Results of cation exchangers against heavy metals in the field. Int. Symposium on Crop Protection, Gent 3. 5. 1977

3. KLOKE, A.: Discussion paper on the subject: "Tolerierbare Gehalte verschiedener Elemente in Böden" für die BGA-Kommission "Landw. Nutzung von Abwasser und ähnlichen Stoffen", Fassung vom 19. 2. 1974, Berlin 1974

4. ROHDE, G.: Bekämpfung der Rieselmüdigkeit. Z. ges. Hyg. 11, 569-583, 1965

5. SANTILLAN-MEDRANO, J.; JURINAK, J.J.: The chemistry of lead and cadmium in soil; solid phase formation. Soil Sci. Soc.Am. Proc. 39, 851-865, 1975

6. SORAUER, P.: Handbuch der Pflanzenkrankheiten; 1. Band, Die nichtparasitären Krankheiten; 2. Teil. Paul Parey Verlag, Berlin und Hamburg 1969

7. TIETJEN, C.: Qualitätsanforderungen der Landwirtschaft an Müllkompost. AID-Information für die Wirtschaftsberatung 26, Nr. 4, 1-12, 1977

8. VAN ASSCHE, C.; DE MEY, W.: Fytiatrie ten opzichte van de zware metalen met kationenwisselaars. Genootschap plantproduktie en ekosfeer, K. VIV., Gent 1975

9. VAN ASSCHE, C.; JANSEN, G.: Anwendung von selektiv wirkenden Kationenaustauschern auf mit Schwermetallen kontaminierten Böden. Landwirtsch. Forsch., Sonderh. 34/II, 215-228, 1978

10. VETTER,H.; FRÜCHTENICHT, K.; MÄHLHOP, R.: Immissionen und Nahrungsqualität. Landwirtsch. Forsch. Sonderh. 30/I., 29-39, 1974

11. VETTER, H.: Private information at the 88th VDLUFA-Congress, Oldenburg 20. - 25.9. 1976

Table 1
Plant trials using substrates with added compost (ratio 1:1); Lettuce crop weights as percentages of the control weight (without added compost)

	control	MKK	MKK+I	MK	MK+I	KSK	KSK+I
sandy soil	100	69,8	82,2	40,2	59,5	7,9	72,2
loamy soil	100	73,8	110	63,3	95,8	35,4	105,4

MKK = compost produced of refuse with sewage sludge
KSK = compost produced of sewage sludge
MK = compost produced of refuse
+I = ion exchange resin

Table 2
Plant trials using substrates with added compost (ratio 1:1);
Analyses of lettuce

Figures in mg/kg dry substance

	Fe	Zn	Cu	Mn	B	Pb	Cd	Cr	Ni
sandy soil control	206	52	4,8	28	43,4	8	*	*	1
MKK	104	260	14,8	18	99,8	8	2	*	3
MKK+I	64	156	14,4	26	94,8	5,5	*	*	*
KSK	900	720	22,2	530	121,6	5,0	9	7	35
KSK+I	152	196	14,4	76	72,6	7,0	5	*	2
MK	292	295	18,0	30	118,8	8,0	1	*	3
MK+I	76	156	14,8	36	115,0	4,5	1	*	*
loamy soil control	128	49	4,8	48	73,6	4,5	*	*	2
MKK	108	230	15,8	24	86,2	4	2	*	*
MKK+I	98	106	14,8	26	74,4	4	*	*	2
KSK	156	780	16,4	680	118,8	4	12	4	46
KSK+I	118	305	14,9	170	60,6	6	5	*	1
MK	108	230	14,8	42	102,4	4,5	*	*	*
MK+I	118	150	15,6	68	112,4	6	*	*	2
GD 5 %	2,9	4,9	0,3	6,9	3,9	1,9	0,8	-	1,4
1 %	4,2	6,9	0,4	9,7	5,5	2,7	1,2	-	2,0
0,1 %	5,9	9,8	0,6	13,8	7,9	3,8	1,7	-	2,9

Explanation of MKK, KSK, MK and +I see Table 1.
* = below the detection limit

Table 3
Heavy metal contents of the compost used

mg/kg dry substance (WESTERHOFF method)

compost	Pb	Ni	Cd	Cr	Cu	Zn	Fe	Mn
MKK	369	32	4	33	239	1293	4530	398
KSK	16	249	44	105	110	4570	2100	1107
MK	403	20	5	14	151	1416	1297	578

Explanation of MKK, KSK and MK see Table 1.

Table 4
Heavy metal adsorption by Lewatit OC 1029 in a 4-year field trial
Amount of ion exchange resin applied: 100 equivalent %, calculated on the total Pb, Cu and Zn content of the soil at the beginning of the trial

	Adsorbed quantity (mole%) calc. on total content in soil (WESTERHOFF method)				Degree of exhaustion of Lewatit OC 1029 in % of total capacity			
Test period (months)	3	11	26	47	3	11	26	47
Al	-	4	7	3	-	17,3	18,5	13,3
Ca	-	-	-	-	48,8	34,0	23,2	21,1
Cd	-	34	44	46	-	<0,1	<0,1	<0,1
Cu	31	44	56	56	12,6	18,1	23,1	23,2
Fe	2	3	4	4	0,8	1,1	2,0	1,7
Mg	-	-	-	-	2,2	1,1	2,1	2,6
Mn	24	11	6	3	0,6	0,3	0,2	0,1
Na	-	-	-	-	-	<0,1	-	-
Ni	-	26	33	55	-	0,2	0,3	0,5
Pb	9	16	19	15	0,5	0,8	1,0	0,8
Sr	-	-	-	-	-	<0,1	-	-
V	-	-	-	-	-	0,1	-	-
Zn	37	51	56	89	8,6	11.9	12,9	23,0

Table 5
Characteristics of Lewatit OC 1029 used in a 4-year field trial after treatment of the resin with hydrochloric acid (reference form = H-form)

Test period (months)	Cation exchange capacity (TC) meq/g	Decrease of TC %	C/N	O_2/N	Water content % by wt.
0	6,57	-	15,14	2,01	47,1
3	6,35	3	15,54	1,97	46,1
11	6,18	6	15,02	1,89	48,0
26	6,16	6	15,28	1,99	46,2
47	6,04	8	15,32	2,09	48,7

Figure 1

Parsley, sown, field trial after 140 days

I = control without Lewatit resin

IV = with Lewatit OC 1029/1030 (1:1), 100 equivalent %, calculated on the content fo Cu, Pb and Zn in the soil

Figure 2

Carrots, sown, field trial after 140 days

I = control without Lewatit resin

IV = with Lewatit OC 1029/1030 (1:1), 100 equivalent %, calculated on the content of Cu, Pb and Zn in the soil

Figure 3

Salvia, planted, field trial after 116 days

I = control without Lewatit resin

IV = with Lewatit OC 1029/1030 (1:1), 100 equivalent %, calculated on the content of Cu, Pb and Zn in the soil

Figure 4

Tagetes planted with compressed soil pots, field trial after 133 days

I = control without Lewatit resin

IV = with Lewatit OC 1029/1030 (1:1), 100 equivalent %, calculated on the content of Cu, Pb and Zn in the soil

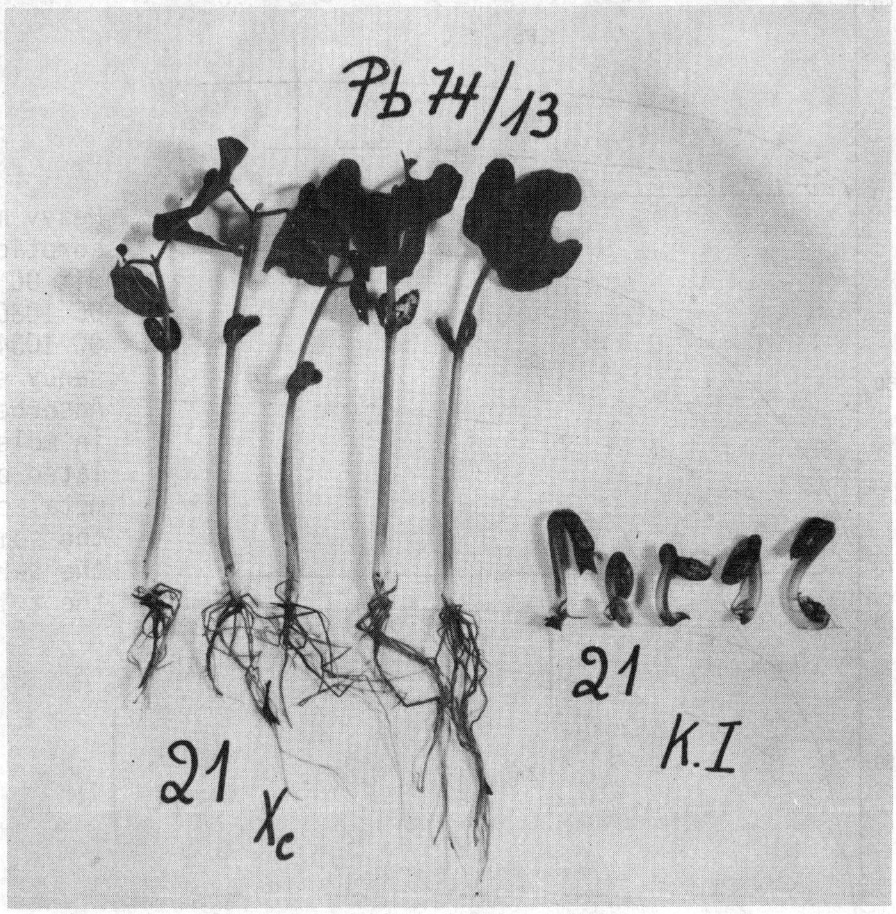

Figure 5

Beans, seedling trials in the greenhouse after 22 days

KI = control I without Lewatit resin

X_c = with Lewatit OC 1029, 140 equivalent %, calculated on the content of Cu, Pb and Zn in the soil

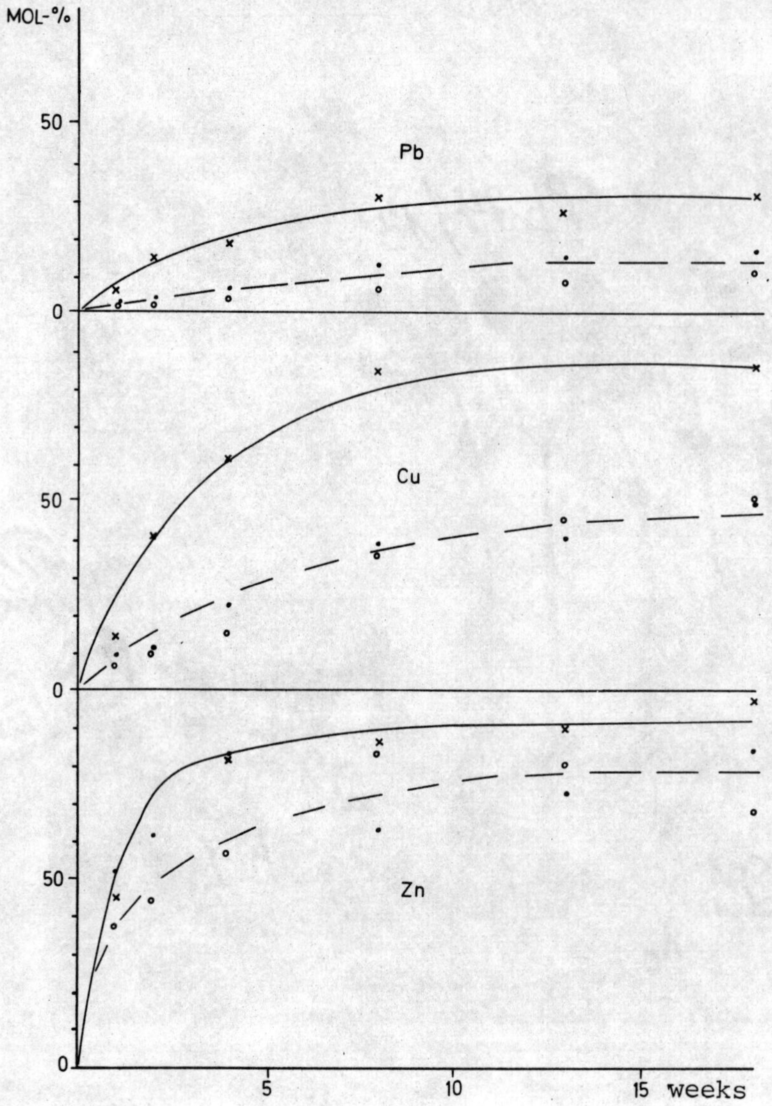

Heavy metal adsorption by Lewatit OC 1029 (˙), OC 1030 (o) and OC 1034 (x) in a sandy soil. Adsorbed quantity in mole %, calculated on the heavy metal content of the soil, versus the duration of the trial

Figure 6

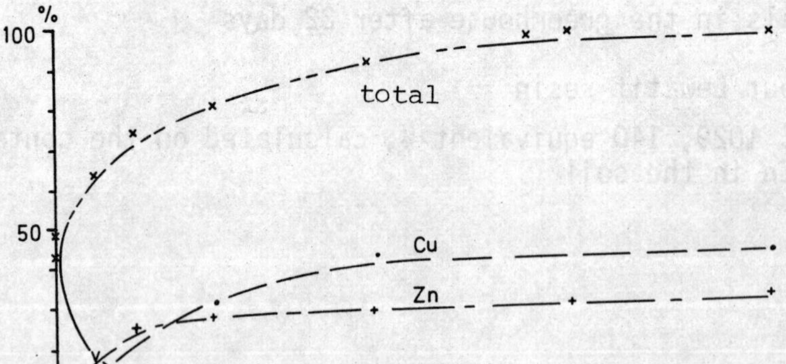

Heavy metal adsorption by Lewatit OC 1034 in a sandy soil. Amount of heavy metals adsorbed given as a percentage of the total capacity of the cation exchange resin, versus the duration of the trial

Figure 7

BACKGROUND LEVELS OF HEAVY METALS IN SOILS AND PLANTS

Iain Thornton, Applied Geochemistry Research Group,

Department of Geology, Imperial College, London.

The main sources of metal contaminants in soils and plants are from metalliferous mining and smelting, other industrial emissions and effluents, urban development, vehicle emissions, dumped waste materials, contaminanted dusts and rainfall, sewage sludge, pig slurry, composted town refuse, fertilizers, soil ameliorants and pesticides. These anthropogenic sources of metals may contribute towards problems affecting crop and animal health and production, the quality of food crops and waters for human consumption, and possibly the health and well-being of the community. Severe contamination can question the use of land for farming and amenity purposes.

Against this backcloth and in the general political and social scene of environmental pollution, where emotion sometimes has precedence over fact, it is essential to have accurate information on the naturally occurring amounts of metals present in order to make a reliable assessment of the significance of contributions from man and industry. The normal abundance of an element in earth material is commonly referred to by the geochemist as background, and for any particular element this value is likely to vary considerably according to the nature of the earth material in which it occurs (10). The establishment of background values for metals in soils and plants has for some time been recognised as part of the geochemists role in pollution studies (4).

Natural levels of metals in soils and their parent materials

The natural ranges in concentrations of most metals in soils are wide. The main sources are the parent materials from which the soils are derived. These are usually weathered bedrock or overburden transported by wind, water or glaciation. 95 per cent of the earth's crust is made up of igneous rocks and 5 per cent sedimentary rocks; of the latter about 80 per cent are shales, 15 per cent sandstones and 5 per cent limestones (21). Sedimentary rocks tend to overlie the igneous rocks from which they were derived and are thus more frequent at the surface. The abundance of heavy metals in igneous and sedimentary rocks is shown in table 1. The weathering of the minerals in igneous rocks and thus the availability of the constituent trace elements and heavy metals varies appreciably and has been discussed by Mitchell (22). Of the sedimentary rocks, sandstones are composed of minerals that weather with difficulty and in general contain only small amounts of metals. Shales may be of organic or inorganic origin and usually contain larger amounts of metals. It is seen that black shales in particular are enriched in Cu, Pb, Zn, Mo and Hg (table 1).

Soils tend to reflect the chemical composition of their parent materials, though the actual total contents and forms of heavy metals in soils are modified to varying degrees by pedogenetic processes, which may lead to the mobilisation and subsequent redistribution of metals both within the soil profile and between neighbouring soils (21, 26). Soils developed from the weathering of coarse grained sediments such as sands and sandstones and from acid igneous rocks such as rhyolites and granites usually contain smaller amounts of heavy metals than those derived from fine grained materials such as clays and shales and from basic igneous rocks.

For example, the average copper content of British surface soils has been recorded as 20 ppm (26). However the range in apparently uncontaminated soils is wide and largely reflects the nature of the parent material. Data for a large number of surface soils from England and Wales show total copper levels ranging from as little as 2 ppm in soils derived from Pleistocene sands to 60 ppm or more in those from some marls, shales and clays (34). The total copper content of soils developed from individual beds of the old Red Sandstone formation in South Wales is generally lowest in the coarser grained materials with concentrations increasing with decreasing grain size (37).

Anomalous amounts of metals which are potentially toxic to plants and/or animals may be derived from metal-rich source rocks and from mineralised materials (table 2). For example, calcareous soils derived from inter-bedded shales and limestones of the Lower Lias formation in Somerset, containing 20 ppm Mo or more, have been associated with molybdenosis in grazing cattle (8, 16, 17). Soils derived from ultra-basic rocks containing nickel-rich ferromagnesium minerals provide a further example of a natural source of metal excess. In parts of Scotland such soils under poor drainage conditions may give rise to nickel toxicity in cereal and other crops (22).

Soils in mineralised areas in Britain frequently contain large amounts of one or more of the elements Cu, Pb, Zn, Cd and As. These soils may be developed over or in the dispersion halo around ore-bodies and mineral veins or from transported overburden containing mineralised materials. The majority of mineralised areas in Britain have been worked and surface soils are usually contaminated to varying degrees by mining, processing and smelting operations. It is often difficult to distinguish whether anomalous concentrations of metals in the soils of these areas are due to natural weathering of underlying mineralised materials or to inputs from mining activities.

The distribution, forms and mobility of heavy metals in the soil are affected to varying degrees in the course of soil formation and ageing by the processes of gleying, leaching, organic matter accumulation and podzolisation. Metals are often leached from the surface horizons of podzols and enriched in the B horizon. Distribution can also be affected by soil reaction (pH) and redox potential (Eh).

The soil-plant-animal relationship

The availability of metals and their uptake by plants is related both to their total concentrations and forms in the soil and to a number of physical and chemical factors operating at the soil-root interface. For example, the solubility and availability of manganese and molybdenum is greater in poorly drained than in freely drained soils. Soil pH can have a marked effect on availability with manganese, cobalt, cadmium and zinc uptake decreasing and that of molybdenum usually increasing with rising soil pH. Intake into plant roots and translocation from root to shoot is also related to the presence of other major and minor elements in the soil. The role of soil micro-organisms on heavy metal distribution and availability is as yet little known and must surely warrent investigation in the immediate future.

Different plant species, and indeed specific cultivars, regulate metal uptake at both the soil-root and root-shoot interfaces to varying degrees. Clover tends to contain larger amounts of trace elements and particularly molybdenum than grasses. Where soils contain excess molybdenum, the clover content of the sward should be kept low to keep intake by grazing livestock to a minimum. Accumulator plants may contain very large amounts of specific elements; Astragalus spp. found in some arid areas of America can contain several thousand ppm Se compared with local grasses growing on the same soil with only a few ppm. Se accumulator plants have not yet been found in Britain, though plants accumulating large amounts of As have been reported on mine-waste in south-west England (24). Metal tolerant ecotypes of common grass species that have adapted to metal-rich soils and waste materials take up different amounts of lead and zinc to cultivars on uncontaminated land. These tolerant plants will also be found on undisturbed mineralised soils.

In general the trace element content of the whole plant tends to increase with increasing maturity, and seasonal influences are also important (9). Lead was found to be present in higher concentrations in grass in winter months than in summer on a soil of normal lead content in Scotland (23). Similar effects have more recently been found in mineralised areas of Derbyshire, where seasonal differences in the lead content of pasture were far greater than those attributable to the lead contents of the soils (fig. 1) (33).

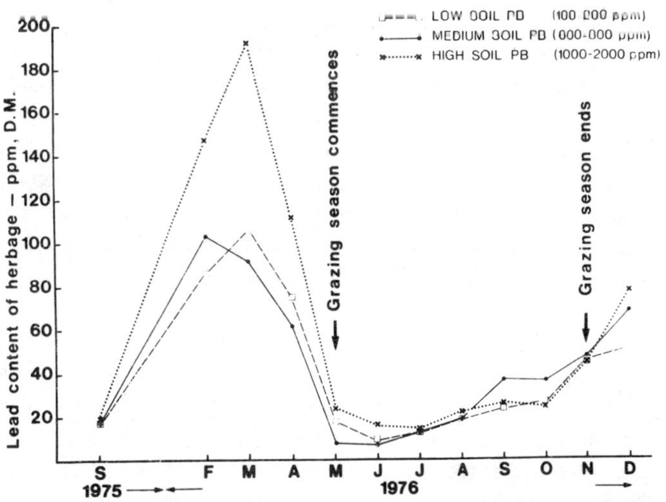

Fig 1: Seasonal variations in the Pb contents of pasture herbage growing in contaminated and uncontaminated soils in Derbyshire

As illustrated in a recent publication (28), individual metals are taken up and translocated in different amounts, usually depending on whether uptake is active or passive. Zinc is an essential element readily taken up and translocated to the shoots of grasses. Amounts of zinc in the shoot reflect the soil content. Lead on the other hand is not required for plant growth, and there are barriers to uptake at both the soil-root and root-shoot interfaces (14). Lead in grass does not usually reflect the soil content. Arsenic in pasture herbage increases by only a small amount over a wide range of soil concentrations, and the amount of copper in the aerial parts of pasture herbage does not significantly rise with increasing soil content. It has been shown that copper accumulated in the roots of ryegrass is not available for transport to shoots, being mostly associated with the cell wall fraction in the root (13).

The relationship between the soil:plant system and the grazing animal is also complex and depends on such factors as the composition of the diet, selectivity of grazing, and the form and availability of ingested metals. Ingested soil ranges from 1 to 10 per cent of the dry matter intake of cattle and may be as high as 40 per cent in sheep under winter conditions. Ingested soil is thus considered to be an important pathway of metals, which are not readily taken up by grass, into the animal (32).

Regional metal maps and geochemical surveys

For agricultural use, maps showing the regional distribution of heavy metals would ideally be based on soil and/or vegetation samples. Such maps have been compiled in New Zealand (7), the whole and parts of the United States (15,25) and Ireland (3), based on the careful selection of relatively few samples and a large degree of geographical extrapolation.

In Britain this approach has so far proved impracticable because of the cost and time necessary to account for local differences due to the complex geology and soil types. In spite of several ad hoc surveys of relatively small areas, undertaken in the course of soil survey and advisory work, there are few systematic data on either total or 'available' levels of trace elements in soils. One of the most comprehensive collections of data showing mainly normal or background levels of both total and 'available' trace elements in soils is that compiled in connection with the 'Survey of Fertilizer Practice' by the Agricultural Development and Advisory Service, based on a random sample of approximately 1500 farms (2).

Geochemical reconnaissance surveys undertaken in Northern Ireland, England and Wales by the Applied Geochemistry Research Group (35, 36) and in Scotland by the Institute of Geological Sciences (12) are based on the systematic sampling of sediments taken from tributary drainage. This primary reconnaissance method is based on the premise that active stream sediment represents a composite sample of the erosion products of rock, overburden and soil upstream from the sample site. The Wolfson Geochemical Atlas of England and Wales includes maps for As, Cd, Co, Cr, Cu, Fe, Mn, Mo, Ni, Pb, Sn and Zn (36). These maps reflect the composition of soils and their parent materials, providing valuable background data on the regional distribution of a number of heavy metals and metalloids. In addition they focus attention on particular anomalous areas where patterns of above-normal metal concentrations reflect contamination on a regional scale from past and present mining and other industrial activity.

Examples of background and anomalous levels

(a) Lead The geochemical reconnaissance map for lead (fig. 2) shows that around 97 per cent of England and Wales is represented by class limits ranging up to 160 ppm Pb in stream sediments (36). This is a good reflection of natural levels in uncontaminated soils which have been reported as 2 to 200 ppm Pb (11) and 10 to 150 ppm (30). However in the metallogenic province of the Derbyshire lead-zinc mining area, clearly shown in figure 2, the natural background levels of lead in surface soils (0-15 cm) are higher, ranging from 75 to 350 ppm Pb, with soils over limestone 125 to 215 ppm Pb, Millstone Grit 75 to 180 ppm Pb, and Coal Measures 115 to 300 ppm Pb (6). In this area soils contaminated by historical mining and smelting contain from several hundred to several thousand ppm Pb, usually exceeding 1000 ppm Pb within 500 m of old surface workings, spoil heaps and smelter sites. Using a 'threshold' value of 350 ppm, it has been estimated that contaminated soils extend over some 250 km^2 of agricultural land (6). The above normal background levels of lead in soil in this area may either reflect minor mineralisation in the country rock or widespread low level contamination arising from air-borne dispersion of metal-rich dusts and fumes from the transport and smelting of ores over several hundred years.

This particular situation serves as a good example of the difference that can at times exist between a 'local' compared with a 'national' background range for a particular metal. Thus the whole of the Derbyshire area could be described as anomalous in comparison to the majority of England and Wales; on the other hand that area specifically contaminated by mining and smelting must be seen in relation to the local background.

The normal background level of lead in grass has been reported as 2 to 3 ppm (d.m.) (4, 20), though ranges of 5 to 40 ppm Pb have been determined in herbage over a wide range of soil Pb in the mineralised areas of Derbyshire and Halkyn Mountain, with little or no relationship between the amounts in soil and the aerial parts of herbage (fig 1). The lead content of food plants, including vegetables, is usually less than 10 ppm (d.m.).

(b) Cadmium Over 80 per cent of England and Wales is represented by the lowest class of less than 1 ppm Cd in stream sediments (fig. 3). This corresponds to normal concentrations in rocks and soils. The map also highlights anomalies related to three main sources of the metal: (a) areas associated with sulphide mineralisation contaminated by base-metal mining activity, including parts of South-west and Central England and Wales; (b) areas contaminated by present-day and recent industrial processes including those near Bristol and the Swansea Valley affected by recent smelting of lead and zinc; (c) areas underlain by cadmium-rich marine shales of Carboniferous age (contain-in up to 100 ppm Cd) in which stream sediments and soils derived from the rock are enriched in the metal. Some soils in each of these types of anomaly have been shown to contain up to 20 ppm Cd or more.

An interesting example of soils markedly enriched in cadmium (together with zinc and lead) is found in the vicinity of Shipham, Somerset, where agricultural soils reclaimed from and near old zinc mines contain from 30 to 800 ppm Cd (18). Amounts of zinc are also very large frequently exceeding 1 per cent. Background levels of Cd in soils developed from the host rock, dolomitic conglomerate of Carboniferous age, usually exceed 10 ppm Cd, possibly due to finely disseminated mineralisation. Cadmium is also enhanced in soils on surrounding higher land underlain by Carboniferous limestone to the north-west and Old Red Sandstone to the south-east of the mines, possibly as a result of air-borne contamination arising from calcining and transporting the ore (31). Cadmium rarely exceeds 3 ppm (d.m.) in the shoots of pasture plants even on the most heavily contaminated soils at Shipham, where it is suggested that cadmium uptake may be restricted by the large amounts of zinc in the soil (19). Background levels in pasture herbage usually range from 0.1 to 1.0 ppm Cd (d.m.), and a similar range is found in most food plants (5).

(c) Molybdenum The Wolfson Geochemical Atlas shows about 90 per cent of England and Wales to be represented by classes with less than 2 ppm Mo in the stream sediment (not illustrated). This corresponds to normal or background concentrations in rocks and soils. The majority of anomalous areas are underlain by marine black shales ranging from Cambrian to Recent age. Soils derived from these rocks contain from 1 to 100 ppm Mo, compared to most soils developed on other parent materials with less than 2 ppm Mo (27). It is now considered that soils with 5 ppm Mo or more may support pasture herbage containing sufficient Mo to reduce copper absorption and utilization in grazing livestock. Molybdenum-induced copper deficiency in cattle is widespread and may result in loss in production and growth retardation. This natural enhancement in Mo must be taken into account in assessing possible contamination from industry and from the disposal on land of waste materials such as fly ash and sewage sludge.

Normal pasture plants contain 2 ppm Mo (d.m.) or less; herbage with 3 ppm or more may be harmful to cattle. Values of 3 to 10 ppm Mo (d.m.) are frequently found in those areas underlain by black shales (29).

CONCLUSIONS

Information on background levels of heavy metals in soils and plants is important in order to assess the degree and extent of contamination from anthropogenic sources. Natural metal values in uncontaminated material are a pre-requisite to the compilation of baseline data in the context of a 'before and after' situation and to the determination of environmental impact. Background data for metal in soils can vary widely as a result of the different chemical composition of parent materials. Amounts of metals in plants show less variation, but reflect species, season, and physical and chemical conditions at the soil:root interface.

A knowledge of background or natural levels of metals in soils is useful in deciding the types of areas in which metal-rich wastes, including fly-ash and sewage sludge, can safely be applied to land. In addition background data can influence farming practice, where the availability of essential or potentially toxic elements can be increased or decreased by the use of fertilizers and liming materials.

In the consideration of background and anomalous values it is important to realise that metals in soils and plants do not occur in isolation, and that multi-element associations and possible interactions are frequently present but not always recognised.

It is essential that future research be aimed at providing reliable information on the forms of heavy metals in the soil and in components of the food chain, and on their availability at the soil:plant, plant:animal and soil:animal interfaces. In this respect little is known of the possible role of micro-organisms in relation to metal 'availability' both in the soil and in the gastro-intestinal tract.

ACKNOWLEDGEMENTS

The examples cited comprise part of the continuing programme of research in environmental geochemistry by AGRG, funded by the Agricultural Research Council and RTZ Services Ltd.

REFERENCES

1 Antonovics, J., Bradshaw, A.D. and Turner, R.G. 1971 Heavy metal tolerance in plants. Advances in Ecological Research 7, 1-85.

2 Archer, F.C. 1979 Trace elements in soils in England and Wales. In Proceedings of ADAS Conference on Inorganic Pollution and Agriculture, London, 1977 (in the press).

3 Brogan, J.C., Fleming, G.A. and Byrne, J.E. 1973 Molybdenum and copper in Irish pasture soils. Ir. J. agric. Res. 12, 71-81.

4 Cannon, H.L. and Anderson, B.M. 1971 The geochemists involvement with pollution problems. In Environmental Geochemistry in Health and Disease, ed. H.L. Cannon and H.C. Hopps, 155-177. Geol. Soc. Am. Inc. Memoir 123, Boulder, Colorado.

5 Commission of the European Communities 1978 Criteria (Dose/ Effect Relationships) for Cadmium. Pergamen Press, Oxford, 202p.

6 Colbourn, P. and Thornton, I. 1978 Lead pollution in agricultural soils in the southern Peak District. J. Soil Sci., 29, 513-526.

7 Department of Scientific and Industrial Research 1967 Soil Bureau Atlas. DSIR, Wellington, New Zealand.

8 Ferguson, W.S., Lewis, A.H. and Watson, S.J. 1943 The teart pastures of Somerset. I. The cause of teartness. J. agric. Sci., Camb. 33, 44-51.

9 Fleming, G.A. 1965 Trace elements in plants with particular reference to pasture species. Outlook on Agriculture 4, 270-285.

10 Hawkes, H.E. and Webb, J.S. 1962 Geochemistry in Mineral Exploration. Harper and Row, New York and Evanston.

11 HMSO 1974 Lead in the Environment and its Significance to Man. Pollution Paper No. 2, London:HMSO, 47p.

12 Institute of Geological Sciences 1978 a and b, 1979 Geochemical atlases of Shetland, Orkney and Caithness and South Orkney.

13 Jarvis, S.C. 1978 Copper uptake and accumulation by perennial ryegrass grown in soil and solution culture. J. Sci. Fd. Agric. 29, 12-18.

14 Jones, L.H.P. and Clement, C.R. 1972 Lead uptake by plants and its significance for animals. In Lead in the Environment, 29-33 Institute of Petroleum, London.

15 Kubota, J. and Allaway, W.H. 1972 Geographic distribution of trace element problems. In Micronutrients in Agriculture, ed. J.J. Mortredt., 525, Soil Science Society of America, Madison, Wisconsin

16 Le Riche, H.H. 1959 Molybdenum in the Lower Lias of England in relation to the incidence of teart. J. Soil Sci. 10, 133.

17 Lewis, A.H. 1943 The teart pastures of Somerset. II. The relation between soil and teartness. J. agric. Sci., Camb. 33, 52-57.

18 Marples, A.E. 1979 The Occurrence and Behaviour of Cadmium in Soils and its Uptake by Pasture Grasses in Industrially Contaminated and Naturally Metal-Rich Environments. Unpub. PhD. Thesis, Univ. of London.

19 Marples, A.E. and Thornton, I. In press Cadmium in soils and pasture grasses in parts of England.

20 Mengel, K. and Kirkby, E.A. 1978 Principles of Plant Nutrition. International Potash Institute, Worblaufen-Bern.

21 Mitchell, R.L. 1964 Trace elements in soils. In Chemistry of the Soil, 2nd edn. (ed. F.E. Bear), pp. 320-368, New York:Reinhold Pub. Co.

22 Mitchell, R.L. 1974 Trace element problems in Scottish Soils. Neth. J. agric. Sci. 22, 295-304.

23 Mitchell, R.L. and Reith, J.W.S. 1966 The lead content of pasture herbage. J. Sci. Fd. Agric. 17, 437-440.

24 Porter, E.K. and Peterson, P.J. 1975 Arsenic accumulation by plants on mine waste (United Kingdom). The Science of the Total Environment, 4, 365-371.

25 Shacklette, H.T., Hamilton, J.C., Boerngen, J.G. and Bowles, J.M. 1971 Elemental Composition of Surficial Materials in the Conterminous United States. Geological Survey Professional Paper 574-D.

26 Swaine, D.J. and Mitchell, R.L. 1960 Trace element distribution in soil profiles. J. Soil Sci. 11, 347-368.

27 Thomson, I., Thornton, I. and Webb, J.S. 1972 Molybdenum in black shales and the incidence of bovine hypocuprosis. J. Sci. Fd. Agric. 23, 871-891.

28 Thoresby, P. and Thornton, I. 1979 Heavy metals and arsenic in soil, pasture herbage and barley in some mineralised areas in Britain: significance to animal and human health. In Trace Substances in Environmental Health - XIII. Columbia:University of Missouri (in the press).

29 Thornton, I. 1977 Biogeochemical studies on molybdenum in the United Kingdom. In Molybdenum in the Environment, Vol. 2, 341-369, ed. W.R. Chappell and K.K. Petersen. Marcel Dekker, New York and Basel.

30 Thornton, I. 1979 Geochemical aspects of heavy metal pollution and agriculture in England and Wales. Proceedings of ADAS Conference on Inorganic Pollution and Agriculture, London:HMSO (in the press).

31 Thornton, I., Abrahams, P. and Matthews, H. 1979 Some examples of the environmental significance of heavy metal anomalies disclosed by the Wolfson Geochemical Atlas of England and Wales. In Proceedings of International Conference on Heavy Metals in the Environment. London 1979 (in the press).

32 Thornton, I. and Alloway, B.J. 1974 Geochemical aspects of the soil-plant-animal relationship in the development of trace element deficiency and excess. Proc. Nutr. Soc. 33, 257-266.

33 Thornton, I. and Kinniburgh, D.G. 1978 Intake of lead, copper and zinc by cattle from soil and pasture. In Trace Element Metabolism in Man and Animals - 3, ed. M. Kirchgessner, Institut fur Ernahrungsphysiologie, Technische Universitate Munchen Freisung - Weihenstephan, 499.

34 Thornton, I. and Webb, J.S. 1975 Distribution and origin of copper deficient and molybdeniferous soils in the United Kingdom. Proc. Copper in Farming Symposium. London:Copper Development Ass.

35 Webb, J.S., Lowenstein, P.L., Howarth, R.J., Nichol, I. and Foster, R. 1973 Provisional geochemical atlas of Northern Ireland. Appl. Geochem. Res. Grp. Tech. Commun. no. 60.

36 Webb, J.S., Thornton, I., Howarth, R.J., Thompson, M. and Lowenstein, P.L. 1978 The Wolfson geochemical atlas of England and Wales. Oxford University Press.

37 Wood, P. 1975 Regional geochemical studies in relation to agriculture in areas underlain by sandstones. Unpub. PhD Thesis, Univ. of London.

	Basaltic Igneous	Granitic Igneous	Shales and Clays	Black Shales	Limestones	Sandstones
As	0.2-10 2.0	0.2-13.8 2.0	- 10	- -	0.1-8.1 1.7	0.6-9.7 2.0
Cd	0.006-0.6 0.2	0.003-0.18 0.15	0-11 1.4	<0.3-8.4 1.0	- 0.05	- 0.05
Cr	40-600 220	2-90 20	30-590 120	26-1000 100	- 10	- 35
Co	24-90 50	1-15 5	5-25 20	7-100 10	0.1	0.3
Cu	30-160 90	4-30 15	18-120 50	20-200 70	- 4	- 2
Hg	0.002-0.5 0.05	0.005-0.4 0.06	0.005-0.51 0.09	0.03-2.8 2.5	0.01-0.22 0.04	0.001-0.3 0.05
Pb	2-18 6	6-30 18	16-50 20	7-150 30	- 9	<1-31 12
Mo	0.9-7 1.5	1-6 1.4	- 2.5	1-300 10	- 0.4	- 0.2
Ni	45-410 140	2-20 8	20-250 68	10-500 50	- 20	- 2
Se	- 0.05	- 0.05	- 0.6	- -	- 0.08	- 0.05
Zn	48-240 110	5-140 40	18-180 90	34-1500 100	- 20	2-41 16

Adapted from table compiled by M. Fleischer and H.L. Cannon (Cannon et al., 1978 In Geochemistry and the Environment, Vol. 3, 20. National Academy of Sciences, Washington D.C.).

Table 1: Range and mean concentrations of some metals and metalloids in igneous and sedimentary rocks (ppm)

Element	Normal range in soil (ppm)	Metal-rich soils (ppm)	Sources	Possible effects
As	<5-40	up to 2500	Mineralisation	Toxicity in plants and livestock; excess in food crops
		up to 250	Metamorphosed rocks around Dartmoor	
Cd	<1-2	up to 30	Mineralisation	Excess in food crops
		up to 20	Carboniferous black shale	
Cu	2-60	up to 2000	Mineralisation	Toxicity in cereal crops
Mo	<1-5	10-100	Marine black shales of varying age	Molybdenosis or molybdenum induced hypocuprosis in cattle
Ni	2-100	up to 8000	Ultra-basic rocks in Scotland	Toxicity in cereal and other crops
Pb	10-150	1% or more	Mineralisation	Toxicity in livestock; excess in foodstuffs
Se	<1-2	up to 7	Marine black shales in England and Wales	No effect
		up to 500	Namurian shales in Ireland	Chronic selenosis in horses and cattle
Zn	25-200	1% or more	Mineralisation	Toxicity in cereal crops

Table 2: Trace elements in soils derived from normal and geochemically anomalous parent materials

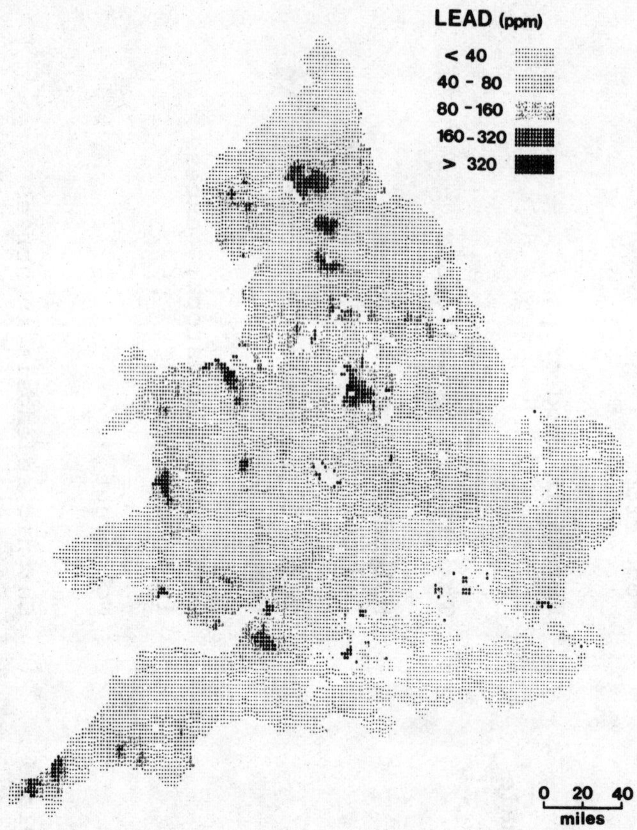

Fig. 2: Map showing the distribution of lead in stream sediment in England and Wales

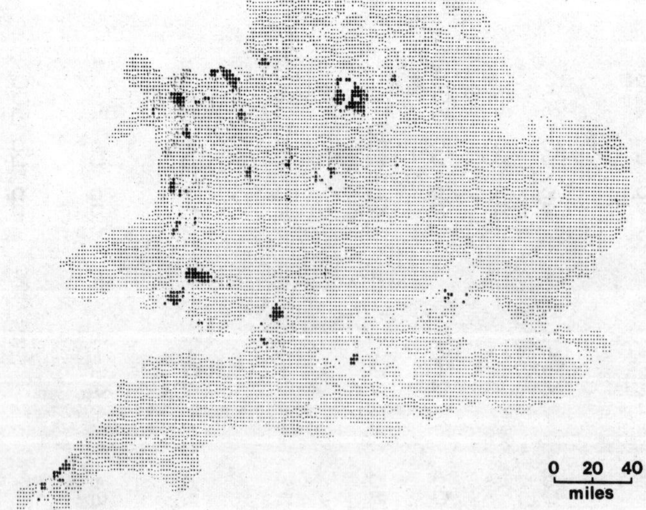

Fig. 3: Map showing the distribution of cadmium in stream sediment in England and Wales

METAL CONCENTRATIONS IN SOIL AND CROP PHYTOTOXICITY

J H WILLIAMS

Agricultural Development and Advisory Service, Wolverhampton

SOURCES OF METAL CONTAMINATION

Sources of metal contamination include heaps of spoil around old mine workings such as occur in Cornwall, West Devon, Derbyshire, Cumberland, Westmorland and parts of wales. Such sources include copper and nickel in Devon and Cornwall, lead and zinc in Wales, Derbyshire, Somerset and the north of England. More widespread damage can occur during periods of very heavy rain when small streams carry debris from abandoned workings onto agricultural land downstream. Dumps of waste in close proximity to smelting works and foundries often renders the ground sterile but this is not normally used for agriculture. There is today a much greater urgency to reclaim such tips either for agricultural production or amenity purposes. The lower Swansea valley is an example where reclamation of such tips has been carried out (17).

Repeated heavy dressings of sewage sludge over a long period of time and the use of sludges from industrial areas containing large quantities of toxic metals have resulted in a build up of one or more metals in the soil.

On a market garden holding in Somerset, the soil, which had received an annual dressing of 50 t/ha of sewage sludge for 30 years, was found to contain 5000 mg/kg total Zn of which 1000 mg/kg was soluble in 0.5 M acetic acid as well as 500 mg/kg Cu, 1500 mg/kg Pb, 150 mg/kg Cr and 50 mg/kg Ni. Vegetable crops were severely damaged and the productivity of the holding seriously reduced.

Emissions from chimneys in vapour form from smelting works and foundries can cause contamination of soils and crops as a result of aerial deposition over many years in the past. Soil profile samples from such sites revealed high levels of extractable zinc and copper especially in the upper horizons of undisturbed land. Land in-filled with industrial waste and covered with only a few inches of topsoil has been known to result in failure of grass to establish.

METAL CONTENT OF SOILS

When considering the degree of contamination of soils by metals it is necessary to know the natural background levels present in normal agricultural soils against which to judge the level of contamination. Ranges and the common levels found in soils are shown in Table 1.

Excessive amounts of some of the chemical elements may be toxic to plant growth and could severely restrict crop yields whilst others may affect the quality of foodcrops for human or animal consumption. The elements which are potentially toxic to plant growth are zinc, nickel, copper, boron and possibly cadmium whilst cadmium, lead and mercury have given rise to some concern as environmental pollutants since they do present a health risk to both man and animals. Other potentially toxic elements but which possibly present a lesser health risk include arsenic, chromium, fluorine, molybdenum and selenium. Some of the elements listed are known to be essential for plant growth and/or stock health in trace amounts, eg. copper, boron molybdenum, zinc and selenium but the optimum range is often narrow and a slight imbalance could aversely affect stock health or crop production without visually manifesting itself.

The geological parent material is the criterion which determines soil metal content. The natural range of many metals in soils is very wide. Fine grained sediments and basic igneous materials usually contain higher concentrations than sandstones or acid igneous parent materials. Naturally occurring levels of some elements may be deleterious to crops and stock eg. the nickel rich ferromagnesian minerals in soils from the basic igneous rocks in parts of Scotland have, if poorly drained, given rise to nickel toxicity in agricultural crops (13). Soils associated with Namurian black shale formations may give rise to excess molybdenum in the diet of grazing cattle resulting in hypocupraemia or low blood copper in the animal (9, 15). Soils in Britain which contain abnormally high concentrations of the heavy metals copper, lead, zinc and cadmium probably result from mining or smelting operations or from heavy applications of sewage sludge from industrial areas. The purpose of the present paper is to present some published data including the results of investigations conducted by ADAS Soil Scientists, and discuss how the harmful effects of metal pollutant concentrations on crop and animal production might be overcome.

AVAILABILITY IN SOIL AND TOXICITY TO PLANTS

Zinc

Zinc has received more attention and blame for causing crop phytotoxicity than any of the other metals. Toxicities of zinc have been ascribed to galvanised wire fencing and troughing, to that associated with deposits of spoil surrounding, or downstream from, old lead mines, and to heavy dressings of sewage sludge applied in the past. Numerous instances of damage to crops from sewage sludge applications in which the zinc extractable by 0.5 M acetic acid varied between 70 and 1500 mg/kg have been recorded. Some examples of the effects of soil metal levels on crop growth are shown in Table 2. Growth of all crops was affected if the extractable zinc content was greater than 100 ppm.

In a series of microplots at the Wolverhampton ADAS Regional Centre, zinc as the sulphate was added to soil and a range of crops grown. The additions were made over the period 1948-50 and the plots cropped without further addition up to 1961. The rates went up to 800 mg/kg of added Zn and the figures for extractable zinc obtained in 1961 using 0.50 M acetic acid are shown in Table 3. Toxic effects were noted in crops grown on those plots receiving the two higher levels of Zn addition equivalent to 150 and 300 mg/kg of acetic acid soluble zinc. There is also ample evidence that liming, to increase the pH over 7.0, reduces the harmful effects of zinc on crops and also some indication that organic matter alleviates the effect of this metal. The analysis of crops grown on Zn - contaminated soils shows that the element is translocated into the aerial parts of the plant.

Copper

Copper is an essential element for both animals and plants and excess can give rise to toxicity very readily in lambs and in clover which is one of the most sensitive crops. Like zinc, the toxicity of copper is also influenced by soil pH and organic matter content. The element would appear to be adsorbed strongly onto organic matter since large applications of copper contaminated sludges have not given any significant increases in copper uptake by cereals and vegetable crops. Richardson (14) found only slightly higher concentrations of copper in lettuce plants grown on soil which had a history of sludge applications than those grown on similar soil which had not received sludge. Copper is less readily translocated within the plant than zinc but it can accumulate in roots. In cereals for example it can produce multiple, branched, swollen roots similar to the symptoms produced by cereal cyst eelworm. In a series of plots at Wolverhampton, similar to those described above for zinc, copper was applied at concentrations up to 800 mg/kg of soil over the period 1948-50. Figures for copper extractable with 0.5 M acetic acid for the 3 highest levels of application are shown in Table 4. The soil possessed a pH of 6.5 and toxicity symptoms were observed at the two highest levels of added copper (80 and 250 mg/kg of extractable copper). Other studies at this centre were concerned with the effects of copper at rates of 100, 200 and 500 mg/kg on soils adjusted to pH 5.7 and 6.4 on a range of crops. Some results for mustard and oats are shown in Table 5. The toxic effect of copper was reduced by raising the pH but at 500 mg/kg of total Cu some damage occurred at both pH levels.

Nickel

Compared with zinc and copper, nickel is injurious to crop growth at much lower concentrations causing induced iron deficiency (7). Normal soils contain between 4 and 200 mg/kg Ni but levels may be much greater (up to 2000 mg/kg) in soils derived from serpentine. The proportion soluble in 0.5 M acetic acid or 0.05 M EDTA is very much less and is a reasonable guide to the possibility of damage to plants. Plant species vary in their susceptibility to nickel toxicity; barley, wheat, ryegrass and beans are the least sensitive whilst oats, turnips, swedes, cabbage, kale and beet are much more sensitive, beet being the most sensitive of all.

Pot experimental work at ADAS Wolverhampton and Wye demonstrated the modifying effect of soil pH on the toxicity of nickel to oats and mustard. The effects of increasing nickel concentration in soil and soil pH on the yields of these two crops are shown in Table 6. The effects of nickel were very toxic at pH 5.7 even at only 50 mg/kg added. At pH 6.4 the yield of oats were still depressed at 50 mg/kg of nickel in soil but mustard remained unaffected up to 100 mg/kg Ni.

Cadmium

Soils generally contain between 0.1 and 2.0 mg/kg cadmium with a common level of 0.1 mg/kg. A recent survey in this country of the cadmium content of 700 soil samples taken in connection with the Survey of Fertiliser Practice showed that most soils contained less than 1.0 mg/kg which compares with the levels reported from Canada. Sewage sludges from non-industrial areas contain less than 15 mg/kg Cd in the dry solids. Some sludges from industrial areas may contain in excess of 100 mg/kg Cd. Where sludge has been applied for a long period of time or at heavy rates of application, soil cadmium contents significantly greater than 1.0

mg/kg are found, eg. 26 mg/kg Cd in soil at one sewage farm site and over 100 ppm in low lying patches or small areas where sewage sludge irrigation has been practised. On allotments and market garden land where sludge has been used over many years as a soil conditioner, levels up to 76 kg/mg Cd have been reported in the topsoil. High levels of cadmium have also been found in soils adjacent to old zinc mine spoil such as the instance which has been highlighted recently in Somerset. Whilst several hundred parts per million of total Cd have been found in the soil, crop growth is not affected. The mineral form in which the cadmium is present in mine spoil is obviously much more insoluble in soil solution than the form in which it is present in sewage sludge.

The application of inorganic salts of cadmium to soil can significantly increase the concentration of the element in crops. Radish tops have been found to accumulate cadmium to a greater extent than the roots. Bingham et al (2) found that a concentration of 20 mg/kg Cd in the edible tissue of radish was associated with a 25% reduction in yield in studies which used cadmium sulphate spiked sludges. Vaidyanathan (private communication) studied the response of perennial ryegrass and lettuce grown under glasshouse conditions in sandy loam and clay loam soils with high and low organic matter contents and treated with cadmium (0, 25, 50 mg/kg) chloride and zinc (0, 50, 100 mg/kg) sulphate. Increasing the cadmium additions to soils resulted in nearly proportional increases in metal uptake by both plant species. Ryegrass dry matter production was unaffected by the cadmium uptake but lettuce growth was checked and retarded significantly in all cadmium treated soils. Uptake was not very much influenced by organic matter, clay or accompanying zinc additions. Tissue concentrations of cadmium in lettuce (average 44 mg/kg) were approximately 4 times greater than in ryegrass (average 10.5 mg/kg).

Soils containing up to 16 mg/kg cadmium as a result of sewage sludge treatment are reported to have produced lettuce containing 27.5 mg/kg Cd in the dry matter. Lettuce have consistently been found to accumulate cadmium to a greater extent than any other crop. Webber (16) found that 5 mg/kg Cd as the chloride added to soil resulted in a reduction in the yield of lettuce. Cereal grains do not appear to accumulate excessive amounts of cadmium when grown in sludge amended soils but high levels have been found by Bingham et al (2) when the cadmium has been added in the inorganic form.

It would appear probable from the evidence that the cadmium which is adsorbed by soil organic matter remains essentially available for plant uptake since Bunzl et al (4) found that it is readily desorbed off peat. This would explain why crops grown with spiked sludges accumulate more cadmium than those grown with naturally high cadmium sludges.

Lead

The normal lead content of soils ranges from 2-200 mg/kg, the highest levels in this country being found in soils derived from limestone. When lead is equilibrated in well-limed soils it becomes very insoluble and is not readily taken up by roots and translocated within the plant. The element has not been shown to be phytotoxic to crop growth but it is both toxic and cumulative in animals and man. The main risks from lead in soils contaminated with mine spoil or sewage sludges are from the surface contamination of crops and the ingestion of lead contaminated soil or sludge by children or animals.

The addition of soil or lime to sludge decreases the solubility of lead in acetic acid but solubility in EDTA is considerably enhanced. The author compared the solubility of lead from 33 sludge treated soils in 0.5 M acetic acid and 0.05 M EDTA. The latter extracted 9-10 times more lead than acetic acid. Bunzl (4) reported that lead was rapidly adsorbed by peat and that desorption was very slow compared with cadmium.

Mercury

Mercury is known to undergo biological tranformation to methylmercury. These alkyl compounds are more mobile than elemental mercury and more lipid soluble. They are therefore of greater toxicity since 90% of any methylmercury present in food is absorbed. Soils generally contain about 0.03 mg/kg whereas plants contain less than 0.02 mg/kg. Sewage sludges can contain up to 100 mg/kg Hg and a number of organisms have been shown to be able to methylate mercury. It is feasible that methylmercury could be present in sludge treated soils and, if so, it could be that the mercury is more available to humans and animals than was thought. However, since the alkyl metal complexes are volatile they could serve to reduce the levels present. The extent to which mercury is taken up by plants is small but investigations into its uptake by crop plants have been few and little is known of its reactions with soil components.

Vaidyanthan (private communication) studied the effects of organic amendments in the form of dried sewage sludge on the uptake of mercury applied as mercuric chloride by ryegrass. The mercury content of the rye-grass was considerably increased when mercuric chloride at 25 and 50 mg/kg Hg was added alone. The sludge amendment ($\equiv$ 125 t/ha) reduced the uptake of mercury applied very appreciably. When the mercury salt was mixed with the sewage sludge before adding to the soil there was an even greater reduction in metal uptake.

Molybdenum

Molybdenum differs from other metals in that it becomes more available to crops as the soil pH increases. Liming, which could well be adopted as a remedial measure for soils contaminated with other metals, would certainly increase the uptake of molybdenum. It is not phytotoxic to crops but consumption by animals of grass containing slightly inflated levels of molybdenum or the direct ingestion of contaminated soil could induce hypocupraemia in the stock. Some soils derived from Lower Lias clay, Namurian black shales and marine alluvium are naturally high in molybdenum. Liming such soils can result in high molybdenum levels in the grass. Concentrations in excess of 2.5 mg/kg in the dry matter can induce copper deficiency and copper supplements need to be administered. Some sewage sludges may contain high levels of molybdenum and in one instance in Lancashire, treated soil was found to contain 90 mg/kg total Mo and the grass contained 30 mg/kg. Cattle in the area suffered from copper deficiency. No studies have been conducted on the effect of molybdenum in sewage sludges on uptake by grass.

Combined Activity of Zn, Cu and Ni

Crop damage can rarely be attributed to the effects of a single element especially if sewage sludges are responsible for the contamination of land. The zinc equivalent concept was developed to give guidance on the safe use of sewage sludge on land. It only took account of zinc, copper and nickel assuming that copper is twice and nickel eight times as toxic as zinc and also assumes the toxicities of these three metals to be additive. It was proposed that 560 kg/ha of zinc equivalent would be an acceptable safe level for a soil maintained at pH 6.5 or above. It can be criticised on the grounds that, except for pH, other soil factors such as organic matter content, and CEC are not taken into account and neither is the cropping sequence. Cereals and grass for example are able to tolerate higher metal concentrations than sensitive vegetable crops such as red beet and lettuce. Leeper (8) and Chaney (5) have suggested that toxic metal additions, expressed as zinc equivalent, should not exceed 5% of the cation exchange capacity of the soil before sludge addition. More recently the results of field experiments with metal contaminated sludges were presented by Marks, Williams and Chumbley (11) at the Conference on Inorganic Pollution and Agriculture. Nickel was by far the most toxic metal for all crops. The toxicities of copper and nickel, based on metal added in the sewage sludge, relative to that of zinc for red beet on the sandy soil was 1.3 and 2.2 for copper and nickel respectively. On the basis of the extractable metal contents of the soils, the relative toxicities were 0.55 and 3.3 for copper and nickel respectively for red beet on the same soil. In general the crops were less tolerant of metal concentrations on the sandy textured soil at Luddington EHS with its lower organic matter content and base exchange capacity. For red beet, nearly two thirds of the variation in crop yield was accounted for by multiple regressions of total metal addition or extractable soil concentrations of zinc, copper and nickel on crop yield. These three metals are the most common phytotoxic contaminants of sewage sludge and since a large part of the variance is explained by these three metals it would appear that the zinc equivalent concent is of use in assessing safe metal additions to soil. It cannot possibly be precise but it does provide a simple working guide to permissible amounts of sludge that may be applied to land. As such it is more likely to be used in practice than more complicated guidelines. However, there is now a need to revise the actual relative toxicities of the three elements based on the experimental work described above. For guidance, the normal and toxic concentrations of zinc, nickel and copper in soils and plants are given in Table 7.

FACTORS AFFECTING METAL AVAILABILITY TO CROPS

In establishing guidelines for safe metal concentrations in soil and minimising the uptake of these metals by crops, account has to be taken of the various factors which influence availability to crops. The all important factor which controls the availability of metals is soil pH and the safe levels of zinc, copper and nickel proposed assume that the pH is maintained at 6.5 or above. If high concentrations of metals are already present it may be advisable to maintain a pH of 7.0 or higher. Other important factors include cation exchange capacity, organic matter content (both of which are interrelated), and plant species.

Soil reaction

The availability of most elements is reduced at higher pH levels, eg Zn, Cu, Ni, Pb, Cd, B and F. The exceptions are molybdenum and selenium, their uptake being enhanced by pH's greater than 6.5. Anderson and Nilsson (1) showed how the concentration of cadmium in fodder rape was reduced as the soil pH was raised from 4.8 to 6.0 or 7.2. The amounts of cadmium extracted with 1.0 M ammonium acetate were also reduced as shown in Table 8.

In a series of replicated micro-plots laid down at Woodthorne, Wolverhampton in 1963, Williams (18) showed how the effects of high rates of zinc and nickel applied in sewage sludge, on a number of vegetable crops were very much reduced on raising the pH to 7.0 or over. Effects on crop yields, expressed as % of control, are shown in Table 9.

Liming the soil to pH 7.0 or 7.2 resulted in appreciable reductions in metal toxicity. At least 4 times the quantities of zinc and nickel were required to produce the same reductions in yields of red beet and onions when the pH was raised by 0.8 units. At pH 7.2 yield of swedes was unaffected by as much as 2560 and 1140 kg/ha of zinc and nickel respectively whereas half this quantity of metals significantly reduced the yield at pH 6.7. At pH 6.5 and 7.0 the reductions in yield of lettuce were similar up to 1280 kg Zn and 570 kg/ha Ni; twice this quantity of metal gave little further reduction in yield at pH 7.0 whereas at pH 6.5 there was near complete crop failure. A number of workers have confirmed the effect of pH in reducing the availability and uptake of lead, cadmium, zinc and nickel. Copper uptake by ryegrass however was hardly affected on raising the pH from 5.3 to 6.3 or over by the addition of calcium hydroxide (3).

Organic matter

MacLean (10) compared the uptake of zinc and cadmium by lettuce grown in soils with organic matter levels of 2.2 and 16.0%. Cadmium concentration was lower but zinc concentration was higher in the lettuce grown in the organic soil. Haghiri (6) found that peat additions to raise the organic matter content up to 7.0% reduced the cadmium content of oats from 13.7 to 7.9 mg/kg. He concluded that this was probably due to the much higher cation exchange capacity of the peat amended soils. The CEC was raised from 17.1 to 30.5 meq/100 g as the organic matter increased from 2.0 to 7.0%. Leeper (8) suggested a decreasing order for the tendency of metals to be chelated by organic matter viz. $Cu^{2+} > Ni^{2+} > Cd^{2+} > Zn^{2+}$. However it is not simply adsorption but rather the strength of the adsorption or, the ease of desorption, which controls the concentrations in soil solution and hence the availability to crop plants.

In studies at Reading designed to investigate ameliorative treatments to reduce the toxic effects of metals applied in previous sewage sludge applications, Richardson (private communication) reported that only liming above the soil pH of 6.5 reduced the content of zinc, cadmium and nickel in lettuce; lead content however was unaffected. Peat addition was without effect on the uptake of Zn, Cu, Ni, Cd or Pb. Copper content of the lettuce was unaffected by double digging, peat, chalk or sulphur additions. Sulphur treatment not unexpectedly increased the content of lead and zinc.

Cation exchange capacity (CEC)

Uptake by plants and safe metal concentrations in soil have often been related to the cation exchange capacity. For a given level of metal addition, the uptake is usually greater from soils with a low CEC. Miller et al (12) showed that the uptake of cadmium was influenced by the CEC of the soil. They related cadmium sorption capacity to the exchange capacity of the soils amended with varying amounts of organic matter.

Phosphorus fertiliser

Addition of lead to increase the soil content up to 44 mg/kg increased the lead content of maize dry matter from 2 to 93 mg/kg after 30 days. When phosphorus was added to soil to raise the bicarbonate soluble P from 22 to 111 kg/ha, the lead content of the shoots fell to 33 mg/kg (19). The effect of applied phosphorus was much greater on the shoot rather than the root content suggesting that the phosphorus affected translocation of lead rather than uptake into the root.

SUMMARY

Metals which have accumulated in soil as a result of contamination with mine spoil or sludge disposal are not lost through leaching down the profile or through crop removal and remain there almost indefinitely. Some like zinc, copper and nickel may be taken up by crops and prove phytotoxic to their growth. Others like lead, cadmium and molybdenum do not usually affect crop growth but may be harmful to the health of stock which consume the crop. The extent to which the metals are taken up depends on factors such as soil pH and organic matter and/or cation exchange capacity and different crops vary in their capacity to take up metals from soil. Leafy vegetables have the highest contents of the mobile elements zinc and cadmium whilst in grain their contents are usually low even though the soil contains high levels.

Lead and copper on the other hand are either rendered unavailable to plant roots or tend to be retained in the roots preventing translocation to the aerial parts of the plant.

Another problem arising from reclamation of metal contaminated land to agriculture intended for grazing or conservation in the ingestion of soil by the grazing animal or soil contamination of the cut grass. The dry matter consumed by the animal may contain as much as 10 or even 15% of the contaminated soil. The intake of elements such as lead, copper, mercury or fluorine, which are not readily taken up by plants, by soil ingestion may present a serious health hazard to stock. Although a great deal of research has been done on the subject in recent years there are still many questions left unanswered when it comes to recommending safe metal additions to soil and to providing sound advice on the reclamation of contaminated land to render it safe for stock and crop production. When all the factors are considered the problem is obviously a complex one and any guidelines which are set at present can only be tentative to be updated upon as our knowledge of the behaviour of metals in soils, crops and animals improve.

TABLE 1. Normal ranges and common levels of metals found in soils and plants

Element	Concentration in soils (mg/l)				Concentration in Plants (mg/kg) (Dry Matter Basis)
	TOTAL		0.05 M EDTA Extract-able	0.5 M HAc * Extract-able	
	Range	Common level			
Zn	10-300	30	1.5-15	1.0-10	20-100
Cu	2.0-100	20	0.5-10	0.1-1.0	5.0- 15
Ni	4-200	20	0.2-5.0	0.2-5.0	1.0- 10
Pb	2-200	10	1.0-10	0.2-4.0	0.1- 10
Cd	0.1-2.0	0.1	0.02-0.2	0.02-0.2	0.2-0.50
Hg	0.01.0.3	0.03	-	-	0.01-0.03
Cr	5.0-3000	100	0.1-4.0	0.01-4.0	0.20-1.0
Mo	0.2-5.0	2	0.02-1.0	$<$ 0.02	0.5- 50
Se	0.1-2.0	0.2	0.01-0.5	-	0.05-0.15
As	0.1-40	6	-	-	0.10-5.0
Co	1.0-40	10	0.05-4.0	0.05-2.0	0.05-0.50
B	2.0-100	10	-	0.2-4.0+	10.0-40.0

*30 minute shaking period
+ Water Soluble B

TABLE 2. Extractable zinc levels in soil related to crop growth

CROP	Extractable Zn (mg/kg) in soil*	
	Poor Growth	Satisfactory Growth
Chrysanthemum	93	37
Wallflowers	154	56
Sugar Beet	96	84
Bedding Plants	150	100

* Extracted with 0.5 M acetic acid.

TABLE 3. Zn addition and extractable zinc content of soils.

Zn addition in 1948-50 (mg/kg of soil)	Extractable Zn* in 1961 (mg/kg)
280	90
360	150
800	300

*Extracted with 0.5 M acetic acid.

TABLE 4. Cu addition and extractable copper content of soils.

Cu addition in 1948-50 (mg/kg of soil)	Extractable Cu* in 1961 (mg/kg)
280	60
360	80
800	250

*Extracted with 0.5 M acetic acid.

TABLE 5. Effects of copper on crop yields (fresh weight in g/pot).

Crop	Soil pH	mg/kg of total Cu in soil		
		100	200	500
Mustard	5.7	26.0	5.6	1.2
	6.4	24.4	23.6	6.4
Oats	5.7	19.4	17.1	7.3
	6.4	19.1	19.7	16.4

TABLE 6. Effect of added nickel and soil pH on the yield of oats and mustard. (fresh weight in g/pot)

Nickel added (mg/kg Ni)	OATS		MUSTARD	
	pH 5.7	pH 6.4	pH 5.7	pH 6.4
0	20.3	19.8	24.8	25.6
50	17.1	16.8	17.2	23.2
100	5.9	14.6	0.8	25.8
250	2.5	13.8	0.4	8.0

TABLE 7. Normal and toxic levels of Zn, Cu and Ni in soils and crops.

Element	Extractant	Concentration in soils (mg/l)		Concentration in plants (mg/kg)	
		Normal	Toxic*	Normal	Toxic
Zn	0.05 M EDTA	1.5-15	130-260	20-100	> 500
Cu	"	0.5-10	50-100	5-15	> 25
Ni	"	0.2-5.0	20-35	1-10	> 50

*Some sensitive crops grown on light textured soils at slightly acid pH's will be affected at the lower end of the toxic range whilst tolerant crops grown at pH 6.5 or higher will only be affected at the upper end of the range.

TABLE 8. Effect of soil pH on the concentration of cadmium in fodder rape and on the cadmium extracted from soil by 1.0 M neutral ammonium acetate.

Cd applied (kg/ha)	Cd in plant dry matter (mg/kg)			Extractable Cd in soil (mg/kg)		
	pH 4.8	pH 6.0	pH 7.2	pH 4.8	pH 6.0	pH 7.2
0	637	121	105	13	9	6
1.75	719	604	384	103	127	119

TABLE 9. Effect of zinc and nickel on yield of vegetable crops grown at different soil pH's (% of control yield).

Crop	Soil pH	Control Yield (t/ha)	Metals applied in kg/ha		
			Zn - 640 Ni - 285	Zn - 1280 Ni - 570	Zn - 2560 Ni - 1140
Red Beet	6.2	39.6	13.9*	7.1*	3.3*
1972	7.0	34.1	100	97.4	47.2*
Onions	6.2	4.9	67.3	22.4*	6.1*
1973	7.0	6.3	103	74.6	61.9*
Swedes	6.7	35.4	87.8	29.7*	0.0*
1974	7.2	35.9	117	113	136
Lettuce	6.5	21.6	100	70.0*	3.7*
1975	7.0	30.0	87.7	76.3	71.6*

*Significantly different from control at p = 0.05 (horizontal comparisons only).

REFERENCES

1. Anderson, A and Nilsson, K O. Ambio, 1974, 3, 198.
2. Bingham, F T, Page, A L, Mahler, R J and Ganje, T J. J. environ. Qual., 1975, 4, 207.
3. Bolton, J. Environ. Pollut., 1975, 9, 295.
4. Bunzl, K, Schmidt, W and Sansoni, B. J. Soil Sci., 1976, 27, 32.
5. Chaney, R L. USDA, 1974.
6. Haghiri, F. J. Environ. Qual., 1974, 3, 180.
7. Hunter, J G and Vergano, O. Ann. App. Biol., 1952, 39, 279.
8. Leeper, G W. Contract No. DACW 73-73-C-0026, Dept of the US Army Corps of Engineers, 1972, pp.70.
9. Le Riche, H H. J. Soil Sci., 1959, 10, 133.
10. MacLean, A J. Can. J. Soil Sci., 1976, 56, 129.
11. Marks, M J, Williams, J H and Chumbley, C G. ADAS Conference on Inorganic Pollution and Agriculture 1977. Paper SS/OC/77/19.
12. Miller, J E, Hassett, J J and Koeppe, D E. J. Environ. Qual., 1976, 5, 157.
13. Mitchell, R L. Netherlands J. agric. Sci., 1974, 22, 295.
14. Richardson, S J. ADAS Conference on Inorganic Pollution and Agriculture 1977. Paper SS/OC/77/20.
15. Thompson, I, Thornton, I and Webb, J S. J. Sci. Fd. Agric., 1972, 23, 878.
16. Webber, J. ADAS Conference on Inorganic Pollution and Agriculture 1977. Paper SS/OC/17/18.
17. Weston, R L, Gadgil, P D, Salter, B R and Goodman, G T. 1965. 5th Symp. Br. Ecol. Soc.
18. Williams, J H. ADAS Conference on Inorganic Pollution and Agriculture 1977. Paper SS/OC/77/17.
19. Zimdahl, R L and Foster, J M. J. Environ. Qual., 1976, 5, 31.

A STRATEGY FOR EVALUATION OF SOIL COVERING TECHNIQUES TO REDUCE TRACE METAL UPTAKE BY SOFT FRUITS AND VEGETABLES.

N.W. LEPP and MARY R. HARRIS

Dept. of Biology, Liverpool Polytechnic, Byrom Street, Liverpool L3 3AF.

INTRODUCTION

Trace metal contamination of soils may present a major problem in the reclamation of land prior to a change of usage. Particular problems occur when the intended new usage involves low-rise housing with garden space; the major fear being that home-grown produce may contain excessive trace metal levels which become incorporated into human food chains at above-normal rates. Problems of this nature can arise in the fringe areas around major urban centres, where re-housing projects are located on the sites of former sewage farms. These sites have often been in use for several decades prior to the change in usage, and, in consequence, may be heavily contaminated with a variety of trace metals. The variety of metals encountered, and the considerable variation in their levels, pose several problems for potential reclamation schemes; optimization of treatments may prove difficult and expensive.

One approach that many believe to have some merit as a standard practice is to cover the contaminated soil with varying depths and types of overlying material. The potential advantages of such an approach are:

1. Simplicity
2. Potential immobilization of metallic contaminants
3. Combination with other amelioration methods.

Its disadvantages include:

1. Potential for upward movement of soil metals through evaporation, activity of soil organisms or cultivation
2. Availability and cost of suitable covering materials.

As such an undertaking would be expected to provide amelioration for a considerable period of time, a short-term experimental approach is of doubtful value. Many factors must be monitored, over an extended time-scale, before suitable guidelines for future use can be developed. The project described below has been initiated as one such survey, designed to measure both soil and crop response to a variety of covering treatments over an existing contaminated site.

THE BEAUMONT LEYS PROJECT

Beaumont Leys, located some 5 km north-west of Leicester, was the site of a former sewage farm. This operated for several decades, prior to its obsolescence and subsequent change of use. This site presently forms a major development site for the City of Leicester, and when completed, will form an integrated community consisting of housing, industry, education and public open space; with a linear park on the north-west boundary, adjacent to Rothley Brook. A comprehensive account of the history and development of this site is given by Heeps and Pike (this volume).

Previous soil surveys (Pike et al 1975a,b) have shown the presence of 'hot spots', where soil trace metal levels reach excessive values. These correspond to processes operating within the former sewage farm; filter beds, sludge-spreading areas and effluent irrigation areas. One major cause for concern are the high cadmium levels found at such sites, although other trace metals are also present at high levels. Cadmium is well known to accumulate in edible portions of several crop plants (Chaney and Hornick 1978) and establishment of agricultural or domestic crops on cadmium-contaminated soil should be viewed with great caution.

The site provides an ideal opportunity to evaluate the effectiveness of soil covering techniques as a method to reduce trace metal uptake into fruits and vegetables. In particular, the opportunity to assess metal uptake into domestic, as opposed to agricultural crops, can be taken. It should be stressed that measures to reduce possible cadmium contamination of domestic crops at Beaumont Leys to an acceptable level have already been taken (see Heeps and Pike, this volume), and that this project has been designed to provide information for future occasions when such sites are developed.

The design of the project has concentrated on domestic crops, as in the future, such sites will probably be used, in part, for housing. Bearing this in mind, the selection of test crops has concentrated on those commonly grown by householders, as well as taking into account known metal accumulation by such crops. The inclusion of soft and top fruit reflects the absence of information on metal accumulation in such plants.

In designing this project, several factors have been taken into consideration. Firstly, any remedial treatments must be effective over time-periods consistent with the life time of the housing; this may be between 60-80 years. The assumption is that covering techniques will provide amelioration over such a time-span, but this is by no means certain. Hence the adoption of a long-term approach (at least six growing seasons) was considered desirable.

The second factor concerns the actual execution of the project. One may use a restricted number of covering treatments and a wide range of test plants, which would give a wide-ranging picture, but with little replication. Alternatively the variety of covering treatments can be increased, the numbers of crops can be drastically restricted, and some replication can be made. Each approach has its merits, and, after careful consideration, the wide-ranging alternative was adopted. The reasons for this decision were as follows. The chosen cover materials are widely available. The variation in metal content of the site soil would have required an excessive degree of replication. The range of crops grown domestically is very wide, and a wide-ranging test would provide information on some of this variety of crops. The alternative approach (wide variety of cover, fewer crops, some replication) has been adopted on a parallel project, and the results from the two exercises will be evaluated together to produce a fully comprehensive picture.

The experimental site - characteristics, covering and planting

An area of land (0.6 ha) was identified as having an elevated (20-40 ppm) cadmium content from the overall Beaumont Leys survey. This was part of an arable field, directly adjacent to Beaumont Lodge Farm. Full details of the characteristics of this site are given in Table 1.

The covering strategy adopted was as follows. Soil cover depths, over the contaminated top soil were 1.2, 0.6, 0.3 and 0.15 m respectively. At the two greatest depths, 1.2 and 0.6 m, three different cover treatments were used. These were:

1. Top soil (Cd content < 1 ppm) for the entire depth (specified top soil).
2. Type (1) subsoil (Quarry Fines) for the lower half of the covering layer, then specified top soil for the remainder.
3. Type (2) subsoil (as raised sand and gravel) as in (2), with an identical top soil cover.

The specified top soil alone was used at the lower (0.3, 0.15 m) cover depths.

Standard experimental plots were designed for use with all covering treatments. Each plot was 32 m long and 4 m wide, subdivided into 32 x 2 m^2 sub-plots. Figure I illustrates such a plot, and gives details of the basic plot construction. Within each main plot, four treatments were superimposed on the main cover depth variable. Half the plot was limed. Crushed dolomitic limestone was incorporated into the contaminated top soil at the rate of 2 tons/acre prior to the covering treatment. In addition, one limed and one unlimed group of sub-plots was designated to receive fertilization treatment equated to 'good husbandry', whereas the two other plot were designated to receive no fertilization. Thus, in each main plot, 4 different groups of sub-plots were present. In each of these groups, eight standard crops were established. These are listed in Table 2, together with reasons for varietal selection.

In addition to the main covering treatments, several other standard plots were established.

1. Additional 32 x 4 m plots:
   (a) 0.5 m 'low' cadmium soil, overlying subsoil, then covered with 0.3 m specified top soil. (Low cadmium soil: total Cd content < 5.0 ppm).
   (b) as above, but 'medium' cadmium soil used. (Medium cadmium soil: total Cd < 15 ppm).
   (c) 0.8 m specified top soil overlying subsoil - 'control' bed.
   (d) 0.65 m site soil ('high' cadmium) - Beaumont Leys control.
2. 10 x 10 m^2 allotment plots.
   (a) 2 plots with 0.3 m specified top soil covering; one limed, one unlimed.
   (b) 1 plot with 0.8 m contaminated site soil.
   (c) 1 plot with 0.8 m specified top soil overlying subsoil.

This gives a total of 16 plots. The layout of the plots within the site is shown in Figure 2, and a summary of the type and depth of soil covering on each plot is given in Figure 3.

The allotment plots were included to allow trials of a wider range of crops than those used in the main experimental beds. As the project develops, these beds will yield valuable additional information on

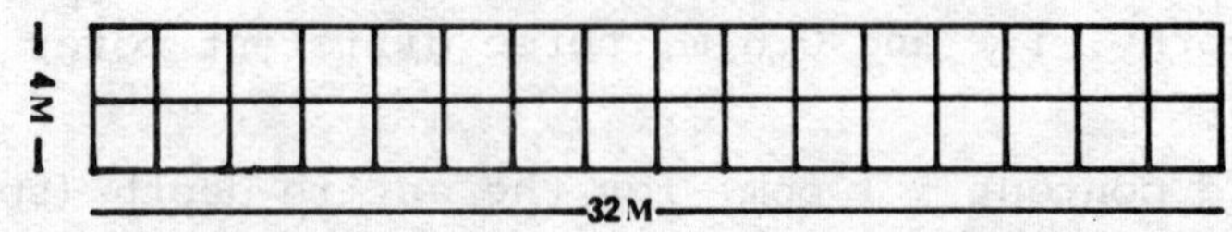

Figure 1. Experimental Plot Each cell, 2 x 2 m. Constructed from concrete slab fencing, lined with 500 gauge polythene sheeting.

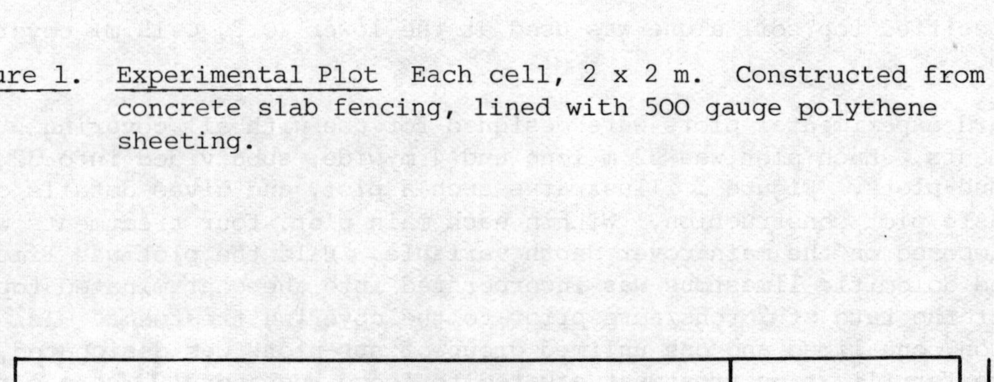

Figure 2. Layout of the experimental site

Plots A-I, N-P experimental plots; plots J-M allotment plots; x - site office; Y - polyhouse. Soil cover depths for each plot are given in Figure 3.

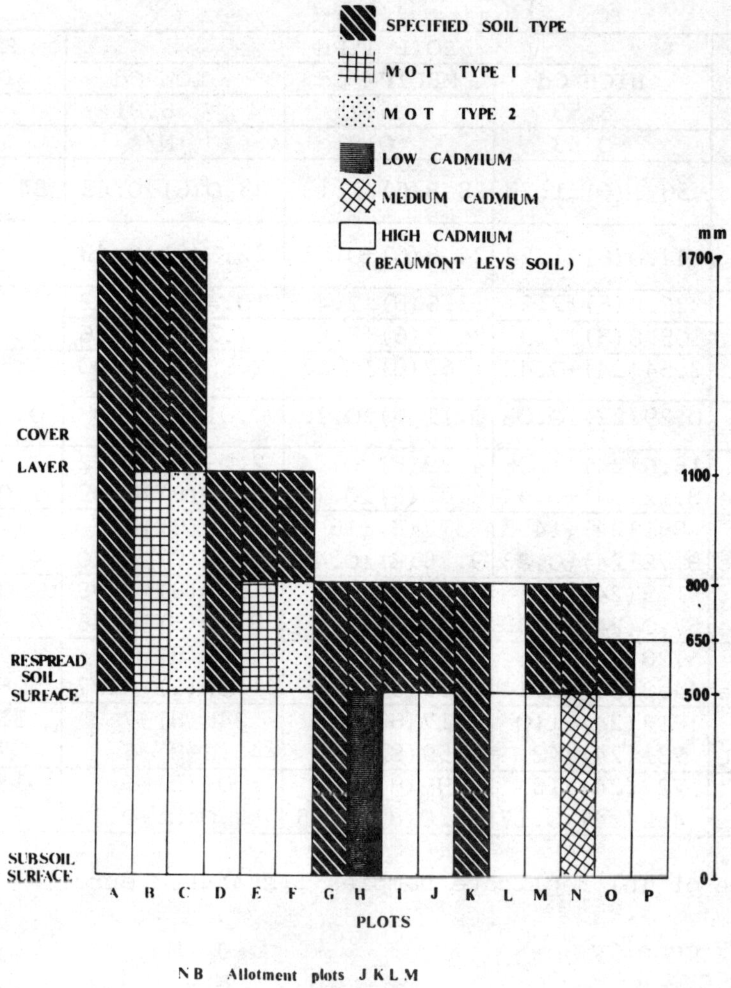

Figure 3. Soil cover depth and type: For location of experimental plots in the overall site, see Figure 2.

TABLE 1
Chemical and physical characteristics of the main soils used for the experimental plots.

PARAMETER	SOIL TYPE			SPECIFIED TOP SOIL
	HIGH Cd	MEDIUM Cd	LOW Cd	
pH	6.55	6.75	6.91	7.02
Bulk Density	0.53	0.81	N/A	0.83
Cation Exchange Capacity	56.1(6)±1.73	28.8(6)±1.13	23.0(6)±0.43	34.7(6)±0.70
Soil Moisture Content	44.0(6)±1.24	24.6(6)±0.51	22.5(6)±0.43	27.1(6)±1.06
L.O.I.	26.6(6)±2.3	13.6(6)±2.4	10.4(6)±0.52	5.0(6)±0.45
Inorganic-P	105.8(6)±7.3	36.3(6)±3.8	27.2(6)±0.76	20.1(6)±2.73
NO_3-N	2.54(24)±0.42	1.68(8)±0.16	2.99(8)±1.00	1.59(8)±0.19
NH_4-N	0.29(22)±0.08	0.11(8)±0.10	0.01(8)±0.0	0.03(8)±0.02
Cadmium total	18.6(24)±1.06	4.29(8)±0.34	2.09(8)±0.12	1.71(8)±0.25
" available	8.12(24)±0.44	2.43(8)±0.21	0.90(8)±0.09	1.00(8)±0.11
Lead total	881(24)±14.5	317(8)±16	95(8)±8	173(8)±15
" available	8.78(24)±0.33	9.25(8)±0.49	0.30(8)±0.20	4.30(8)±0.30
Nickel total	179(24)±7.6	46(8)±2.1	47(8)±0.98	36(8)±2.60
" available	57.0(24)±3.39	9.25(8)±0.89	8.10(8)±0.46	4.80(8)±1.06
Copper total	728(24)±39	136(8)±8	47(8)±3	71(8)±8
" available	44.8(24)±2.1	12.9(8)±0.85	1.00(8)±0.27	3.65(8)±0.73
Zinc total	1812(24)±110	417(8)±25	246(8)±7	353(8)±24
" available	731(24)±42	159(8)±8	44.3(8)±6	74(8)±13
Chromium total	722(24)±13	196(8)±11	101(8)±5	137(8)±15
" available	4.5(24)±0.17	3.8(8)±0.15	2.0(8)±0.11	2.2(8)±0.13

Table gives mean of (n) replicate samples ± standard error.

Units are:		
	Bulk density	g cm^{-3}
	C.E.C.	meq 100 g^{-1}
	Soil moisture content and L.O.I.	%
	Inorganic-P, NO_3-N and NH_4-N	mg 100 g^{-1} d.wt.
	All metal levels	µg g^{-1} d.wt.

cadmium uptake in many different types of garden crop.

The plots were constructed from concrete slab fencing, lined with 500 gauge polythene sheeting. Prior to plot construction, the site was levelled to the subsoil surface, and the resultant stockpile respread to ensure a more homogeneous distribution of contamination. The filling of the plots was carefully supervised, and the deeper plots were filled cell by cell with the appropriate soil type.

Broad,terram-based paths were laid between beds to allow easy access for future cultivation. Figure 4 shows the site before and after construction.

Crop rotation and layout

The test crops were allocated to each individual sub-plot on a random basis (Gemmell 1977). Four of the sub-plots were designated as permanent (currants, grass, raspberries, apples). Strawberries would be rotated on a 3 year cycle, and the three vegetable plots were to be rotated annually.

Figure 4(a) Experimental site prior to construction

Figure 4(b) Experimental site after construction completed

TABLE 2
Experimental Crops

CROP	CULTIVAR(S)	REMARKS
Strawberry	Royal Sovereign	Virus-Resistant
Lettuce Leek	Webbs Wonderful Musselburgh	Known Cd accumulators. Both popular varieties
Potato	Arran Pilot	Second Early Variety
Red Beet Parsnip	Detroit Asmer Improved Marrow	Root crops
Grass	Mommersteeg M22 Seed Mix	Used for amenity grass at Beaumont Leys
Redcurrant Blackcurrant	Laxtons No. 1 Wellington XXX	Popular varieties
Raspberry	Malling Jewel	Popular variety
Apple	James Grieve	Self-fertile

Main experimental programme

The major aim of the project is to assess the effectiveness of soil covering treatments on subsequent trace metal uptake and accumulation in edible plant parts. To achieve this end, regular sampling and analysis of all plant material will be undertaken, with special, but not exclusive, emphasis being placed on the edible portions. The long-term nature of the project will enable a full picture of changes in metal availability and accumulation patterns to be constructed.

In addition to the main plant experiments, numerous subsidiary projects, not directly related, but highly relevant to the main schedule, will be initiated.

Subsidiary experimental programme

Four subsidiary experimental programmes will be initiated, to provide a fuller insight into all aspects of the reclamation procedure. These are:

1. Continuous monitoring of soil moisture movement.
2. Studies on soil fauna.
3. Studies on soil microflora.
4. Monitoring changes in soil chemistry in relation to covering treatment and crop growth.

Project 1 will be continuous throughout the life of the experimental work. This involves making a permanent bore-hole to the water-table, and monitoring changes in this on a regular basis. In addition, an array of soil tensiometers will be established, in vertical files, in each 1200 mm plot. The tensiometers will be at 1400, 1200, 60 and 45 mm respectively, and will provide indications of the efficiency of both moisture break materials, as well as allowing comparisons of the suction exerted by the various soil types.

Projects 2-4 will be undertaken in two parts. Initial, base-line surveys, undertaken within the first year of site establishment, will establish initial populations of organisms, and soil chemical status. After 3 growing seasons, studies will commence to investigate the nature and magnitude of changes in these factors that have occurred with time.

CONCLUDING REMARKS

The project described above has been specifically designed to produce answers of great relevance to contemporary land reclamation projects. At present, no conclusions or comments are forthcoming, but it is hoped that after two growing seasons, a picture of the relative effectiveness of the various treatments will start to emerge. The introduction of a wide programme of related, relevant projects should result, in the final analysis, in the production of a more integrated picture of the consequences of a reclamation project of this type than reported so far.

In conclusion, the aims of the project are simple, but the results, in terms of total environmental response, may have far-reaching consequences.

REFERENCES

Chaney, R.L. and S.B. Hornick (1978). Accumulation and effects of cadmium on crops. Proc. 1st. Int. Cadmium Conf., San Francisco 1977: ILZRO. pp. 125-140.

Gemmell, R.P. (1977). Colonization of industrial wasteland. Edward Arnold, London. 80 pp.

Pike, E.R., Graham, L.C. and M.W. Fogden (1975a). An appraisal of toxic metal residues in the soils of a disused sewage farm. I. Zinc, copper and nickel. J. Assn. Pub. Anal. 13: 19-33.

Pike, E.R., Graham, L.C. and M.W. Fogden (1975b). An appraisal of toxic metal residues in the soils of a disused sewage farm. II. Lead, cadmium and arsenic, with notes on chromium. Ibid. 13: 48-63.

ACKNOWLEDGEMENTS

This project is undertaken for, and is funded by, the Department of the Environment. We would like to thank Mr. M.A. Smith of the D.O.E. for his unflagging endeavours in initiating and guiding this project, and Mr. K. Heeps and Mr. I. McLean of the City Engineers Department, Leicester City Council, for invaluable assistance and advice in establishing the experimental site.

Project 2.3 will be undertaken in two parts [illegible]

[illegible]

[illegible]

In conclusion, [illegible]

REFERENCES

[illegible]

[illegible]

[illegible]

[illegible]

ACKNOWLEDGEMENT

[illegible]

CADMIUM AND LEAD CONTAMINATION OF SOILS AND VEGETABLES IN RELATION TO HISTORIC BASE METAL MINING.

Brian E. Davies and Helen M. White

Department of Geography, U.C.W., Aberystwyth, Dyfed, Wales.

INTRODUCTION

During the nineteenth century Great Britain was a major source of base metal ores. The minerals of greatest interest were galena (PbS) and sphalerite (ZnS) which occur primarily as pipes or lodes in Lower Palaeozoic sediments and the Carboniferous Limestone. Derelict lead mines, waste heaps and trial cuts are scattered throughout western counties of England and Wales and in the Derbyshire peak District but most of the productive mines were concentrated in several mining districts. These are listed in Table 1. together with their production figures and the approximate size of each mining district.

AREA	Ore production: 1850-1947 tons x 10^3 PbS	ZnS	% total PbS	ZnS	Approximate Area km^2	Total ore Production per km^2
N. Pennines	1696	267	39.0	25.7	2850	0.69
Lake District	183	38	4.2	3.6	1500	0.15
Derbyshire	678	80	15.2	7.6	300	2.53
Shropshire	237	20	5.4	1.9	290	0.89
Clwyd	658	342	15.2	33.0	260	3.85
Punlumon	479	151	11.0	14.5	1040	0.61
Mendips	100	50	2.3	4.8	135	1.11
Cornwall & Devon	317	89	7.3	8.9	2240	0.18
TOTALS	4348	1037	100.0	100.0	8615	0.63

Table 1. The major British mining fields : production statistics compiled from Dunham (1959) and Green (1958).

For most of the production period the ore preparation and concentration processes were both inefficient and uncontrolled by any environmental protection legislation. As a result of public disquiet the Rivers Pollution Commissioners heard evidence in 1872 that lead mining was causing considerable river pollution leading to the death of stock and failure of crops in floodplain land. They remarked (Frankland and Morton, 1874) that mining was carried on in an extremely slovenly and wasteful manner and that reckless waste of valuable materials was permitted in many mines.

It is possible to estimate approximately these losses of lead to the environment. Jones (1972) reported that the proportion of waste rock to ore at the Goginan mine near Aberystwyth was around 25:1 (the fines from the spoil there contain between 0.8 and 1.5% PbS); Frankland and Morton (1874) reported up to 25% losses during the ore separation stages; Willies (1971) suggested that cupola lead smelters were 63% efficient; Lewis (1967) discussed the Gadlys smelter in Clwyd and found that 13% of lead metal was lost during desilvering. By aggregating these losses we estimate that between 1.0 and 1.6 tonnes of lead were lost to the environment for every tonne reported in the official mineral statistics and the losses of zinc are presumed to be similar.

The Rivers Pollution Prevention Act of 1876 required the miners to pass their effluents into settling pits and while this proved of great benefit the peak of the industrial activity and its environmental consequences had already passed. Today, the once polluted rivers have nearly all recovered and the only overt signs of the past are derelict buildings and spoil heaps barren of vegetation. However, this paper is not concerned with these areas of dereliction but rather with the legacy of heavy metal contamination in apparently productive fields and gardens in lead mining areas.

CADMIUM AND LEAD IN SOILS

Early work had established that many fields in West Wales, Cornwall and Devon were contaminated by Pb, Zn, Cu and Cd, the latter element being a common impurity of sphalerite. The main distribution processes were atmospheric transport of smelter smoke and the deflation of fine-textured spoil, fluvial redistribution of solutes and sediments in river floodplains and gravity-induced movement of sediments downslope from waste heaps. This work has been reviewed by Davies (1979).

Assuming that the intensity of pollution is related to the total ore production compared with the size of the mining field then, from Table 1, Derbyshire, Clwyd and the Mendips appear likely to be the worst affected. An interesting feature of both Clwyd and Derbyshire is that their sphalerites are rich in Cd. El Shazly et al. (1956) reported that 12 samples from Halkyn and Minera (Clwyd) contained 0.5 - >1% Cd and a specimen from the Mill Close mine (which accounted for 30% of Derbyshire Zn production and 60% Pb production) near Matlock contained 1% Cd. In the Mendips the chief ore was smithsonite ($ZnCO_3$) for which no data are available but a soil sample from Shipham taken in 1972 contained 330 mg. Cd/kg. Thus, cadmium contamination appeared of possible special interest in these areas. On the other hand, the dominant country rock was limestone, in each case, which implied that fluvial dispersal of contaminants would be small. The investigations were extended to these three areas and soil samples (0 - 15 cm) were collected from agricultural fields and analyzed for several heavy metals: summary data for Cd and Pb are presented in Table 2.

The Halkyn values are derived from a systematic survey (1 x 1 km grid) which has been described by Davies and Roberts (1978). The Minera figures are based on an unpublished survey (0.5 x 0.5 km grid; B.E. Davies) as are those for the Mendips (2 x 2 km; B.E. Davies and R.C. Ginnever). The Derbyshire sampling was carried out along two transect lines, 1000 m apart, at Castleton and another near Darley Dale. The sample interval was 500 m (B.E. Davies and L.J. Roberts, unpublished). Data for the Tamar Valley (Davies, 1971) and the Ystwyth Valley (Alloway and Davies, 1971) are included for comparison. As a rough guide it can be assumed that a normal soil rarely contains more than 100 mg Pb/kg or 1.0 mg Cd/kg.

TOTAL METAL CONTENT : mg/kg DRY SOIL

AREA	Number of Samples	Pb Mean	Pb Range	Cd Mean	Cd Range
Derbyshire	25	2414	99-19693	8.6	2.4 - 67.0
Halkyn	260	886	35-47995	6.1	0.4 - 540.0
Minera	23	870	74- 6944	1.9	0.5 - 8.5
Mendips	69	390	31-11236	4.4	0.6 - 116.0
Ystwyth	17	1419	90- 2900	2.5	1.2 - 4.0
Tamar	58	110	20- 522	1.8	0.7 - 4.5

Table 2. Summary analytical data for agricultural soils.

Mean metal values are given in Table 2 but differences in the design of the sampling programmes make direct comparisons between the areas difficult. However, comparisons of the maximum values can be made and it is seen that the limestone areas do emerge as being more severely polluted by both Cd and Pb than the non-calcareous Ystwyth and Tamar Valleys.

Collection of samples on a regular grid allows isoline or contour maps to be prepared and Davies and Roberts (1978) have published such maps for Pb, Zn, Cu and Cd in the soils of the Halkyn Mountain mining area. The maps revealed a very distinctive pattern of metal accumulation in agricultural soils. The highest soil Pb value (47955 mg/kg) was found near the New North Halkyn mine (at SJ 203707) which ceased production in 1958. Pb levels decline progressively away from the mines but certain detached anomalies to the east were identified with the sites of historic smelters. Despite the absence of fluvial dispersal of contaminants soil pollution was widespread and it was calculated that there is 171 km^2 of contaminated land (i.e. > 100 mg Pb/kg soil) in the Halkyn district. Within this affected area there is 44 km^2 in the range 1000 - 10000 mg Pb/kg and 3 km^2 with contents

> 10000 mg Pb/kg. The Cd map was broadly similar and the area contains 69 km^2 with soil contents > 2 mg Cd/kg and 3 km^2 in excess of 100 mg Cd/kg. The work in the Mendips has also revealed a characteristic pattern of soil contamination by Cd and Pb and will be published elsewhere. No map is available for any part of Derbyshire but the transects suggest a similarity with Halkyn Mountain and Colbourn and Thornton (1978) have remarked that "polluted soils cover most of the eastern part of the limestone and areas of 0.5 - 1 km radius around pollution sources on other geological strata".

TOTAL METAL CONTENT : mg/kg DRY SOIL

AREA	Number of Samples	Pb Mean	Pb Range	Cd Mean	Cd Range
Derbyshire	8	4958	363-12837	18.3	3.3 - 38.7
Halkyn	17	2871	920- 5897	9.3	1.6 - 40.0
Minera	6	2453	1221- 3741	11.7	2.4 - 23.9
Mendips	7	1603	597- 2612	69.5	8.8 - 97.9
Tamar	12	260	41- 510	2.7	1.3 - 4.0

Table 3. Summary analytical data for garden soils.

Although the mining industry is largely defunct it does not follow that the towns and villages of the area are necessarily in decline. In many of the more remote areas old cottages have been purchased and renovated for holiday use while elsewhere towns or villages such as Shipham, Halkyn, Holywell or Winster are expanding both through private development and council house estates. In all cases these new housing estates are expanding over contaminated soils while the gardens of the older houses were subjected to pollution during the active mining period. Therefore, the soil investigations have included studies of garden soils for which summary data are given in Table 3.

The data are derived from the same sources as those quoted for Table 1. except that the Mendips values are quoted from Davies and Ginnever (1979). It is clear that the values, while more restricted in range than those for the agricultural soils, are a subset of the figures in Table 1. and that the garden soils are characterised by similar pollution problems.

CADMIUM AND LEAD IN VEGETABLES

The toxic nature of Cd and Pb is widely known and the possibility of excessive amounts being present in human foodstuffs is therefore a matter for concern. In the case of Pb the maximum amount permitted in food is 2 mg/kg on a fresh weight basis. There is no similar limit for Cd but the World Health Organisation has suggested that an individual's maximum weekly intake be limited to 400 - 500 µg; for Pb the tolerable maximum weekly intake is 3000 µg.

Although the soil data reveal a significant pollution problem in mining areas it does not necessarily follow that crops or vegetables grown on these soils would also be contaminated. Only a small proportion of the total amount of any element in soil is available for absorption by roots and actual uptake is modified by soil properties and plant species. Nonetheless, Davies and Roberts (1975) reported that radish grown in certain gardens in Halkyn contained more than the permissible amounts of Pb in the edible root. To investigate this further, trial plots have been established in Clwyd and Derbyshire and several kinds of vegetable grown. Results for three vegetables are presented in Table 4. In order to facilitate interpretation the values have been calculated to a nominal fresh weight basis assuming that the dry matter contents of the vegetables are, lettuce = 5%, carrot = 20% and onion = 10%. The values are presented in ascending order of soil Pb content.

For Pb, half the lettuces and onions and 7/8 carrots have contents equal to or in excess of the 2 mg/kg limit and the order in which they accumulate Pb is carrots > onion > lettuce. In the case of Cd the results may be interpreted by assuming that each vegetable is the sole vegetable consumed for a week.

The Ministry of Agriculture bases its interpretations on the premise that an individual consumes 1470 g food/day of which vegetables account for 350 g, i.e., 2450 g/week. Therefore, a diet of only lettuce would contribute, using the mean Cd content, 1348 µg Cd or 3.4 x the WHO limit of 400 µg Cd; only carrot would contribute 809 µg Cd (2.0 x WHO) and only onion, 515 µg Cd (1.3 x WHO). The order of Cd accumulation is lettuce > carrot > onion. The data do not support any direct link between soil and vegetable metal contents and comparatively low levels of soil pollution can give rise to vegetables of suspect quality.

	Soil			Lettuce leaves		Carrot roots		Onion bulbs	
	pH	Pb	Cd	Pb	Cd	Pb	Cd	Pb	Cd
1	6.5	318	2.3	0.15	0.06	5.52	0.32	0.28	0.04
2	7.0	782	3.7	2.93	0.59	5.80	0.32	0.49	0.19
3	6.5	949	5.0	2.23	0.89	2.26	0.19	1.98	0.13
4	5.4	1970	4.3	1.07	0.99	4.04	0.36	5.65	0.33
5	6.7	2267	18.2	0.31	0.68	1.04	0.13	1.14	0.39
6	7.3	3130	29.2	6.00	0.85	5.40	0.61	3.06	0.29
7	6.5	5680	5.2	1.15	0.18	12.66	<0.05	2.03	0.19
8	7.1	11618	35.1	2.00	0.15	4.08	0.41	2.31	0.09
Mean		3339	12.9	1.98	0.55	5.10	0.30	2.12	0.21

Table 4. Analytical results for garden soils and vegetables. Metal values are given as mg/kg dry soil or vegetable fresh weight.

It is not clear to what extent fertilizers or manures can control the uptake of heavy metals by plants. Liming is commonly supposed to be a useful measure to reduce the availability of Pb but the pH values of the garden soils suggest they have a low lime requirement. Another possibility is that vegetable cultivars might be selected which assimilate smaller quantities of heavy metals and John and Laerhoven (1976) have published results which support a contention that Cd uptake by lettuce is genetically controlled. To investigate differential uptake by lettuce cultivars six varieties were grown on three contaminated soils in pots in a glasshouse. Full details of this experiment will be published elsewhere but summary values are presented in Table 5. From the mean values it is seen that the order for Pb accumulation is (see the table footnote for the meaning of the abbreviations) SB > WD > WW > CH = AYR > PW whereas the order for Cd is accumulation is AYR > SB > CH > PW > WD > WW. But there are significant differences in the order for individual soils, e.g., variety SB accumulated most Pb from soils 1 and 3 but least from soil 2. This experiment suggests that a cultivar which is a weak absorber of Pb is not necessarily also a weak absorber of Cd and, anyway, other (unidentified) factors are more important in practice in controlling heavy metal uptake by lettuce.

LETTUCE VARIETY

	SOIL	AYR	SB	CH	WW	WD	PW
			mg/kg				
Pb:							
1	754	0.38	0.49	0.34	0.24	0.25	0.16
2	850	0.93	0.50	0.79	1.12	1.30	0.75
3	169	0.20	1.10	0.38	0.25	0.26	0.27
Mean	591	0.50	0.70	0.50	0.54	0.60	0.39
Cd:							
1	4.5	0.82	0.65	0.60	0.41	0.51	0.55
2	5.5	1.41	1.13	1.08	0.69	0.98	0.98
3	2.2	0.56	0.35	0.41	0.28	0.33	0.36
Mean	4.1	0.93	0.71	0.70	0.46	0.61	0.63

Table 5. Summary metal data for 5 varieties of lettuce grown in 3 contaminated soils. Metal values are quoted as mg/kg dry soil or lettuce leaf, fresh weight. Varieties were, AYR = All Year Round (cabbage, butterhead); SB = Salad Bowl (perpetual); CH = Constant Heart (cabbage, butterhead); WW = Webb's Wonderful (cabbage, crisphead); WD = Winter Density (cos); PW = Paris White (cos).

CONCLUSIONS

Although visual evidence of dereliction of and damage to land is confined to the immediate environs of derelict base metal mines several surveys have revealed that Cd and Pb contamination of fields and gardens in the mining areas is widespread. The gardens in which experimental vegetables were grown were selected to be representative of the range of metal contents encountered in Clwyd and Derbyshire. The results suggest that even the lower soil Pb and Cd levels can give rise to vegetables of suspect quality. It is not known to what extent individual families rely on home-grown vegetables and it is possible that consumption of foodstuffs derived from elsewhere might considerably ameliorate the problem. Nonetheless, it would be prudent to assume a problem does exist and to investigate further whether Cd and Pb absorption by vegetables might be reduced either through agronomic practices or through the growing of cultivars with a reduced ability to absorb metals.

ACKNOWLEDGEMENT

Part of this work has been supported by the Ministry of Agriculture, Fisheries and Food and by the Welsh Office.

REFERENCES

Alloway, B.J. and Davies, B.E. 1971. Geoderma 5, 197-207.

Colbourn, P. and Thornton, I. 1978. Journal of Soil Science 29, 513-526.

Davies, B.E. 1971. Oikos 22, 366-373.

Davies, B.E. 1979. (in) "Inorganic Pollution and Agriculture". MAFF Technical Bulletin. H.M.S.O. (in press).

Davies, B.E. and Ginnever, R.C. 1979. Journal of Agricultural Science (Cambridge) (in press).

Davies, B.E. and Roberts, L.J. 1975. Science Total Environment 4, 249-261.

Davies, B.E. and Roberts, L.J. 1978. Water, Air and Soil Pollution 9, 507-518.

Dunham, K.C. 1959. (in) The Future of non-ferrous mining in Great Britain and Ireland : a symposium. I.M.M. London.

El Shazly, E.M., Webb, J.S, and Williams, D. 1956. Transactions Institution of Mining and Metallurgy 66, 241-271.

Frankland, E. and Morton, J.C. 1874. Fifth Report Rivers Pollution Commission. H.M.S.O. London.

Green, G.W. 1958. Bulletin Geological Survey 14, 70-90.

John, M.K. and Laerhoven, C.J. van. 1976. Environmental Pollution, 10, 163-173.

Jones, O.T. 1922. Special Report on the Mineral Resources of Great Britain, volume 20. H.M.S.O., London.

Lewis, W.J. 1967. Lead Mining in Wales, University of Wales Press, Cardiff.

Willies, L. 1971. Bulletin Peak District Mines Historical Society 4, 384-394.

OPTIONS FOR TREATMENT OF METAL-CONTAMINATED MINE WORKINGS

M. S. Johnson, B.Sc., Ph.D.

Department of Botany, University of Liverpool, Liverpool L69 3BX

SYNOPSIS

Environmental problems of derelict metalliferous mine workings in England and Wales are examined. The prospects for waste utilization are discussed, together with remedial measures based on vegetative stabilization. Factors to be considered in establishing and maintaining a permanent vegetation cover, and the means by which this may be achieved, are reviewed. The relevance of these techniques to redeveloping contaminated land in the urban environment is discussed.

INTRODUCTION

The contribution of housing and recreational schemes to plans for re-developing derelict land in urban-industrial areas has highlighted environmental quality problems significant to these land use proposals. Pollution of land and water resources within former industrial centres has led to an extensive programme of research into the residual construction, public health and safety hazards of these areas.

From a biological viewpoint attention has been directed towards trace metal contamination of ecosystems with emphasis on the mobility, behaviour and bioavailability of lead, cadmium and mercury residues in the terrestrial environment (eg. Lagerwerff, 1972; Johnson *et al*., 1978). Remedial measures which allow re-use of contaminated sites without restricting land capability must be effective on a permanent basis. However, the recent appreciation of contamination problems in relation to sensitive development proposals for urban land means that the conclusions from treatment trials monitored over several years are not available to developers currently implementing large scale reclamation works. Consequently, it may be advisable to draw on experience gained from similar situations in which trace metal pollution of the soil environment has been a major consideration in treatment policy.

A great deal of information has been gained from re-development works on metal-rich wastes derived from the non-ferrous metal mining and smelting industries of the British Isles (Street & Goodman. 1967; Johnson & Bradshaw, 1977). Active trace element pollution from abandoned mine workings is a significant problem in certain areas of England and Wales noted for base metal mineralizations. Land and water contamination within the Pb-Zn and Cu orefields of Wales has provided sufficient impetus for a detailed survey of over 650 metalliferous workings (Table 1) to identify priority sites for treatment programmes (Welsh Development Agency & Liverpool University, 1978).

TABLE 1

SURVEY OF ABANDONED METALLIFEROUS MINES IN WALES*

County	No. of mine† workings	Environmental impact appraisal ‡ Major	Moderate	Minor/negligible
CLWYD	221	4	31	186
DYFED	248	19	24	205
GWYNEDD	92	5	17	70
POWYS	102	3	16	83

* Welsh Development Agency & University of Liverpool (1978)

† Excluding sites of <100m² surface area

‡ Based on quantitative assessments of the effects on land and water resources, landscape quality and potential hazards.

Redevelopment objectives for these sites are usually environmental protection simultaneous with amenity improvement and recreational use. This requires the establishment of a permanent ground cover vegetation but several revegetation systems have been developed differing greatly in complexity, practicability, cost and indeed effectiveness in the long term.

ENVIRONMENTAL PROBLEMS

Environmental contamination through erosion of trace element-rich spoil and dispersal of mine waters is mainly an historical phenomenon, but one which remains significant to resource management in England and Wales at local and regional levels. Flood deposition of mineralized sediment has severely damaged pastures causing lead poisoning in livestock around Llanwrst in Gwynedd, N.Wales (Thornton, 1969), and geochemical reconnaissance data suggest that in Dyfed alone over 485ha of farmland on alluvial soils have been contaminated by mine waste deposition (Thornton, 1975). Even where transport in surface drainage and watercourses is insignificant, low level contamination of vegetation and surface soil may occur at considerable distances from the source (Roberts & Johnson, 1978) due to wind erosion (Fig.1).

Livestock fatalities caused by ingestion of contaminated vegetation or use of polluted water supplies are now infrequent, but the significance of sub-lethal toxicity to animal production, and of metal contamination of cereal and vegetable crops to food quality has been questioned (Thornton, 1975). Zinc toxicity in arable crops and lead, zinc and cadmium enrichment of vegetables grown experimentally on domestic garden soils of the Flintshire-Denbighshire mining district (now Clwyd) have been reported in recent years (Ball, 1960; Davies & Roberts, 1975); but the exact significance of these accounts of soil and crop contamination to human health is very uncertain.

Continuing metal contamination of the terrestrial environment by fine (< 20μ) and intermediate (0.02-5mm) tailings is encouraged by their low inherent resistance to erosion processes, and by deficiencies in the original disposal policy. In certain situations the environmental hazards of large unstable spoil tips undoubtedly justify rehabilitation works but there are many potential solutions to these problems.

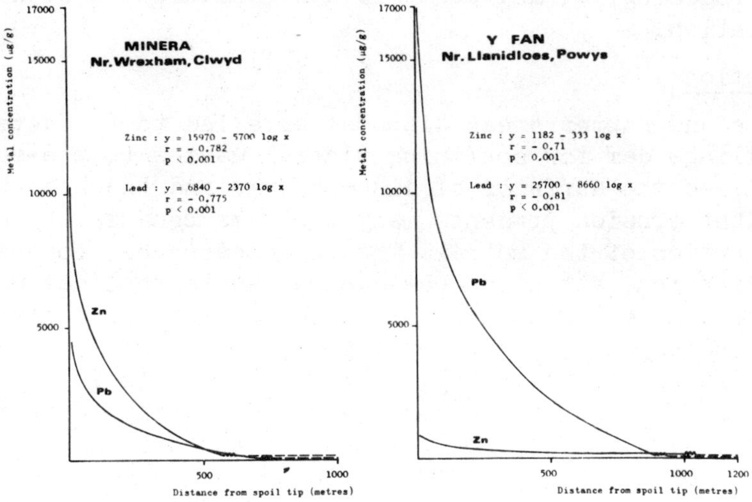

FIG. 1 Lead and zinc concentrations in surface soils around abandoned metalliferous mines in Wales

ALTERNATIVES TO REVEGETATION

Revegetation forms a basic part of landscape redevelopment in the metal mine environment since it is the only means by which the objectives of environmental protection,amenity enhancement and,wherever possible, provision of a definite afteruse can be achieved simultaneously. However modern social and economic pressures have stimulated industry and public authorities to investigate alternative approaches,in particular the prospects for positive utilization and secondary disposal to landfill sites.

Waste Utilization

In recent years there has been a revival of commercial interest in metalliferous mine wastes because of the residual metal values. Responding to changes in mineral prices, small scale reprocessing has developed using advanced ore treatment systems. At present these are limited to large, metal-rich tips in west Wales, Derbyshire and south-west England. Whilst these operations are liable to increase in number, it is unlikely for reasons of site accessibility, tonnage and grade of deposit, transport, planning and environmental considerations,that they will ever be on a significant scale except at the local level.

The prospect of waste utilization making a significant impact on environmental problems through substitution for standard industrial raw materials(eg.in glass and brick manufacturing)is limited worldwide on technical and economic grounds. Constructional use of intermediate tailings is now restricted to re-surfacing of unmetalled roads but even less desirable applications including road gritting,pebbledashing and pathway construction in new housing estates have been known in recent years. Far from solving existing environmental problems such uses extend metal contamination to new and more sensitive locations.

Removal and disposal

Backfilling of abandoned shafts with coarse discard has been practised on a local scale throughout England and Wales, but mainly to eliminate physical hazards rather than as a deliberate disposal policy. Applied to intermediate and fine tailings, such a policy could have a serious impact on water resources. Chemical stabilization of tailings followed by disposal in old quarries and opencast sites has been known but site availability, hydrogeological and economic considerations prevent its widespread application.

Surface stabilization

Limitations on use and subterranean disposal have led to the development of the modern tailings dam for confining mineral wastes to pre-set areas. The need to stabilize the surfaces of disused dams and beached areas of active sites against erosion presents many problems because of the low load-bearing properties of the matrix. Though research has focused on vegetative stabilization, attention has been given to physical and chemical techniques where limitations of scale, access and climate prevail. Methods based on reacting certain chemicals (eg.ligno-sulphonates and elastomeric polymers) with the tailings to form an erosion-resistant surface layer are, at best, temporarily effective and fail to satisfy the redevelopment objectives for the British situation. Physical techniques of surface treatment with brushwood, straw or by erection of windbreaks are also unsuited to the physiochemical and landscape characteristics of most metalliferous workings in England and Wales.

VEGETATIVE STABILIZATION

Plant growth problems

Non-ferrous metal mine wastes are amongst the most hostile environments from a revegetation viewpoint because of the severe limitations to plant growth. The most prominent limiting factor is heavy metal toxicity which stems from the poor recovery of metallic minerals from ore deposits exploited during the main period of mining activity (1750-1900).

Old tailings in the fine and intermediate categories usually exhibit the highest metal concentrations (Table 2), and release of plant-available metal salts by weathering inhibits plant biochemical and physiological functions causing total disruption of metabolic processes. Direct effects on photosynthesis and respiration are accompanied by secondary

TABLE 2

TOTAL METAL CONCENTRATIONS OF ABANDONED METALLIFEROUS SPOIL TIPS † IN ENGLAND AND WALES

County	No. of sites examined	Pb	Zn	Cu	Cd	Authors
CLWYD	13	1770-76000	1200-127000	30-588	34-395	Johnson & Bradshaw,1977
CUMBRIA	7	2070-6370	4690-7370	77-380	0.2-29	"
DERBYSHIRE	16	5760-65250	700-18500	31-230	1-10	Colbourn & Thornton,1978
DEVON & CORNWALL	23*	48-2070	23-1950	65-6140	3-45	Johnson & Bradshaw,1977 Porter & Peterson,1977 Smith,1973
DYFED	13	1670-54000	475-8000	77-560	<1-58	Baker,1978 Johnson & Bradshaw,1977
GWYNEDD	4	220-2000	80-2400	90-2300	8-510	Gregory & Bradshaw,1965
POWYS	6	2030-71200	285-16000	27-285	5-162	Johnson & Bradshaw,1977

† Extractable in concentrated HNO_3 or $HNO_3/HClO_4$; all values in µg g^{-1}

* Total As concentration range : 68-26530µg g^{-1}

problems such as the adverse effects of root growth inhibition on the absorption of water and nutrients. The at best sparse colonization of many British metalliferous tips is caused by copper, lead or zinc toxicity or combinations thereof. However, in terms of remedial works consideration must also be given to trace contaminants of the principal metallic minerals (eg. silver in galena (PbS); cadmium in sphalerite (ZnS)) because of their biological properties and significance in terms of food chain transfer.

Many factors influence heavy metal phytotoxicity. Apart from the chemical form of the metal(s), substrate acidity greatly affects metal availability for root absorption. In alkaline conditions [eg.spoil tips dominated by calcite ($CaCO_3$)] calcium-heavy metal interactions reduce the availability of copper, lead and zinc by precipitation and complexing reactions. Nevertheless cation antagonism rarely compensates to the extent of eliminating phytotoxicity. In acid substrates heavy metal mobility is increased and revegetation problems can be compounded by aluminium and manganese toxicity. Moreover, in the presence of significant quantities of pyrite [eg. iron pyrites (FeS_2)] metal mobility and acidity problems may be exacerbated by weathering reactions which release free sulphuric acid.

Metal mine wastes are characteristically deficient in essential plant macronutrients. Inadequate supplies of nitrogen, phosphorus and potassium stem from the absence of organic (N,P) and mineral (P,K) stores of these nutrients. The lack of organic matter and clay minerals, which together account for the cation exchange capacity of normal soils, also contributes to the low nutrient status through the inability of metalliferous tailings to retain nutrients against leaching.

Whilst chemical problems are the major factors to be overcome, mine wastes also present a range of adverse physical characteristics relevant to revegetation works. Modern tailings share several of the physical properties of older material but the predominance of silt and clay sized particles emphasises the growth limitations of waterlogging due to compaction and surface capping, and the practical problems of mechanised revegetation works. By way of contrast the coarser intermediate tailings and waste rock dumps are periodically subject to drought conditions. Seasonal temperature fluctuations also inhibit plant growth through alternate expansion and contraction of tip surfaces. This causes root tearing and instability because the shallow fissures which develop serve as focal points for erosion.

Land use objectives

Certain land use objectives are more demanding than others and it is important that the technical requirements are aligned with practical and economic constraints when decisions are made about treatment policy. For example, tip stabilization for erosion control and amenity enhancement does not necessarily require the complex treatment measures for residential and intensive recreational land uses. In this respect the metal mine environment constitutes a departure from contaminated land in urban areas even though the principles of revegetation may be common to both situations.

PRACTICAL RECLAMATION SYSTEMS

Tolerant plant populations

Early attempts to revegetate metalliferous spoil tips invariably failed. Direct sowing of agricultural grasses and legumes, transplanting of tillers and large turves as centres for vegetative colonization and seed sources proved totally unsuccessful. A refinement of the turfing system based on a latticework design improved the erosion resistance by acting as a windbreak and an interceptor of surface run-off, but the failure of root growth in the underlying spoil eventually led to dieback of the transplant material.

Stabilization by direct seeding can only be achieved by countering substrate phytotoxicity. The sparse natural vegetation of certain sites has been of outstanding value in this connection. Certain species, although not innately tolerant, have evolved ecotypes which can persist on heavy metal contaminated soils (Gregory & Bradshaw, 1965; Smith & Bradshaw, 1972). These metal tolerant plants also possess physiological and morphological characters which reflect environmental stresses other than toxicity and are more tolerant than normal populations of the low nutrient status and episodal drought conditions of the mine tip environment.

Tolerance is a stable and heritable character which is resilient to the adverse effects of gene flow, and only a minimum of selection has been necessary to produce commercial cultivars from natural ecotypes. Three cultivars, Festuca rubra cv. 'Merlin' (Pb/Zn), tolerant Agrostis tenuis cv. 'Goginan' (Pb tolerant) and Agrostis tenuis cv. 'Parys' (Cu tolerant), are now undergoing commercial seed production. The long term performance of tolerant ecotypes on metal polluted substrates is infinitely superior to that of non-tolerant plant material (Fig. 2).

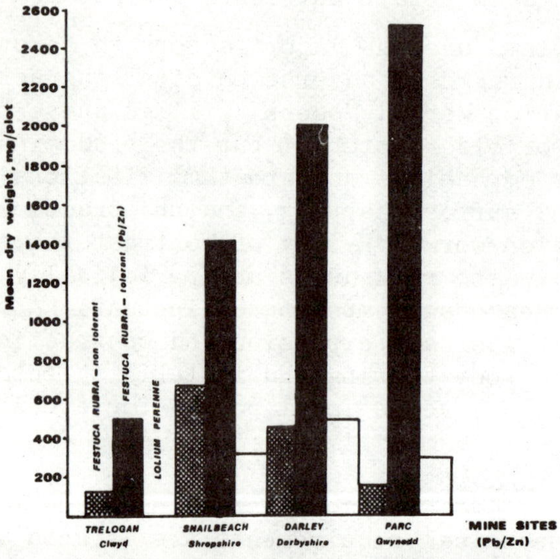

FIG. 2 Yield of tolerant and non-tolerant populations of red fescue (Festuca rubra) and non-tolerant ryegrass (Lolium perenne) on Pb/Zn mine wastes treated with NPK fertiliser, after 12 months growth.

The prospects for large scale use of tolerant populations for tip stabilization are encouraging. Since the technique is based on direct seeding and treatment with inorganic fertilisers (Smith & Bradshaw, 1972), without application of bulky amendments, there are economic advantages as well as the attraction of being able to stabilize inaccessible sites by manual application of seed and fertiliser. There are however certain limitations, notably the specificity of the tolerance mechanism, which mean that existing varieties may not persist on land containing phytotoxic concentrations of metals other than those to which they are adapted. Furthermore, grazing of the sward is not possible, and it is questionable whether swards composed entirely of tolerant plant material could maintain their continuity when subjected to significant recreational pressures.

Traditional surface treatments

Surface amelioration in which the toxic tip is covered by a blanketing layer of innocuous material is by far the most common restoration technique employed at the present time. Amelioration with a 100-500mm layer of topsoil, pulverized fuel ash, sewage sludge or peat and sand has been attempted in several smelter and mine waste situations (Street & Goodman, 1967; Johnson & Bradshaw, 1977). Modifications of this system to improve surface stability and encourage root development involve incorporation of the ameliorant into the surface 100-300mm of waste. Successful restoration has been achieved with these approaches but doubts have been expressed about their long term prospects.

Root development is confined to the innocuous material unless tolerant populations are sown and this can lead to instability if physical detachment occurs at the amendment/mine waste interface. In situations where only a limited amount of regrading is practical, and only a shallow depth of amendment can be applied, frost heaving and contraction on drying can cause massive slippage and initiate erosion (Smith & Bradshaw, 1972). When applied as a discrete surface layer, certain ameliorants have the added problem of encouraging upward mobilization of soluble metal salts and secondary contamination of the amendment. Through the processes of root absorption, translocation, senescence and decay, this can lead to accumulation of metallic residues in upper soil horizons (Johnson et al., 1978).

Modified surface treatment

Because of the pressing need to find effective solutions to the revegetation problems of contaminated land, attention has naturally been turned towards treatment methods which could provide a significant improvement on existing systems. In the metal mine situation, faced with the problems of attaining (1) a permanent, low maintenance vegetation cover capable of accepting recreational pressure, and (2) a sward which is effectively independent of the toxic tailings, several surface treatment trials have been established. These have the objectives of (1) preventing root accumulation of metals and the re-appearance of toxicity symptoms in the long term, (2) eliminating transference of metals between the toxic waste and amendment materials, and (3) providing a vegetation cover in which nutrient cycling and plant growth proceed independently of regular maintenance.

Many of the regression problems experienced with conventional treatments centre on the high nutrient status and physical characteristics of the amendments. The former leads to vigorous growth and rapid root development towards the amendment/waste boundary, and the texture and structure tend to encourage upward movement of metals in the evapotranspiration stream because there is no physical barrier to capillary conductivity. With amendments of high cation exchange capacity, metals may be held tenaciously on the exchange complex so surface accumulation occurs.

Field trials monitored on several counts during the last four years have shown that innocuous mineral wastes such as non-pyritic colliery shale, slate quarry spoil and graded (5-30mm) limestone chippings applied at depths of 150-900mm have considerable potential as alternatives to topsoil (Johnson et al.,1977). The low nutrient status of these wastes can be advantageous in that root development, sward productivity and species composition can be manipulated in line with land use objectives more easily than with topsoil. Moreover, upward movement of metals can be reduced considerably by careful selection and grading of the amendments.

Apart from the physical discontinuity which inhibits capillary movement at the amendment/spoil boundary, the low cation exchange capacity of, for example, colliery spoil tends to minimise retention by providing an environment in which metal movement is in a state of flux. In many cases metals transferred to the amendment profile are not complexed or tightly bound as is the case with topsoil and organic-based materials. Consequently, they are subject to leaching according to seasonal changes in water balance and movement. The potential problems of using non-toxic mineral spoils must however, be recognised. The greater susceptibility of the cover vegetation to drought, the importance of keying the amendment to the supporting substrate for increased stability, and the need for preliminary treatment of the amendment material are important considerations.

In the urban environment, modifications of this system based on the principles of a break layer are being examined. The potential of this approach involving barriers of calcareous gravels and coarse sand or limestone etc., has yet to be established in the redevelopment of contaminated land for housing purposes.

Direct seeding

Active non-ferrous metal mines in England and Wales, the largest of which are working deposits of tin (Cornwall) and lead/fluorspar (Derbyshire), are subject to constraints on positive waste utilization similar to those described previously. The principal impacts on the terrestrial environment are caused by tailings dams which, though small by international standards, are significant from a landscape viewpoint because of their shape, size and location.

This is particularly true in the Peak District National Park where work has been carried out on disused fluorspar tailings dams to stabilize the surfaces against wind re-entrainment, and to minimise aesthetic conflicts with the surrounding countryside. Whilst the scale of these operations presents a series of practical reclamation problems, the efficiency of modern processing methods has reduced one of the major limitations to plant growth, namely metal toxicity (Fig. 3). Considering the scale of revegetation works, the scarcity of suitable amendment material and the low load-bearing properties of the matrix, it is doubtful whether restoration could have been achieved without these beneficial changes in metal recovery. By careful selection of species mixtures and fertiliser treatments, and use of specialized seeding methods and planting systems (Johnson et al., 1976), the environmental impact and landscape conflicts of future fluorspar mining schemes could be reduced considerably.

Combination treatments

It is only during the last few years that the environmental quality hazards posed by specific metalliferous workings have been given full recognition. Recent surveys have been carried out to establish a base-line databank for identifying problem sites (Welsh Development Agency & University of Liverpool, 1978), and remedial measures have already been applied to a derelict lead-zinc working in the Conwy Valley, Gwynedd, north Wales (Parc mine : SH 787603) with a history of environmental problems. Estimates suggest that during the period 1974-78 over 13000

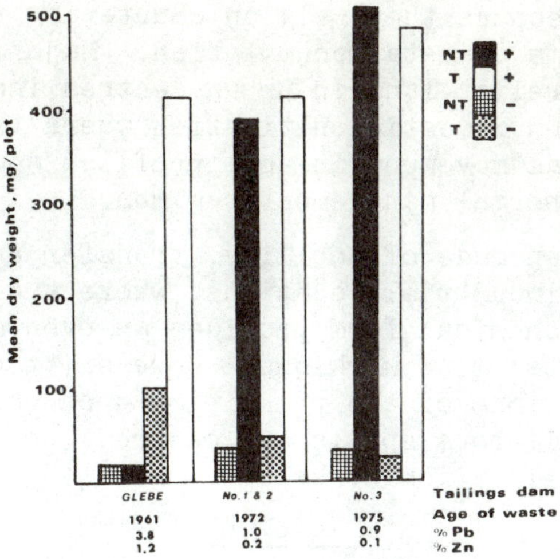

FIG. 3 Growth of lead-zinc tolerant (T) and non-tolerant (NT) Festuca rubra in the presence (+) and absence (-) of NPK fertiliser, on lead/fluorspar tailings of different ages

tonnes of mineralized waste (equivalent to 43t Pb, 204t Zn and 1t Cd) have been eroded episodally from the main spoil deposit of 265000 tonnes. The impact of this dispersal on agricultural pastures flanking the River Conwy, and the remaining hazards to the main river, its estuary and coastal waters, were considered sufficient to justify reclamation works commencing in 1977.

In view of the physicochemical properties of the spoil tip, its location, and the restricted availability of amendment material, no single reclamation system could be guaranteed to meet the engineering and biological requirements of the redevelopment work. Because of concern about the stability of the final landform, which was dictated by the restricted area available for recontouring, neither direct seeding nor surface treatment could alone be contemplated. This was overcome by combining the two approaches so that the risks of erosion pockets and physical displacement were eliminated. Whilst the long term success of this policy cannot be guaranteed, site works having been completed very recently (October, 1978), this combination system offers several technical and practical advantages in the handling of very difficult contaminated mine sites and is likely to be employed elsewhere.

CONCLUSIONS

Recognition of the potential health and environmental hazards associated with different types of metal contaminated land has provided the incentive for a series of revegetation trials during the last decade. The exact objectives of these trials differ for individual areas depending in particular on land use proposals, but in all cases a treatment is required which effectively counters heavy metal toxicity by preventing accumulation of metallic residues in vegetation.

In general the non-ferrous metal mine environment presents the most intensive problems from a plant growth viewpoint but, because of the less demanding afteruses of these sites, their treatment requirements may be no more exacting than for the very sensitive urban-industrial locations where new housing schemes are proposed.

Tolerant plant populations are unlikely to be of great value in the urban environment because they rely on countering rather than avoiding the problems of metal accumulation. Major considerations in the recycling of derelict land in urban centres include intensive recreational uses and domestic cultivation where the objectives must be to eliminate metal movement in soil profiles and to prevent root accumulation under normal management regimes.

Few studies have been made of mobility, transfer and bioaccumulation of metals in urban-industrial locations, where the range of metallic elements and their chemical form provides an even more complex environment than metal mine workings. However, there is reason to believe that adaptations of the principles upon which surface treatment of metalliferous spoils are centred, could form the basis of successful rehabilitation works.

REFERENCES

BAKER, A.J.M. (1978). Ecophysiological aspects of zinc tolerance in *Silene maritima* With. New Phytol., 80, 635-42.

BALL, D.F. (1960). *The soils and land use of the district around Rhyl and Denbigh*. Sheets 95 and 107. Mem.Soil.Surv.Eng. and Wales. London, HMSO.

COLBOURN, P. and THORNTON, I. (1978). Lead pollution in agricultural soils. J.Soil Sci., 29, 513-26.

DAVIES, B.E. and ROBERTS, L.J. (1975). Heavy metals in soils and radish in a mineralised area of Wales, Great Britain. Sci.Tot.Env., 4, 249-61.

GREGORY, R.P.G. and BRADSHAW, A.D. 1965). Heavy metal tolerance in populations of *Agrostis tenuis* Sibth and other grasses. New Phytol., 64, 131-43.

JOHNSON, M.S., BRADSHAW, A.D. and HANDLEY, J.F. (1976). Revegetation of metalliferous fluorspar mine tailings. Trans.Inst.Min.Metall., 85A, 32-7.

JOHNSON, M.S. and BRADSHAW, A.D. (1977). Prevention of heavy metal pollution from derelict mine sites by vegetative stabilization. Trans.Ins.Min.Metall., 86A, 47-55.

JOHNSON, M.S., McNEILLY, T. and PUTWAIN, P.D. (1977). Revegetation of metalliferous mine spoil contaminated by lead and zinc. Environ. Pollut., 12, 261-77.

JOHNSON, M.S., ROBERTS, D. and FIRTH, N. (1978). Lead and Zinc in the terrestrial environment around derelict metalliferous mines in Wales (U.K.) Sci. Tot. Env., 10, 61-78.

LAGERWERFF, J.V. (1972). Lead, mercury and cadmium as environmental contaminants. In *Micronutrients in Agriculture*. pp.593-628. Madison, Soil Sic. Soc. Amer.

PORTER, E.K. and PETERSON, P.J. (1977). Arsenic tolerance in grasses growing on mine waste. Environ. Pollut., 14, 255-65.

ROBERTS, R.D. and JOHNSON, M.S. (1978). Dispersal of heavy metals from abandoned mine workings and their transference through terrestrial food chains. Environ.Pollut., 16, 293-310.

SMITH, R.A.H. (1973). *The reclamation of old metalliferous mine workings using tolerant plant populations* Ph.D.thesis, University of Liverpool.

SMITH, R.A.H. and BRADSHAW, A.D. (1972). Stabilization of toxic mine wastes by the use of tolerant plant populations. Trans. Inst. Min. Metall., 81A, 230-7.

STREET, H.E. and GOODMAN, G.T. (1967). Revegetation techniques in the Lower Swansea Valley. In *The Lower Swansea Valley Project* (ed.Hilton, K.J.), pp.71-110. London, Longmans-Green.

THORNTON, I. (1969). The application of regional geochemical reconnaissance to agricultural problems. Ph.D.thesis, University of London.

THORNTON, I. (1975). Applied geochemistry in relation to mining and the environment. In Minerals and the Environment (ed. Jones, M.J.), pp.1-16, London, Institute of Mining and Metallurgy.

WELSH DEVELOPMENT AGENCY and LIVERPOOL UNIVERSITY (1978). Survey of abandoned metalliferous mines in Wales, Vols.1-5,156 pp.

ASPECTS OF MINE WASTE RECLAMATION IN CENTRAL CANADA

E. M. Watkin, Department of Crop Science,
University of Guelph, Guelph, Ontario, Canada

INTRODUCTION

Central Canada, comprised of two provinces, Ontario and Quebec, is divided into two major geological regions, the Canadian Shield and the Great Lakes - St. Lawrence Lowland. The Shield region dominates the mountain areas of each province. Composed of rocks of Precambrian age it has many natural resources, and for over fifty years has been Canada's leading source of metals. Major minerals being recovered on a commercial basis include copper, lead, zinc, gold, silver, nickel, uranium and iron. The Great Lakes - St. Lawrence Lowland region is underlain by limestone, dolomite, sandstone and shale. Apart from its agricultural importance, it is a large producer of aggregate for the intensively settled areas of the region. Only a few metaliferous occurrences have been found. In the southern portion of Quebec a third geological region occurs, the Appalachian Orogen, which is the northeastern extension of a much longer belt in the United States. Asbestos mining is a very large industrial activity in the region.

Land directly disturbed by hard rock mining activities in central Canada was surveyed by Murray (1977) as part of a country-wide assessment. Some 15,000 hectares, or 57 square miles, are listed as having been disturbed by overburden removal, waste rock and tailings deposition (Table 1) in the region. That area represents sixty-eight per cent of the total area which has been disturbed by mining in central and Atlantic Canada. Assessment of disturbed areas was undertaken using LANDSAT-1 and LANDSAT-2 satellite imagery. Certain limitations of the technique were recognized but the data obtained were considered a relatively acceptable assessment of the extent of disturbed areas larger than ten hectares when surveyed in 1974.

Approximately ten per cent of the land disturbed by mining in central Canada has been reclaimed to varying degrees. Indeed, this proportion is correct for the country as a whole. As pointed out by Ripley, Redmann and Maxwell (1978), some of the land was disturbed long before environmental concern or reclamation were ever considered. In numerous instances reclamation will remain impossible until presently active operations are closed. Nevertheless, these authors feel such figures make it clear that reclamation has not been a high priority in Canada in the past. Canada is a vast country, and there has been little pressure toward efficient use of all land space. Most mining operations occur in areas distant from the country's population centres. Under these circumstances, therefore, the pressure to reduce the environmental impact of mining is much less than in the more densely populated areas of Europe and the eastern United States, where such activity can occur in or adjacent to large metropolitan areas.

Within the past five years there has been an upsurge both in interest and effort towards land reclamation in mined land areas. Prior to this time, most effort had been directed towards the improvement of air and water quality standards in mining regions. It is only within recent years that recognition has come that improvement in water quality around mined areas is also partly dependent upon the control of contamination caused by wind and water erosion from disturbed land areas.

Table 1

Extent of Land (Hectares) Disturbed and Subsequently Reclaimed by Hard Rock Mining in Ontario and Quebec (1)
(from Murray 1977)

	Disturbed		Reclaimed	
	Ontario	Quebec	Ontario	Quebec
Asbestos	77	1,863	2	197
Base Metals - (Cu, Pb,Zn,Ni.)	2,596	2,034	279	62
Gold/Silver	2,714	778	265	238
Iron Ore	2,465	1,686	259	62
Uranium	512	-	13	-
Molybdenum	119	-	0	-
Titanium	-	49	-	0
Provincial Total	8,483	6,410	818	559
Area Total	14,906		1,377	

(1) Includes rock dumps, overburden piles and tailings deposits.

RECLAMATION CASE HISTORIES

A. Wastes Containing Iron Sulphides

Acid tailings seepage was not identified in Ontario until the mid-1960's (Hawley 1972). Subsequently the problem was recognized at other base metal mining areas in Quebec, Atlantic and Western Canada. The seepage of acid solutions from tailings deposits is of greater concern than the acid mine drainage which typifies coal fields in the eastern and mid-western United States. Acidity values of seepage water can be 15-fold those reported for acid mine drainage in the eastern United States (Caruccio - personal communication).

1. Nickel Tailings - Copper Cliff, Ontario

The reclamation program undertaken by Inco Metals at Copper Cliff, Ontario dates from 1917, when disturbed areas around the town were first revegetated. During the 1950's, an intensive program was begun to establish vegetation on extensive areas of tailings. As a result of that program, some 1,000 acres (400 ha) of nickel tailings are now vegetated. In terms of iron sulphide content, the tailings deposited at Copper Cliff are classified as low, as the majority of the acid-producing iron sulphides contained in the original ore are removed during the nickel extraction process and used in sulphuric acid manufacture. Nevertheless, the tailings have to be treated with agricultural limestone (3-5 tons per acre/7-9 tonnes per hectare) before vegetation can be established. Precise details of the procedure for establishing vegetation have been described by Peters (1978).

Originally, vegetated tailings were managed on an agricultural basis with annual cutting of grass for hay. However, the observation that trees voluntarily established in vegetated tailings caused a change of approach in tailings management. Reforestation trials were undertaken to supplement naturally occurring tree establishment. In 1974, plans were drawn up to develop the tailings and adjacent area as a Wildlife Management Area. Planning was completed by 1976 and since that time investigations of various components of the plan have been carried out to determine the final route which will be taken. At present, investigations are being made as to whether or not heavy metals accumulate in the blood or muscles of waterfowl using the area, in small mammals or insects inhabiting it, the pattern of tree root growth and iron pan penetration by roots, as well as bird census, nesting and banding studies (Peters 1978).

2. Copper Tailings - Noranda, Quebec

In 1974, following initial reclamation attempts, Noranda Mines Ltd. (Horne Division) initiated a joint research programme with the University of Guelph to devise reclamation methods for copper tailings areas containing iron sulphides (pyrite and/or pyrrhotite) that fell within their jurisdiction in northwestern Quebec. The programme had several objectives: 1) establishing vegetation on acid tailings (pH 1.5-3.0) to control wind and water surface erosion, minimise seepage by evapotranspiration from the plant canopy and increase the aesthetic appearance of the waste areas; 2) evaluate the effect of vegetation as a means of controlling the oxidation of iron sulphides in fresh tailings (pH 9-10); 3) characterization of tailings composition in order to design reclamation programmes for deposited tailings and future tailings from non-mined ore bodies containing iron sulphides.

Several methods of amending acid tailings were evaluated through growth room and on-site studies. Amendments evaluated included agricultural limestone, rock phosphate, dusts from cement and phosphate fertilizer manufacture (Table 2), anhydrous ammonia and fly ash. The efficacy of the various amendments, as indicated by plant growth responses, was very variable, both between and within tailings areas. Agricultural limestone is now accepted as the amendment material which provides the most consistent amelioration, though the rate of application will vary considerably within a particular site (Table 3). It is now possible to make specific recommendations on the use of agricultural limestone in large scale reclamation projects. A standard form of plant assay test has been developed to determine the agricultural limestone requirements necessary for sustained plant growth to occur on a tailings area (Fig. 1).

The addition of sufficient fertilizer is essential, particularly phosphorus (Table 4). Fertilizer requirements are variable from site to site, as well as within a site, but the minimum requirement determined to date for any site is 1500 lbs/acre 5-20-20 (1680 kg/ha), with or without additional phosphorus. Nitrogen fertilization is kept to a minimum, since the primary constituent of the seeds mixtures utilized is a legume, birdsfoot trefoil (*Lotus corniculatus* L.). Too high a nitrogen rate

Table 2

Shoot Growth of Birdsfoot Trefoil on Site 1 Tailings Samples Treated with Various Amendments

Grams/Pot after 150 Days

Amendment material	t/ac.	Sample Areas 3-5	3-6	3-7	3-8
Agricultural limestone	25	0.07	2.23	2.50	1.87
	40	1.54	1.99	1.90	2.30
	60	1.63	2.23	2.31	1.63
Cement dust	25	0	0	0.14	2.09
	50	0	0	0.18	0.01
	75	0	0	0	0
Phosphate dust	25	0	0	0	0.70
	50	0	0	0.20	1.01
	75	0	0	0.08	0.55
Agricultural limestone + rock phosphate	10-0	0	0.10	0.28	2.36
	10-10	0	2.18	1.75	4.02
	10-20	0.34	2.33	2.80	3.66

Table 3

Shoot Growth (Grams Dry Weight/Pot) of Three Species on Site 2a Samples after 165 Days

Assay Species	Agricultural limestone tons/acre	Quemont slope sample number I	II	III	IV	V	VI	VII
Birdsfoot trefoil	25	2.25	2.30	0.0	1.10	0.0	0.0	0.0
	50	1.80	1.80	0.0	1.59	1.80	1.05*	0.0
	75	1.90	1.85	0.0	1.68	1.37	2.23	0.0
	100	2.40	2.21	0.15*	1.80	1.98	1.32	0.0
	125	2.06	1.81	0.18*	1.32	1.64	1.69	0.01*
Creeping red fescue	25	0.95	0.93	0.0	0.73	0.14	0.0	0.0
	50	1.16	0.92	0.0	0.84	0.71	0.57	0.0
	75	1.00	1.14	0.12*	1.00	0.65	0.59	0.0
	100	0.82	0.93	0.33	0.89	0.66	0.61	0.0
	125	1.13	0.61	0.22	0.95	0.70	0.50	0.16*
Tall fescue	25	1.60	1.48	0.0	1.10	0.0	0.0	0.0
	50	1.13	0.97	0.0	1.05	1.29	0.69	0.0
	75	1.66	1.18	0.68	1.40	1.28	0.85	0.0
	100	1.26	1.13	0.78	1.44	1.22	0.80	0.0
	125	1.24	1.21	0.76	1.23	1.08	0.65	0.14*

* Partly dead, sustained growth not expected.

___ Acceptable level of plant growth.

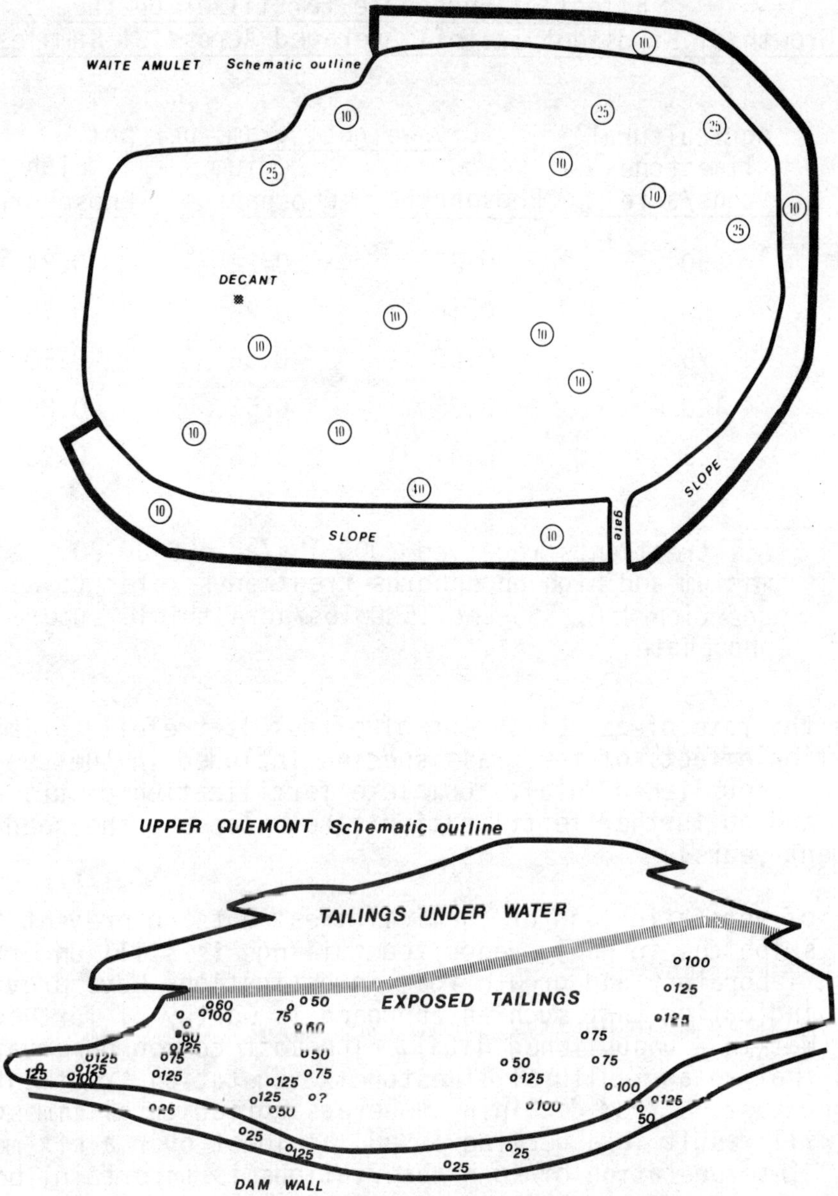

Fig.1 - Schematic outline of two tailings sites for which agricultural limestone requirements have been determined. Each site is approximately 80 acres (32 hectares). All figures in tons per acre.

Table 4

Effect of Phosphate Fertilizer on the
Shoot Growth of Birdsfoot Trefoil Averaged Across 31 Samples from Site 2.

Agricultural limestone tons/acre	Dry weight, grams per pot		
	Low Phosphorus	Medium Phosphorus	High Phosphorus
50	0.01	0.22	0.24
60	0.16	0.25	0.39
75	0.23	0.36	0.50
100	0.36	0.53	0.86
125	0.41	0.65	0.93

All treatments received 2000 lbs/acre 5-20-20. Low, medium and high phosphorus treatments refer to an additional 0, 500 and 1500 lbs/acre triple superphosphate.

reduces the rate of establishment of birdsfoot trefoil, primarily through competition effects of the grass species included in the seeds mixture (Watkin - unpublished data). Complete fertilization occurs at time of seeding and no further fertilization takes place in the seeding or subsequent years.

The use of vegetation and/or chemical treatments to prevent the oxidation of iron sulphides in newly deposited tailings is still uncertain. Nevertheless, laboratory and growth room investigations have provided sufficient indication that such an approach is worthy of further investigation (Watkin - unpublished data). The most common observation made to date is that an agricultural limestone x vegetation treatment will prevent container pH from declining, whereas agricultural limestone, used alone, will result in a decline in pH, at least over a six-month test period. Interpretation of such observations is uncertain, because of the inability to satisfactorily characterise the composition of the tailings samples under investigation. Numerous attempts have been made to determine a relationship between tailings composition, agricultural limestone requirement and plant growth. Over a three-year period, various analyses of mineralogical and chemical components were undertaken (e.g.Anon.1975). None of these have provided a good correlation with degree of plant growth recorded in growth room tests, for example, analysis of water soluble extracts (Table 5).

Uranium

Extensive studies have been made in order to develop revegetation techniques for acid tailings arising from extensive uranium mining in the Elliot Lake region of Ontario. Composted organic wastes, sewage sludges, rock phosphate and agricultural limestone were shown to be suitable amendments in laboratory trials (Murray 1972, Watkin and Winch 1974). Subsequent field trials demonstrated that agricultural limestone was the most efficient and cheapest form of amendment for the highly acid

Table 5

Comparison of Water Soluble Extract Analysis - Tailings Grouped by Limestone Requirement

Tailings Sample Number	Agric. lime-stone require-ments	ppm oven dry material												
		Mg	Mn	Ni	Pb	Cd	Cr	Zn	Cu	K	Na	Fe	Al	Ca
B4	10	300	24	0.9	0.2	0.2	0.3	55	12	0.2	15	1,250	475	2,250
B5	10	1,100	15	0.2	0.2	0.9	-	5	9	3.0	20	1.5	475	2,250
B10	10	1,250	72	1.4	0.1	0.1	0.1	30	5	0.2	10	150	2,000	2,025
B2	30	20	3	0.6	0.2	0.1	0.2	6	17	1.0	25	2,250	55	2,300
B3	30	700	47	1.4	0.5	0.2	1.0	40	18	0.2	15	1,050	1,075	2,203
B6	30	800	47	0.7	0.2	0.1	0.7	25	8	0.5	10	400	1,100	1,600
B9	30	120	7	1.8	0.7	0.2	0.5	75	7	1.0	13	2,000	250	240
B1	50	51	7	1.3	0.7	0.2	0.2	10	9	0.7	25	4,750	125	155
B11a	70	68	3	0.3	0.1	0.1	0.2	4	13	1.8	19	400	128	100
B11	70+?	25	10	0.4	0.2	0.1	0.1	4	4	4	20	175	55	625

tailings materials (pH 1.9-2.3). Coarse and fine tailings (73 and 96 per cent + 300 mesh, respectively) differed in the amounts of agricultural limestone required to neutralize tailings acidity. The former required eight tons per acre (18 t/ha), the latter 32 tons per acre (72 t/ha) (Murray and Moffett 1977). In 1978, two reclamation contracts for some 140 acres (66 ha) were let by one of the area mining companies and by a federal government department. Both contracts called for the establishment of a grass/legume mixture by seeding directly into the tailings, following application of agricultural limestone and fertilizer.

The establishment of a vegetative cover on the uranium tailings at Elliot Lake will obviously reduce wind and water erosion of the various tailings areas' surfaces. Since those tailings contain 85 per cent of the total radioactivity originally present in the ore, reduction in surface erosion is essential if contamination of adjacent land areas is to be prevented. However, vegetation established on uranium tailings may concentrate radioactive isotopes. Moffett and Tellier (1977) compared the level of various radioisotopes in four grasses species grown on uranium tailings for four years with the same species grown in potting soil. Of the radioisotopes studied (uranium, thorium, radium-226, lead-210 and polonium-210) only uranium and radium-226 were significantly higher in the grasses from the tailings than from the control. The concentration levels recorded for uranium and radium-226 were below those permitted by federal government water quality standards.

B. <u>Wastes Not Containing Iron Sulphides</u>

1. <u>Asbestos Tailings, Quebec</u>

The problems associated with the establishment of vegetation on serpentine asbestos mine wastes were reviewed by Moore and Zimmermann, 1977. Inadequate mineral nutrition was considered the main factor preventing plant growth. Because of the high pH (9.0-9.2), toxicity due to the presence of chromium, iron and magnesium was less than anticipated. It was cautioned that, following asbestos waste amendment and plant establishment, a decline in pH could occur with time in the immediate vicinity of the plant root, with consequent detrimental effects.

Field trials conducted over a four-year period showed that applications of high rates (1 kg m^{-2}) and organic waste (4 kg m^{-2}) enabled a satisfactory vegetative cover to be established (Zimmermann and Moore 1977). The addition of the organic waste was considered an absolute necessity. However, at the end of that period, severe nitrogen deficiences were apparent, and fertilizer maintenance treatments would be necessary to ensure continued success.

Logistical problems associated with applying and incorporating organic wastes, seed and fertilizer materials on large conical dumps with angles of rest of 35° - 40° were recognized by the above investigators, as were potential difficulties for the retention and availability of nitrogen on the alkaline wastes. At present both these problems remain unresolved in terms of undertaking a practical, large-scale reclamation program.

2. <u>Iron Ore Tailings, Ontario and Quebec</u>

No published information is available on the revegetation of iron-ore tailings. Several companies have initiated small-scale studies. To date, the results of these would indicate that there should be no biologically-based problems in establishing vegetation on the neutral to slightly alkaline wastes (pH 6.9-8.0). Rather, the major problem is one of logistics, based on waste area size and/or distance from centres capable of providing equipment for carrying out revegetation work.

REFERENCES

1. ANON. 1975. Study of unoxidized tailings for Noranda Mines Ltd. Report No. 17606, Mineral Research Branch, Ministry of Natural Resources, Toronto.

2. MOFFETT, D. and TELLIER, M. 1977. Uptake of radioisotopes by vegetation growing on uranium tailings. Can.J.Soil Science 57, 417-424.

3. MOORE, T. R. and ZIMMERMANN, R.C. 1977. Establishment of vegetation on serpentine asbestos mine wastes, southeastern Quebec, Canada. J.appl. Ecol. 14, 589-599.

4. MURRAY, D.R. 1977. Pit slope manual, Supplement 10-1, Reclamation by vegetation = Volume 2 - Mine waste inventory by satellite imagery; Canada Centre for Mineral and Energy Technology, CANMET Report 77-58, 216 p; Ottawa, Canada.

5. MURRAY, D. and MOFFETT, D. 1977. Vegetating the Elliot Lake uranium mine tailings. J.Soil and Water Conserv. 32(4), 171-174.

6. PETERS, T.H. 1978. INCO Metals reclamation program. Canad. Inst. Mining Bull. December, p. 104-106.

7. RIPLEY, E.A., REDMANN, R.E. and MAXWELL, J. 1978. Environmental impact of mining in Canada. Centre for Resource Studies, Queen's University, Kingston, Ontario, Canada. pp.274.

8. WATKIN, E.M. and WINCH, J.E. 1974. Composted organic wastes,sewage sludges and rock phosphate for the amelioration of acid uranium tailings. Proc. Int. Conf. Land for Waste Management, Ottawa, Canada. pp.48-56.

9. ZIMMERMANN, R.C. and MOORE, T. R. 1977. Study of vegetation of asbestos mine wastes, Eastern Townships, Quebec. Supplementary Report No.3 to the Environmental Control Committee, Quebec Asbestos Mining Association. 22 p.

DISCUSSION

Paper C1

There is a significant positive relationship between levels of cadmium in a soil and in some plants grown on it but this relation is influenced by other factors which have not yet been quantified.

Statistical analysis of the data on cadmium levels had provided a graph with a positive intercept on the y axis of Cd levels in the soil versus those in the crop. Factors, other than the statistical method itself, are aerial deposition of metals and possible errors in analytical methodology when working at these very low levels.

Paper C2

The choice of test crops for use as indicators in garden soils where soil fertility may be suboptimal and the need to repeat field trials over a number of years to allow for varying weather conditions were points raised in discussion.

Dr R D Davis said that league tables of crop sensitivity to particular elements in soils had been prepared and a copy of the table for cadmium is shown. It is based on cadmium concentrations in the edible parts of crops grown on two contaiminated soils. The greatest hazard is presented by lettuce; the least by pea.

In order to assess worst-case conditions, lettuce should be grown as the indicator crop although it would be realistic to grow potato as well because it is a more important plant in terms of usual dietary intake.

Uptake of elements is thought to be greater in glasshouse conditions and, therefore, a single test in the glasshouse would probably take account of the worst possible result that might occur under field conditions. The need is to conduct rapid plant tests which can be used for routine purposes. Nevertheless, the direct information on hazard provided by the plant test as described justifies the extra time needed to carry it out.

Dr B E Davies (Aberystwyth) added that in a recent survey of a large number of garden soils it had been found that both heavy metal content and general soil fertility tended to climb with length of the period under garden cultivation.

League table of crop sensitivity to elements in soil

Cadmium (Zootoxic: edible parts)

TOBACCO	Sensation Hybrid (Ornamental)	(310)
LETTUCE	Tom Thumb	100
LETTUCE	Webbs	77
LETTUCE	Paris White Cos	73
SPINACH	Bloomsdale	58
CELERY	Golden selfblanching	47
CABBAGE	Greyhound	40
KALE	Maris Kestrel	38
MANGOLD	Yellow Globe	15
BARLEY	Aramir	13
BEETROOT	Boltardy	13
TURNIP	Snowball	13
BARLEY	Mazurka	12
WHEAT	Spartacus	11
BARLEY	Abacus	10
WHEAT	Sappo	10
BEETROOT	Detroit	9
TURNIP	Bruce	9
LEEK	Musselburgh	9
TOMATO	Moneymaker	9
OAT	Leander	8
OAT	Trafalgar	8
RYEGRASS	S24	7
RAPE	Orpal	7
PARSNIP	Giant Exhibition	7
SUGAR BEET	Sharpes Klein Monobeet	7
CARROT	Standard Improved	7
RADISH	French Breakfast	6
BARLEY	Julia	6
SWEET CORN	Golden Early	5
BARLEY	Ark Royal	4
SWEDE	Acme	4
COURGETTE	Zucchini	3
POTATO	Desirée	3
MAIZE	Caldera	3
SUNFLOWER	Tall Single	3
FRENCH BEAN	Canadian Wonder	1
PEA	Onward	1

Paper C3

The choice of leeks as an overwintering vegetable in the Thamesmead trials was queried because of the difficulty of getting samples entirely free of contaminating soil if the objective is to measure uptake through the root system. Mr Jones replied that gross surface contamination was removed by washing in de-ionised water. The intention in the Thamesmead field trial is not to investigate root uptake but to determine metal concentrations likely to be ingested through eating crops grown on areas treated thus.

More information was sought on the barrier treatments at Thamesmead, especially on plot site preparation and the soil suction characteristics. In reply, Mr Jones said that compaction had been difficult on the small plots used in this trial but it was hoped to investigate soil water relations in collaboration with staff from Liverpool Polytechnic.

A difficulty foreseen by one questioner was that break barriers with a high cation exchange capacity might get saturated and become ineffective while another thought that efforts to maximise compaction could be counter-productive with certain materials. The use of a 'Terram' sandwich filled with clean aggregate as a capillary barrier was said to have been successful over a four-year period in land reclamation around the coast of the Arabian Gulf and the Red Sea. The matting prevented silting-up of the air-spaces in the aggregate barrier from above and also prevented settlement of the barrier into the ground below.

The value of a soil-marker, such as silicon, as a control in the analytical procedure was mentioned by one questioner. Mr Jones said that none had yet been used in this work but in other research titanium had been determined for this purpose and this would be investigated.

Paper C4

In reply to a question on the economics of using the material described by Dr Heller he said that the relatively high cost of cation exchange resins would preclude their use in agricultural cropping situations but they may be justified in other situations where low concentrations of resins might assist the establishment of self-supporting vegetation as for example in amenity and recreation areas. Their use may also be justified in areas of high population or in the immediate vicinity of metallurgical works and the like where land prices are high or the soil may be severely contaiminated.

Paper C5

In reply to a query on the use of the Wolfson Geochemical Atlas of England and Wales as a guide to background levels of metals in soils, Dr Thornton explained that the atlas was based on stream sediments which are themselves composite samples of the metals existing in the catchment area represented by each sample. The atlas was useful as a primary reconnaisance because it focussed attention on specific areas which could then be delieated for detailed local investigation.

Paper C6

Questioned about the form in which the metals had been added to the test soils and their subsequent incorporation in the soils with adjustment for pH values, Mr Williams replied that metals had been added to the soils in the form of their inorganic salts (sulphates), watered in and thoroughly mixed by cultivations each year from 1948 to 1961 in the case of the results presented for Zn and Cu in Tables 3 and 4. In the pot experiment described in Tables 5 and 6, copper sulphate was ground with a small quantity of the soil and then mixed into the bulk of the soil. In the case of nickel, the chloride salt was used. The pH value of the original soil was 5.7 and a portion of the area was given the equivalent of 5 t ha^{-1} of ground limestone to raise the pH to 6.4.

Paper C7

It was suggested that the Liverpool Polytechnic investigation of trace metal uptake by soft fruit at Beaumont Leys near Leicester should be extended to include brambles which are often the first fruit to colonise waste ground. Dr Lepp agreed and said that it was intended to include them in the fruit plot in the next stage of that trial.

Mean lead contents of 0.79 mg kg^{-1} have been reported recently (J G Farmer, "Lead in wild blackberries from suburban roadsides" J.Sci.Food Agric. 1979, 30, 816-818). These levels do not exceed the MAFF recommended limits of 1 mg kg^{-1} in fresh food and 2 mg kg^{-1} in canned fruit. Apparently little of the contaminating lead is removed by washing the fruit. It has also been reported that blackberries gathered from Beckton Gasworks site had elevated cadmium levels (5 mg kg^{-1} dry weight).

Paper C7 - C10

A questioner regretted that not one of the speakers had given any public health statistics to indicate that heavy metal content of food does, in fact, present a real problem. He then suggested that Dr B E Davies' data on the Cd content of lettuce were perhaps misleading because anyone subsisting solely on a diet of lettuce would die of gross malmutrition before the intake of cadmium reached toxic levels.

Dr Davies replied that in his presentation he had been careful to talk only of potential hazard. In the case of Pb the interpretation is clear. It is an offence to sell food which does not comply with the Lead in Food Regulations and a limit of 1 mg Pb/Kg fresh weight has been set. This provides a useful standard by which homegrown vegetables, not for sale, may also be judged. It is clear from Table 4 in his paper that vegetables grown in those mining areas are often of suspect quality.

There is no legal limit for Cd in food and one has to seek other ways of eveluating Cd levels in vegetables. One such method was given in the paper but it was not his intention to suppose that anyone might subsist entirely on lettuce. Another approach is through an average figure. Dr D G Lindsay, in his paper presented earlier in the conference, had shown that an individual's weekly consumption of vegetables is approximately 2.0 kg and that vegetables contribute to Cd intake. In order to contribute the WHO maximum tolerable weekly intake of 400 - 500 μg Cd, vegetables as the sole source of Cd would have to contain 0.20 - 0.24 mg Cd/kg. Comparing with the higher figure, 5 of 8 plots in Dr Davies trial produced lettuce of suspect quality, 5 of 8 suspect carrots and 3 of 8 suspect onions. This would surely seem to be evidence of a potential hazard.

It is also worth bearing in mind that permitted levels of metal concentrations are more usually decreased not relaxed. Examples of this are the Pb levels in food and in drinking water.

Of course, it is not known whether the potential hazard is a true hazard. Some workers have attempted to associate the prevealence of diseases with environmental trace elements. Stocks & Davies (1964) reported that the ratio of Zn/Cu in soils from North Wales and Cheshire was higher in gardens associated with a death from stomach cancer by a long-term resident. The effect was more pronounced and consistent in soils taken from houses where vegetables were grown. It may be noted that Cd resembles Zn in its plant uptake behaviour. In collaboration with others Dr Davies has been attempting to relate the prevalence of human dental caries with woil chemical composition.

In Cornwall and Devon a higher prevalence of caries in 13-year old children was found where soils and vegetables were contaminated by Pb (Anderson et al, 1976, Davies, 1978). But in Shipham, Cd contamination did not appear to affect caries prevalence (Anderson et al, 1979). The link between disease and soil composition is certainly unproven but there is now enough evidence to support a view that accumulations of heavy metals by vegetables should not be dismissed as of no concern.

Anderson R J, Davies B E, James P M C, 1976 "Dental caries prevalence in a heavy metal contaminated area of the west of England" Brit. Dental J. 141, 311-314

Anderson R J, Davies B E, Nunn J H, James P M C, 1979 "The dental health of children from five villages in north Somerset with reference to cadmium and lead" Brit. Dental J. 147, 159-161

Davies B E, 1978 "Plant available lead and other metals in British garden soils" Sci. Total Envir. 9, 243-262

Stocks P, Davies R I, 1964 "Zinc and copper content of soils associated with the incidence of cancer of the stomach and other organs" Brit. J. Cancer 18, 14-24

Paper C9

A question was asked about breeding tolerant grass species for other than aesthetic purposes. Are there any that withstand intensive pasture management? and Are there any like some maize varieties that do not take up heavy metals and which could be developed? Dr Johnson said in reply that in practice very few species of agricultural importance are found growing in toxic areas and no tolerant material has been found among the better species but it would be worthwhile to continue the search.

Professor Bradshaw added that, in general, tolerant populations seem to take up only about half as much of the trace elements that susceptible species do when growing in the same soil.

Paper C10

The succesful treatment of copper tailings in Canada by applying agricultural limestone was in contrast to UK experience where ryegrass can only be grown on fertilised areas for one or two years after reclamation but Dr Watkin said that pyrite in coal tips should not be directly compared with pyrite in wastes from the processing of massive sulphide ores.

In relation to revegetation, the types of pyrite at Noranda cannot be characterised with any certainty. No colonisation by plants has occurred on untreated areas 20 years after abandonment. Pyrrhotite (an iron sulphide of composition $Fe_{x-1}Sx$) also occurs in numerous massive sulphide ores and is believed to act as a catalyst in pyrite oxidation, as well as oxidising very fast itself, and is a very important factor under the conditions at Noranda and elsewhere in Canada. No signs of regression in vegetation quality have been seen in the areas treated 12 years ago. There is evidence that the existence of a healthy stand of vegetation which includes legumes can itself prevent the oxidation of pyrite in tailings. Where regression has occurred it may have been in areas with only weak vegetative cover, such as an area of grass without legumes.

Hazards to Immediate Workers on Site

L.C. SMITH, HM Senior Chemical Inspector
Health & Safety Executive

INTRODUCTION

Any work activity on contaminated land will attract the requirements of the Health & Safety at Work Act 1974, both with regard to the protection of the workers on site and other persons who may be affected by the work activity. The effect of Health & Safety legislation on work activities on contaminated land is discussed in the paper by Mr Graham Walker, HSE. This paper describes some examples of the casework investigated by HM Factory Inspectorate of HSE, where toxic and fire hazards were encountered in land reclamation and construction work on contaminated land.

My Section in the Factory Inspectorate is concerned with Occupational Hygiene and most of the examples described are work activities on contaminated land where the nature of the contaminant presented a toxic or asphyxiation hazard. In some examples the contaminants presented a toxic and fire hazard and the fire risk to site workers from flammable contaminants is introduced briefly.

The casework selected includes examples of hazards encountered in land contaminated with asbestos, carbonaceous deposits and by seepage of hazardous liquids. The paper does not attempt to deal with all the hazards likely to be encountered in work on contaminated land. Recommendations made by HM Factory Inspectorate are given in the examples to show the measures taken for the protection of site workers. They are specific for each incident and the site conditions and should not be taken as model guidance for all situations.

RECLAMATION OF A WASTE TIP CONTAINING ASBESTOS WASTE DEPOSITS

The health hazards associated with the inhalation of asbestos fibres are nowadays well known. This example illustrates the precautions taken to protect site workers and the public from exposure to asbestos during reclamation work on a waste tip containing asbestos waste deposits.

The tip, which included a hillock of material known locally as the "white heap", was started over 100 years ago by a chemical company dumping waste consisting mainly of calcium compounds. From about 1900 the company began production of insulating materials containing asbestos. Asbestos waste was dumped at the tip and ceased there in the early 1970's.

The land containing the "white heap" was bought by a local development corporation for the purpose of reclamation and redevelopment. Part of the land was to be landscaped for a park and part was to be used for industrial development. Drillings taken at the heap confirmed its history showing it consisted mainly of a core of sulphite cake overlain by calcium carbonate with asbestos contaminated material near the edges. Sampling and analysis of the asbestos contaminated material showed that amosite, chrysotile and crocidolite were all present.

The development corporation proposed to place a contract in 1976 for levelling the heap and depositing contaminated material in a nearby landfill site. Before placing the contract they consulted all interested parties including the Employment Medical Advisory Service and the Factory Inspectorate. Special conditions of contract were drawn up detailing the hygiene measures and working practices to be followed by the Contractor. The Factory Inspectorate recommended that initially the site workers could use positive pressure powered respirators during excavation and disposal operations. It was agreed that a trial run of one day's work would be carried out with the Factory Inspectorate in attendance to determine by personal sampling what levels of airborne asbestos dust would be produced.

The work consisted of digging into the waste heap with a mechanical shovel and loading the spoil into lorries for transport to a landfill site about half a mile away. Precautions were taken to ensure that asbestos contaminated spoil exposed during excavation was disposed of in specially identified areas at the landfill site. All the site workers wore positive pressure dust respirators fitted with high efficiency filters. Personal samples were taken on membrane filters for the site workers including the shovel driver and lorry drivers. Samples of the excavated spoil were also collected for examination. Microscopic evaluation of the personal samples indicated levels of airborne asbestos dust up to about 1.0 fibres/ml in the form of amphibole asbestos. The results for airborne asbestos dust were surprisingly high considering the damp state of the spoil. Significant amounts of amosite and chrysotile asbestos were found in the spoil samples and the presence of crocidolite was confirmed. In the opinion of a Senior Chemical Inspector there were very few areas of the pit, except the most recent, likely to be free of asbestos. In the circumstances, the Factory Inspectorate recommended that site workers should continue to wear positive pressure powered respirators during excavation and disposal operations. The other hygiene measures laid down in the conditions of contract should also be maintained throughout the project.

On the following day when no digging was being done personal samplers were worn by Factory Inspectors while walking round the site. The fibre counts were less than 0.1 fibres/ml indicating that respiratory protection need not be used so long as the tip was not disturbed. However, when the surface of the tip was disturbed, particularly during periods of dry weather, there could be a significant increase in airborne asbestos dust on site. Persons outside the excavation and disposal sites were unlikely to be affected except in two places where site vehicles crossed public highways. There appeared to be no easy solution to this problem which depended very largely on climatic conditions. If the spoil could be maintained in a very damp condition this would suppress dust emissions at the excavation and tipping sites and would also obviate the need for sheeting down lorries. The Factory Inspectorate recommended that the excavated area of the heap should be kept thoroughly wet by use of water sprays or sprinklers if necessary; the approaches to the public highway crossing points should also be kept damp; and other working practices laid down in the conditions of contract to prevent the spread of asbestos contaminated material should be strictly observed.

The development corporation is to be commended for its approach to this project. The precautions and working practices adopted for the protection of site workers and the public from exposure to asbestos, and which were mainly laid down in special conditions of contract before the commencement of work, are summarised as follows:-

a) No employment of persons under 18 years of age on site.

b) Protective clothing provided for use by site workers consisting of:-

   i) Polyester cotton overalls close fitting at the neck, wrist and ankles.

   ii) Polyester cotton cap.

   iii) Rubber boots.

c) Facilities provided for vacuum cleaning protective clothing at the end of each day's work and for regular laundering, at least weekly, of dirty clothing. Dirty clothing sealed in strong plastic bags clearly marked "Asbestos Contaminated" and sent to a laundry capable of carrying out correct washing procedures.

d) Showers and lockers for site workers personal and protective clothing provided. Site workers instructed to shower at the end of each day's work and to leave their protective clothing in the lockers provided for that purpose.

e) A personal issue of a positive pressure powered respirator to each site worker for use during excavation hauling and tipping operations.

f) Periodic sampling by HM Factory Inspectorate or an organisation engaged by the development corporation for the determination of asbestos emissions produced during site operations.

g) Security measures provided to prevent entry by the public during site operations.

h) Moisture content of the spoil maintained within specified limits dousing the material within the area of operations.

i) All vehicles and plant thoroughly washed and cleaned before leaving the site. (In addition to removing deposits from wheels and exterior of vehicles the cab should be cleaned.)

j) No overloading of vehicles carting excavated material and material spilt on the haul route removed immediately.

k) At the end of each day's work all asbestos contaminated spoil covered by uncontaminated material.

TUNNELLING IN LAND CONTAINING DEPOSITS OF CARBONACEOUS WASTE

The following examples illustrate the hazards from toxic and oxygen-deficient gases encountered in tunnelling through land contaminated with coal waste or decaying organic matter. Slow oxidation of buried coal waste or the microbiological decomposition of buried organic matter generates a gaseous mixture deplete in oxygen and rich in carbon dioxide. This gaseous mixture will seep into any tunnels, shafts or pits dug into gas-laden ground. Falling atmospheric pressure will increase the rate of seepage raising the carbon dioxide concentration and lowering the oxygen content of the air in the confined excavated space. The tunneller in these conditions risks being overcome by toxic concentrations of carbon dioxide and/or asphyxiation through oxygen deficiency. Exposure to an atmosphere containing about 3% carbon dioxide leads to laboured breathing and headache, these symptoms becoming severe at 5 to 6%. Exposures to 10% result in headache, visual disturbance and tremors and may be followed by loss of consciousness. Death has been known to occur at a probable concentration of 22%. Oxygen deficiency in the absence of toxic gases or vapours produces symptoms when the concentration falls below about 16% (the flame in a flame safety lamp is extinguished when the concentration falls below 17%). At 12% oxygen content abnormal fatigue upon exertion and disturbed respiration may occur. Collapse and loss of consciousness occur at below 10% oxygen.

In this accident investigated by HM Factory Inspectorate, two men collapsed and died in a sewer tunnel being driven through filled ground. A site survey carried out prior to the start of work indicated the presence of slag, coal waste and other industrial waste in the fill which lay above coal measures. The tunnel 1m in diameter was being excavated by miners at a depth of 9 to 10m between 2 vertical shafts. At the time of the accident it had been driven about 60m leaving another 25m to be excavated. There was no forced ventilation in the tunnel, no system of monitoring the atmosphere and no system of communication between the miner and the surface.

There had been no excavation work in the tunnel for five days when two engineers descended the working shaft to check the alignment of the tunnel. After about half an hour, when the engineers did not return to the surface, two colleagues descended to the tunnel to investigate and became seriously affected after brief exposure to the atmosphere in the tunnel. One man collapsed and was dragged out by his colleague. The two engineers were later brought out by the fire service, one man being found dead and the other dying shortly afterwards.

Over a period of 1 to 4 days after the accident the Factory Inspectorate carried out atmospheric sampling within the working shaft and tunnel. A Chemical Inspector wearing self-contained breathing apparatus (2 hour oxygen set) entered the tunnel with the assistance of the fire service. Tests of the atmosphere were made with oxygen analyser, flammable gas detector and gas detector tubes. A 60m long sample tube was installed near the tunnel face to enable samples to be taken from outside the tunnel.

At the bottom of the working shaft the oxygen content was 21% and there were insignificant concentrations of toxic and flammable gases. Air samples taken from the tunnel two days after the accident indicated oxygen concentrations around 12.5% and carbon dioxide concentrations around 4.5%. Carbon monoxide was detected at 10 to 20 ppm. Other toxic gases e.g. hydrogen sulphide, hydrogen cyanide and flammable hydrocarbon gases were not detected in significant quantities. Two days later the oxygen level had risen to 16% and the carbon dioxide level had fallen to 2.5%.

The excavated material appeared to include slag, clinker, ash and coal waste. Samples on analysis indicated a combustible content ranging from 2.5 to 85%.

After the accident the contractor drilled a series of bore holes along the line of the tunnel still to be excavated. Analysis of the gases in the holes indicated carbon dioxide and carbon monoxide concentrations up to 20% and 20ppm respectively. Temperatures in the bore holes ranged from 10 to 15°C. The warm temperatures plus the presence of carbon monoxide, indicated in the opinion of a mining consultant the presence of a "hot spot" some 30 to 50m from the bore holes.

Local meteorological records showed that there had been a marked decrease in atmospheric pressure during the night preceding the accident. Under these conditions gases entrained in the fill material would have been drawn into the tunnel.

By extrapolating the tunnel air sampling results back to the day of the accident it was concluded that the carbon dioxide concentration in the tunnel most probably exceeded 4.5% and could well have approached the 20% concentration found in the bore holes. The oxygen concentration would have been correspondingly reduced and may have been too low to support human life. These conclusions were confirmed by the pathologist's report which attributed the mens' death to asphxia due to lack of oxygen and the inhalation of a gas or gases.

It was the opinion of the Factory Inspectorate that tunnelling could continue providing the following safeguards were rigorously implemented:-

a) Effective forced ventilation should be provided. Air should be blown into the tunnel at the face at a rate of about 0.24 m^3/s. The foul air from the tunnel should be extracted from the shaft at the tunnel entrance at a rate exceeding 0.24 m^3/s.

b) A permanent sample tube should be installed up to the tunnel face. Instruments at ground level should be used to monitor the tunnel atmosphere for oxygen and carbon dioxide concentration. The miner should carry a portable oxygen alarm.

c) Some communication should be established between the surface and the tunnel heading.

d) No person should enter the shaft or the tunnel until it had been established by testing that the air was respirable. If faults developed in the safety equipment entry should be forbidden and persons in the tunnel immediately called out.

In this type of small diameter tunnelling an electrically-driven trolley on rails is moved up behind the miner at the tunnel face to carry away excavated material. Rescue was complicated by the presence of the trolley which had to be removed before one man could be reached. The men were found about 37m inside the tunnel, and the firemen, wearing 45 minute self contained breathing sets used up nearly all their air supply before reaching the surface. This incident brought to light the difficulties of rescuing persons overcome by harmful gases in long small diameter tunnels. In the opinion of the Chemical Inspector investigating the accident rescue in this type of tunnel would not be possible at distances exceeding 75-90m.

The following fatal accident occurred during the preparation of a site for landscaping. The site was a disused and filled in canal which after draining some years previously had been used as a waste tip. A storm water drain, consisting of a 9" diameter semi-porous pipe with catch pots and access shafts at about 100m intervals had been laid along what had been the base of the canal for about a total length of 750m. The access shafts were about 6m deep.

Contractors were levelling the site with soil and one of the manholes to an access shaft, which was in a shallow depression, had to be raised about 1.5m. It was intended that the work should be done from outside the manhole. The manhole cover had been opened only once some years previously since construction of the drain system 12 years ago.

On the morning of the accident a bricklayer and a labourer prepared to raise the level of the manhole. Shortly after opening the manhole cover the bricklayer entered the access shaft. After a few minutes he was seen by the foreman to start to climb up from the shaft bottom, collapse and fall back. The labourer entered the shaft to attempt a rescue and also collapsed. The foreman and another man made a second rescue attempt but fortunately for them they both felt the effect of "fume" before reaching the base of the shaft and were able to climb out. The bodies of the two men were later recovered by firemen wearing breathing apparatus.

The shaft was sealed after the accident and on the following day the Factory Inspectorate took atmospheric samples from inside the shaft through a sample tube. On analysis the oxygen concentration of the samples was 10% or less and the carbon dioxide concentration was 5 to 6%. Methane was present at 0.3% and no highly toxic gases were detected.

It was concluded from the high carbon dioxide concentration and the presence of methane in the samples that the decay of organic matter in the drainage system and surrounding fill had depleted the oxygen of the air in the shaft.

The men who died were overcome in minutes of entering the shaft indicating the presence of an asphyxiating atmosphere deficient in oxygen. Although the storm water drain was open at both ends there was insufficient natural air movement to ventilate it.

The manhole could be raised without anyone entering the shaft and the Factory Inspectorate recommended that the work should be carried out under strict supervision to ensure that no person entered the shaft for any reason without wearing breathing apparatus. Before the start of each day's work and at regular (hourly) intervals the atmosphere in the vicinity of the workmen's breathing zones should be tested for excessive amounts of carbon dioxide. The Local Authority were to ensure that no members of the public had access to any part of the drainage system.

These examples also illustrate the danger of attempting to rescue persons overcome in confined spaces without proper rescue equipment and careful preparation.

TOXIC AND FIRE HAZARDS ENCOUNTERED IN LAND-CONTAMINATED BY SEEPAGE OF HAZARDOUS LIQUIDS

There are basically two causes of land contamination arising from the seepage of hazardous liquids:-

a) deliberate contamination from dumping of liquids on waste ground, or waste tips;

b) accidental contamination by leakage from containers or pipelines.

A water immiscible liquid may seep down to water saturated ground where it may spread out and travel underground above the water table. The liquid may seep into an excavation site, sometimes a long way from the original point of release. If highly toxic or flammable liquids are encountered in tunnelling the work may only be able to proceed by open trench cutting to prevent the build-up of dangerous vapour concentrations. If the risk to site workers or to other persons not involved in the site operations is extremely high the excavation may have to be abandoned.

HM Factory Inspectorate were called to investigate an incident which occurred during the construction of a trunk sewer system close to the course of a river. The sewer was being driven partly in tunnelling operations and partly by excavations where the ground fell away and there was inadequate cover for a tunnel.

Along part of the line of the sewer the land had been levelled and had been used for railway sidings. In this section there was insufficient cover for a tunnel and an open trench about 4m deep had to be cut through underlying rock by a mechanical excavator. While the excavator was working at the rock surface, after the top soil had been removed, there was a small explosion and a flash as apparently some gas was ignited. Work on the trench was stopped while the contractor's engineers took samples and with the help of a chemist from a neighbouring factory the gas was identified as carbon disulphide.

The source of the carbon disulphide contamination was most likely from spillages which seeped into the ground whenthe land was used for railway sidings. Some years ago the factory used to receive deliveries of carbon disulphide in rail tankers, the liquid being pumped from tankers to storage tanks on the opposite bank of the river. Leaks of carbon disulphide had probably occurred from equipment used during the transfer operation.

The quantity of carbon disulphide which had seeped into the underlying rock and its spread were unknown. In view of the highly toxic and flammable risk to the site workers and additionally the risk of igniting a large pocket of gas, which could damage a gas main laid within a few metres of the line of the excavation, adoption of a more cautious method of excavation was advised by the Factory Inspectorate.

After discussion with the contractor of the possible methods it was agreed to investigate the possibility of diverting the course of the sewer to bypass the contaminated area. No one could be certain that no more gas would be encountered even when excavating well clear of the original course and excavation should in any case proceed with caution. Trial pits would be dug, using an excavator with a long arm, along the line of the proposed diversion route. The excavations would be carried out carefully, removing small quantities at a time, while tests for carbon disulphide would be carried out at frequent intervals as the work proceeded.

In this next incident workers were exposed to a toxic risk by inhalation and/or skin contact from coal tar residues seeping into an excavation for a storm water drain. A large diameter concrete storm water drain was being constructed below the bed of a tidal creek flowing into the river Thames. For part of its course the creek ran adjacent to the site of an old gasworks. Construction involved excavating an open trench about 6m deep by 3 m wide, laying the pipe sections in concrete then back filling.

As construction reached the vicinity of the old gasworks site it was noticed that, in addition to surface water, what appeared to be coal tar residues were seeping into the excavation at the gravel-clay interface about 2m above the bottom. The contractor had samples of residues analysed and they were found to consist of a mixture of cresoles.

Apart from pumping out accumulated water and residue further excavation work was suspended pending advice from the Factory Inspectorate. At the time of the Factory Inspector's visit there was a substantial deposit of residue on the soil at the bottom of the excavation.

Factory Inspectorate recommendations for the protection of the men working in the excavation are summarised as follows:-

a) Impermeable protective clothing, gloves, boots and face shields should be worn.

b) Accidental splashes of residue on skin should be removed immediately by washing with soap and water. (Good washing facilities including showers were provided on the site).

c) Dilution air should be blown through flexible trunking as close as practicable to the working position at the bottom of the excavation to minimise the risk of cresole vapour accumulation.

d) Periodic tests should be made to check the effectiveness of the dilution ventilation.

While construction engineers were tunnelling under a runway at a major airport for the installation of a new fuel line aviation fuel vapour was detected. The work was stopped by the Factory Inspectorate. Tunnelling was abandoned and excavation was allowed to continue by open trench cutting. It was concluded that over many years seepage from leaking fuel lines and spillages had percolated through the sub soil to the water table under the airport.

There is no way of knowing how far immiscible liquids may travel underground above the water table. A refinery was founded in the mid 19th century for the refining of gaswork liquors and over the years there have been a number of complaints of hydrocarbon liquids seeping into excavations, basements and ground floor levels of neighbouring premises. The refinery now has been closed for some eight years but seepage of hydrocarbons still occurs in low lying areas. A neighbouring engineering firm had extended their plating bath pit and were installing equipment when the vapour from a pocket of liquid ignited. On analysis the liquid was found to consist of a mixture of aromatic and parafinic hydrocarbons of flash point minus 2.3°C. The liquid had apparently seeped in after a heavy storm which raised the water table. The Factory Inspectorate advised lining and sealing the pit to prevent further ingress of hydrocarbon liquids.

SUMMARY OF SAFETY MEASURES FOR THE PROTECTION OF SITE WORKERS

These examples of casework investigated by HM Factory Inspectorate illustrate the essential need for prior knowledge of the nature of the contaminant and the hazard it presents. The benefit of prior knowledge of the hazard is shown in the asbestos example where a safe system of work was drawn up before the work commenced. Forewarning of the hazard plus a safe system of work would avoid the serious consequences which could occur through blundering into a dangerous working situation.

A safe system for work on contaminated land may include:-

a) Monitoring of the atmosphere within the breathing zones of the workers to ensure that throughout the progress of their work they are not exposed to harmful concentrations of airborne contaminants or to an asphyxiating atmosphere.

b) Monitoring of the general atmosphere in an excavation to ensure early detection of points of seepage and dangerous gas concentrations.

c) The provision and use of protective clothing where it is essential to protect the worker from skin contact with the contaminant and/or to prevent contamination of his personal clothing. Special facilities for cleaning, laundering and storing protective clothing may be required.

d) The provision of washing facilities, which may include showers, convenient to the place of work for the protection of the worker against contaminants toxic by skin absorption and/or ingestion.

e) The provision and use of respiratory protection, where through the nature of the site operations the concentration of an airborne contaminant in the breathing zone of the worker cannot be controlled to a safe level by ventilation.

f) The maintenance of an adequate supply of fresh air to the working face of a confined excavation such as a deep trench or tunnel to ensure:

i) there is sufficient oxygen for respiration;

ii) dangerous gas concentrations are diluted to a safe level.

Forced ventilation is generally used for the control of the working atmosphere in a confined space, but for some applications a combination of local exhaust and forced ventilation may be necessary. In tunnelling it is normal practice even when no particular gas or dust problem is expected, to provide ventilation at a rate of 0.25 to 0.3 m^3/sec of air per m^2 of working face.

In practice dilution of a toxic/flammable gas to a safe respirable level i.e. Threshold Limit Value, will ensure that the concentration is far below the lower explosive limit (LEL). There may be circumstances where an asphyxiant/flammable gas, such as methane, may be present in an excavation at dangerous concentrations. The atmosphere at the working face should be monitored to ensure that the flammable gas concentration is maintained well below its LEL. In such circumstances it is good working practice to maintain the flammable gas concentration at below 5% LEL. Even when there may be no forewarning of a hazard from site sampling atmospheric tests should be carried out throughout the progress of work in deep excavations. It is suggested the tests include the determination of oxygen deficiency; carbon dioxide, flammable gas (methane), carbon monoxide, and hydrogen sulphide concentrations. If there is any suspicion, for example by smell, of the presence of other contaminants further tests should be carried out. These may be done by general and specific gas detector tube tests.

ACKNOWLEDGEMENTS

The author wishes to thank his colleagues in the HSE for their advice and the information used in the preparation of this paper.

BIBLIOGRAPHY

1) Guidance Notes from the Health and Safety Executive

MS 13 Asbestos
EH 10 Asbestos hygiene standards and measurement of airborne dust concentrations
EH 15 Threshold limit values for 1978
GS 5 Entry into confined spaces

2) Waste Management Paper No 18, Asbestos Wastes, DOE

3) Safety in Wells and Boreholes, The Institution of Civil Engineers, 1972

4) Safety in Sewers and at Sewage Works, The Institution of Civil Engineers in collaboration with the Ministry of Housing and Local Government, 1969.

5) BS5573 : 1978, Code of practice for Safety precautions in the construction of large diameter boreholes for piling and other purposes.

6) Work with Asbestos Insulation and Sprayed Coatings. Draft Approved Code of Practice and Draft Guidance Notes Consultative Document, Health and Safety Commission.

[illegible]

[illegible]

ACKNOWLEDGEMENTS

The authors [illegible] in the preparation of this paper.

REFERENCES

1) [illegible]

2) [illegible]
10 [illegible]
and Conditions [illegible]

3) [illegible]

4) [illegible] 1972.

5) [illegible] 1969.

6) [illegible]

7) [illegible] Commission.

'NON-NUCLEAR' RADIOACTIVE WASTES

Mr A Windsor

Radiochemical Inspector, Department of the Environment

Non-nuclear Radioactive Wastes

The schedule to the Radioactive Substances Act 1960 defines the levels at which the concentration of naturally occurring radioactive elements renders material to be radioactive. The Act requires all premises holding radioactive substances to be registered and it requires any disposals of radioactive waste to be formally authorised. An authorisation is drafted as necessary on behalf of the Secretary of State on the recommendations of the radiochemical inspector, the inspector having consulted professional associates in the local and water authorities as he thinks relevant. The drafted authorisation is then sent formally to the local authorities for comment before final modification and issue.

In making recommendations for the disposal of radioactive waste an inspector considers each case on its merits. If one if involved in the clearance of a site that has become contaminated with radioactive waste as a result of some process that has been performed there, one must ask oneself whether or not it is desirable to clear the site, what the final objective should be and what the technical options are to achieve the objective. One must also consider the acceptability of the options to the operators, landowners, and representatives of the communities involved.

Radioactive materials differ from other contaminants in that they can give rise to enhanced gamma background radiation levels. In addition they can enter into food chains and water supplies as a result of their normal chemical and/or physical properties and in some cases concentrate in particular organs of the body which then become irradiated. In the case of naturally occurring radioactive elements in the uranium and thorium decay chains further complications arise as the chemical and physical properties of the elements vary during the decay sequence. In addition it should be appreciated that nothing can be done to change the radiation and half life characteristics of the elements involved.

There have been a number of industrial processes in this country over the past hundred years that have been involved unwittingly in the processing of materials that happen to contain elements in the naturally occurring uranium and thorium series. Apart from the production of radium as a source of radiation, none of these processes has intentionally involved their radioactive properties, although some of them (such as luminising) have taken advantage of the energy associated with the radiations. The inheritance has been a number of contaminated land areas varying in size from a small spoil heap to several acres.

The principal decay series producing these various problems is that starting with thorium 232, although that including radium 226 has also been involved. Waste is usually in the form of derelict buildings, contaminated industrial land, waste spoil etc. Quantities vary from thousands of cubic feet to tens of cubic feet on a particular site. Concentrations can be quite low. A concentration of 10^{-4} microcuries of thorium 232 per gramme in soil will produce at least a twenty fold increase in background radiation levels. Although thorium does not appear to be taken up in plants, it can migrate slowly in soils and appears to concentrate slightly in brick. At such low p.p.m. the chemical form of the parent can be difficult to assess and the meaning of such terms as soluble and insoluble is indeterminate in the complex of soil chemistry.

In the years immediately preceding the promulgation of the Radioactive Substances Act a review of the many processes involving materials contaminated with thorium and uranium was completed. The sites associated with the work were identified and their occasional clearance has been taking place under the provisions of the Act ever since. Ideally the waste should be removed or buried so that it does not enhance background radiation levels where people live or work or where land may be developed and the material re-exposed, and it should not be disposed of in an environment where it can realise any potential it may have to enter into food chains or water sources. If at all possible conventional local disposal routes should be used. If the waste is too radioactive, removal to the specifically controlled landfill site operated by BNFL may be possible. Otherwise routes may include disposal as uncontained bulk waste at sea, disposal in special containers at sea or indefinite controlled storage - if it is philosophically possible to class that as disposal.

The choice between these options should be taken by balancing the cost against the net benefit from the disposal. However the technical aspects of radioactive waste are frequently less significant than the political ones. For example if one can locate a suitable landfill site within trucking distance, nobody in local life can afford to identify with the use of the site for 'nuclear' waste. Conflicting information and advice is available from all as the effects of exposure to radiation at slightly enhanced background levels is bound up in the problems of measuring small statistical variations in large samples and predicting these effects in both exposed individuals and future generations.

The basic UK recommendation to which we are obliged to conform is that sites should be cleared to a level of 10^{-5} uCis/gm. However if it is not possible to do this because of local restraints, the resulting action may not be the best radiological solution although it may well be the best practicable solution. However it is most unlikely that any course of action taken would result in the appropriate dose equivalent levels being exceeded, and in general terms exposure levels are kept as low as reasonably achievable. Some examples will be presented which will range from very little being done to change radiation exposure levels, to a clearance programme involving hundreds of thousands of pounds which has nevertheless prevented the aim of land reclamation due to public opposition.

FIRE AND EXPLOSION HAZARDS WITH CONTAMINATED LAND

K N Palmer

MA, CChem, FRIC, CEng, FInstF, FInstPet, FIFireE

SUMMARY

Problems can arise with contaminated land containing combustible materials, from either fire aspects or from the generation of flammable gas, particularly methane, by microbial action. The amount and nature of the combustible material can vary widely but one case is quoted in which the calorific value of the organic matter exceeded that of peat.

Smouldering is the type of combustion usually encountered underground in contaminated land and the available scientific information on smouldering is reviewed. The smouldering may be initiated by various ignition sources, but the possibility of self-ignition is given particular attention because it is an inherent property of the material.

Generation of flammable gas is a hazard when buildings are constructed on contaminated land, and there may be environmental nuisance arising from the odour of co-products. Existing information on the explosion problem is reviewed and the areas where further work is needed are indicated.

Finally consideration is given to the mitigation problems associated with fires and generation of flammable gas. The current methods of fighting underground smouldering fires are described; they are labour intensive and slow. The need for further developments in techniques is clear.

INTRODUCTION

Contamination of land may arise from various causes. Where quarries, pits etc are infilled with material brought from elsewhere the intention may be to provide made-up ground on which buildings or other facilities may be installed. It is in the making-up process that the potential hazards begin. Local cleansing authorities and private organisations commonly use such methods to dispose of domestic and industrial waste, which may well contain organic and combustible material with the associated risk of fire and explosion. Where a waste tip is operating the life of the tipping may be several decades and the composition of the material being tipped will reflect the changes in composition of the waste over relatively long periods. There may also be problems with the adequate documentation of types of material being tipped. Made-up ground can also be generated adjacent to an industrial operation where material is being excavated. After sorting,the unwanted residue is deposited frequently above ground level as a tip heap. The material in the heap may contain combustible constituents so again the problem of fire can arise.

Another source of contaminated land is the industrial site where over long periods material soaks into the ground and eventually builds up to a significant concentration. The situation is clearly most likely to arise in old industrial areas, which have become enveloped in the suburban spread from city centres. Ultimately when the industrial site is abandoned, there would be strong pressure for the re-use of the derelict land and decisions are necessary as to any measures necessary to overcome potential problems. Problems such as the chemical attack on the foundations of buildings, the possibility of toxic substances being present in the ground which could eventually find their way into home-grown food, and sub-ground movement due to settling of loose infills are all beyond the scope of this paper. Here concentration will be made on the fire and explosion hazard associated with combustible infills likely to be considered where any major re-development takes place.

The fire hazard arises because of the combustible material present in the ground, together with the small supply of air available to support combustion underground. Because the air supply is small the rate of combustion is likely to be slow and therefore glowing or charring is more likely to take place below ground than flaming. If the burning should break through to the surface, or is initiated there, then of course more air is available and flaming can take place. However below the surface smouldering is the form of combustion most likely to be encountered and temperatures up to red heat, say, at least 600°C, are quite possible.

The sources of ignition to initiate the smouldering are various. Firstly the material may be actually burning when it is first deposited, as is the case with hot ashes in domestic and industrial refuse. Secondly the smouldering may be initiated by a fire at surface level arising from the burning of vegetation or of bonfires to dispose of refuse. Alternatively the smouldering may be initiated below ground level by localised heating, eg due to electrical cables. Another method is the self-heating and ignition of materials which slowly react with the limited supply of air and eventually reach a temperature at which smouldering will propagate. In this context it is worth noting that although the supply of air may be very limited underground, by definition the heat insulation is very good there. Thus, although the rate of heat generation may be low, it can be conserved and over a long period the temperature can build up to that at which a propagating form of combustion can start. In any event because of the various types of ignition source available it is not possible to assume that ignition will not occur.

Because of the presence of organic matter in the contaminated land there is a possibility of microbial reactions taking place. These reactions can be broadly classified as anaerobic fermentation and the gases generated will tend to prevent the very slow natural ingress of air to levels below ground. The principal gases produced are likely to be carbon dioxide, methane, with possibly some nitrogen due to the limited ingress of air, with some minor constituents such as hydrogen sulphide which because of their smell and toxic properties are environmentally objectional and give rise to complaint. An explosion hazard can arise because methane is a combustible gas and if the mixture of gases can accumulate within a closed volume, eg under the foundations of a building, or in a closed basement, then there is a possibility of build-up of concentrations sufficiently high for the mixture to become ignitable. The result of ignition is the rapid propagation of flame which raises the temperature of the gas and unless relief venting is available the pressure will rise. In a closed compartment sufficient venting area is not available and, as pressures as low as 15 kN/m^2 are sufficient to cause serious structural damage to buildings, measures of explosion protection or prevention are clearly necessary.

With both smouldering combustion and methane generation the level of the water table is of great significance. Clearly if the water table is permanently within a short distance of the surface then prolonged burning or microbial reaction are less likely. If the water table is permanently well below surface levels so that the combustible material is only affected by surface water draining downwards then conditions for burning and microbial reaction are encouraged. Complex situations can arise where the water table fluctuates seasonally since the time required for burning or reaction to become a serious problem may be longer than that between cycles of flooding. On the other hand they may not, and under these conditions there is a possibility of permanent burning or microbial action.

One of the simplest means of assessing the problem on contaminated land is to make borings and take samples which can subsequently be investigated in the laboratory. Data from one such investigation, on a projected housing site, is available[1]. A large number of samples were taken and were analysed for organic content and for calorific value. From samples all over the site the content of organic matter varied from above 50 per cent down to about three per cent, which is the order of value associated with natural uncontaminated soil. The calorific values, as measured by bomb calorimetry, for the soils followed a comparable range of variation. However, a particularly interesting finding was that after allowance was made in the samples for the organic content the calorific value of the organic content itself was fairly constant at around 24.5 MJ/kg. This may be compared with a value of 17.5 MJ/kg for dried peat, indicating that the calorific value of the organic matter in the samples was substantially higher than that of peat. Even though the samples were not constituted solely of organic matter, the calorific value was sufficiently high for there to be concern about the possibility of smouldering combustion. The characteristics of smouldering therefore become of considerable practical importance.

COMBUSTION PROCESSES

Information is available on the basic properties of smouldering[2], although the investigations were carried out in other contexts. Two experimental regimes were investigated, that where the smouldering propagated along a 'train' of material in the open, either in still air or under an air draught, and a second regime where smouldering was initiated within a heap and the time for burning to progress to the surface was studied.

Table 1

Smouldering rates of beechwood sawdust in still air (cm/h)

Mean particle diameter mm	Moisture content %	Depth of dust layer (mm)				
		3.0	8.0	10.0	16.5	24.0
0.48	9.4	ns	ns	ns	5.4	5.1
0.27	9.8	ns	ns	ns	5.5	-
0.19	9.4	ns	ns	5.0	5.8	-
0.14	9.8	ns	6.3	5.7	6.0	-
0.12	9.7	ns	6.7	-	6.1	-

ns indicates smouldering not sustained

From the experiments with burning in trains in the open it was readily shown that the rate of propagation of smouldering was slow in still air, of the order of cm/h (Table 1). This conclusion held for a range of dusts of vegetable origin. The smouldering rates were markedly increased, sometimes by an order of magnitude, by an applied air flow. A simple approximate relation was found between the air velocity and the smouldering rate of fine dusts but this broke down with coarser materials. If the dust was sufficiently coarse, flaming could develop from the smouldering, the transition occurred at lower air velocities as the particle diameter was increased. In addition smouldering rates of fine dusts in still air varied with the depth of dust provided that the depth was not less than a critical minimum value for smouldering to be sustained. With fine dusts this minimum value was of the order of millimetres. With coarse materials it could be several centimetres. The minimum depth of dust layer necessary for sustained smouldering depended upon the particle size of the dust and the air flow but, like the smouldering rate, it appeared to be independent of the thermal conductivity of the base upon which the dust rested. Not all combustible dusts would undergo smouldering, in particular those which liquefied under the action of heat tended either to burn with flame or to extinguish. Dusts of cellulosic origin, and many coal dusts, smouldered without difficulty and in the present context this ability to smoulder is very relevant.

Experiments were also carried out in which smouldering was initiated at the base of small heaps or at the centre of the base of open topped cubical boxes of size up to 90 centimetres. No indication was found that the depth could be so increased that smouldering would not be sustained. The time of propagation of the smouldering from the base to the top of the deposit of dust was approximately in proportion to the square of the depth and for most of the period of burning there was no evolution of smoke or steam. For a fine mixed wood sawdust about 24 hours were necessary for smouldering to propagate from the base to the top surface of the dust within a box of side 30 centimetres.

All the above work was carried out with combustible dusts singly, and not mixed with non-combustible components. A very modest amount of information is available on the effect of adding calcium carbonate to wood sawdust smouldering in open trains, where two indications, rather than conclusions, can be drawn. These were firstly that the amount of calcium carbonate necessary to be mixed with the sawdust to prevent smouldering was relatively high, say at least 80 per cent of the mixture. Secondly the addition of smaller amounts of calcium carbonate in fact accelerated the smouldering rate. This initially surprising fact may be explained on the basis that the smouldering rate is governed by the transport of oxygen to the smouldering zone, and the rate will not be greatly affected by the admixture of non-combustible with the fuel. On the other hand the density of fuel in the train is reduced by the addition of the non-combustible material so that a greater volume can be burnt with the same amount of air. Thus the propagation rate may in fact increase. Eventually the heat sink effect of the diluent would be expected to become important but this appears to require relatively large amounts of diluent. No information is available on the effect of non-combustible diluents to smouldering initiated at the base of heaps and this is clearly an area, in the present context, where a great deal more information is urgently required. Two factors are important. The extent to which inert diluents must be added to prevent such smouldering needs to be determined, so that consideration could be given to measuring calorific values from actual ground and then being able to decide whether underground smouldering was possible. Secondly the effect of diluents on the burning rate needs to be understood because the faster the rate at which the smouldering can propagate underground then the wider the area over which fire-fighting and other emergency actions become necessary. This second aspect is considered in more detail below.

Although contaminated land may be vulnerable to external ignition sources such as matches, fires etc there is a possibility of self-heating leading to ignition underground. Self-ignition can occur because if the combustible material is slowly reacting with air, exothermically, then the rate of heat release will increase exponentially with temperature according to the Arrhenius equation. Heat losses from underground will depend directly on the temperature difference between the point at which the reaction is taking place and the surface. Thus depending upon the chemical constants in the equation a temperature may be reached at which the rate of generation of heat is greater than the rate of heat loss. At this point the temperature of the system rises, the reaction accelerates and the losses become progressively relatively less. Under these conditions the reaction rate can increase sufficiently to lead to the initiation of smouldering which is ultimately controlled by the rate at which air can penetrate to the reaction zone. There is a considerable literature on the physics and chemistry of self-heating and self-ignition and it is certainly relevant to the present context. The only proviso is that although self-ignition is a slow process it is likely to show itself during the first few years of formation of the contaminated land, rubbish tip, etc rather than become a problem many decades after the deposit has become established. However because of the slow rate of smouldering following self-ignition the effects of the burning may themselves take a long time to show on the surface. Building on contaminated land which has only been in position for a few years would raise the question of the possibility of self-ignition.

The generation of flammable gas mixtures due to microbial action on organic matter below ground level which subsequently escape to the atmosphere or within buildings can be covered to some extent from existing literature. Thus detailed information on the percentage of gases such as carbon dioxide and nitrogen which need to be mixed with methane and air to prevent explosion is readily available[3]. One problem is that the composition of the gas mixture produced is not controlled and information is needed on the ratio of methane to non-combustible gases present before firm conclusions as to possible ignition and explosion hazards can be drawn. If the ratio of the non-combustible gases to methane is 90 per cent or less by volume then the subsequent mixture in air could be explosible. The explosion characteristics of mixtures of methane, non-combustible gas and air in the context of the relief of explosion pressure in buildings are not well documented and further work in this area may be necessary in order to give informed guidance. Clearly the first line of defence will be ventilation, to remove flammable atmospheres as they penetrate into vulnerable areas, secondly some form of back-up protection such as explosion relief venting may be necessary in all but the simplest cases. The ignition energy of gas mixtures is so low that prevention or explosion by eliminating ignition sources, in ordinary industrial or domestic premises, is not practicable. Removal of the explosible mixture, or the safe handling of any explosion that does occur are therefore the prime methods of protection.

METHODS OF MITIGATION

With the fire hazard of contaminated land the methods of mitigation are either taken before ignition or after the outbreak of fire is noticed.

In practice relatively little is customarily done to prevent the initiation of fire. Mention has already been made of the effects of variation in water table which can naturally give some protection but generally measures are not taken in advance. This may well be the result of relatively little work having been done to define what such measures should be, and much more thought needs to be given to this point in the future. Protection against ignition by surface fires is assisted by the practice of covering the land with a layer of top soil, say, one metre in thickness, but this top soil is deposited on the land for other reasons. In any case it would only be a partial protection against surface fires since the occupier can in fact excavate readily to a depth of one metre for his own purposes and unknowingly lose the protection against ignition. However, this could well be an exceptional situation so that the presence of the top soil is certainly helpful in preventing surface ignitions. It does not of course significantly help to prevent the spread of fire underground should smouldering be initiated and the control of such fires requires direct action.

The control and extinction of underground fires in waste ground has long been known to be a difficult problem and in many cases a compromise solution has been agreed. This is to attempt to control and reduce the burning rate, without necessarily extinguishing the combustion. The immediate problem is then solved since smoke and noxious fumes may no longer be evolved but the problem is liable to recur at intervals. The methods of attacking such smouldering fires involve either injecting extinguishing materials from the surface, through cracks or possibly bore holes, or to undertake substantial earth moving activities in order to isolate the burning region. Both methods are slow and labour intensive, but on waste ground this is not necessarily a great disadvantage. However where the burning is progressing under the foundations of buildings, then there is a need for more rapid action and neither method is particularly attractive. Further development work in this area is urgently needed so that in event of burning penetrating beneath buildings a reasonably effective and quick, preferably economic, method of attack can be employed. An alternative suggestion is that of fire protection by screening the area upon which buildings are situated by means of a non-combustible 'fence' below ground level. If effective this would prevent burning from more remote areas penetrating into the area upon which the building stood, but would not of course deal with ignitions which occurred within the fenced area. That such ignitions do occur is known for instance as a result of the burning of refuse in the backyard or even as a result of heat from a boiler being conducted through the floor slab of the basement of the building and heating the contaminated ground beneath to such an extent that self-ignition eventually developed.

Dealing with the generation of flammable gases produced by microbial action also presents some difficulties because the gases generally do not diffuse upwards over the whole surface area but tend to concentrate and move up through cracks in the ground. Although methane is less dense than air, carbon dioxide is more dense, so that the mixture can have a density either less than or greater than that of air. This means that natural buoyancy may not be available to disperse the gases once released to the atmosphere. Generally the protection is to attempt to drain the evolving gases into some form of ducting so that they can be dispersed away from the environs of buildings or where people may object to noxious smells. It is not practicable to await the termination of the generation of the gases as this could persist over decades and at present there is no general treatment for the injection of agents into the ground to suppress microbial processes. However, with more knowledge of gas composition from actual sites the extent to which natural buoyancy can be used to disperse the vapours and typical rates of flow of gas could be determined. It is clearly important to prevent the access of these gases into buildings and although impermeable floor slabs help considerably their protection can be nullified by the introduction of service ducting carrying electric cables, telephones, water pipes etc unless these are properly sealed to prevent the passage of gas through them. The techniques for protection are available but more stringent codification could be helpful in ensuring that they were more effectively used.

CONCLUSION

The fire and explosion hazards arising from contaminated land are of particular importance where the land is being used for building purposes. More information is needed to enable an assessment of whether the calorific value of the ground as deduced from analysis of samples could be used as a guide as to whether or not there is a fire hazard. If there were some threshold of calorific value above which smouldering could occur, and below which it did not, under the conditions of practical cases, then attention could be concentrated on those sites of calorific value above the threshold. More information is needed on the burning characteristics underground, of the possibilities of self-ignition, and on improved methods of fighting fires once ignition has taken place.

Flammable gases produced by microbial action are likely to be mixed with non-combustible gases produced by the same processes and although information is available on the amount of non-combustible gas needed to prevent the methane from burning, the lack of control of the composition of the gas mixture at the point of generation means that flammable mixtures may be produced. In order to prevent damage to buildings adequate ventilation must be installed and some consideration given to the need for protection against explosion. Both aspects are in need of further investigation particularly that where complex mixtures of gases, combustible and non-combustible, may be present.

The current methods of dealing with fire and explosion hazards are in need of further investigation in order to improve their effectiveness and also enable design data to be laid down. Once fire has started underground serious problems can arise if it propagates below buildings and the present methods of fighting such fires are in need of improvement.

ACKNOWLEDGEMENT

This paper forms part of the work of the Fire Research Station, Building Research Establishment of the Department of the Environment. It is contributed by permission of the Director, BRE.

REFERENCES

1. STARK, G W V. 1978. Private communication.

2. PALMER, K N. Smouldering combustion in dusts and fibrous materials. Combust Flame 1957, 1 (2) 129-154.

3. COWARD, H F. and JONES, G W. US Bureau of Mines Bulletin 503. Washington DC, 1952.

Gas generation and the problems it creates in land reclamation

J. Bromley

INTRODUCTION

For a long time it has been known that the 'Will-o-the-Wisp' flames on decomposing refuse and the bubbles of 'marsh gas' have been due to the microbiological generation of methane. However, recently because of a number of fires and explosions of methane on reclaimed landfill sites attention in many countries has turned to its safe venting and to its possible collection and utilization. An Intergovernmental methane task force meeting, sponsored by the United States Environmental Protection Agency and the United States Department of Energy, Division of Building and Community Services considered the hazards and opportunities of methane in landfill at Denver in March 1979[1] and a meeting in Switzerland in November 1978 surveyed the European approaches to the problem[2].

The CH_4/CO_2 mixture which is formed by anaerobic decomposition of refuse has a calorific value around 450 B.T.U./cu.ft. which is similar to 'coal gas' but by removal of the 35% of carbon dioxide it can be upgraded to almost pure methane with a calorific value of 1000 B.T.U./cu.ft. for possible injection into a natural gas grid system[4].

The extent of the problem

Much of the derelict land that has become available for reclamation in the U.K. is on old landfill sites. These were often situated from 1900 to 1950 in relatively shallow mineral extraction holes on farmland around the dense centres of our larger cities. However with the expansion of the towns these old landfill sites have now become surrounded by urban and light industrial developments.

Consequently they have been used as sports fields, parkland and sometimes for light industrial building or even housing. The degree of awareness of the problems associated with the construction of buildings on such sites has been variable. Often the sites are shallow, the refuse was incinerated as it was deposited and little biodegradable material remains, but on the more recent sites, when incineration was prevented and good topsoil was provided, methane build up has been appreciable.

More recently, it has become difficult to obtain sites for landfill near to, or midst, housing and many of the large cities 'transfer' their refuse from the street collection vehicle to liner trains or barges for deposition at sites such as Sutton Courtenay, Stewartby and Pitsea in quantities exceeding 1000 tons/day to each site. These large often deep sites are usually away from town centres and they are not likely to be reclaimed for housing or light industry but will probably be returned to agricultural use. The 'problems' of methane penetration into buildings will be replaced by the possible advantages of collecting the methane in order to utilise it.

The 30×10^6 t/annum of domestic and industrial refuse deposited in 10 metre deep layers at a density of around 0.8 tonnes/m^3 covering an area of 40×10^6 m^3 (40,000 hectares) over the 1960-1970's will tend to be replaced in the 1980-1990's by 40×10^6 t/annum in 30 metre deep sites at densities, after compaction and settlement, of 1.0 tonne/m^3 covering the smaller area of 13×10^6 m^3 (13,000 hectares). The deeper sites will be more likely to produce methane at higher rates and the flow rate of methane out of the reduced surface area of a smaller number of large sites will tend to present a different 'problem' in the future - indeed a challenge to recover and use the gas economically.

One further reason for a change in the nature of the gas problem has already been mentioned, namely the tendency to allow refuse to self ignite in the 1900's to 1950's. The ash from domestic fires was often deposited hot in the 'dustbin' and constituted a major component of refuse. Refuse was not given a daily cover of top soil and bulldozers and compaction equipment were not used. Any methane that was generated was ignited by the 'tip'.

Some of the sites being considered for reclamation have a history of this kind but other more recent sites have never been on fire, are well covered and they have therefore a tendency to cause gas to move laterally along fissures in chalk, sand lenses in clays or even culverts and service ducts to locations remote from the landfill. They therefore might present a greater potential hazard.

Bromley and Parker[3] have recently reviewed the problems associated with methane generation from landfill sites.

<u>The generation of gases in reclaimable land</u>

Before discussing landfill sites in more detail it is perhaps important to recognise that there may be other possible gas sources as follows:-

(a) 'Marsh gas'. A marshy or peaty base to a reclaimed area may generate inflammable gas mixtures which will penetrate through any infill to give methane release at the surface. Marsh gas is usually a CO_2/CH_4 mixture similar to the gas obtained from anaerobic decomposition of refuse.

(b) Old coal mine workings, surface coal seams or rock fissures connected to coal seams. The gas that is released from coal contains no CO_2 and may be distinguished from landfill gas by gas analysis.

(c) Gas mains, fuel pipe lines or tanks. The fracture of pipelines, the spillage of fuels or the dumping of solvents or hydrocarbon wastes can give rise to 'gas' problems on reclaimed land. The gas chromatographic analysis of a sample of the vapour should enable the true source of the gas to be ascertained - see below.

(d) Inert refuse containing gypsum or plasterboard may generate hydrogen sulphide under anaerobic conditions.

(e) Domestic and industrial refuse. As well as generating CH_4 and CO_2 other vapours may be released if the codisposal of solvents etc with the refuse has been practised on an appreciable scale. From 1972 onwards it should be possible to ascertain what materials of this kind have been deposited at a given site from the Deposit of Poisonous Waste Act notifications.

Gas identification

The instruments commonly used are of two general types. Firstly the explosiometer type which work on a catharometer principle and are really only detecting the presence of gases with different thermal characteristics from air. They are calibrated in % Lower Explosive Limit (LEL) usually for methane in air which has a Lower Explosive Limit of 5%. Thus 10% LEL means 0.5% methane (or other gas with similar thermal properties in air. These instruments are safe to use in areas where explosive gas mixtures might be found. Secondly there are the flame ionisation type of instruments. These record the presence of ions in a hydrogen/air flame and must not be used in areas where explosive gas mixtures are likely to occur. They have the advantage that they are much more sensitive and can therefore be used for a detailed examination of the routes by which methane is percolating from refuse into a building.

The following table shows the typical analysis of landfill gas compared with natural gas.

Percentages	CH_4	CO_2	H_2O	Other Hydrocarbons
Landfill Gas	55-65	35-40	1-5	1
Natural Gas	93.6	0.2	-	Ethane 2.9 and as below

The composition of the hydrocarbon fraction is as follows:-

ppm	Acetylene	Ethylene	Ethane	Allene *	Propane	Isobutane	n-butane	1 butene	Unsaturated C4S	Isopentane	n-pentane
Landfill Gas	<1	20	22	50	141	122	184	18	65	83	196
Natural Gas	-	-	19,000	-	5,000	1,000	1,000	-	-	-	-

* or Acetaldehyde

The gas generation process

The microbiological processes by which methane is produced in anaerobic digester systems have been extensively studied[5]. Farquhar and Rovers[6] have studied the factors influencing landfill gas production. Evolution increases with increasing water content[7] but may be inhibited by a great excess of water. Anaerobic methane production rate is a maximum at 30-35°C but may be inhibited by a great excess of water. Temperatures up to 60°C are sometimes attained during an initial more exothermic aerobic stage which produces CO_2 and no methane. The optimum pH is 7.0.

Four stages are usually postulated:-

Stage I aerobic decomposition in which entrained oxygen is converted to CO_2. Heavy compaction and good daily cover and rapid infill of deep layers reduce the importance of this stage. It has normally finished before reclamation starts but may recur if the refuse is exhumed during construction work.

Stage II anaerobic decomposition commences as the oxygen becomes depleted. CO_2 and even hydrogen are produced and the nitrogen from the air is progressively displaced.

Stage III anaerobic methane production rises to a peak and carbon dioxide concentrations fall. Nitrogen concentrations drop to zero.

Stage IV a slow decline in methane production rate occurs over many years depending on the extent to which the gases can escape from the site either vertically through the top cover or horizontally into surrounding rock structures. Gas compositions at this stage are usually in the methane 50-70% range, carbon dioxide 50-30% range. The landfill sites normally considered in reclamation schemes have reached this fourth stage.

The problems of gas generation

Awareness of the possible problems caused by gas generation is very variable. Some reclamation schemes and building projects have taken expert advice in the planning, operational, constructional and post construction phases, whilst others appear to have been unaware that any problems might exist. It is hoped that public and professional awareness will increase and balanced, environmentally safe, and cost effective approaches to the various problems will soon be applied whenever required as a result of the increased attention this topic is now receiving.

The types of 'problem' will now be briefly reviewed.

Fires

Methane/CO_2 gas mixtures from landfill will burn continuously if conditions are satisfactory. In the U.K. a simple burner system has been in operation since 1972 at Merstham G.L.C. landfill site. Gas is collected from a number of holes drilled into the refuse and delivered via a manifold to a simple burner system. A pilot light system has been used to keep the flame alight.

If the gas is not collected in this way internal fires may sometimes develop within refuse. These will be aided if fresh oxygen is drawn into the site. This is more likely to occur in sites with a steep gradient where a fire at the top end draws air in at the lowest point.

If the burners are correctly designed they should not present any risk of explosion. Great care must be taken to ensure that the flame is not drawn back during conditions of low flow so that it reaches zones where the 65% methane gas has been diluted down into the explosive limit range of 5-15%. Flame traps must be installed to ensure this does not happen.

Explosions on sites[8]

When methane concentrations in air are in the 5 to 15% explosive limit range there is a danger of explosion from any source of ignition. The explosive wave will travel along voids and may dislodge dust and start fires where oxygen ingress is possible to any inflammable materials.

The exact value of the explosive limit range for CO_2/CH_4 mixtures diluted by air is not known. It is usual to take the 5-15% figure for CH_4/air mixtures and consider the CH_4 content only of the landfill gas when assessing the potential hazard.

Explosions within the body of a landfill are not likely to occur since the energy within the advancing explosion front will be adsorbed by the solid material of the fill. If voids are created by drilling, or shafts are constructed for leachate removal, or refuse is disturbed during excavation of trenches for services, there is always a possibility of a methane/air explosion. Ignition can be caused by electrical equipment, vehicle ignition systems, tools or excavators hitting stones or rocks, welding equipment or cigarettes.

Sudden changes in atmospheric pressure or temperature can cause quite rapid expansion of gas into previously safe areas. Frequent monitoring of working areas may therefore sometimes be necessary.

Explosions in buildings

The reclamation schemes of the past have frequently included the construction of lightweight industrial buildings on the completed landfill, or the building of a sports pavilion, or of a housing estate on adjacent land. Sometimes landfill sites have been sited near to existing buildings and gas from the 'reclaimed' landfill has migrated to the buildings.

Mobile homes, parked vehicles, landfill operators huts can also act as collection vessels for landfill gas as it moves vertically upwards from cracks in top cover or up soakaway pipes or car park drainage systems.

The following possible routes for the gas into void spaces, buildings or vehicles have been observed. This list does not claim to be exhaustive but it gives an idea of the careful investigational work that is required to find out why an accident occurred and also of the factors that must be considered in building design etc.

1. Cracks or joins in concrete rafts

Differential subsidence of the decomposing refuse under a raft may cause cracks to occur. Gas coming up such cracks has been ignited by welding equipment and blow lamps[1].

2. Permeation of concrete rafts

It is common practice to drive piles down to bed rock and link the piles at the surface with strong cross members and then lay a raft over the resulting bearing structure. This leaves an inverted set of voids when the refuse settles which will fill with CH_4/CO_2 gas. It will try to find weak spots in the raft or will eventually spill round the sides in its escape to the atmosphere.

The weak spots in such a raft are often found to be:-

i. where services such as electricity, gas, telephone, water, drains, penetrate the raft

ii. at the floor screed/wall joints

iii. at low points in the floor such as drains or showers, gulleys in kitchen floors

iv. at places where nails, screws, bolts, rawplugs, have been driven into the floor

v. at door posts inset into floor structures

vi. at points where stress bearing roof structures join the piles

3. Drainage systems

(1) Sewage systems

Although sewage gas is present in large town sewers where the flow of organic wastes is large it is not usual to find high concentrations in the small domestic waste systems around a house where the amount of organic material is small. High methane concentrations in sewers (above 20% methane in air) usually reach the sewer because 'T' junctions and inspection plate systems are exposed to landfill gas. Sometimes joints in sewage pipes fracture as the landfill subsides or when heavy vehicles drive over the unconsolidated surface.

(2) Soakaway systems

Soakaways for surface run-off water and for water from gutters are frequently built in landfill refuse. The pipes delivering water to the soakaways act as paths for methane to return to the buildings.

Blocked gutters will cause the gas to come out of pipe joints below overhanging roof structures. The gas may enter roof voids and travel from room to room above ceiling level.

4. Inverted catchment features

Overhanging roof structures, sometimes with lips on, will catch upwards moving methane gas streams spilling from under a concrete raft or from vent pipes.

Surface cladding to buildings may be provided with a water deflecting skirt to keep water away from damp courses. Such skirts will pick up gas spilling from the edge of a concrete raft.

Movement of gas within buildings

Air ventilation ducting and ceiling voids can cause gas to pass from room to room. Electric cable ducting systems also will act as a pathway and incidentally lead gas to switches which are potential ignition sources for the gas mixture.

Other explosive situations

Vehicles, caravans, mobile offices, weighbridges can all act as collection systems for methane on landfill sites. Changes in ventilation can of course rapidly increase or decrease the hazard in any given situation. The closing of windows and doors will often rapidly allow explosive gas situations to develop.

Asphixiation

It unfortunately is necessary to point out that CO_2/CH_4 gas mixtures will not support life and shafts for leachate removal, culverts, old tunnels on or even near to landfills should be filled in or sealed off. People have been killed entering such systems in the U.K.

Vegetation

The revegetation of landfill sites has been surveyed in an article by F.B. Flower[20]. It has been shown that plants such as tomato are not affected when their roots are exposed to CH_4 but CO_2 has a marked effect. A study on the effectiveness of various techniques for preventing landfill gas from reaching the roots of plants and shrubs which has been in progress for two years has shown that black gum was the most tolerant and honey locust the least tolerant of the 19 species under study. Extra soil mounds on the top cover and lined and ventilated trenches backfilled with suitable soil, gave the best results. Adequate irrigation was essential.

Previous work had shown that methane could penetrate several hundred feet from landfill sites and damage crops. The practice of providing good daily cover and compacting the top cover, enhanced lateral movement of gas and the high compaction oftem made the development of good root structures very difficult.

Smell

When gas evolution occurs from refuse other volatile constituents are driven out together with the methane and carbon dioxide. Hydrogen sulphide is usually considered to be one of the most important odiforous constituents of landfill gas. Its concentration is reported to be zero at the Palos Verdes landfill site[14]. However it is usually considered that compounds other than H_2S cause most of the unpleasant odour. Gas chromatographic mass spectrometric work is in progress in an attempt to identify the exact nature of the compounds.

The smell can possibly be overcome by the following techniques.

1. by burning off the methane when hopefully the unpleasant smelling compounds will also be destroyed. If they are not destroyed they may rise in the plume of hot burnt gas and be dispersed several hundred feet above the site.

2. by lowering the water table. This has been carried out on one site where free sulphide ion concentrations in excess of 500 ppm have since fallen to levels around 7 ppm and so far no smell complaints have been received. Gypsum waste is known to have been deposited at this site.

3. by using a masking odour. This can hardly be considered a long term option for a reclamation scheme, but may be a useful technique during a construction phase when sulphide containing refuse might be disturbed. However it is generally considered unwise to consider the construction of buildings on sites where much sulphide is likely to be present. The admixing of different refuse horizons with differing pH and sulphide contents may result in considerable smell problems.

4. putting much deeper top cover on the site. This solution will result possibly in a build up of smelly compounds within the refuse to higher concentrations but should result in a lower flow rate of the smell from the site and a reduction in public nuisance.

<u>Poisonous gas release</u>

The only known cases of this have not been during or after reclamation schemes but rather during landfilling operations when hydrogen sulphide has been the toxic gas responsible for the deaths. However the possibility of the sudden release of toxic hazardous or poisonous gases during the reclamation of derelict land is always present. The Hazardous Materials Service at Harwell has been called upon to assist a number of incidents of this kind. Sometimes extensive civil engineering works through old landfill will require the continuous presence of a chemist capable of giving technical advice when gas or solvent clouds are released and arranging for the safe removal and destruction of any potential sources of gas such as corroded gas cylinders. These are usually removed to a safe location and detonated with explosives.

Sometimes it is suspected that there may be release from arsenical compounds or spent cyanides on land that is to be developed. These rather special cases need individual assessment and no general rules can be given except that it is always best to err on the safe side and obtain professional advice if any uncertainty exists. Remedial measures can be very expensive.

<u>Techniques used to avoid gas problems during land reclamation</u>

The obvious solution is of course to operate the site so as not to generate gas. This however does not seem to be a realistic option and it is generally accepted that gas will arise with modern methods of landfill technology. The period over which it is generated can be considerably extended if the refuse is kept dry by the use of gradiented compacted top cover allowing good run off of rainfall.

The reclaimer used to be the person who purchased land in order to reclaim it for some better end use. Reclamation today is a vital part of the mineral abstraction process and must be considered in detail during the initial planning process. Top soil is saved, the mineral abstracted, landfilling follows and then the site is 'restored'. The problems of gas evolution are therefore dealt with as part of an overall plan leading to final site restoration.

The techniques for dealing with landfill gas have been summarised in an article by Stone[4] and by Moore and Igbal[21]. The tragic loss of three lives at the North Carolina National Guard Armory in 1969 due to the lateral movement of methane a distance of 7 metres into a building, alerted landfill operators to the need to build into their sites measures to prevent lateral migration.

The installation of control measures during landfill site infilling is cheaper and more effective than the drilling of shafts and building of grout walls in finished landfills. In the U.K. many current landfill operations are incorporating gas control measures and there are also old sites where it is necessary to install gas control measures to alleviate situations that have developed because no precautions were taken.

Two main techniques are used. Ventilation trenches or wells to vent the gas, or grout or plastic impermeable walls to prevent gas migration.

Ventilation trenches. Trenches filled with hardcore or large aggregate are constructed around the landfill to allow the warm gas to rise rapidly and be diluted in a safe manner at the surface with the air. The trenches can easily go to the base of the fill if they are constructed during the landfilling operation but it may not be possible to install them cheaply to the base fill more than 8 metres deep once the fill is completed. Such trenches may not prevent migration into surrounding rock structures unless the rock is first sealed with a clay seal.

Ventilation shafts or wells within the refuse. These can be constructed as the landfill depth increases to relieve the pressure of the gas that builds up over a large well covered site. (The pressures are rarely more than 1 or 2 cm water gauge). Construction of shafts in completed fills is difficult and expensive. Naylor et al.[9] give the 1977 costs of a 10m deep percussion drilled 15 cm hole in refuse at £400. 20, 40 and 60 metre holes would cost £700, £1200 or £2300 respectively. The venting of gas via a number of vertical shafts or wells does lend itself more readily to the collection of the gas for its utilisation. In the United States several landfill sites now have burner systems installed which flare off the gas continuously. Local customers can frequently be found who will purchase the gas or it can be used directly to generate electricity[10].

Grout walls and impermeable plastic barriers. These are difficult to install and unless they are very carefully constructed they may not be perfectly continuous and they will leak. P.J. Jarrett[11] comments that present day clay injection technology for creating barrier systems is costly and it is difficult to achieve the uniform compaction density required for such a system.

The low gas pressure generated within landfills have already been mentioned. It is common practice to apply a suction to the wells when the gas is being collected for its use as a fuel. Typically suctions of about 4 inches water gauge are used. A pressure of about 7 inches water gauge is usually used to drive the gas along the main to the processing plant or user. Great care must be taken to ensure that the suction is not too great or air will be sucked into the landfill. This will result in some nitrogen appearing in the landfill gas and its calorific value will be reduced. Pressure transducer techniques for measuring the gas pressures have been developed by Charles Kunz[12]. The flow of methane into half an inverted oil drum imbedded into the surface can be used to estimate the flux of methane out of a landfill surface[13].

The utilisation of gas from landfill sites[22]

The developments in the utilisation of gas from landfill sites will have repercussions on the reclamation of landfill sites in the future. There will be a tendency for sites to be constructed so as to maximise gas production over as few years as possible and for all larger sites to have some form of gas control system.

It has been estimated that there will be several hundred sites in the United States feeding gas directly to local consumers with the minimum of treatment. As it is cheaper to transfer the energy as electricity rather than as gas many experts consider that on site electricity generation will eventually become common place. It is not anticipated that there will be more than half a dozen sites where it is economically feasible to upgrade the gas to pipeline standards although this technology has now overcome many of its teething problems. The Reserve Synthetic Oil plant at Palos Verdes landfill in California[14] has changed from its original absorbent system using molecular sieves and carbon, which suffered excessive acid corrosion problems, to one using a conventional triethylene glycol dehydration system. This unit suffers from the formation of polyglycols and research is in progress to prevent their formation in future units. Very low concentrations of certain unspecified acid constituents cause stress corrosion cracking of the carbon steels. A caustic wash system has been installed to remove them and new absorbents for CO_2 adsorption are being developed.

Despite these setbacks and a net energy efficiency estimated by Eberhart[15] to be less than 70% on stream, availability is reported by Collins to be 90% and commercial feasibility is assured[14]. Scepticism still exists and the amount of gas that can be converted economically to High BTU pipeline quality gas is small[16].

Giuliana[17] has reported results of one year's operation of a gas extraction system on a 40 acre area of the Fresh Kills Landfill site on Staten Island. Four wells only 40 feet deep each recovered 175 cfm of gas, and Constable, Farquahar and Clement in a recent report on Gas Migration and Modeling[18] compare measured gas flow rates at a number of sites in Southern California. Pohland[19] points out that if leachate is recirculated over refuse most of the gas production can be made to occur in the first three months after refuse is deposited.

CONCLUSION

The generation of methane by the decomposition of domestic refuse in old landfill sites can lead to dangerous situations in which explosive concentrations of methane in air build up particularly in buildings on, or adjacent to, the refuse deposit. However the techniques of venting the gas and of preventing its migration are now understood and can provide solutions to the problems that arise.

It will probably become common practice at many future landfill sites to install migration control equipment and at many sites it will be economic to use the gas as an energy source. The upgrading of the gas to pipeline standards has been demonstrated in full scale plant but it is generally agreed that this process will not be economically viable at most locations. Direct use of the CO_2/CH_4 will be more attractive.

Great care should be taken in the details of the siting and detailed design of buildings in or near landfills to ensure that the buildup of explosive concentrations of methane in air within the buildings is made very unlikely.

REFERENCES

1. Proceedings of a working symposium on methane from landfills: hazards and opportunities. March 21-23 (1979) held at Colorado Women's College, Denver. Chairmen R.C. Eberhart and J.P. Morgan.

2. Internationale Studientagung. Sickerwasser und Gase aus geordneten Deponien. November 8-10 (1978) held at Krattigen, Switzerland.

3. Bromley, J. and Parker, A. Methane from Landfill Sites. International Environment and Safety. June (1979).

4. Bowerman, F.R., Rohatgi, N.K., Chen, K.Y. and Lockwood, R.A. A case study of the Los Angeles Palos Verdes Landfill Gas Development Project. EPA-600/3-77-047 (1977).

5. Anaerobic Biological Treatment Processes. Advances in Chemistry Series Volume 105. American Chemical Society ed. F. Pohland.

6. Farquahar, G.J. and Rovers, F.A. Gas production during refuse decomposition. Water, air and soil pollution 2 (1973) p.483-495.

7. De Walle, F.B., Chian, E.S.K. and Hammerberg, E. Gas Production from Solid Waste in Landfills. J. Env. Eng. Div. ASWA. June (1978) p.415-432

8. Zborovszky, Z. The Flammability of Gas Mixtures. Paper presented at Symposium reference (1) (1979).

9. Naylor, J.A., Rowland, C.D, Young, C.P. and Barber, C. The Investigation of Landfill Sites. Water Research Centre Technical Report TR91 (1978).

10. M.J. Blanchet et al. Treatment and Utilisation of Landfill Gas, Mountain View Project Feasibility Study, USEPA Report SW-583 (1977).

11. Jarrett, P.J. Paper presented at Midlands Geotechnical Society on the Engineering Behaviour of Industrial and Urban Fill. University of Birmingham, April 23-25 (1979).

12. Kunz, C. and Lu, A.H. Transducer Measurement of Landfill Gas Pressure. Paper presented at Symposium reference (1) (1979).

13. Kunz, C. and Lu, A.H. Flux Box Measurements of Methane Emanation from Landfills. Paper presented at Symposium reference (1) (1979).

14. Collins, R.H. Upgrading Gas to Pipeline Specifications. ISWA Journal No. 28/29 p.1-3, March/April (1979).

15. Eberhart, R.C. et al. Preliminary Study of the Potential for the Utilization of Landfill Methane in Baltimore County, Howard County and Montgomery County, Maryland, John Hopkins University. Report CPE 7803, December (1978).

16. Anon. A.G.A. Monthly p.17. September (1978).

17. Giuliana, A.J. Methane Recovery from a Shallow Landfill, Experience at Fresh Kills, Staten Island N.Y. Paper presented at Symposium reference (1) (1979).

18. Constable, T.W., Farquahar, G.J., Clement, Gas Migration and Modeling. In Press (1979).

19. Pohland, F.G. Leachate Recycle as a landfill management option. Paper presented to ASCE Conference, San Francisco, July (1979)

20. Flower, F.B., Leone, I.A., Gilman, E.F., Arthur, J.J. A Study of Vegetation Problems Associated with Refuse Landfills, EPA/600/2-78-094 (1978).

21. Moore, C.A., Igbal, S.R. Design Criteria for Gas Control Services. 3rd Annual Municipal Solid Waste Research Symposium, September (1977). EPA-600-77-026.

22. R.C. Eberhart Ed. "Proceedings of Symposium on the Utilisation of Methane Generated in Landfills" JHW/APL. Maryland March 9-10 (1978).

CONTAMINATED LAND - EFFECTS ON WATER RESOURCES

C.P. Young and C. Barber
Water Research Centre

1. INTRODUCTION

Land contaminated by the deliberate disposal of waste materials, or by the accidental discharge of potentially polluting substances, may threaten the quality of water resources. In those cases where mobile components or breakdown products of the contamination enter the aqueous phase to form leachates, pollution of surface- or groundwaters may arise because of migration of the leachate away from the contaminated site. In a limited number of cases, the component of this contamination may be immobilized in the soil and not enter the aqueous phase.

The nature of the groundwater contamination at a given site will be related to the type of waste disposal or accidental discharge, whilst the variation and extent of the contamination will depend on local hydrological, hydrogeological and geochemical conditions. Such conditions may be altered by disturbances such as land drainage during reclamation.

2. REVIEW OF PROBLEMS

2.1. The pollution encountered at contaminated land sites arises from either deliberate waste disposal practices or from accidental spillages and leaks of contaminating substances. Certain potential contaminants may be strongly retained within the wastes, or in soils on which the wastes are deposited. Heavy metals retained in soils may produce phytotoxic effects, whilst essentially immobilized hydrocarbons may lead to the risk of explosions or fire in excavations rendering such sites unsuitable for direct agricultural or building use respectively. However, the more mobile contaminants, including some degradation products of otherwise immobile substances, are likely to enter the aqueous phase and give rise to contamination of either ground or surface-waters. Surface water is generally accessible for quality monitoring and contamination of this resource may be identified and rectified with relative ease. By comparison, groundwater contamination is less easily detected because of the smaller number of suitable monitoring points and is considered more likely to produce long-lasting problems. Reviews of water resources contamination resulting from previous land use have been produced recently (1-5), mainly dealing with problems encountered in the United States, although experience suggests that such problems are universal. This section provides examples from the published literature, and from the Water Research Centre's investigations, of groundwater contamination from a variety of land uses, covering the more important types of contaminated land.

2.2. Industrial sites and waste disposals

The disposal of plating effluents containing hexavalent chromium (Cr VI) and cadmium in lagoons was found to have polluted a gravel aquifer at Long Island, New York (6, 7). In the United Kingdom, contamination of groundwater in the Triassic Sandstone aquifer by hexavalent chromium has been reported (8) at distances of up to 250 m from a site where chromite ore sludge is deposited. The disposal of liquid industrial wastes at the Rocky Mountain Arsenal, Colorado, has given rise to extensive contamination of a shallow aquifer at up to

5 km from the site (9). The frequency of such problems is indicated by a survey of 50 industrial waste disposal sites in the United States (10), 43 of which proved to have led to local groundwater contamination. In general, groundwater contamination has been found to be most extensive at those sites where liquid waste or sludges are exclusively discharged or where where liquid disposals on to solid wastes have exceeded the absorptive capacity of the solid wastes (8).

2.3. Hydrocarbon spills

The spillage of hydrocarbons may lead to contamination of both land and water resources. It has been estimated that at least 1500 m^3 of hydrocarbon products are lost annually in the United Kingdom as a result of accidental spillages or leakage (11), some of which have resulted in short-term taste and odour problems in groundwater supplies (12). The long-term effects of hydrocarbon contamination of groundwater are difficult to determine on present evidence and studies are continuing to rectify the situation. The problem is exacerbated by the very low concentrations of soluble hydrocarbons which give rise to detectable, and objectionable taste and odours in water supplies and to the difficulties and costs of remedial measures to rehabilitate contaminated aquifers. As an example, following the accidental spillage of 50.5 m^3 of aviation kerosene on to the Chalk aquifer in June 1978 (11), only 15% were recovered by scavenge pumping and the use of oil-absorbent rope mops by March 1979. When deposited in association with domestic wastes, oils have been found to be strongly adsorbed to the wastes, limiting their potential to migrate to groundwater, but appear to remain undegraded at least 10 years after disposal (8). Where aerobic conditions prevail, more rapid degradation of hydrocarbons may occur.

2.4. Disused gas-works

Groundwater pollution at such sites appears to have arisen principally from the storage or disposal of waste products on site. Phenolic substances, complex cyanides and thiocyanate were found in solid samples from one site, with phenols and cyanide present in groundwater beneath the site, the former being detectable up to 400 m distance in the direction of groundwater flow (8). Low concentrations of phenols in groundwater from a borehole at the site of an old gas works which processed whale oil, have been detected 120 years after production ceased (13) and attributed to the leaching of tarry substances present within the Chalk adjacent to the borehole.

2.5. Sewage of effluents and sludge

The effects of disposal of secondary sewage effluent at a rate of between 400 and 800 m^3/day, over a period of about 25 years, has been investigated (14). Percolation of the effluent through a 10 to 15 metre thick unsaturated zone above the water table led to an almost complete removal of potassium and phosphorus from the effluent, and a complete removal of viruses within 335 metres of the site. Biodegradation within the unsaturated zone was indicated by large reductions in Total Organic Carbon (TOC) concentrations, but the same processes did not appear to affect significantly the total nitrogen content of the infiltrate. Both nitrate, and the conservative contaminant chloride, were attenuated in in the saturated zone in the direction of groundwater flow by dilution due to dispersion. Low concentrations (0.5 mg/l total) of halogenated hydrocarbons, produced by chlorination of the secondary effluent during treatment, were detected at the water table beneath the disposal area but were found to be reduced to levels of 1 μg/l within 0.5 km of the site. Over 99% of pathogenic bacteria (*E. coli* and faecal streptococci) were removed during infiltration, but small numbers were recorded in samples taken up to 2.0 km away. At other sites leakage from sewers has caused short-term pollution of groundwater by bacteria and chloride (15). The pollution of groundwater from septic tank drainage has been reported from permeable gravels at Long Island, New York (7), whilst leakage from cesspools was identified as a source of Chalk groundwater contamination on the Isle of Thanet in 1886 (16).

Some 80% of the sludge arising from sewage treatment at inland sewage works is disposed of to land, about half to landfills and the remainder to act as a nutrient supplement or soil conditioner to agricultural land. The potential phytotoxic and health effects of the agricultural use of sludges has been recently reviewed (17), taking particular account of the risks arising from pathogens and heavy metals in sludges derived from combined industrial and domestic sewage treatment. The investigation of agricultural sites on the Chalk (18), which had received digested sludge as the principal source of plant nutrients indicated that heavy metal ions are strongly retained within the soils and are not leached in significant quantities to the water table, but that high rate of sludge application may increase the rate of nitrate leaching to groundwater.

2.6. Solid domestic wastes

Groundwater pollution from the disposal of solid domestic wastes has been investigated and reviewed recently by several research programmes (8, 9, 19). Leachates within the wastes may be highly polluting with respect to organic (BOD, COD, TOC) and inorganic (NH_4-N, Fe, Mn, Cl, Na, K, Zn) constituents (Table 1). These liquids are most polluting during the early stages of breakdown of the organic fraction present in the wastes, but tend to decrease in strength with time as the wastes stabilize. Leachates from partially stabilized wastes may remain polluting with respect to the concentration of ammonia (Table 1). A number of examples of contamination of both surface- and groundwaters near domestic waste landfills have been reported (8, 9, 19), although the extent of pollution varies widely, being dependent not only on the wastes and method of emplacement (20) but also on the local hydrology.

2.7. Mine wastes

The dominant mining activity in the United Kingdom which produces potentially polluting wastes is coal mining. Approximately 50 million tonnes of argillaceous spoil are produced annually, a high proportion of which is disposed of as landfill (21). Such wastes are frequently rich in disseminated pyrite which, on weathering, gives rise to mineralised leachate rich in chloride, sulphate, iron and manganese. An example of groundwater contamination by leachate from a colliery spoil heap in the Doncaster area has been reported (22).

The drainage water pumped from mines may be highly mineralised. Discharge of mine drainage on to the Chalk aquifer in Kent over the period 1907 to 1973 is estimated to have added a total of 318,000 tonne of chloride to the Chalk groundwater system, giving rise to chloride concentrations in excess of 750 mg/l over 13 km^2 of an important aquifer (21).

3. FACTORS INFLUENCING THE CONTAMINATION OF WATER RESOURCES

3.1. Chemical factors

The composition of percolating water within a contaminated site can vary considerably according to the chemical conditions present. Important chemical variables which affect the solubility of specific contaminants are the redox conditions within a site, the pH value of percolating water, and particularly for heavy metals, the presence of complexing ligands in the aqueous phase. Soil microbial activity may also play a significant role, by conversion of insoluble material to soluble products.

The mobility of soluble substances in the subsurface environment will be affected by complex physical, chemical and biological processes which interact to attenuate movement, for example by filtration of suspended material, by chemical precipitation, sorption or removal of organic and inorganic substances by soil microorganisms, reducing their effect on water resources.

The removal of cations by a range of soil types has been investigated in several studies with particular reference to the percolation of leachate from landfill sites containing domestic wastes (31-34). In general, the degree of attenuation of contaminants is greatest in soils rich in clay minerals and in those containing hydrous iron oxides. Fuller and Kate (32) found that Cr, Ni and Hg were generally most mobile in soils, Pb and Cu were least mobile, whilst attenuation of As, Be, Cd, Se, V and Zn varied with chemical conditions (soil pH, concentration of metal). The movement of leachate through clays, investigated in another study (33) showed little attenuation of chloride, sodium and organic substances (COD). Potassium, ammoniacal-N, magnesium and silicon were moderately attenuated, whilst heavy metals (e.g. lead, cadmium, mercury and zinc) were least mobile. Leachate percolating through clays under anaerobic conditions solubilised calcium (by ion exchange) and iron and manganese by reduction of oxide coatings on clay mineral surfaces.

3.2. Hydrological and hydrogeological factors

In order for ground or surface water contamination to occur, a leachate must not only be generated by solubilisation of components of the waste, but also migrate from the contaminated area to a water resource.

For solid wastes or contaminated land lying entirely above the water table, the potential quantity of leachate which will be generated within a given time is equivalent to the net infiltration from all sources during the same period. A comprehensive method for estimating rates of leachate production, based on short or long-term accounting periods for the water balance has been proposed by the United States Environmental Protection Agency (23) and extended by Naylor et al (24) to take account of the additional hydraulic load which may be applied as a result of the co-disposal of liquid and solid waste, or from the run off of surface water from surrounding areas on to contaminated land. If long-term average infiltration values are substituted in the balance, changes in moisture content storage of the soils and wastes may be assumed to be zero, thus simplifying the computations. However, experience suggests that leachate production rates vary seasonally and estimates made on a weekly or monthly basis provide a more realistic assessment.

The initial water content of solid wastes is often below their field capacity at the time of disposal. Between 60 and 125 mm of infiltration have been found necessary before the initiation of continuous drainage from a metre thickness of domestic wastes. If the rate of deposition of fresh wastes is such that their accumulated moisture deficits exceed the rate of effective rainfall, continuous leachate production may be delayed considerably. Capping of such sites, to reduce or eliminate infiltration may be possible but could give rise to very slow stabilisation of the waste mass.

At landfill sites, or contaminated lands where the hydraulic conductivities of the underlying strata are lower than the mean rate of net infiltration, leachate will accumulate within the site and may discharge horizontally as interflow or surface seepages. In general, such sites are underlain by fine grained rocks such as clays, shales, mudstone and marls, but recharge rates augmented by liquid waste disposals may lead to the ponding of leachate on formations with relatively high hydraulic conductivities (8). When the hydraulic conductivity of the underlying rock exceeds the net infiltration rate, permanent accumulation of leachate within the wastes does not occur, but leachate migrates vertically through the unsaturated zone, to the water table. The leachate may move either by intergranular seepage, between the constitutent grains of the rock or sediment, or by fissure flow through fractures. In the former case the rate of movement through partially gas-filled spaces is relatively slow, affording a maximum opportunity for chemical and biochemical processes of leachate attenuation, as well as effective filtration of particulate contaminants. In contrast, fissure flow is rapid, with a minimum rock/leachate contact and reduced opportunities for attenuation. Similarly, horizontal flow velocities within the saturated zone are generally significantly lower in intergranular aquifers than in fissure-flow systems, where velocities of the order of kilometres per day may be achieved. The storage coefficients of intergranular flow systems are often in the range 10 to 25%, whilst that of fissure flow aquifers may not exceed 1 or 2%. As a result, groundwater level fluctuations in fractured rocks are frequently rapid, responding to individual infiltration periods, with annual flucturations that may be measured in tens of metres. In such situations, wastes or contaminated ground lying many metres above the summer water table level may be inundated by rising groundwater levels in the winter period, with subsequent flushing out of leachate. By comparison, the natural groundwater level fluctuations of intergranular flow aquifers are muted, being buffered from sudden recharge by the temporary storage available in the unsaturated zone (Fig. 1). Sand and gravel beds and poorly consolidated sandstones, such as parts of the Lower Greensand formation, are examples of dominantly intergranular flow aquifers, whilst fractured granite or massive limestones, such as the Carboniferous Limestone in South Wales, are representative of fissure-flow aquifers.

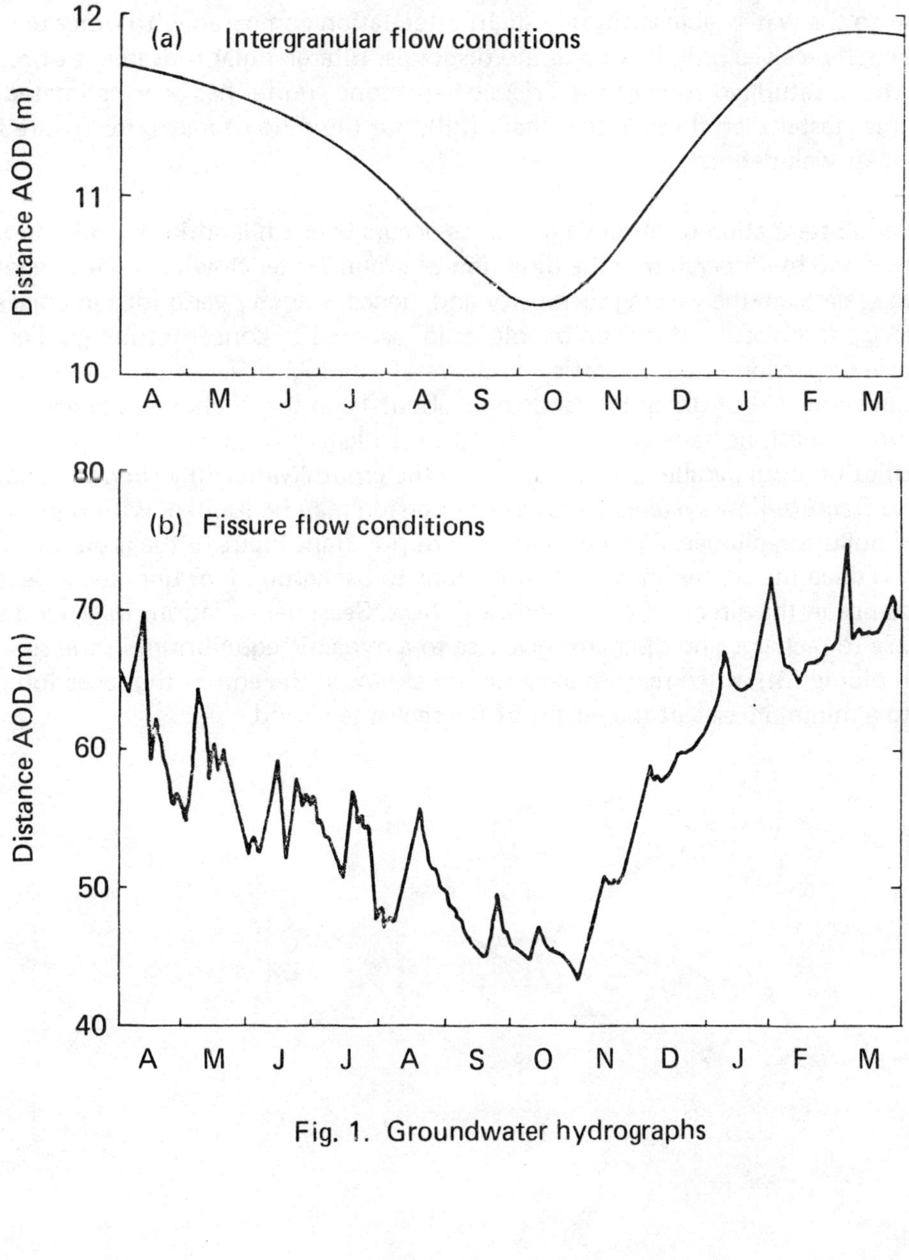

Fig. 1. Groundwater hydrographs

The principal aquifers in the United Kingdom, the Chalk and Triassic Sandstone, exhibit a range of behaviours between the two extremes, the ratio of intergranular to fissure flow being dependent on a number of factors including lithological variations and the local degree of fissuring due to tectonic deformation and human activities such as mining. The movement of water and solutes through the unsaturated zone of the Chalk is the subject of considerable hydrogeological study. The Upper Chalk characteristically possesses a porosity of between 35 and 50%, with main pore diameters of less than one micron (0.001 mm) (25), giving rise to intergranular saturated hydraulic conductivities of the order of 10^{-6} cm/sec. The rock is traversed by frequent,near vertical fractures, which are estimated to account for between 1 and 3% of the bulk porosity (25), but which raises the bulk hydraulic conductivity to between 10^{-3} and 10^{-2} cm/sec. Arising from the investigation of the movement of plant nutrients leached from agricultural land, Young et al (26) have postulated that under natural infiltration rates a high proportion of recharge to the Upper Chalk, both water and solutes, moves slowly through the matrix of the unsaturated zone, at velocities estimated to be in the range 0.5 to 1.5 m/year, but that perhaps 15% of recharge takes place rapidly via the fissures. The implications for leachate movement from the Chalk landfill sites receiving only natural rates of infiltration are important, in that a high proportion of contaminants may move slowly, allowing a maximum time for attenuation mechanisms to operate. The hypothesis suggests that a greater proportion of recharge occurs rapidly via fissures at enhanced rates of infiltration and some evidence of this was recorded at a landfill site on the Upper Chalk in Suffolk (27), where high rates of liquid waste disposal to part of the landfill appeared to have encouraged leachate migration to the water table with only slight attenuation compared with other parts of the landfill which received only limited liquid disposals. Intergranular movement of recharge through the unsaturated zone of the Triassic Sandstone aquifer has been estimated to be several times faster than through the Chalk (28), but the ratio of matrix to fissure flow in that aquifer is less well defined.

Dilution and attenuation of miscible leachates occurs beneath landfills by mixing with groundwater flows and by dispersion in the direction of groundwater flow (29). Dispersion arises on the micro-scale from the varying tortuosity and, hence, velocity variations in adjacent flow paths and from molecular diffusion of solutes in response to concentration gradients. On the macro-scale dispersion is due to vertical and lateral lithological variations in the aquifer. Dispersion constants (α) of 0.6 m for sandstone, about 10 m for Chalk and between 15 and 40 m for fractured limestone have been reported (8), the higher values indicating greatest attenuation by dilution both parallel and transverse to the groundwater flow (Fig. 2). In poorly interconnected fracture-flow systems transverse dispersion may be limited, with only slight lateral spread of pollution plumes. At a constant rate of pollutant input to the groundwater, dispersion may reduce the contaminant concentrations to background, or undetectable levels at some distance in the direction of groundwater flow. Seasonal variations in groundwater flows in response to recharge and discharge give rise to a dynamic equilibrium in the size of the pollution plume (3), which reaches a maximum extent at the end of the recession period and shrinks to a minimum size at the height of the recharge period.

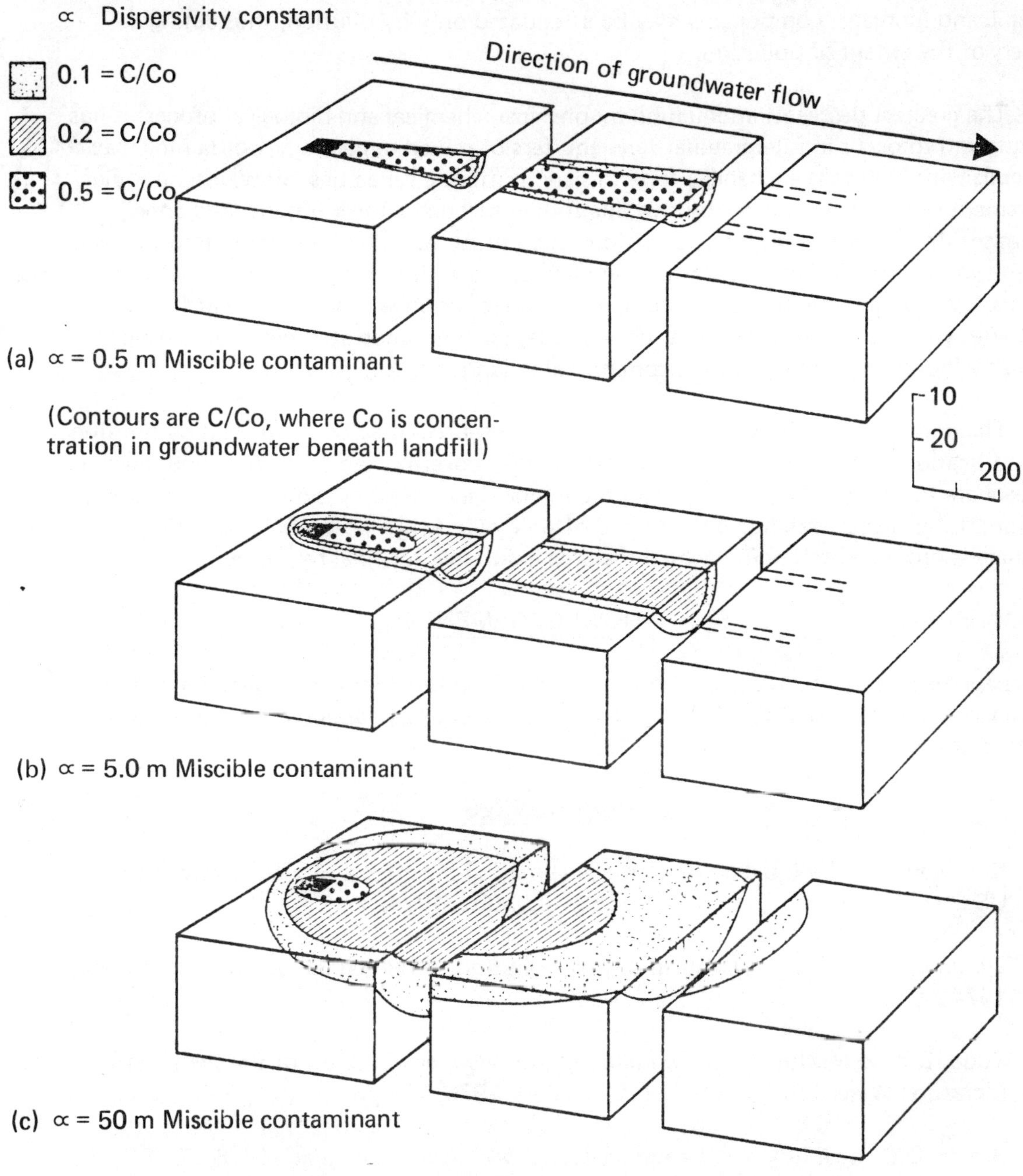

Fig. 2. Steady state distribution of pollutants in an isotropic, homogenous aquifer due to dispersion. Longitudinal and transverse dispersion assume equal. (after DOE 1978 (8))

4. SUMMARY AND CONCLUSIONS

4.1. Field investigations of existing areas of contaminated land and laboratory and pilot-scale experiments on leachate production and attenuation have shown that physical, chemical and biological processes may act to ameliorate the effects of contamination on water resources.

4.2. These processes may operate on both the contaminated land and during migration of the leachate through underlying or adjacent strata. Removal of particulate contaminants, biodegradation of labile organic components and sorption, precipitation and complexing of organic and inorganic substances may attenuate leachate, but certain persistent organic and inorganic components may be attenuated only by dilution, thus acting as tracers of the extent of pollution.

4.3. The greatest degree of attenuation by physical, chemical and biological processes has been found to occur in intergranular flow aquifers of mixed mineralogy, containing clay minerals with high base exchange properties and buffering capacities. Slow, intergranular movement of water and solutes by two phase flow through a thick unsaturated zone optimises the opportunities for attenuation. In contrast, rapid leachate migration through fissures in massive, non-porous strata offers few opportunities for attenuating mechanisms, other than limited dilution, to operate. Attenuation will be minimal at fissure flow sites with a shallow water table, or where seasonal groundwater level fluctuations regularly inundate part or all of the contaminated land.

4.4. The impact of landfill site leachate on water resources may be controlled by limiting the infiltration rate, in the form of low permeability covering, etc., thus reducing the rate of leachate formation. This management technique may, however, slow the rate of stabilization of the wastes, extending the period of time over which potentially polluting leachate is produced and before which the area is suitable for final reinstatement.

5. ACKNOWLEDGEMENTS

This paper is published with the permission of the Directors of the Medmenham and Stevenage Laboratories of the Water Research Centre and with approval of the Department of the Environment.

6. REFERENCES

1. Karubian, J.F., Polluted groundwater: estimating the effects of man's activities. US Env. Prot. Agency, report EPA/680/4074-002. 1974.

2. Summers, W.K. & Spiegel. Groundwater Pollution - a bibliography. Ann Arbor Press, 1974.

3. Todd, D.K. & McNulty, D.E.O. Polluted groundwater - a review of the significant literature. Water Infer. Center Inc. New York 1975, 179 pp.

4. Braids, O.C. & Gillies, N.P. Groundwater. *J. Wat. Poll. Centr. Fed.* 1978, 50 (6), 1406-1414.

5. Geraghty, J.J. & Miller, D.W. Status of groundwater contamination in the US. *J. Am. Water Wks Assoc.* 1978, 70 (3), 162-167.

6. Lieber, M., Perlmutter, N.M. & Fraventhal, H.L. Cadmium and chromium (VI) in Nassau County Groundwater. *J. Am. Wat. Wks. Assoc.* 1964, 56 (6), 9 pp.

7. Perlmutter, N.M. & Lieber, M. Disposal of plating wastes and sewage contaminants in groundwater and surface water, South Farmingdale Massapequa Area, Nassau County, Long Island, New York. US Geol. Survey Water Supply Paper 1879-G. 1970. 74 pp.

8. Department of the Environment, Disposal of hazardous wastes in landfill sites, Final Report of the Policy Review Committee 1978. HMSO.

9. NATO-CCMS. Disposal of hazardous wastes: Landfill. US Dep. Commerce, Nat. Tech. Inf. Service Rept. PB-276811. 1977.

10. Miller, D., Braids, O. and Walker, W. The prevalance of subsurface migration of hazardous chemical substances at selected industrial land disposal sites. US EPA, Rept. EPA 68-01-2966. 1977.

11. Hoare, M.J., Eastwood, J.W. and Brown, S. Groundwater contamination: combatting problem on site. *Surveyor*, 24 May 1979. 12-13.

12. Smith, H.E. A water supply polluted by aviation fuel. *Proc. Soc. Wat. Treatment and Exam.* 1962. 11 (1) 32-33.

13. Woods, E.C. Pollution of groundwater by gasworks waste. *Proc. Soc. Wat. Treatment and Exam.* 1962. 11 (1).

14. Baxter, K.M. and Edworthy, K.J. The Impact of Sewage Effluent. Recharge on Groundwater in the Chalk for an area in south-east England. Paper presented at International Symposium on Artificial Groundwater Recharge, Strasbourg. 14-18 May, 1979.

15. Shepherd, J.M. Pollution of groundwater supplies by sewage. *Proc. Soc. Wat. Treatment and Exam.* 1962. 11 (1) 12-16.

16. Whitaker, W. The Water Supply of Kent, with records of sinkings and borings. Mem. Geol. Surv. England and Wales. 1908. 399 pp. HMSO.

17. Standing Committee on the Disposal of Sewage Sludge. Report of the Working Party on the Disposal of Sewage Sludge to Land. DOE/NWC. Standing Technical Committee Report No. 5. 1977. Department of the Environment, 1977.

18. Edworthy K.J., Wilkinson, W.B. and Young, C.P. The Effect of the Disposal of Effluents and Sewage Sludge on Groundwater Quality in the Chalk of the United Kingdom. *Prog. Wat. Tech.* 1978. 10 (5/6) 479-493.

19. Zanoni, A.E. Potential for groundwater pollution from the land disposal of solid wastes. *CRC Critical Reviews in Env. Control.* 1973. 3 (3) 225-260.

20. McCabe, D.J. Gas and leachate from baled and shredded solid waste. Conf. on Land Application of Residual Materials. Eng. Foundation Conf., Easton, Maryland. Sep. 1970. ASCE, New York.

21. Headworth, H.G. and Wilkinson, W.B. Measures for the protection and rehabilitation of aquifers in the United Kingdom. *Proc. Water Research Centre Conference.* Groundwater Quality - Measurement, Prediction and Protection. Reading, England. 6-8 September 1976. pp. 760-817.

22. Nicholls, G.D. Pollution Affecting Wells in the Bunter Sandstone. pp.116-125. in Groundwater Pollution in Europe. Ed. J.A. Cole. Proc. Conf. Water Research Association, Reading, England. Sep. 1972. Water Information Centre Inc. Port Washington, New York.

23. Fenn, D.G., Hanley, K.J. and de Geare, T.V. Use of the Water Balance Method for Predicting Leachate Generation from Solid Waste Disposal Sites. Report SW-168. 1975. Office of Solid Waste Management Programs. US Environmental Protection Agency, 60 pp.

24. Naylor, J.A., Rowland, C.D., Young, C.P. and Barber, C. The Investigation of Landfill Sites. Technical Report No. TR 91, Water Research Centre, Oct. 1978. 68 pp.

25. Ballif, J.L. Porosite de la Craie. Appreciation de la taille et de la repartition des pores. *Ann. agron.* 1978. 29 (2), 123-131.

26. Young, C.P., Oakes, D.B. and Wilkinson, W.B. The Impact of Agricultural Practices on the Nitrate Content of Groundwater in the Principal United Kingdom Aquifers. Paper presented at IIASA Conference "Environmental Management of Agricultural Watersheds", Smolenice, Czechoslovakia, 23-27 May, 1979 (in press)

27. Young, C.P., Barber, C., Naylor, J.A. and Maris, P. Investigation of a site at Culford Road, Ingham, near Bury St. Edmunds, Suffolk. Programme of Research on the Behaviour of Hazardous Wastes in Landfill Sites: WLR Technical Note No. 22. 1978. Department of the Environment, London.

28. Brereton, N.R. and Wilkinson, W.B. Flow Mechanisms Governing the Movement of a Pollutant in a Groundwater System. *Proc. Water Research Centre Conference.* Groundwater Quality - Measurement, Prediction and Protection. Reading, England. 6-8 September 1976. pp.6-8.

29. Oakes, D.B. Use of idealized models in predicting the pollution of water supplies due to leachate from landfill sites. *Proc. Water Research Centre Conference*. Groundwater Quality - Measurement, Prediction and Protection. Reading, England. 6-8 September 1976. pp.611-624.

30. Millbank, P. Results of long-term monitoring of industrial wastes at a site in Kent. *Surveyor.* 6 August 1976. 28-30.

31. Leighton, I.W. & Blanc, F.C. Removal of cations from leachate by interaction with subsurface soils. *Boston Soc. Civ. Eng.* 1973. 60 (4) 145-162.

32. Fuller, W.H. Investigation of landfill leachate pollutant attenuation by soils. USEPA Report No. EPA-600/2078-158 1978, 218 pp.

33. Griffin, R.A. and Shimp, N.F. Attenuation of pollutants in municipal landfill leachate by clay minerals. USEPA, Report No. EPA-600/2-78-157.. 1978. 146 pp.

34. Farquhar, G.J. &Rovers, F.A. Leachate attenuation in undisturbed and remoulded soils in Genetelli, C.J. & Cirello, J. (Eds.) Gas and leachate from landfills: formation, collection and treatment . USEPA Report No. EPA-600-9-76-004. 1976. pp.111-122.

Table 1. Composition of typical leachates from landfills (in mg/l except pH)

	Fresh Refuse	Aged Refuse
pH	7.1	7.1
COD	11 600	125
BOD	7 150	5
TOC	4 440	40
NH_3-N	340	40
NO_x	0.5	2
Ortho-P	0.1	0.02
Cl^-	2 100	225
SO_4^{2-}	460	-
Na	2 400	240
Mg	390	90
K	490	75
Ca	1 150	280
Fe	160	0.1

MINE AND QUARRY WASTE - A STABILITY PROBLEM

SYNOPSIS

W. JOHN SEARLE

HM INSPECTORATE OF MINES AND QUARRIES

Waste - General.

Some 150 million tons of waste are deposited on land annually in the United Kingdom of which 100 million tons is mine and quarry waste; and 50 million tons is all other waste such as domestic, trade, commercial, and factory waste.

The deposition on land of mine and quarry waste is controlled in respect of STABILITY, following the Aberfan disaster, by the Mines and Quarries(Tips)Act 1969, and to the detailed requirements of the Mines and Quarries(Tips)Regulations 1971. The 1969 Act and the 1971 Regulations are administered by HM Inspectorate of Mines and Quarries.

The deposition of all other waste, with the principal exception of radioactive waste, is controlled in respect of POLLUTION by the Control of Pollution Act 1974 wherein it is defined as 'controlled' waste. Controlled waste may not be deposited on land without a licence to do so issued by a Waste Disposal Authority who will impose conditions on the licence to ensure that the waste will not pollute the environment. The waste disposal operation is subject to the provisions of the Health and Safety at Work Act 1974 which is administered by HM Inspectorate of Mines and Quarries where the controlled waste is being deposited in quarries or mines; and by HM Inspectorate of Factories where it is being deposited on factory premises or elsewhere.

Mine and Quarry Waste - Deposition.

Mine and quarry waste is generally concentrated in very large tips of solid waste and/or in lagoons of liquid waste close to its point of extraction. In comparison with most 'controlled' wastes referred to above, mine and quarry waste is inert but has a high density. So, in general, whereas 'controlled' waste presents a POLLUTION hazard; mine and quarry waste presents a STABILITY hazard.

In order to ensure the stability of solid tips of mine and quarry waste and the stability of bunds retaining lagoons of liquid waste as required by the Mines and Quarries(Tips)Act 1969; all large waste tips and retaining bunds must be pre-designed and sited by persons qualified to do so; and a Report detailing the proposals must be submitted to HM Inspectorate of Mines and Quarries 30 days before the tip or bund is started.

Solid waste tips must be deposited on a site which is capable of taking the imposed waste loading without failure or undue settlement. The tip must be compacted and profiled such that it is inherently stable and will not collapse, move, or crack open. The surface of the tip must be properly drained so that rainfall cannot penetrate into the body of the tip and cause instability. The site of the tip must be adequately drained to prevent the accumulation of water around the tip and must have a proper outfall to carry all water to natural drainage. Finally the tip must be inspected weekly and any signs of bulging, cracking, movement, water accumulation, or any other sign of instability must be reported and rectified immediately; and every two years a full report on the stability of the tip must be prepared by a person qualified to do so.

Similar design, construction, siting, drainage, inspection, and reporting procedures are required for bunds retaining lagoons of liquid mine and quarry waste,

Mine and Quarry Waste - Records.

All plans, maps, inspection records, and reports regarding the stability of large tips of solid mine and quarry waste and bunds retaining lagoons of liquid waste are retained by the mine or quarry during its active life; and are deposited with HM Inspectorate of Mines and Quarries when the mine or quarry is abandoned.

There are similar statutory retention and deposition arrangements for the weekly record of mine and quarry waste. This record contains the nature, quantity, and location of waste deposited on all large tips of mine and quarry waste and in lagoons of liquid waste.

Before commencing reclamation work on any mine or quarry waste tip or lagoon enquiries should be made to ascertain whether any of the above records are available. The records will be useful in planning any reclamation because they will indicate how the deposit was designed and constructed and will record the nature and variation of the waste. Any records of instability, movement, or dangerous occurrence will give a warning as to what precautions are necessary to avoid instability hazards during reclamation works.

Mine and Quarry Waste - Reclamation Hazards

The combustion, heating, and gas emission hazards in coal mine waste; and the potential hazards in the chemical properties of some other mine and quarry waste which must be considered in regard to reclamation are adequately dealt with in other papers. This paragraph is therefore confined to instability hazards.

The potential instability hazards within large accumulations of mine and quarry waste are too often not fully appreciated until they cause a sudden disaster as was caused at Aberfan; and more recently at Collonne-Ricouart in France; near Liege in Belgium; and at Buffalo Creek in the United States.

Instability in mine and quarry waste is in general caused by random, unplanned tipping and it is for this reason that legislation to control mine and quarry waste tipping was introduced into the United Kingdom following the disaster at Aberfan.

The first step in planning reclamation of mine and quarry waste areas is therefore to ascertain whether the waste was deposited in accordance with existing legislation; and if so, to obtain and study the various plans, maps, and records prepared and retained in accordance with that legislation. These documents should give adequate information in regard to the construction and any historical record of instability to enable reconstruction works to be planned which will avoid any danger from instability.

It may well be, however, that the waste was deposited prior to existing tip and lagoon stability legislation; in which case it is more likely to present a stability hazard during reclamation for the following main reasons:-

1. Solid mine and quarry waste was generally loosely tipped; and slips were often encouraged in order to reduce the height of the tip and provide more tipping space. Each slip of a large mass of waste produced a slip plane or slickenside along which the waste could again slide if disturbed during reclamation.

2. Lagoons of liquid waste were often formed in pockets on solid waste tips and later covered over. On poorly drained cohesive waste these buried lagoons build up a pressure which can burst out and create a hazard during reclamation work.

3. Many old mine and quarry waste tips were not surface graded to shed rainwater and ponds have formed on the surface where water has accumulated and gradually seeped into the body of the tip causing toe bulging and surface slides often repeated a number of times and producing failure slip planes within the tip.

4. Because of restricted tipping area some old mine and quarry tips have been built with very steep side slopes resulting in tension cracks along the top which fill with water and cause instability.

These few examples give a brief indication of the instability problems which can be found on old mine and quarry waste which was deposited prior to the existing stability legislation. The unknown stability situation within such waste is a real danger and every effort should be made to assess the real situation by means of an appropriate site investigation in order that reclamation works can be undertaken safely.

@@@@@@@@@@@@@@@@@@@@@@@@

STRICTURES AND STANDARDS FOR WATERS FOR AMENITY PURPOSES

Kenneth Guiver, Esq. Chief Scientific Officer

Southern Water Authority

Water quality standards are not laid down in the United Kingdom generally, for the various uses of water, including recreational use. Nevertheless, there is a considerable amount of information on the quality of water needed for fisheries and expert advice can be obtained on the subject. An idea of water requirements for other uses is available from various sources.

There is a need to take into account other factors such as depth of water, planting of trees etc., in constructing a lake on reclaimed land. Prevention of pollution is important and it is advisable to consider ways of mitigating it at an early stage of any development.

Introduction

Rain, falling on land, either percolates into the groundwaters, or runs fairly directly from the surface into local streams if the soil and substrata are more impermeable. The same thing will happen on reclaimed land and the quality of the water that runs off will depend on the material that has been used to reclaim the land and the period during which the rainwater is in contact with that land.

If there is any pond or impoundment of water on the land surface itself, the contaminants in the water will have a marked effect on whether it becomes an eyesore with possible septic conditions giving rise to local smell complaints, toxic with other chemicals such that it cannot support even the lowest forms of life, or a clean and clear stretch of water that can be used for various purposes, and enhance the beauty of the surroundings. Although little has been published, particularly in the United Kingdom, on the quality of water that is required for various amenity uses other than fisheries, it is possible to give some guidance on what is likely to occur and action that can be taken to prevent significant pollution. The early installation of protective measures against possible pollution, can mean that high standards of water quality are obtained to meet most purposes, and certainly add to the aesthetic value of a pond or lake.

The pollution of water from reclaimed land

To prevent any pollution of rainfall run-off from reclaimed land, it is probably best to completely seal the surface of the land if there is any chance of the run-off being adversely affected by the reclamation material. Then the water can flow directly into any impoundment or lake created, and is generally of the same quality as rainfall itself. A covering of soil with proper underdrains, above the sealed surface of the reclaimed land will allow the growth of grass and plants etc. The water itself might contain some chemicals such as fertilizers leached out from the soil itself, but will not be contaminated with any toxic materials from the reclaimed land. Weed growth may occur in the water due to run-off from the soil containing fertilizers, but this can generally be controlled, particularly if the water is of adequate depth, and excellent fisheries and general amenity waters can be produced.

The most economic ways of sealing a land surface will depend on local circumstances. If clay can be obtained near at hand, it can be used successfully. Any fine grained material with a low permeability can be equally good, particularly if underdrainage of the overlying top-soil provides a preferential water flow to a surface stream or lake. Puddled chalk has been found to have satisfactory low permeability if properly prepared and of adequate depth.[1] In the absence of such material, synthetic rubber liners have often been used although somewhat expensive for large areas.

In many circumstances, however, it will not be possible to seal the surface of reclaimed land, and water run-off may then dissolve materials from the land such that they can cause problems in ponds etc. If land on which domestic refuse has been tipped for instance is rehabilitated, the run-off is likely to contain organic compounds from the decomposition of the rubbish such that the water will be likely to become anaerobic when impounded, and to have a high ammoniacal nitrogen content which can be toxic to fish and insects. The leachates from such refuse tips even in the absence of toxic industrial waste, are difficult to purify, and the mere fact that in many instances they are anaerobic, leads to additional problems with the low pH of the waters dissolving iron and manganese from the tipped material and local strata. Thereafter, when the water is aerated, the iron is precipitated out as an objectionable brown semi-colloidal material.

If anaerobic conditions develop in accumulations of leachate from refuse tips, they can be overcome by intensive aeration or the addition of chemicals which can provide oxygen such as sodium nitrate, but satisfactory purification is difficult and the waters are never likely to be of an adequate quality for amenity purposes without considerable further improvement, either by dilution with clean waters or by removal of contaminants in some other way in a separate purification pond. Percolation of water through some distance of clean soil or sub-strata can of course give rise to considerable purification with removal of metals by adsorption onto soil particles. There may also be advantages in creating the water impoundment some distance away from reclaimed land in order that other groundwaters are collected to improve the water quality.

The amount of water that can be impounded and the area from which it is collected must obviously be dependent on the local geological characteristics of the area, including the permeability of the various strata surrounding the reclaimed land, the catchment area concerned and the amount of rainfall. As mentioned above, sealing of the reclaimed land to prevent rainfall run-off leaching out pollution, may be of benefit in those instances where the reclaimed land is likely to give rise to problems, but this may reduce the amount of rainfall run-off that can be collected from the area. Reclamation of areas such as gravel pits and land where there is no likelihood of pollution of rainfall run-off, is obviously a much easier proposition, in terms of water quality. In these cases there may be difficulties in maintaining a permanent pond or lake unless the final land surface intersects the normal water table, or the pond has its base sealed.

Standards of Water for Amenity Purposes

The greatest amount of information available on the needs for water quality for amenity purposes, probably arises with regard to fishing. In so far as fish have to live their whole lives in water, a great deal of research work has been done on chemicals which are toxic to them. Even potable water supplies, although they require standards to prevent problems with human consumption, do not often need more stringent conditions than those for fish. In fact fish are generally encouraged in most rivers to act as biological indicators in order to warn potable water abstraction points, when problems might arise. There are chemicals which have different toxicity characteristics between animals and fish, but, in the present absence of satisfactory chemical methods for continuous monitoring water quality, fish can still provide a useful function like the miner's canary.

To some extent the toxicity characteristics of chemicals are dependent on the particular species of fish that are present in waters. Trout are normally found to be more sensitive for instance than carp and other coarse fish. (This factor is recognised in a recent EEC Directive concerned with the quality of river waters for fish. Higher dissolved oxygen concentrations are required for trout and other salmonid fish when compared with the coarse fish (cyprinids)).[2] It is proposed that rivers will be designated as either salmonid or cyprinid waters, and will then have to meet certain standards for a large number of chemical constituents and physical conditions.

Ammonia is another potentially toxic constituent of waters. It can be derived from effluent discharges or the decomposition of organic matter. The toxicity of ammonia will depend on other factors such as the pH and temperature of the water since it is the unionized ammonia concentration which is of main concern. The EEC Directive thus gives limits for both total ammoniacal nitrogen and unionized ammonia in designated rivers. Not all the limits in the Directive are however primarily concerned with toxicity to fish. There are other factors taken into account such as the concentration of chemicals that might taint fish flesh which could be used for human consumption. The Directive also gives certain limits for nutrients such as nitrates and phosphates which have no toxic effect on fish, but which can give rise to weed and algal growth problems which have an indirect effect on fisheries.

The Anglian Water Authority has recently put forward various standards for rivers in its region to be used for various purposes, such as potable supply, industrial supply, fisheries etc. It has taken account of various EEC directives in producing the standards, but as far as fisheries are concerned has also set out limits for ammoniacal nitrogen dependent on temperature and pH. In addition it has taken the first steps in the U.K. to define standards of water quality for amenity purposes.(3) Different limits are set for general amenity and high amenity waters, with figures in each case being stipulated for maximum permissible values and maximum desirable values of various constituents. The maximum permissible values are defined such that higher concentrations of a particular constituent are likely to give rise to dangerous conditions. The maximum desirable concentrations are somewhat lower and are to be used as guidelines or working levels which should be maintained at all times if possible.

For waters that are going to be used for bodily contact sports such as swimming or water skiing, it might be considered that some form of bacteriological standard might be desirable. Such standards are in fact set out for bathing waters in an EEC Directive that was passed in 1975.(4) The standards are 10,000 total coliforms per 100 ml and 2,000 Escherichia Coli per 100 ml, but it is only intended that they should be applied to "designated" waters that are used by large numbers of bathers. It is intended that the directive shall apply to both fresh and saline waters, but as far as the U.K. is concerned, it appears unlikely that any inland waters will be designated.

In fact there are many rivers which are taken for potable supply which do not meet such standards due to the run-off from agricultural land and effluent discharges etc. For potable supplies there is a need to sterilize the water using chlorine gas or other chemicals, but this should not be necessary, and could be highly undesirable in recreational waters. Swimmers and water skiers do not normally swallow large volumes of the water they are using for their sports and the dangers of excess chlorine with its high toxicity to fish and other organisms can be a much greater hazard than the risk of public health problems. Self purification takes place in any mass of water that is left in contact with air without additional pollution being added to it, and the ultra-violet action of sunlight acts as a strong sterilizing agent to kill bacteria in waters, provided that turbidity does not restrict penetration of the light.

<u>Water testing and treatment</u>

No matter how much information is already available on the land surrounding a pond or lake, it is always worthwhile having a full analysis taken before deciding on any use of the water. Depending on the size and depth of the water, it may be necessary to take a series of samples in order to have a fully representative picture of the water quality. Samples should also be taken of any streams feeding into the lake, particularly from reclaimed land. If heavy rainfall means that additional streams develop, they should all be tested to ensure that no toxic material is being washed out of the surrounding land.

For an interpretation of the results of analysis it will probably be necessary to obtain expert advice. Any of the Regional Water Authorities within the United Kingdom should be able to offer advice and in many instances even be able to assist with the analysis. The Authorities have both chemists and biologists who are able to comment on the quality of waters and assess potential uses. The staff of the Water Research Centre at Stevenage have access to most world-wide information on the toxicity of various chemicals to fish, and carry out considerable experimental work on the subject themselves. Further information on various chemicals can be obtained from United States publications by the Environmental Protection Agency.(5)

Consent from the Regional Water Authorities in this country is required under the Salmon and Freshwater Fisheries Act 1975, for any introduction of fish to new lakes or other waters. It needs to be established that there is no risk of any disease being introduced with the fish, and that they will not be affected by the new water. It is also important to consider whether there will be subsequent variations in water level in the pond or lake. Uniformly shallow waters are not generally successful as fisheries and should never be used for trout stocking, because trout are more susceptible to the high temperatures that might occur. The most satisfactory conditions for fish to thrive are where there are some deep sections of lakes and there is no chance of the water level dropping such that the lake surface is reduced and the fish are crowded in a shallow pond of hot and muddy water.

Most trout fisheries in this country are maintained by artificial stocking rather than by natural propagation of fish and account must be taken of the costs of buying the fish in deciding how much to charge for fishing. Coarse fisheries are more likely to be self-sustaining since the fish are not generally removed. Even then it may be of considerable benefit to provide facilities for coarse fish to spawn by sinking bundles of sticks into the water and ensuring a good plant growth without making fishing impossible because there is too much growth. Some shading of part of the water by surrounding trees is often also worthwhile provided it does not interfere with the angler when he wishes to cast his line.

In many cases, before obtaining representative analyses of water samples it will be worthwhile leaving the water for a time to see how it develops of its own accord. This will allow time to see what variation in depth and condition of the water arises from rainfall. Providing there is no toxicity and everything possible is done to keep the water as clean as possible, it will also give an indication of what natural growth might develop. Self-purification and oxygenation of the water can be encouraged by the introduction of some water plants. With reasonably large bodies of water it may even be worthwhile depositing sand and shingle material on the bank of the water most remote from the prevailing wind direction, in the anticipation that circulation of the water through the sand will assist with filtration and purification of the water which can be of benefit.

Protective measures to limit pollution of the pond should be provided. Any incoming streams that are known to suffer from oil slicks or other surface material should have an interceptor constructed on them even if it is only the provision of a simple undershot weir. It may be necessary to allow some form of settlement pond or interceptor on minor streams, which can be readily emptied to prevent the main pond becoming silted and shallow. If trout or coarse fish are stocked and there is evidence of deoxygenation of the water either from organic material or possibly the die-back of weed growth, it may be worthwhile considering the installation of floating aerators which can oxygenate and circulate the water. In isolated instances of deoxygenation of water, it may be satisfactory to pump out and spray back over the lake surface using a simple pump from the power take-off of a tractor. In cases where ponds become seriously silted and weed growth is a problem, it is possible to use aquatic herbicides under expert guidance for control. It is generally preferable in the long-term, however, to carry out major dredging to deepen such waters rather than depending on the temporary palliative of weed removal.

General Considerations of Amenity Use

The length of this paper does not allow comment on the many aspects which need to be taken into account when considering the use of a lake for a particular recreational use. Although fishing can be practised on a relatively small area of water, most active sports require much more in order to be practicable. Yachting requires space and an open area where the winds can provide the strength and variability that make the sport exciting. Rowing requires a straight stretch of water of adequate length to build up the strength of the crew. Both require that weed in the water is kept to a minimum.

For general advice on potential recreational use of water it is suggested that contact is made with the Water Space Amenity Commission at St. James, London. The various Regional Sports Councils around the country may also be able to offer advice on particular sports.

References

1. "Promising Future of Chalk needs D.o.E.Investigation", Surveyor, 19th January, 1978. page 14.

2. Commission of the European Communities. Council Directive on the quality of fresh waters needing protection or improvement in order to support fish life 78/659/EEC. Official Journal No. L222/1, 14th August, 1978.

3. Price, D.R.H. and Pearson, M.J. "The Derivation of Quality Conditions for Effluents Discharged to Freshwaters". Inst. Water Pollution Control 78. 1979. 1. Page 118.

4. Commission of the European Communities. Council Directive concerning the Quality of Bathing Water. Official Journal No. L31, 5th February, 1976.

5. The Environmental Protection Agency. "Water Quality Criteria 1972".

Reclamation of Agricultural Land used for Disposal of Sewage and Sewage Sludges.

A.A.Wood and A.H.Ross

Thames Water, Chiltern Division.

SYNOPSIS

Manor Farm, Dorney, Bucks, was a sewage farm from 1879 until about 1938. From that time until the present, farming has continued in conjunction with sludge disposal.

Analysis of top soil showed that the land was unsuitable for general agricultural purposes if the recommendations of M.A.F.F.(A.D.A.S) and D.O.E./N.W.C. were to be followed.

Wide variations in heavy metal content were detected, on a field by field basis. Three fields were surveyed in depth and the results from the two at the extreme ends of concentration ranges are dealt with in some detail.

The need for reclamation having been established the question arose as to what options might be exercised to achieve that objective. Four such options are currently being investigated and on completion of the soil renovation stage on selected areas cultivation will recommence. Subsequently, a repeat, in depth survey of the soil will be performed and uptake of metals into various crops will be ascertained in order to establish which option produced the best results from a crop quality aspect. It is also intended that effects on ground water quality will be monitored as some methods of renovation may well be expected to increase the potential release of trace elements by leaching.

HISTORY OF THE PROBLEM.

In the latter half of the nineteenth century the disposal of sewage in inland towns of any size was commonly performed by irrigation over land. Frequently, this operation was practised in conjunction with farming since it was known the sewage had a manurial value.[1]

Sewage farms inevitably tied up relatively large areas of land and with the population growth which was occurring the sewage farms generally became subject to overloading. In many instances the overloading was detrimental both to the farming activity and to the sewage treatment function.

With the development of relatively intensive methods of sewage treatment such as the percolating filter and the activated sludge process, sewage farming, as the sole method of treatment decreased fairly rapidly from about the turn of the century.

In many cases the newly constructed sewage works were built on part of the sewage farm land. Because of the intensive nature of the new processes however, the major part of the sewage farm became redundant and therefore available for other use. The fate of such land varied from place to place, sometimes being sold off partly or entirely to agriculture, housing or industrial development, or other purposes. In a number of cases though, the local authority retained the land with a tenant farmer or continued to farm the land directly. In the latter case part of the farm might be used for treatment of storm-water by irrigation and the remainder used for sludge disposal.

As a consequence of the previous use of the land for sewage treatment the nature of the soil, and its chemical composition could be very different from that of soil in fields adjacent, which had not been so used.

GROWTH OF SLOUGH

In the pre-World War I era, Slough was a small country town with a population of only a few thousands. The principal industries at that time appear to have been agriculture and brickmaking.

During the first World War an extensive motor vehicle servicing depot was established by the War Department on the western extremity of the town. At the same time, a small sewage treatment works was constructed at Cippenham to deal with the sewage generated by the depot. The Cippenham Works, covering an area of about 10 hectare was designed to deal with the sewage from a population of 10,000.[2]

Sewage from the rest of Slough had been pumped to a 10 hectare sewage farm at Dorney since 1879. As growth of local population took place additional farmland was purchased and by 1921 the farm covered about 50 hectares.[2]

Shortly after the first World War had ended the War Department depot at Slough became vacant and it was then purchased and development as an industrial estate commenced. From that time, and until the recent past Slough was a 'boom-town' with industry and population growing at such a pace that the sewage treatment facilities were soon outstripped.

In 1930 - 31 the Urban area was increased fourfold and the Council acquired the Cippenham Sewage Works which adjoined the sewage farm. Thereafter the sewage works and sewage farm were operated in conjunction one with the other, but in increasingly overloaded conditions. A completely new works was obviously required and constructional work commenced in 1936 and was completed in 1938.[2]

SLOUGH SEWAGE FARM

Manor Farm, Dorney, now covers approximately 68 hectare and incorporates the whole of the pre - 1938 sewage farm apart from the land on which the present treatment works lies. The present area also includes two fields which represent late additions, i.e. since 1945. One of these fields became part of the farm as an indirect result of the construction of the M4 motorway. This field (Plot 4) is one of the two fields considered in detail later.

During the inter-war period while Slough was 'booming' the whole of its sewage, and industrial effluent was treated at Cippenham and Dorney. Unfortunately, though it might be thought that the growth of Slough was paralleled by the New Towns development after the second World War this is not so. The New Towns were provided with up to date sewage treatment facilities which kept pace with the population growth and industrial expansion. The New Towns were able to control trade effluents from the outset by virtue of the existence of the Public Health (Drainage of Trade Premises) Act, 1937. While Slough sewage farm was being operated as such it was subject to increasing overload with sewage containing substantial volumes of trade effluents which were not, and indeed could not, at that time, be effectively controlled.

From 1938 to 1966 the sewage farm changed its identity, largely, to that of a conventional farm although still under the same management as the sewage treatment works. Mixed farming was practised with rotation between cereals, root crops and grazing. From 1950 - 60 crops such as potatoes, mangolds, beet, oats, barley and kale appear in the records. Fields adjacent to the Roundmoor Ditch were still in use occasionally for storm-water irrigation and the rest of the farm was used for disposal of sludge, virtually all the sludge produced being disposed of to the farm.

In the mid 1960's a decision was taken that the farm would be used solely for rearing of beef cattle and simultaneously some old sludge lagoons in the farm hinterland were allowed to dry out and subsequently returned to grazing meadows.

At the same time as the changeover to beef-rearing took place it was decided that the ever increasing volumes of digested sludge being produced could no longer be dealt with adequately by lagoon storage and disposal on Manor Farm.

The solution of the problem was seen to lie in the provision of additional land and purchase of road going tankers designed to carry and distribute liquid sludge direct on grassland. Two additional farm units were purchased by the Corporation; 30 hectare at Knowl Hill a few kilometres west of Maidenhead and 26.5 hectare at Hitcham a few kilometres to the north west of Cippenham Treatment Works. Ex - W.D. tankers were purchased and a workable system for sludge disposal was established.[3] From that time until the present the amounts of sludge disposed of on Manor Farm have been more closely related to the needs of the crop (grass and hay) and have been therefore substantially less, on an annual basis, than was the 'norm' prior to 1966.

Monitoring of treated sludge had already been instituted at that time, on a regular basis and shortly afterwards, an atomic absorption spectrophotometer was secured for use in the sewage works laboratory. From then on the daily monitoring of the sludge was supplemented by monitoring of soils on land receiving sludge whether the land was owned by the Corporation or not.[4]

Heavy metal concentrations in the top-soil of Manor Farm proved to be generally very much higher than the normal background level for the area. There was, however, no evidence to suggest that these, elevated, metal levels were affecting the growth of grass, or the health of the cattle reared on the farm.[5]

RESEARCH AND REHABILITATION.

Immediately prior to the reorganisation of the water industry in 1974 some thought had been given to the future of Manor Farm. When Thames

Water took over the running of Slough Sewage Treatment Works and the associated farms further consideration was given to the question of what should be done with a farm whose topsoil was unfit for agricultural purposes, in the light of A.D.A.S.10.[6] The more recent guidelines embodied in the D.O.E./N.W.C. Report No. 5[7] add weight to the argument that some means should be sought whereby the contaminated land might be renovated.

In May 1977 a meeting was held between senior officers representing Operations and Scientific Services. The officers were drawn from Regional, Divisional and Group levels. The situation was discussed in some depth and at that time it was agreed that deep ploughing might prove to be a viable solution to the problem.

The deep ploughing operation was subsequently made the subject of a research project and this work was approved by the Authority in July, 1977.

The aim of the project is, 'To assess the effectiveness of deep ploughing techniques on metal concentrations in soils and subsequent up-take by grass (and possibly other crops) with particular reference to cadmium.'[8] The Research Requirements Group after considering the merits of the project stressed that the work should be carried out with the full co-operation of A.D.A.S.

A further meeting was called to discuss and arrange the programme of work and a Senior Soil Scientist from M.A.F.F. (A.D.A.S.) attended the meeting and gave valuable advice.

Agreed Programme

1) Three fields; plots 2,4, and 8 were initially selected for profile investigation.
2) Sections of the fields would be treated in a variety of ways depending on the results of the profile investigation and on the advice of A.D.A.S.
3) Following the treatments the fields would be restored to grass.
4) After an acceptable sward had been established, profile investigation would be repeated in order to determine the effect of the various treatments on metal distribution.
5) Ground-water samples would be taken from beneath the fields in an attempt to determine changes in quality arising as a result of the renovation process.
6) After a season or two in grass a variety of crops would be grown in each field. Particular regard would be had to the yield and quality of the crops grown and it would be essential to determine whether or not the metal content of the crop would render it unfit for human consumption.

Modes of Treatment

Subject to the results of profile investigation the treatment of sections of each field would be as follows:-
1) Left undisturbed as a control.
2) Top-soil inverted to a depth of up to 1 metre, using specialised ploughing equipment.
3) Mixing by ploughing to approximately 0,7 metres.
4) Skimming of surface soil, followed by:-
   a) Inversion as in 2 above.
   b) Mixing as in 3 above.

5) Overtopping with uncontaminated soil if available on or near site.

Profile Investigation - Sampling and Analysis.

Ten, equally spaced, positions on each field were selected as sampling points and marked with stakes. A mechanical excavator (J.C.B.) was used to open up the ground at each position to dimensions of 2 x 1 metres and 1 to 1.2 metres deep. Thames Water and A.D.A.S. Soil Scientists working together with small hand trowels then obtained discrete samples from the vertical sides of the trenches in 7.5cm bands down to 15 cm from the surface and 15cm bands down to 90cm from the surface. Particular care had to be exercised in order to avoid contamination of each sample with soil from other levels. Seven samples were obtained from each trench. The size of each sample was approximately 15cm x 15cm x band height (7.5 or 15cm) and weighed 500 - 600 gm (for a 15cm band). The samples were stored in polythene bags prior to analysis.

Each individual sample, after initial preparation[9], was analysed for copper, zinc, cadmium, nickel, lead, chromium and manganese. (Chromium and manganese are less significant than the other metals and are not referred to again in this contribution).

The prepared soil was first digested with nitric acid followed by 4:1 nitric/perchloric acid mixture and the metals determined using a Pye Unicam SP 2900 Atomic Absorption Spectrophotometer with deuterium background correction.

A summary of results of analysis for heavy metals is given in table I. The mean, and 90% confidence limits are given for the ten samples from each depth band.

Samples were also taken for pH value, and available phosphorus, potassium and magnesium [10]. In the case of these determinands, three depth bands were selected:- 0-15cm;30-45cm and 60-75cm. Samples from several points were composited prior to analysis the results of which are given in table II.

The extractable 'zinc equivalent' metals (Zn,Cu,Ni) were also determined on eight samples from each field for the upper three bands.[6] These results are summarised in table III. For comparison purposes the total zinc, copper and nickel concentrations are also given for the corresponding, eight samples.

For the sake of brevity results from plot 8 are not included in any of the tables. The level of contamination in that plot lies between the levels found in plots 2 and 4.

Progress to-date.[11]

All three fields had sections of top-soil stripped off in the latter quarter of 1977. This operation was carried out using a box-grader drawn by a tracked vehicle. The top-soil removed was stock-piled on site and subsequently removed to be used for landscaping, primarily on industrial sites and on motorway embankments.

Provided there are suitable outlets for the top-soil this is obviously an ideal and practical method of rehabilitation. Given that the extent of contamination downwards has been determined it is a relatively simple matter to skim off the required depth of top-soil, the only unsurmountable obstacle to this procedure being presented by the presence of an unfavourably shallow, top-soil layer.

The success of the experimental top-soil removal was such that during the summer of 1978 approximately 6 hectare of land was skimmed and the soil removed at no cost to the Authority. Further work of this nature is planned for 1979.

The effect of this operation can easily be seen from results tabulated below:-

Metal concentrations mg/kg in upper 15cm of plots 2 & 4 before and after treatment by skimming.							
Total metal		Cu	Zn	Cd	Pb	Ni	pH
Plot 2	before	658	1560	127	735	198	6.0
	after	18	119	2.6	56	31	6.2
Plot 4	before	76	246	13	72	29	6.5
	after	16	61	1.0	31	18	7.1

Further progress with the project was, unfortunately delayed due to difficulties in locating, in this country, a plough capable of inverting the soil to the depth required. It was known that the Forestry Commission used ploughs capable of working to a depth of about 0.9m[12]. One such plough was kindly made available, on loan, by the Forestry Commission but proved to be unsuitable because inversion was minimal.

Considerable time was spent in tracking down the plough required but eventually after many false starts a successful route of discovery opened up early in the summer of 1978. This route commenced with a contact in the National College of Agricultural Engineering and then on via a Field Officer at M.A.F.F. Cambridge to a farm near St.Ives, Huntingdon. Arrangements were made to inspect a plough there which was described as being of the type we were seeking. This turned out not to be so, the plough being a deep-mixer but the contractors who owned it knew where there was a plough of exactly the type required. On inspection that plough did prove to be precisely the one we were looking for and although it had lain, unused, in the open, for over ten years it was obviously capable of being put back into working order. The plough was purchased and subsequently refurbished and became available for use in the late spring of 1979. So far as is known this plough, which was manufactured in the Netherlands, is the only model of its type in this country. It is designed to plough a single furrow to a depth of approximately 1.4m. The beam supports two plough shares, the working depth of the forward share being independently variable (within limits) of the working depth of the after, main-share. The working depth of the main share is controlled by a shoe supported by the beam. The plough is raised and lowered, from and to its working position by hydraulic rams controlled from the tractor. The function of the forward share is to take a slice of top-soil, up to about 20cm in thickness and deposit this layer in the base of the previous furrow. The main after - share, cuts and inverts the soil to the required depth.

A trial section was ploughed, on Plot 4, in the late spring of 1979. A crawler tractor developing about 460 h.p. was used and the operation of the plough proved to a depth of about 1.25m. At this depth the tractor (relatively new and in first class condition) had little in hand. It is understood that working on peaty fenland with a clay sub-soil, the plough was hauled by two D8 tractors when working to 1.4m. The plough and tractor together can only be operated successfully on a field of adequate dimensions.

Due to the waterlogged nature of the ground on the low-lying plot 2, it was not possible to complete the trial on that plot during the spring. Two furrows were cut however before the trial was suspended until ground conditions improved.

Rehabilitation by means of overtopping with uncontaminated soil was not, on further consideration, seen to be a viable proposition. However, some relatively uncontaminated material did become available on site as a result of civil engineering work so the opportunity was taken to deposit this material on one section of plot 2. This will allow crop quality on this section to be compared with that from other sections at a later date.

SOIL DESCRIPTIONS 1977.

Plot 2. Variable levels with three basically differing profiles.

1) Slightly higher ground with shallow sandy loam onto gravel by between 30 and 50 cm. Water table more than 1 m down.
2) Very organic sandy loam top-soil to 30 cm onto very organic silty clay loam to about 60 cm. This band was grey blue in colour and anaerobic. Gravel beneath 60 cm usually with a water table by 70 cm.
3) Very organic sandy loam top-soil onto the silty clay loam layer by 20cm and gravel by 70cm. Water table at 50 to 70 cm.

Plot 4. Almost without exception this was a well drained brown very fine sandy loam to more than 1 m. Gravel was encountered in some areas at between 60 and 80cm. There were occasional bands of loamy fine sand and additionally some bands of about 30 cm thickness of heavier silty clay loam. These latter were brown in colour, well structured and well drained and would not impede roots.

CONCLUSIONS.

Various methods of renovating contaminated agricultural land are currently being evaluated at Manor Farm, Dorney. The methods of skimming and deep inversion appear promising with the first of these methods having greater merit. That method is not appropriate however if contamination has been dispersed to any depth, e.g. by conventional ploughing.

The success or otherwise of the varying modes of treatment will only become apparent after the land has been put back into cultivation and various crops grown.

Since the land at Manor Farm is unlikely to be of unique character the results of the current exercise may well prove of value in similar situations elsewhere in the country.

ACKNOWLEDGEMENTS.

The Authors wish to thank Mr.E.C.Reed, Director of Operations, Dr.M.C. Dart, Director of Scientific Services and Mr.A.E.Collins, Divisional Manager, Chiltern Division, Thames Water Authority, for permission to publish this paper.

Thanks are also due to Mr.R.McD.Daniels, Regional Farms Manager, Thames

Water Authority, Mr.Selwyn Richardson, Soil Scientist A.D.A.S. for their valuable assistance and advice on operational aspects and finally to those staff at Slough Sewage Treatment Works, particularly Mrs.C.M.King, without whose efforts this contribution would not have been possible.

Any views expressed are those of the authors and not necessarily the opinions of the Authority.

REFERENCES.

1) Stanbridge.H.H. History of Sewage Treatment in Britain. Vol.5 Land Treatment. I.W.P.C.
2) Finch J. Description of the Sewage Purification Plant (Slough) 1963. Slough Corporation.
3) Claydon M.B. Et.Al. Disposal of Municipal Sludges to Agriculture. Proceedings of Symposium on Disposal of Municipal and Industrial Sludges and Solid Toxic Wastes 1973. pp.74 et.seq. I.W.P.C.
4) Annual Report of the Director of Water Pollution Control, 1970-71 p.18. Slough Corporation.
5) Department of Health and Social Security. The use of 'Cinagro' as an Organic Fertilizer - A Critical Appraisal in Relation to Possible Environmental Hazards and Human Health.
6) Ministry of Agriculture Fisheries and Food. Permissible Levels of Toxic Metals in Sewage Used on Agricultural Land. A.D.A.S. Advisory Paper No. 10, 1971.
7) Standing Committee on the Disposal of Sewage Sludge, Report No 5, 1977. Department of the Environment/National Water Council. ISBN 0 904561 22 4
8) Thames Water, Research and Development Programme for 1979/80. Department of the Environment/National Water Council.
9) Standing Committee of Analysts. The Sampling and Initial Preparation of Sewage and Waterworks Sludges, Soils, Sediments and Plant Materials Prior to Analysis, H.M.Stationery Office 1977. ISBN 0 11 751251 6.
10) Ministry of Agriculture, Fisheries and Food. Technical Bulletin 27. The Analysis of Agricultural Materials. H.M.Stationery Office 1973 (Revised 1979).
11) Thames Water, Progress in Research 1977/78 p.16.
12) Forest Ploughs. Forestry Commission Leaflet 70, 1978. H.M. Stationery Office. ISBN 0 11 710222 9.

TABLE I

DEPTH (CM.)	TOTAL METALS mg/kg DRY SOIL									
	PLOT 2					PLOT 4				
	COPPER	ZINC	CADMIUM	LEAD	NICKEL	COPPER	ZINC	CADMIUM	LEAD	NICKEL
0–7.5	987±164	2410±283	204±25	599±176	239±29	131±27	398±78	22±4	108±16	37±4
7.5–15	329±91	709±137	50±13	872±308	156±29	22±3	94±16	3.3±1.3	36±2	20±2
15–30	21±6	206±73	5.9±1.8	84±45	47±15	12±1	49±4	0.3±0.1	30±4	15±2
30–45	14±4	88±15	1.2±0.5	34±16	29±3	13±1	53±6	0.1±0.05	14±2	19±2
45–60	17±5	71±12	1.2±0.5	32±11	35±3	12±1	52±5	0.1±0.05	13±2	23±4
60–75	10±3	44±15	0.3±0.15	17±5	22±5	13±1	50±5	0.1±0.05	12±2	26±4
75–90	9±3	34±12	0.4±0.15	17±7	28±13	10±1	50±8	0.1±0.05	9±2	25±4

TABLE II

AVAILABLE NUTRIENTS EXPRESSED AS mg/kg DRY SOIL (EXCEPT pH)

DEPTH (CM.)	PLOT 2				PLOT 4			
	pH Value	PHOSPHORUS	POTASSIUM	MAGNESIUM	pH Value	PHOSPHORUS	POTASSIUM	MAGNESIUM
0 - 15	6.0	126	391	144	6.5	129	115	84
30 - 45	6.2	24	124	280	7.1	56	87	51
60-75	5.3	35	85	110	7.4	41	72	51

TABLE III

TOTAL AND EXTRACTABLE METALS (MEAN OF 8 RESULTS)

		PLOT 2			PLOT 4		
		0 - 7.5cm	7.5 - 15cm	15 - 30cm	0 - 7.5cm	7.5-15cm	
ZINC	TOTAL	2481±349	728±177	210±95	362±83	99±19	
	ACETIC ACID EXTRACTABLE	956±101	216±82	42±35	196±56	34±14	
	% EXTRACTABLE	39±4	30±6	20±5	54±6	34±9	
NICKEL	TOTAL	246±36	149±36	49±19	35±4	20±2	
	ACETIC ACID EXTRACTABLE	109±12	60±15	11±4	18±4	.6±2	
	% EXTRACTABLE	44±3	40±5	23±5	51±6	28±6	
COPPER	TOTAL	1064±168	322±117	21±7	119±30	22±4	
	EDTA EXTRACTABLE			10±3		15±4	
	% EXTRACTABLE			48±7		66±6	
ZINC EQUIVALENT				153±63		107±30	

METAL CONTAMINATION AND THE SIGNIFICANCE OF THE METAL OF METAL TOLERANCE

A.D. Bradshaw, M.A., Ph.D., F.I. Biol.

Department of Botany, University of Liverpool, England.

SYNOPSIS

Tolerance to heavy metals occurs very commonly in populations of plants growing on metal contaminated ground. It is due to normal evolutionary processes which can take place very rapidly and in very localised areas. As a result the presence of a vegetation cover on contaminated land does not mean that there is no toxicity. Metal tolerant populations can be valuable in very toxic areas as a stabilising and reclamation technique, and in less heavily contaminated areas as a means of reducing the metal content in vegetation which might be used for grazing.

EVOLUTION OF METAL TOLERANCE

Heavy metals are extremely toxic to plants. Concentrations of a few parts per million can kill plants completely. The actual level depends on the metal concerned, copper and nickel being among the most toxic. The primary effect of heavy metals is on root growth : this leads to the subsequent death of the whole plant. The levels of metal in solution which cause a 50% reduction in root growth are remarkably low (Table 1).

Table 1. Concentrations of metals causing 50% reduction in the root growth of ryegrass (Lolium perenne) (data of W.H. Wong)

copper	0.02	lead	1,7 ppm
manganese	0.05	cadmium	2.1
nickel	0.18	chromium 6+	3.8
aluminium	0.93	mercury	6.6
zinc	1.3	iron	9.3

In normal soils or in mine wastes a large proportion of the metal is unavailable, due to complexing with organic matter, to presence as insoluble compounds, or to being included in minerals. As a result metal toxicity symptoms are usually only found in plants when the total concentrations of individual metals in soil or wastes is over 1000 ppm. Such levels are high : but, unfortunately, polluting industrial operations such as metal refineries and normal metalliferous mining has produced innumerable situations both in British Isles and overseas where levels of 1000 ppm have been exceeded. Modern extraction techniques have reduced metal concentrations in tailings to the point where toxicity is not apparent, but older wastes are usually very toxic (Johnson et al. 1976).

In these heavily contaminated habitats the effects of metal pollution are very apparent. In the most toxic areas the ground will be completely devoid of vegetation. In less toxic areas certain characteristic species are to be found which in former times have been used as indicators of the presence of metals (Antonovics et al. 1971). In the least toxic areas an almost normal range of species may be found making it very difficult to tell without soil analysis that there is any contamination at all.

Described in this way the effects of metal contamination appear simple and easy to understand. However the whole matter is complicated by the occurrence of metal tolerance in certain species. When samples of seed of a common grass species such as *Agrostis tenuis* (common bent grass) are sown on metal contaminated soil the seeds germinate but then die. However a very few seedlings (about 0.1%) continue to grow and appear unaffected by the metal. If the metal tolerance of the plants is examined by a rooting test it is found that these plants have the capacity to root in solutions containing metal which are completely toxic to seedlings from the same population which has been grown on normal soil (Walley et al. 1974). It is possible therefore to select for a metal tolerance within the species which is quite outside the average behaviour of the seedlings of the species.

These selected individuals, if interbred, produce metal tolerant offspring. We can therefore cause artificially the evolution of metal tolerance. Does the same occur in nature ? If the populations of species found occurring naturally on metal contaminated land are tested for metal tolerance, it is always found that they are much more tolerant than populations of the same species taken from normal soils. This tolerance is specific to individual metals. We now know that metal tolerance has evolved in a wide range of species throughout the world and it is the best example of the way in which evolution can occur in plants (for reviews see Antonovics et al. 1971 and Bradshaw 1975). A film has been made for the Open University by the British Broadcasting Corporation (shown at this conference).

SIGNIFICANCE FOR METAL CONTAMINATED LAND

From the point of view of understanding plant growth on metal contaminated land the evolution of metal tolerance has several significant aspects.

1. The occurrence of vegetation on metal contaminated ground

The evolution of metal tolerance confers on plant populations of certain species the capacity to grow on very heavily contaminated soil, containing for instance 0.5% Cu or 2% Pb or 2% Zn. These levels would be completely lethal to normal populations of the same species or to other species not found on contaminated soils. The presence of plants growing on an area known or suspected of being metal contaminated is therefore not an indication that the area is not toxic.

At the same time the absence of plants from a particular part of a contaminated area are not necessary evidence of extremely high levels of toxicity. Plant growth on inhospitable substrates is often determined by the lack of normal plant nutrients (Bradshaw and Chadwick 1979). On metalliferous mine wastes in England and Wales the presence of vegetation cover is determined by levels of soil phosphorus and not metal level (Smith and Bradshaw 1980). This suggests that metal tolerance allows species to cope with a wide range of metal levels. So in fertile, nutrient-rich, sites which are metal contaminated a deceptively vigorous growth of species may occur having the appearance of normality, in fact permitted by the evolution of metal tolerance.

2. The rapid evolution of tolerance

It could be imagined that evolution takes a long time, so that vegetation growth on recently metal contaminated sites would not be assisted, and therefore complicated, by the evolution of metal tolerance. However the artificial selection experiments suggest that metal tolerance might be able to evolve within a single generation. Recent investigations of the populations of Agrostis stolonifera (creeping bent grass) around a copper refinery in Lancashire (Wu et al. 1975) and of Silene vulgaris (bladder campion) around a zinc smelter in Westphalia (Ernst 1975) show that substantially tolerant populations can evolve in 3-5 years, although fully tolerant populations may take 10-50 years to develop.

It appears that in some situations the evolution of tolerance can occur so rapidly that the effects of increasing metal contamination can be masked, for instance, under a galvanised wire fence (Snaydon in Bradshaw et al. 1965). The anti-fouling effects of copper paints on the hulls of ships may be overcome by the evolution of copper tolerance in the fouling alga Ectocarpus siliculosus (Russell and Morris 1970).

3. The geographical localisation of tolerance

Most plants are self sterile and pollinated by wind or insects. As a result cross pollination must occur over distances of many metres. Many metal contaminated sites are only one hectare in extent or less. Does this mean that evolution of metal tolerance is prevented because plants in the contaminated area are constantly being cross pollinated with non-tolerant plants in the surrounding normal soils ?

Detailed investigations show that evolution of metal tolerance can readily occur on very small contaminated areas, of less than 100 m^2, and that at the boundary of contaminated areas there is a very sharp transition over a few metres between metal tolerant and normal populations (Jain and Bradshaw 1966). Cross pollination does occur but there is strong natural selection acting against it, so that non-tolerant individuals are eliminated from metal contaminated areas, and tolerant individuals are eliminated from normal uncontaminated areas. The latter is due to competition from other plants against which the tolerant individuals cannot survive because of their slower growth (McNeilly 1968).

4. The use of tolerant populations for the stabilisation of metal contaminated ground

Metalliferous wastes can cause considerable environmental problems. Because they are usually bare of vegetation they can be eroded by wind and water, particularly in storm periods, and fine particles containing high levels of metals can move onto surrounding farm land and into water courses where it can travel great distances (Australian Government 1974). At the same time percolation of water through metalliferous wastes can carry soluble metal salts into neighbouring streams.

A vegetation cover will eliminate or reduce drastically all these sources of pollution, and a variety of ways have been developed for achieving this (Johnson 1979, Bradshaw and Chadwick 1979). The use of metal tolerant populations is one method that has great potential since it allows plants to be established directly into metalliferous waste. Owing to the plant nutrient deficiencies in such wastes fertiliser must be added at time of sowing and further dressings may have to be provided subsequently. But with the addition of fertiliser metal tolerant populations will grow extremely successfully on heavily contaminated wastes (Smith and Bradshaw 1972).

The advantage of the technique is its simplicity and low cost. It is ideally suited to situations where immediate stabilisation of a very toxic waste is required. It is necessary to identify suitable plant material. So far three cultivars (cultivated varieties) are available commercially, of which "Merlin", a lead and zinc tolerant cultivar of Festuca rubra (red fescue), has already been used with great success on a large scale in the United States as well as in Britain. For other metals and in other climates other cultivars must be developed : "Parys" Agrostis tenuis is tolerant to copper, and in Rhodesia populations of Cynodon dactylon, Sporobolus virginicus and other grasses are available adapted to the high salt levels common in metalliferous tailings deposits in arid climates (Hill 1977). Where metal tolerant plant material is required it should be very easy to develop appropriate tolerant cultivars within five years providing that a systematic collection, assessment and breeding programme is followed (Bradshaw et al. 1978).

5. Control of metal movement into vegetation

Tolerant plants growing in a heavily metal contaminated medium take up metal into their roots and shoots, as much as 1-5,000 ppm of dry weight. As a result such vegetation cannot be used for grazing unless consumption of the contaminated herbage by individual animals is strictly controlled.

However in less heavily contaminated areas the mechanism of metal tolerance in tolerant plants does cause a considerable restriction of the movement of metals into the foliage of plants. This can be important for areas where agricultural use is contemplated (Smith and Bradshaw 1972) (Table 2). Laboratory experiments have demonstrated that metal tolerant plants possess a mechanism whereby heavy metals entering the plant are complexed within the cells of the root so that the subsequent movement of metals is restricted. This mechanism is much less well developed in non tolerant plants. In areas of low level contamination by metals control of metal levels in herbage by the use of appropriate tolerant cultivars seems a promising technique since the differences are even more marked than in Table 2. Genotypes differing in their metal uptake are being identified even in species where full heavy metal tolerance is not found (John and Van Laerh ven 1976).

Table 2. Mean heavy metal content of shoot material of tolerant and non tolerant populations of two grass species growing on a range of lead/zinc wastes for at least 12 months (data of R.A.H. Smith)

Species	population	total content (ppm)	
		Pb	Zn
Festuca rubra	tolerant	397	683
(on 6 calcareous wastes)	non tolerant	502	1184
Agrostis tenuis	tolerant	3790	236
(on 3 acidic wastes)	non tolerant	5361	467

CONCLUSIONS

Plant evolution is a process which until recently was thought to be so slow that it could be disregarded except over geological periods of time. It is now known to be extremely rapid and pervasive, affecting all aspects and characters of plants. Evolution in plant due to processes of natural selection is almost as spectacular as that found in bacteria.

Nowhere is this more obvious than in metal contaminated habitats where the results are not only of theoretical but also considerable practical significance.

REFERENCES

ANTONOVICS, J., BRADSHAW, A.D. and TURNER, R.G. (1971). Heavy metal tolerance in plants. Adv. Ecol. Res. 7., 1-85.

AUSTRALIAN GOVERNMENT JOINT TECHNICAL COMMITTEE, (1974.) Mine waste pollution of the Molonglo River, Final Report on Remedial Measures. Canberra : Australian Government Publishing Service.

BRADSHAW, A.D. (1975). The evolution of heavy metal tolerance and its significance for vegetation establishment on metal contaminated sites. In Heavy Metals in the Environment (Ed. T.C. Hutchinson) 599-622. Toronto Univ. Press, Toronto.

BRADSHAW, A.D. and CHADWICK, M.J. (1979). The Restoration of Land, Oxford : Blackwell.

BRADSHAW, A.D., HUMPHREYS, M.O. and JOHNSON, M.S. (1978). The value of heavy metal tolerance in the revegetation of metalliferous mine wastes. In Environmental Management of Mineral Wastes, ed. G.T. Goodman and M.J. Chadwick. 311-334. Alphen aan den Rijn : Sijthoff and Noordhoff.

BRADSHAW, A.D., McNEILLY, T. and GREGORY, R.P.G. (1965). Industrialisation, evolution and the development of heavy metal tolerance in plants. In Ecology and the Industrial Society, ed. G.T. Goodman et al. 327-343. Oxford : Blackwell.

ERNST, W. (1975). Physiological and biochemical aspects of metal tolerance. In Experimental Studies on the Biological Effects of Environmental Pollutants, ed. T.A. Mansfield, 115-133. London : Cambridge Univ. Press.

HILL, J.C., (1977). Establishment of vegetation on copper - , gold - and nickel - mining wastes in Rhodesia. Trans. Inst. Min. Metall. 86A, 135-145.

JAIN, S.K. and BRADSHAW, A. D. (1966). Evolutionary divergence among adjacent plant populations. I. The evidence and its theoretical analysis. Heredity, Lond. 21, 407-421.

JOHN, M.K. and VAN LAERHOVEN, C.J. (1976). Differential effects of cadmium on lettuce varieties. Environ. Pollut. 10, 163-173.

JOHNSON, M.S. (1979). Options for treatment of metal contaminated mine workings. In Reclamation of Contaminated Land. Society of Chemical Industry 1979. (these proceedings.)

JOHNSON, M.S., BRADSHAW, A.D. and HANDLEY, J.F. (1976). Revegetation of metalliferous fluorspar mine tailings. Trans. Inst. Min. Metall. 85A, 32-37.

McNEILLY, T. (1968). Evolution in closely adjacent plant populations. III. Agrostis tenuis on a small copper mine. Heredity, Lond., 23, 99-108.

RUSSELL, G. and MORRIS, O.P. (1970). Copper tolerance in the marine fouling alga, Ectocarpus siliculosus. Nature, Lond. 228, 288-289.

SMITH, R.A.H. and BRADSHAW, A.D. (1972). Stabilisation of toxic mine wastes by the use of tolerant plant populations. Trans. Inst. Min. Metall. A.81, 230-7.

SMITH, R.A.H. and BRADSHAW, A.D. (1980). The use of metal tolerant plant populations for the reclamation of metalliferous wastes. J. Appl. Ecology. in press.

WALLEY, K.A., KHAN, M.S.I. and BRADSHAW, A.D. (1974). The potential for evolution of heavy metal tolerance in plants. I. Copper and zinc tolerance in Agrostis tenuis. Heredity, 32, 309-319.

WU, L., BRADSHAW, A.D. and THURMAN, D.A. (1975). The potential for evolution of heavy metal tolerance in plants. III. The rapid evolution of copper tolerance in Agrostis stolonifera. Heredity, 34, 165-187.

GEOTECHNICAL PROBLEMS OF BUILDING ON DEEP FILL

J A Charles, MSc(Eng), PhD, ACGI, C.Eng, MICE
Building Research Establishment

SYNOPSIS

Reclamation of land that has been used for industrial purposes often involves construction on deep fills. This type of situation occurs with old docks that are backfilled, with tips containing industrial and domestic waste and with backfilled opencast mineral workings. Building on such sites may involve severe geotechnical problems. A number of field investigations are currently being carried out by the Building Research Establishment. Various ground treatment methods have been used in attempts to improve the load carrying characteristics of different types of fill and their effectiveness has been assessed.

INTRODUCTION

Reclamation of land that has been used for the deposition of industrial and domestic waste, for other industrial purposes or for mineral workings often involves construction on fill. The term 'fill' describes a wide range of materials which occur under many different circumstances. Fill can be composed of either natural soil and rock or of domestic and industrial wastes. It may have been placed in thin layers and compacted under controlled conditions or it may have been placed by uncontrolled end tipping in high lifts either in the dry or through water. The fill may be a surcharge above existing ground level or the backfill for an excavation. The depth of fill found on different sites varies greatly. Depending on the nature of the fill, a wide variety of problems may be encountered in developing a site (as the papers presented at this Conference indicate), but only problems of a geotechnical character are considered here.

Where a fill that has been loosely placed and which is liable to settlement is only shallow, the simplest solution to the geotechnical problems may be to either remove the fill or to take footings down through the fill into the underlying natural ground. This assumes, of course, that the natural ground is itself an adequate foundation material. With somewhat deeper fills piling may be an economic solution to the foundation problems. There may however still be problems where services enter buildings and piling can be difficult in some fills due to obstructions. Where the fill is deep and the structure to be built on it is comparatively small, it will usually be necessary to found the building on the fill.

The geotechnical problems that arise where construction takes place on fill have been studied at the Building Research Establishment for some years. A digest was produced in 1949 describing the general principles of building on fill and two years later Meyerhof (1951) published a paper on the subject which is still widely quoted. In more recent years interest in this topic has increased and further investigations have been carried out, particularly with a view to assessing the effectiveness of various ground treatment methods on fill. A new digest has recently been produced (Building Research Establishment, 1979). Recent conferences also reflect the increased attention being paid to the problems involved in developing reclaimed land; the Land Reclamation Conference held at Grays, Essex in 1976, the American Society of Civil Engineers' Conference on Geotechnical Practice for Disposal of Solid Waste Materials in 1977, the International Conference on the Use of By-Products and Wastes in Civil Engineering held at Paris in 1978, and the Midland Geotechnical Society Symposium on the Engineering Behaviour of Industrial and Urban Fill in 1979.

DEEP FILLS

In any consideration of geotechnical problems that may arise when construction takes place on a filled site, it is useful to make a clear distinction between sites where filling has yet to begin, and consequently control over the type of fill and method of placement can be exercised (new fill), and those sites where uncontrolled filling has already taken place (existing fill). In the former situation it should be possible by proper control over the filling to ensure that there are few problems in subsequently building on the fill. It is situations that fall into the latter category that are most likely to give rise to difficulties. A wide variety of human activities have left sites covered by deep fill and a few of these situations will be briefly described.

(i) Opencast Mining

The ratio of overburden to extracted mineral is large in opencast coal mining in Great Britain. Typically it may be necessary to remove 20 cubic metres of overburden to obtain one tonne of coal and coal is being extracted by opencast working at an annual rate of over 10 million tonnes in Great Britain. Usually the overburden is replaced in the excavation without systematic compaction and consequently opencast coal mining leaves large depths of loose fills. (When it is known that a site will be needed for building development after opencast mining is complete, it may be possible to arrange for the backfill to be replaced in thin layers with compaction. However, this may be expensive.) 30 metres is a fairly typical depth for an opencast working, but sites can be much deeper and at Westfield in Scotland it is planned to work to a depth of 220 metres (Leigh and Rainbow, 1979). As the fills at these sites are composed of natural soils and rocks there are no inherent problems due to the nature of the parent material. Difficulties in building on such sites are largely due to the uncompacted state of the fill. Excavation of overburden for opencast mining can be by scraper, face shovel or drag line or a combination of these and so the fills at different sites can vary widely in their condition as well as in their composition. During opencast mining at a site part of the fill may be temporarily surcharged by a spoil heap while another area may be the site of a washery plant and consequently large variations can occur in the state of the fill at a particular site.

(ii) Domestic Refuse

Disposal of domestic refuse in Great Britain has been mainly by landfill methods. In the years 1976-77 a total of 21 million tonnes of wastes were disposed of by the Waste Disposal Authorities in England alone and over 80 per cent of the waste was landfilled (Chynoweth, 1978). As a consequence of this practice, over the years large areas of ground have been covered with considerable depths of domestic refuse. When such sites are considered for development some years after filling has been completed, the age of the refuse is of great significance. Not only has older refuse had more time for decay and decomposition of organic matter to take place, but it is inherently a much better material. Changes in the composition of domestic refuse over the last forty years are described in BRE Digest 168 (1974). Typically in 1935 it contained almost 60 per cent by weight of ashey material whereas recently the proportion has been as low as 20 per cent. Over the same period there has been a large increase in the proportion of paper and some increase in the quantity of plastics, metal and glass. This data is based on average values from towns in Britain and clearly great variations may be found in particular situations. Landfill practice has also changed greatly over the years and there are currently significant variations in the method of placement and the type of compaction plant used on landfill sites. It should be expected that where a fill contains significant quantities of domestic refuse of recent age, large settlements will occur over many years as decay and decomposition occur.

(iii) Industrial Wastes

The disposal of industrial wastes also has resulted in many sites being covered with considerable depths of fill material. Colliery spoil (the waste product of mining coal) is derived from the rocks (mainly siltstones and mudstones with seat-earths) adjacent to the coal seams. It has been estimated that there is approximately 3,000 million tonnes of colliery spoil in Great Britain and that it is being produced at an annual rate of 50 million tonnes (Gutt et al, 1974). Until fairly recently colliery spoil was loose tipped and some heaps became ignited. However, most colliery spoil is unburnt. Many recent land reclamation schemes have involved colliery spoil; old spoil heaps are re-contoured with their tops removed and sometimes excess spoil is used to fill adjacent quarries or clay pits. Colliery spoils have been used extensively for highway earthworks. When the spoil is placed in thin layers under controlled conditions and adequately compacted problems should be few. Development of existing spoil heaps where the fill has been placed by end tipping will present more difficulties.

Another industrial waste which has been used quite widely as a fill material is pulverised fuel ash (PFA), the waste from coal burning power stations. PFA consists mainly of minute glass spheres in the silt size range and can be compacted to form a lightweight fill.

(iv) Old Docks

The redevelopment of dock areas may present particular difficulties. In principle an old dock can be drained and then backfilled with selected fill in properly compacted layers. In practice this may not be feasible due to problems with dewatering and instability of the dock walls. As a consequence old docks are frequently backfilled by end tipping through water and a loose fill results. Where there is a significant depth of silt at the dock bottom, end tipping may succeed in displacing much of the silt, but it is likely that some pockets of silt will be trapped within the backfill. Currently large areas of London's dockland are being redeveloped and a number of old docks are being backfilled in London and in other ports in Great Britain.

SETTLEMENT OF FILLS

Construction on deep filled sites may involve a number of geotechnical problems. Where the fill has been placed above natural ground level, the stability of the slopes at the edges of the filled ground should be considered. The bearing capacity of the fill also should be examined. However, the major geotechnical problem when building on deep fill is likely to be the long-term settlement of the fill. This settlement may be attributed to several causes such as self-weight of the fill, increased loading on the fill or wetting up and inundation of a loose fill.

(i) Settlement due to self-weight of the fill

In many situations self-weight of the fill will be the principal cause of long term settlement. A particular case is the consolidation of a well compacted wet cohesive fill which may result in significant settlement over a considerable period as excess pore water pressures slowly dissipate. The inclusion of drainage layers within such a fill can speed the process up and so reduce problems that it may cause.

With poorly compacted fills, primary compression will occur almost immediately a load is applied to the fill. However, some long term compression will occur under conditions of constant effective stress and moisture content and this can be termed 'creep' settlement. For a wide variety of fill materials creep compression occurs approximately linearly with the logarithm of the time that has elapsed since the fill was placed. This type of behaviour results in a fairly rapid decrease of settlement rate with time. A parameter, α, can be defined as the percentage compression of the fill occurring during a log cycle of time (Sowers et al, 1965), ie α represents the percentage compression occurring between (say) 3 months and 30 months or 1 year and 10 years after placement of the fill. For a well-graded rockfill placed in layers and well compacted, α may typically be as small as 0.2 per cent. Surface settlement monitored at a restored opencast mining site indicated $\alpha = 0.74$ for the uncompacted backfill (Kilkenny, 1968). It is noteworthy that α values for domestic refuse can be much larger. Measurements made in the USA indicate that α values are likely to range between 2 and 10 per cent depending on how favourable conditions are for decay (Merz and Stone, 1962; Sowers, 1973; Yen and Scanlon, 1975; Rao et al, 1977). Figure 1 shows creep compression plotted against the logarithm of time and illustrates the wide variations in creep rates for different types of fill materials. It should be emphasised that the use of an α parameter to predict the likely settlement of a fill can only be valid where conditions in the fill remain unaltered. A change in stress or moisture content can lead to much greater movements occurring.

(ii) Settlement due to Applied Loads

The compressibility of different fill materials under increased loading varies widely. For a particular fill material the compressibility will be dependent on the initial degree of compaction, the moisture content and the stress level. Laboratory tests carried out on a well graded, saturated, sandstone rockfill in a 0.6 m diameter oedometer have shown that when the vertical stress on the rockfill was increased from 85 kN/m^2 to 170 kN/m^2 the coefficient of compressibility (m_v) varied between 51×10^{-6} m^2/kN and 331×10^{-6} m^2/kN depending on the initial compaction (Charles, 1973). When the vertical stress was increased from 1380 kN/m^2 to 2760 kN/m^2 m_v was much smaller and did not depend very much on initial compaction. It varied only between 13×10^{-6} m^2/kN and 17×10^{-6} m^2/kN. Thus at the lower stress range an increase of stress of 100 kN/m^2 would

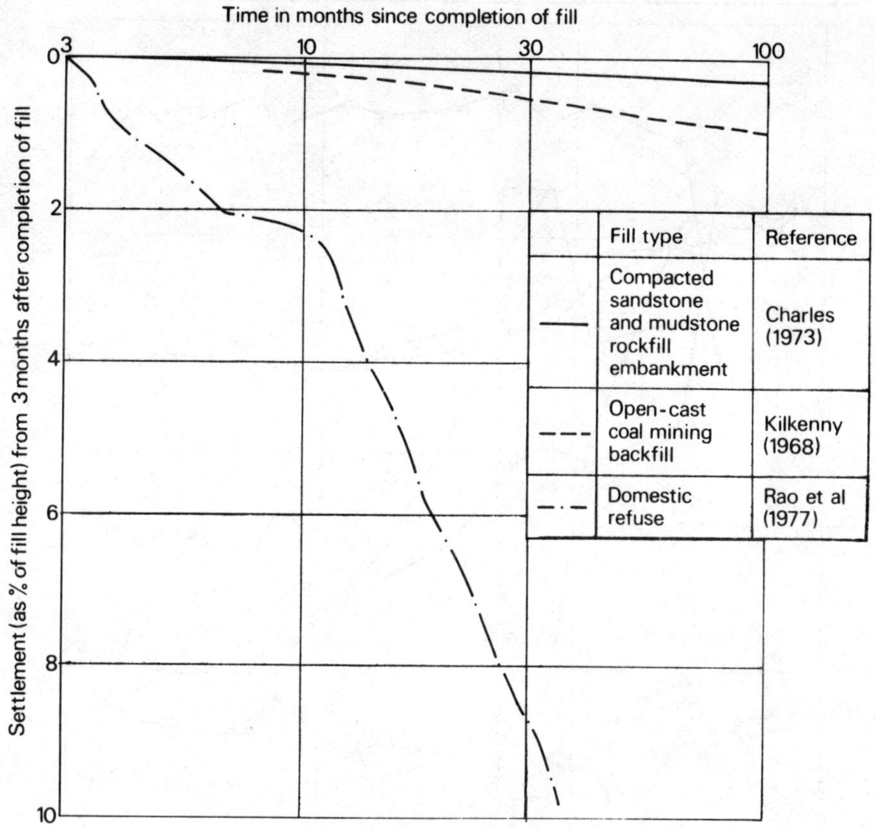

Fig 1 Settlement rates of different types of fill (vertical compression plotted against the logarithm of time)

produce between 0.5% and 3.3% vertical strain (depending on initial compaction) and at the high stress range an increase of vertical stress of 100 kN/m^2 would produce only about 0.15% compression. Field observations have indicated that a freshly dug colliery spoil placed in 2 m lifts with little compaction can have a coefficient of compressibility of the order of 500 x 10^{-6} m^2/kN at low stresses, which means that an increase in stress of 100 kN/m^2 could produce about 5 per cent compression. There is evidence that at low stresses the compressibility of domestic refuse can be much greater than this (Sowers, 1973; Rao et al, 1977).

(iii) <u>Collapse Settlement due to Inundation</u>

Loose unsaturated fill materials are usually liable to collapse compression on inundation. If inundation occurs within a fill during or subsequent to construction on the fill, then serious problems may arise due to collapse settlements. An investigation was carried out at a restored opencast coal mining site at Horsley to investigate the effect of a rising ground water table on the 70 m of uncompacted mudstone and sandstone backfill (Charles et al, 1977). Collapse compression at some locations and depths within the fill was locally as large as 1.5 per cent. Figure 2 illustrates the manner in which vertical compression occurred at different depths as the water table rose up through those particular levels within the fill. Guest (1975) has described damage to houses built on a relatively shallow restored opencast coal mining site in the Midlands. An investigation by the Building Research Establishment suggested that water penetrating at depth into the fill through drain trench excavations probably was largely responsible for the damaging movements.

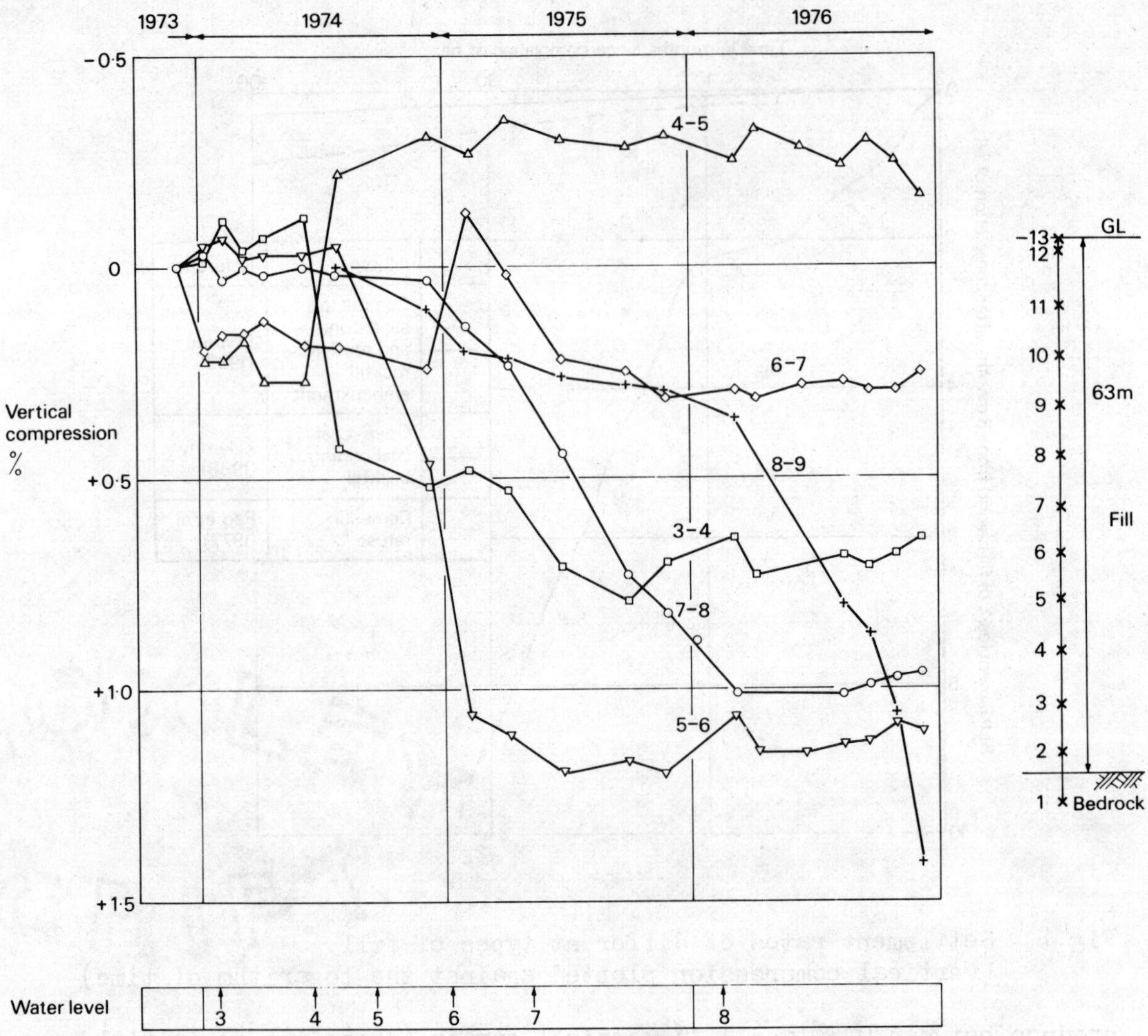

Fig 2 Effect of rising ground water table on an opencast coal mining backfill

Having given consideration to some causes and likely magnitudes of settlements of deep fills, it is well to remember that it is differential settlement rather than total settlement that leads to distortion and damage to buildings and roads built over deep fills. Differential settlement is likely at the edge of filled area and at any other part of the site where the depth of fill is changing rapidly. Even with a constant depth of fill, large differential settlements may occur if the fill is very variable. Thus the poorer types of fill, placed without control, are not only liable to large total settlements, but also the variability within such fills makes it probable that there will be large differential movements.

SITE INVESTIGATION

Construction on an existing filled area should be preceded by careful investigation of the site and it is clearly desirable to integrate the geotechnical site investigation with the investigation of other relevant factors, eg combustibility, gas generation, toxicity, chemical attack. Information about the history of the site and the filling that has taken place there should be sought and evaluated (Dumbleton, 1979). The boundary of the filled area should be established. A major problem with existing filled sites is the possibility of extreme variability and the site investigation should include a number of trial pits so that a comparatively large quantity of the fill is examined. With variable fill material, small scale laboratory tests are of limited value whereas a programme of field measurements can yield much valuable information.

A programme of field measurements is most likely to be useful if it is closely related to those aspects of the behaviour of the fill giving most cause for concern. These could include long term creep settlement of the fill under its own weight, compression under structural loading or collapse settlement on inundation. Creep movements can be investigated by monitoring the movements of surface settlement stations by precise levelling over an adequate period of time. Usually this should be at least one year and preferably longer. Large scale loading tests are a practical way of assessing the behaviour of a fill under structural loading, but it should be remembered that in most instances settlement under working load over a lengthy period will be of much more interest than the results of a rapidly carried out bearing capacity test. Simple techniques used by the Building Research Establishment to provide constant loads on fills over long periods have included small surcharges of fill and rubbish skips filled with sand. Where the possibility of collapse settlement on inundation arises, a field inundation test may be useful. This can be carried out by excavating a trench in the fill, filling it with water and monitoring the movement of adjacent settlement stations by precise levelling.

GROUND TREATMENT METHODS

Where buildings are to be founded on deep fill and the site investigation suggests that significant differential movements may occur over the area of a building, several different approaches to the problem may be considered. The foundations may be designed to either withstand or tolerate the ground movements. Reinforced concrete edge beams that can span areas of loss of support may be used. Flexible structures such as timber framed houses, which can accept large movements, can be adopted. In some situations an alternative approach of improving the load carrying characteristics of the fill by the use of some ground treatment method prior to development of the site may be an attractive solution. Several of the available methods are briefly described.

(i) Refilling in thin layers with Compaction

When the main problem with a fill is its loose uncompacted state, the simplest solution may be to excavate the fill and replace it in thin layers with adequate compaction. The Department of Transport Specification for Road and Bridge Works (1976) gives a good guide to the layer thickness and type of compaction suitable for different fill types. With this approach it is possible to examine all the fill and remove any unsuitable material. Where the fill is very deep it may not be practical or economic to excavate the full depth of fill and a decision has to be made about the minimum depth of fill that should be recompacted. This is unlikely to be less than 4 to 5 metres and in some situations it could be much deeper. This treatment method may sometimes be overlooked by builders and architects, but it should be remembered that an earthmoving operation of this type can be quite effective and relatively economic.

(ii) Pre-loading with a Surcharge

Another ground treatment method involving bulk earth moving is the pre-loading of filled ground with a surcharge of fill. As a loose fill will compress immediately under load, the surcharge need not be left in position for an extended period. On a large site a small area can be surcharged and the surcharge fill moved around the site in a continuous earth moving operation. Usually the surcharge will be designed so that it applies stresses to the filled ground greater than any that will be applied during the subsequent development of the site.

The treatment of an uncompacted 24 m deep cohesive fill left by opencast ironstone mining has been described by Charles et al (1978). Figure 3 shows that the settlement produced by a 9 m high surcharge of fill occurred mainly as the surcharge was being placed with little subsequent movement. The compression of this cohesive fill can be attributed to reduction of air voids in the fill rather than the squeezing out of water from the clay lumps themselves. That the settlement occurs immediately as the surcharge is placed is an important consideration as surcharging is sometimes rejected as a ground treatment method because it is thought to be a lengthy process. This is not the case for uncompacted fill materials.

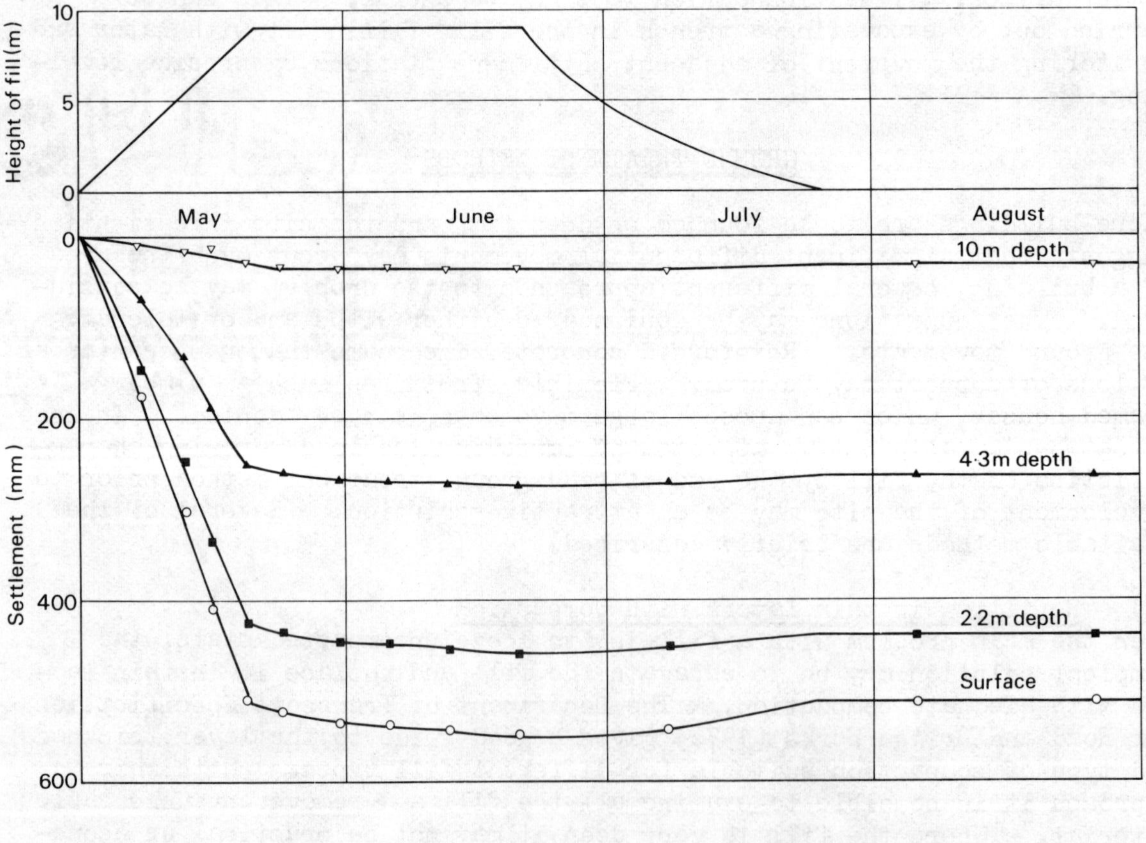

Fig 3 Effect at different depths within a 24 metre deep opencast ironstone mining backfill of pre-loading with 9 metre high surcharge

Tomlinson and Wilson (1973) described how an uncompacted colliery waste fill was treated by pushing a 5 m high shale surcharge across the site with a large bulldozer. Recently old docks at Hull (Victoria Dock) and Methil (No 3 dock) have been backfilled by end tipping into water and then treated by small 2 m high surcharges of fill.

(iii) Dynamic Consolidation

In this ground treatment technique, deep compaction of soils is attempted by repeated impacts of a heavy weight onto the ground surface (Menard and Broise, 1975). The Menard system of dynamic consolidation as it has been applied in the United Kingdom has typically involved dropping a 15 tonne weight from heights of up to 20 metres. Primary tamping has usually consisted of repeated impacts at a number of points on a fairly widely spaced grid. At two sites the Building Research Establishment carried out investigations to determine the depth within the fill to which the treatment was effective. At site A the 24 m deep fill was predominently cohesive in its upper part (Charles et al, 1978) and at site B the 6.5 m deep fill was 40 years old domestic refuse (Charles, 1979). Figure 4 shows the enforced settlement measured at different depths within these two fills treated by dynamic consolidation. It is seen that the ground treatment technique (in which there was an average energy input of 2800 kN.m/m^2 at site A and 2600 kN.m/m^2 at site B) effectively compacted the fills to depths of about 5 metres. With clay fills the total input of energy required to compact the soil will be largely a function of the undrained shear strength of the soil (Charles, 1978).

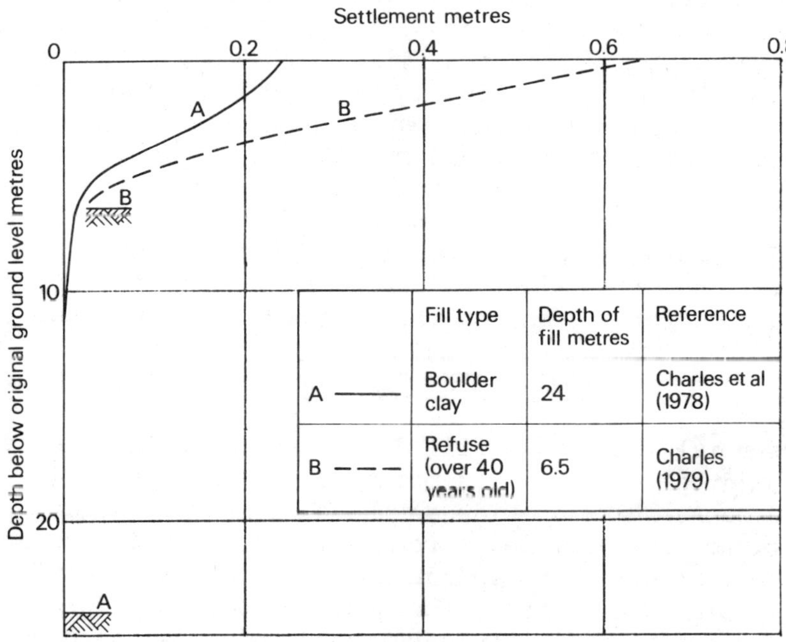

Fig 4 Dynamic consolidation of two fills - effectiveness at depth

(iv) Vibro Techniques

This type of process was developed in Germany in the early 1930's to compact loose naturally occurring sands. The basic tool is a cylindrical poker which contains in its bottom section an eccentric weight. Rotation of the weight results in vibrations in a horizontal plane being transmitted to the soil and the poker penetrates into the ground. This leads to the densification of a granular soil. In a cohesive soil little compaction is achieved, but the long cylindrical hole produced by the poker vibrator can be backfilled with granular fill. The cohesive soil is consequently effectively reinforced by stone columns. The method can be used in fill materials in a similar way to its use in natural soils. Moseley and Slocombe (1978) have described the use of this technique on an opencast mining backfill.

(v) Inundation

It has been pointed out that inundation of a loose unsaturated fill material can cause significant settlement to occur. When this happens after construction on filled ground has taken place, it can be a serious problem. However, pre-inundation of a site before development can form a method of ground treatment. In some situations inundation may occur naturally as a ground water table re-establishes itself after opencast mining as at Horsley (Charles et al, 1977). An attempt to inundate via surface trenches at the restored opencast ironstone mining site at Snatchill, Corby, was not particularly successful (Charles et al, 1978). In the top 5 metres of fill the collapse compression was generally between 1 and 2 per cent, although a compression strain of 6 per cent was measured at one location. However, subsequent movements have been quite large over a lengthy period in the area of ground treated by inundation at Corby.

Any ground treatment method that is considered for use on a filled site should be examined in the light of the full range of problems existing at the site. In some situations a particular approach may be beneficial in more than one respect. For example, compaction of a colliery spoil will not only improve its geotechnical properties, but may also eliminate the possibility of combustion within the fill. However, in other situations different problems may have conflicting requirements. For example, from a geotechnical standpoint a sensible way of dealing with an old industrial waste fill might be by excavation and refilling with adequate compaction; however such treatment could lead to mixing and reactivation of chemicals with associated ground movements. Another example of this latter type of situation could arise with the use of piles for a structure founded on an old refuse tip. The piles might be adequate to carry the structural loads but could also form paths for the escape of methane gas.

CONCLUDING REMARKS

1. It is essential that every filled site which is considered for development should be judged on its merits as the load carrying properties of fills have a range quite as wide as those of natural soils. At one end of the spectrum, for example, a rockfill placed in layers and suitably compacted may form an excellent foundation material while at the other extreme a recently placed domestic refuse may be completely unsuitable. Consequently an existing fill should be thoroughly investigated prior to construction on the site. The type of structure to be built on the site and its foundations should be carefully considered in relation to the site conditions.

2. A principal deficiency of many existing fills is that they have been placed without systematic compaction. As a result of this they are in a loose state and there is the possibility of great variability across the site. Large differential settlements may occur. It may be possible by the use of a suitable ground treatment technique to densify the fill and thereby greatly reduce subsequent settlement.

3. When a ground treatment technique is used, it is desirable to carry out a post-treatment programme of testing to confirm its effectiveness. As a major problem with many fills is the possibility of extreme variability, it is clear that large scale field tests are preferable to laboratory tests on small, probably unrepresentative, samples. The test programme

is most likely to be useful if it is closely related to those aspects of the behaviour of the fill giving most cause for concern. These could include creep settlement of the fill under its self-weight, compression under structural loading or collapse settlement on inundation.

4. On many deep fill sites there will be a variety of problems besides those of a geotechnical nature. In this situation it is obvious that the various problems should not each be considered in isolation. At the site investigation stage it is desirable to have an integrated approach and any solution that is considered should be examined in the light of the full range of problems.

ACKNOWLEDGEMENTS

The paper is published by permission of the Director of the Building Research Establishment. The helpful comments made by Dr J B Burland are gratefully acknowledged.

REFERENCES

BUILDING RESEARCH ESTABLISHMENT (1974): Digest 168: Refuse Handling: 1. HMSO, London.

BUILDING RESEARCH ESTABLISHMENT (1979): Digest 222: Fill and Hardcore. HMSO, London.

CHARLES, J A (1973): Correlation between laboratory behaviour of rockfill and field performance with particular reference to Scammonden dam. PhD thesis, University of London.

CHARLES, J A (1978): Methods of treatment of clay fills. General Report, Conf on Clay Fills, Instn of Civ Engrs, London, pp 315-321.

CHARLES, J A (1979): Field observations at a trial of dynamic consolidation on an old refuse tip in the east end of London. Symp on The Engineering Behaviour of Industrial and Urban Fill. Midland Geotech Soc, Birmingham.

CHARLES, J A, NAISMITH, W A and BURFORD, D (1977): Settlement of backfill at Horsley restored opencast coal mining site. Conf on Large Ground Movements and Structures. UWIST, Cardiff, pp 229-251. (Also Building Research Establishment Current Paper 46/77).

CHARLES, J A, EARLE, E W and BURFORD, D (1978): Treatment and subsequent performance of cohesive fill left by opencast ironstone mining at Snatchill experimental housing site, Corby. Conf on Clay Fills, Instn Civ Engrs, London, pp 63-72. (Also Building Research Establishment Current Paper 79/78).

CHYNOWETH, D (1978): Waste disposal statistics. County Councils Gazette, vol 70, part 10, pp 268-273.

DEPARTMENT OF TRANSPORT (1976): Specification for Road and Bridge Works. HMSO, London 194 pp.

DUMBLETON, M J (1979): Historical investigation of site use. Conf on Reclamation of Contaminated Land, Soc Chem Industry, Eastbourne.

GUEST, J E (1975): Contribution to discussion: Conf on Settlement of Structures, Cambridge. Pentech Press, London, p 662.

GUTT, W, NIXON, P J, SMITH, M A, HARRISON, W H and RUSSELL, A D (1974): A survey of the locations, disposal and prospective uses of the major industrial by-products and waste materials. Building Res Establishment

Current Paper 19/74.

LEIGH, W J P and RAINBOW, A K M (1979): Observations of settlement of restored backfill at opencast mine sites. Symp on The Engineering Behavior of Industrial and Urban Fill. Midland Geotech Soc, Birmingham.

KILKENNY, W (1968): A study of the settlement of restored opencast coal sites and their suitability for building development. Bulletin no 38, Dept of Civ Engg, Univ of Newcastle-upon-Tyne, 33 pp.

MENARD, L and BROISE, Y (1975): Theoretical and practical aspects of dynamic consolidation. Geotechnique, vol 25, no 1, pp 3-18.

MERZ, R C and STONE, R (1962): Landfill settlement rates. Publ Works, vol 93, no 9, pp 103-106, 210, 212.

MEYERHOF, G G (1951): Building on fill with special reference to the settlement of a large factory. The Structural Engineer, vol 29, no 2, pp 46-57. Discussion: vol 29, no 11, pp 299-305.

MOSELEY, M P and SLOCOMBE, B C (1978): In-situ treatment of clay fills. Conf on Clay Fills, Instn Civ Engrs, London, pp 165-168.

RAO, S K, MOULTON, L K and SEALS, R K (1977): Settlement of refuse landfills. Conf on Geotechnical Practice for Disposal of Solid Waste Materials. Univ of Michigan, Am Soc of Civ Engrs, New York, pp 574-598.

SOWERS, G F (1973): Settlement of waste disposal sites. Proc 8th Int Conf on Soil Mech & Found Engng, vol 2.2, pp 207-210.

SOWERS, G F, WILLIAMS, R C and WALLACE, T S (1965): Compressibility of broken rock and the settlement of rockfills. Proc 6th Int Conf on Soil Mech & Found Engng, vol 2, pp 561-565.

TOMLINSON, M J and WILSON, D M (1973): Pre-loading of foundations by surcharge on filled ground. Geotechnique, vol 23, no 1, pp 117-120.

YEN, B C and SCANLON, B (1975): Sanitary landfill settlement rates. Proc Am Soc Civ Engrs, vol 101, no GT5, pp 475-487.

SEWERAGE : WITH PARTICULAR REFERENCE TO THE BECKTON DEVELOPMENT IN EAST LONDON

H.G. LONGLANDS THAMES WATER AUTHORITY

G.H. TOWNSEND MASON PITTENDRIGH AND PARTNERS

BACKGROUND

The specific case history which forms the basis of this paper is located within the Beckton Development Area, in the London Borough of Newham, some 11 km. east of the centre of London. Whilst the location falls within the overall Docklands Redevelopment Area, the Beckton scheme largely comprises virgin marshland and what was Europe's largest coal gas production works, with its attendant railway sidings and coal storage areas, together with old small housing areas. Significant to the particular problems recounted in this paper was a very large plant which processed the by-products of the gas production, particularly the large tip of waste which was allowed to accumulate. This tip is locally known as the Beckton Alps and contains the most horrifying agglomeration of chemical compounds.

The previous system of surface water drainage depended on small diameter pipelines and open channels across the marsh areas. The contract on which these problems arose was for 4km. of surface water sewers which has cost some £7m. when completed in 1978.

The development, when completed, will provide for some 8,000 new houses, large industrial areas and a completely new shopping area incorporating what may be the future terminus of the new Jubilee underground rail line. The total extent of the area is some 3km. x 2 km.

The problem area was between the East Ham Manor Way and the Beckton Alps. See plan in Appendix 2.

EXTERNAL CONSTRAINTS AND INFLUENCES

To the outsider looking at the problem for the first time, especially in hindsight, the first question asked is "Why was a major concrete structure such as this pipeline sited so close to the tip?". The basic answer is that the overall plan for the area has zoned the land to the west of East Ham Manor Way for housing and a large sewer laid through this area would pose an unacceptable constraint on the layout of dwellings: straight rows are to be avoided for aesthetic reasons. Whether that constraint would have been dominant had the future problems been fully anticipated is a matter for conjecture. The land to the east of the road is zoned for industrial use and the presence of the sewer was more acceptable.

Another major restraint which was relevant during construction was that the material in the waste tip was regarded as so toxic that the relevant waste disposal authority, the Greater London Council (GLC), placed an embargo on the carting away of the material from the site to a tip elsewhere. This prohibition was later extended to the contaminated material excavated from the trench.

PRE CONTRACT INVESTIGATIONS

The route having been fixed, the next obvious question is "Why were the extreme problems encountered not predicted before construction started?". The answer lies in a combination of overwhelming pre contract evidence that the ground alongside the tip was not contaminated and an imperfect understanding of the localised geological conditions.

The consulting engineers had been employed by Newham L.B. since the mid 1960's in connection with the preliminary works for the whole of the Beckton Development Area. As part of this work, a complete grid of boreholes had been sunk to determine the pattern of ground strata. At a later stage, further holes were bored along the line of sewer and extensive chemical analyses of samples were undertaken.

Of boreholes taken in 1975, four were along the relevant length. The concentration of sulphates expressed as SO_3 in the water samples ranged from 150 to 360 p.p. 100 000. Reference to the Building Research Establishment (BRE) recommendations (1) revealed that in their view a dense, cement rich concrete utilising sulphate resisting portland cement would stand up to concentrations of 250 - 500 p.p. 100 000 and that therefore pipes made from such a concrete would be more than adequate to withstand chemical attack from the sulphates. The pH value of the samples lay between 6.4 and 7.3, which again was within the range stipulated by BRE of 6 to 9.

The actual borehole logs made reference to "some coal tar contamination (made ground)" and "possible coal tar contamination".

MANNER IN WHICH PROBLEMS DEVELOPED

The problem arose when, upon opening up the thrust and reception pits for the thrust bore under East Ham Manor Way, large quantities of tarry substances were carried in with the ground water. This was being pumped from the excavation to the open watercourses that constituted the surface water drainage system at the time.

The land drainage pollution control staff of Thames Water were alarmed at the probable consequences and immediately banned further discharges to the watercourses and suggested that discharges should be made to the foul sewers. The local sewers were not suitable to receive large volumes of water, nor to accept quantities of strong pollutants.

The only alternative was to discharge to the nearby main Northern Outfall Sewer leading to the Beckton Sewage Treatment Works. It was necessary, however, to analyse the contaminants in order to assess its suitability for acceptance at the Beckton Works, and it was these tests that showed that sulphates, cyanides and phenols were present in varying degrees of concentration. The most alarming of them was the level of sulphates that peaked at 10 000 p.p.m. during the course of regular testing and recording over a few weeks.

A representative sample of those obtained is given below and was adopted as the standard which any protection system would have to meet:-

Soluble sulphates 5000 - 6000		Tar Oils	850 ppm
Ph 6.5 - 8.0		Napthalene	195 ppm
Phenols	875 ppm	Pitch	990 ppm

As far as the concrete was concerned, the most serious constituent were the sulphates with their well known and documented effects and which on their own can be dealt with by a variety of traditional materials and methods most of which, however, are based upon bituminous compounds. Unfortunately these are particularly susceptible to cresols and phenols.

Finding a solution to this problem had to be achieved against the background of possible contractual claims for delays amounting to at least £75,000 per week and, depending upon the supply of 1000 metres of 3.2 m. dia. pipes for the remainder of the contract, could have been £120,000 (1976 prices).

The internal diameters, total lengths and weights of pipe involved in the area of the heavily contaminated ground were as follows:-

1800 mm.	38 metres	6.48 tonnes per unit
2000 mm.	137 metres	7.6 tonnes per unit
2400 mm.	388 metres	9.4 tonnes per unit

By virtue of the type of pipe, the outer concrete surface was rough and undulating since it was applied by a guniting type of process.

DETERMINATION OF POSSIBLE PROTECTIVE MEASURES

Since the pipes had already been ordered and were due for delivery in accordance with the Contractor's programme, initial consideration of protective measures had to be devised for them, as distinct from devising alternative methods of construction using different materials that would in themselves be resistant to the chemicals present. This restriction was somewhat alleviated subsequently when it was agreed that the 388 m. of 2400 mm. dia. pipe would be laid elsewhere as an extension to the Beckton Surface Water Sewerage System which had been programmed for construction at a later date.

Apart from chemical consideration. other problems had to be borne in mind - (i) the size and weight of the pipe would affect the treatment process, transportation and frictional forces between pipe and gravel bed during the laying process and (ii) the impracticability of detecting any damage around the bottom of the pipe after it had been laid on the gravel bed.

A large number of alternatives were considered. The two which were adopted are described below, the others are listed in the Appendix.

Concrete Surround Protected by an External Membrane

The wrapping of the pipes in a membrane was considered, with the whole being surrounded with mass concrete (alternative 8 in the Appendix). The actual method adopted for part of the works was to reverse the protection, i.e. the membrane was wrapped round the concrete surround.

Advantages were:-

a. Straightforward construction technique.
b. Back up system of concrete avoided total reliance on the integrity of the membrane.
c. Similar technique applied to manholes.

Disadvantages were:-

a. Long trench required to be opened.
b. Special measures required to prevent damage of membrane.
c. No positive test available to ensure 100% continuity of protection.

Since the membrane provided the chemical protection, various materials were investigated:

a. POLYTHENE SHEETING. Whilst impermeable to sulphates, ICI would give no assurances if phenols present.

b. P.V.C. For it to be used as a flexible sheeting, polyvinylchloride requires the addition of plasticiser to the raw compound, this then requiring a suitable backing material for mechanical protection. That recommended by Dupont was Hypalon. These materials are available intimately bonded by the suppliers to form a single sheet, at present manufactured in widths only 1.65 m. wide. No special problem was envisaged in site jointing and the material itself is completely impervious to phenols and sulphates. There was, however, insufficient material readily available in the U.K. on the scale required.

c. BUTYL RUBBER. Butyl rubber as polyisobutylene is commonly used for tanking and lining reservoirs and it is not readily attacked by alkalis and other chemicals harmful to natural rubbers. Phenol attack is, however, manifested as a softening. It is manufactured in large widths and jointing is straightforward.

d. P.V.C. NITRILE. This material is also commonly used for tanking around oil storage installations to prevent accidental leakage contaminating the surrounding ground. It is superior to butyl rubber since under the presence of phenols it experiences only a slight loss of mechanical strength. It can be reliably welded and jointed and is readily available in large sheets. It was possible to get written confirmation that it had a minimum life of 20 years in the application being considered and was the one adopted.

<u>Polyurethane Coating reinforced with G.R.P.</u>

This was a process offered by Colebrand Ltd. utilising one of their own products. The external surface of the pipe was coated with solventless polyurethane applied by airless spray in one coat to a total thickness of approx. 2 mm. At the spigot and socket this was reduced to 1 mm. This gave the chemical protection required. Before applying the polyurethane the surface of the concrete was prepared by sand blasting and priming with an epoxy resin primer, over which was sprayed aluminium leaf to allow for spark testing of the polyurethane coat. Mechanical protection for transporting, handling and laying was achieved by applying a 3 mm. coat of G.R.P. by spray over the barrel of the pipe.

Advantages included:

a. Applicable to either factory or site operation.
b. High level of quality control and guarantee of 100% protection due to ability to apply spark test.
c. Separate chemical and mechanical protection meant that damage to mechanical cover did not necessarily rupture the chemical

cover, and would only be localised.
d. Little variation to the pipe laying techniques required.

Disadvantages included a need for special equipment to suspend and rotate pipes during treatment, and pipe storage under cover prior to treatment.

CONSTRUCTION

General problems working on site in contaminated ground

These manifested themselves in a number of ways which are dealt with in order of significance.

a. HEALTH AND SAFETY HAZARDS. When working in an excavation into which chemicals are flowing at the rate experienced on this contract, it is only on the larger battered excavations that it is possible to avoid the use of masks or forced ventilation. On smaller excavations, e.g. manholes, it was necessary to force large volumes of air by means of ducting into the excavation without which men complained of dizziness, headaches of increasing severity and ultimately stomach upsets with vomitting.

Protective clothing was necessary to avoid skin complaints and in addition it was found necessary to use barrier cream to combat the effects of vapour. Not unnaturally, incentives had to be paid to men to work in these conditions. The oily nature of the contamination was such as to increase the hazards created by slippery surfaces and great care had to be exercised in handling any item of equipment used in the construction, in ascending and descending ladders and walking on staging.

b. EFFECTS ON CONSTRUCTION. As has already been mentioned, discharging water from excavations to natural watercourses was prohibited and this had to be taken to foul sewers. It was perhaps convenient that the Northern Outfall Sewer was at hand which would cope with fairly large columns of water, it consisting of 5 No. 9 ft. dia. barrels at this point. However, indiscriminate discharge of pollutants could not be allowed even into this sewer and a sophisticated system had to be devised to enable construction to proceed.

A proper appreciation of the problem requires a description of the geological structure. This consists of made ground overlying alluvial clay under which was peat overlying gravel. The latter stratum was the limit required to be considered since the excavation was not taking place beyond its thickness. It was found that the phenolic contamination originating from the waste products tip had penetrated the topsoil and made ground but was fairly effectively sealed off from the lower strata by the clay layer. Thus when a hole was dug the contaminant could be seen running into it from a high level. It was a prime concern to reduce to a minimum the extent to which the water bearing gravel and peat would be polluted by the uncontrolled release of contaminants into the excavations.

A shallow ditch was therefore dug in the clay between the tip and the main excavation to act as an interceptor. Unfortunately, the natural fall of the ground away from the Northern Outfall Sewer meant that this ditch had to be dug in a series of steps to enable water to be pumped in three stages up the slope. Water from the main excavation was pumped into this ditch and thence to a large filter bed. The filter media consisted of clinker overlying 75 mm. gravel rejects, the feed was by perforated tubes from the pumped discharge. Land drains beneath the filter media led the effluent into a channel, equipped with baffle boards, and thence into a sump where dosing could take place before pumping to the Northern Outfall Sewer. In the event, however, it was found preferable to dose the interceptor ditch and for this purpose lime and jizer were used. Changing the filter media at regular intervals ensured that the system remained effective in preventing high concentrations of contaminants entering the Northern Outfall Sewer.

c. EFFECTS ON PLANT. Pumping the effluent from the ditches involved a continuous problem of pump performance and maintenance: to clear blockages and repair damaged seals, which perished under the action of the creosote. To ensure that the interceptor ditch remained effective, not one but two standby pumps had to be provided due to the frequency of breakdown.

Description of Construction

a. WRAPPED PIPEWORK. 38 metres of 1800 mm. dia. pipe and 137 metres of 2000 mm. pipe were treated in this way, it being particularly applicable to sheet piled construction, which these sections were. The sequence of construction following excavation to formation was as follows:-

1. Lay concrete raft to brace bottom of piles and produce satisfactory working surface.
2. Lay P.V.C. nitrile sheet across the bottom and turned a short way up each side. The sheeting was sandwiched between two layers of Fipec to prevent mechanical damage.
3. Instal pipes on cradles to line and level.
4. Weld side sheets to upturns and suspend from top of excavation between Fipec protection.
5. Cast 250 minimum thickness of high workability concrete around the pipes. Proportions of the mix per cubic metre were as follows:-

S.R. Cement	390 Kg.	Zone III landbased fines	515 Kg.
12 mm. Lytag	750 Kg.	W/C ratio	0.8
		Plasticiser (Flocrete N)	975 Ml

6. Weld top P.V.C. nitrile sheets to side sheets.
7. Backfill.

b. WRAPPED MANHOLES. Manholes were treated in basically the same way as pipes except that the additional operations of shuttering and reinforcement meant that the sheeting was at risk for longer and the Fipec protection thus much more important.

c. PRECOATED PIPEWORK. Pipes were precoated by Colebrand Ltd. at the Newport Pagnell factory of Rocla Pipes Ltd. A temporary scaffolding and polythene sheet building was erected into which were installed four large cantilever brackets onto which the pipes were loaded by fork lift truck. Rollers on the brackets driven by an electric motor allowed the pipe to be rotated so that an operator standing on a scaffold was able to treat the pipe from one position. To ensure that a satisfactory standard of spark testing was maintained, an independent specialist was appointed to make random checks.

Once treated the pipes were handled in the normal way after the curing period had elapsed and no damage was sustained to the coating during transport.

Special jointing rings had to be produced for these pipes for two reasons:

1. Change from the normal neoprene to butyl rubber.
2. Reduced cross section to allow for the fact that the joint gap in the pipes was reduced by 2 mm. as a result of the polyurethane coating.

Initial difficulties in laying arose while jointing when the polyurethane, which was of a deliberately resilient character, tended to roll up in front of the ring and frustrate complete entry of the spigot into the socket. The formulation was therefore quickly changed to give a harder membrane at spigot and socket only and thereafter laying and jointing proceeded as normal.

d. SPRAYED MANHOLES. Two manholes were treated using the polyurethane protection and were highly successful, the protection being applied after structural concrete work was complete.

In the case of the base slab polyurethane was applied direct to the good quality blinding layer and then covered, except for a bonding area around the perimeter, with G.R.P. sprayed on to give the necessary protection to the membrane during subsequent activities.

Where polyurethane continued on from P.V.C. nitrile sheeting a complete seal was obtained by spraying onto the sheeting, to which it bonded intimately. Although it would have been possible, spark testing was not used on the insitu work as it was desirable to cut down the exposure time of men to the unpleasant conditions.

e. ADDITIONAL PRECAUTIONS. Since it appeared from the evidence of the excavations that the Northern Outfall Sewer to the north, East Ham Manor Way to the west and Winsor Terrace to the south were the boundaries of the polluted area it was essential that all measures possible should be taken to confine the pollutants within it. Therefore to protect the pipe and also ensure a cut off where it entered or left these locations, by means of thrust boring in each case, resin grouting was carried out to slightly overlap the P.V.C. nitrile or polyurethane protection. Diagonal and vertical lances ensured complete encirclement of the pipe by the grout.

CONCLUSIONS

Many conclusions can be drawn from this case history. Among the more important points the authors would put forward are the following:-

a. In any large scale redevelopment the sewers tend to be the first construction works, and to be the deepest and most extensive. The problems will frequently be of an unexpected and of a more extreme nature than any succeeding work will throw up. As a corollary the cutting of sewer trenches tends to alter the whole hydro-geological balance.

b. Nobody ever disbelieves pre contract site investigations to the extent that would have been necessary on this project to have foreseen all the problems.

c. When a solution is being sought, no channels should be left unexplored: the Colebrand solution was only arrived at after a discussion over a beer about chimney problems!

d. Developers and planners (Newham London Borough in both roles in this case) need to be restrained from dictating the position of important installation until they have been made aware of the full implications: following these troubles, the site of the future foul sewage pumping station has been moved into the residential area to the west of East Ham Manor Way, even though great opposition was met from a planning point of view.

ACKNOWLEDGEMENTS

Acknowledgements in this case take two forms.

There are the formal acknowledgements to Mr.E.C.Reed, Director of Operations, Thames Water, and Mr. G.R. Iley, Director of Engineering and Surveying, London Borough of Newham, for permission to present this paper. Thames Water being the clients and Newham the agent council for these works which were financed by Thames Water.

In addition, many people devoted much time and thought to the problems and the Authors particularly wish to acknowledge the substantial assistance given by (i) the staff of Newham Council who, having employed consulting engineers, still put in considerable efforts at all stages of the problem; and (ii) the scientific staff of Thames Water in relation to the treatment and disposal of the ground water, and who analysed innumerable samples so that the extent of the problem could be assessed.

The Authors must emphasise that the views expressed in this paper are their own, and do not necessarily reflect those of Thames Water or Newham London Borough.

REFERENCE

1. Building Research Establishment Digest. No. 174 February 1975. 'Concrete in sulphate bearing soils and groundwaters'.

APPENDIX 1

ALTERNATIVE MEANS OF PROTECTION.

The large number of alternatives which were not adopted, but which nevertheless were considered, are listed below. The order of listing is not of significance:

1. Painting the outside of the pipe Paints were not recommended

because of their lack of mechanical resistance to abrasion, poor adhesion in wet conditions and lack of a suitable product to withstand phenol attack.

2. Bitumen reinforced with fibreglass   Spraying or painting of the pipe with a thick layer of bitumen and laying on fibreglass strands to increase resistance to abrasion was demonstrated on site.

   The major disadvantages included incompatibility of bitumen with phenols and the likelihood of damage, particularly to spigots and sockets which would have to remain untreated by fibreglass.

3. Wrapping with Densotape   This entailed the use of a strip of compound continuously wound round the pipe, covered with a secondary protective plastic tape. The nature of the pipe with its undulating rough surface gave rise to doubts as to whether complete protection could be achieved due to rucking; it was also incapable of being turned into the socket. A trial proved the point and also demonstrated the practical impossibility of wrapping the joint and manholes insitu.

4. Epoxy Coatings

   a. Normally applied by spraying with a two component mix. Whilst site construction techniques would remain unchanged, they are all expensive because of the material costs, special equipment and rigorous conditions for application. In general, not all the solvents evaporate, and since the residues are moisture absorbent the coatings would not be stable under conditions of permanent ground water.

   b. One particular brand of these coatings was thoroughly investigated, this was Shell Composite's Monoguard; a coal tar epoxy reinforced with glass fibre. Such coatings are highly resistent to impact and an experimental pipe dragged over the ground showed a high reistance to abrasion on a test far more severe than was likely to be met in practice. This particular product was acceptable, in fact it has since been used on another pipeline nearby, but was ruled out on cost grounds on this particular contract.

5. Mass Concrete Surround   Sulphate resisting cement can give protection up to 5000 mg/1 SO_3. Consideration was given to the provision of a thick sacrificial concrete surround, present data indicating that the rate of attack in the high concentrations of SO_3 would be of the order of 150 mm. in 30 years. Local attacks could be more destructive and there is no knowledge of the effect that the other chemicals present might have on the decaying process. There was not therefore sufficient confidence in the adequacy of the proposal to warrant serious consideration of the cost implications. Had the pipes been of normal reinforced concrete construction this may have been regarded as a possible method, but since they were prestressed the consequences of pipe failure were regarded with alarm.

6. Mass Concrete Surround using Lytag Aggregate   The advantage seen here was that the high workability inherent in the use of this aggregate, and the consequent extremely uniform results obtained, would have reduced the chances of local attack through a faulty or

honeycombed area. These improvements over No 5 did not however provide complete re-assurance and the matter was not pursued.

7. PFA Surround This could have been used as a conditioned PFA or as part of a designed mix, the protection afforded by conditioned PFA relies on the possolanic properties. Problems would arise in compacting the material round the pipes. The alternative would be partial replacement for cement or as a partial substitute for both sand and cement to give a concrete having a low permeability. However, the chemical composition of PFA varies considerably and could have created problems in quality control. No specific data is available concerning its improvement in resistance to sulphate attack and no data at all about its resistance to the chemicals present in this case.

8. Insitu Wrapping of Pipe within a Concrete Surround This method envisaged the construction of a continuous membrane around the pipe itself, jointing being affected by glueing or welding.

The advantages were:-

a. The total integrity of the concrete surround was not being relied upon to give complete protection, secondary protection being afforded by the membrane.
b. No damage likely to be caused by backfilling.

The disadvantages included:-

a. Difficulty in providing a slippery surface so that the membrane was not torn during pipe laying.
b. Any rupture of the membrane could cause extensive attack since ground water could travel between the membrane and the pipe.
c. Difficulty in moulding the membrane around pipe joints and bonding onto manhole wrapping.

9. Insitu Reinforced Concrete Construction An alternative of constructing an in-situ reinforced concrete culvert was considered as opposed to pre-cast concrete pipes. Extra thickness of sacrificial concrete was considered, as was a protective surrounding membrane, but the whole concept was eliminated on cost grounds.

10. Mass Concrete Construction Similar comments to the previous method apply.

11. Steel Pipes suitably protected Paint systems on steel pipes have become extremely sophisticated and almost certainly such a pipe in conjunction with a Viking Johnson type of coupling would have met the case. First stage estimates showed, however, that the basic price was too high.

12. "Pipe" made up from Cast Iron Segments These were envisaged as being used in open trench but the basic cost of the segments was too high.

13. Pipes made from H.D.P.E. This was not pursued after initial investigations showed limitations on pipe size.

14. Pipes made with G.R.P. The products of two manufacturers were considered: Redland's Fibaflo and Johnston's Armflow. The resins used for their standard product were not suitable to resist

the chemicals present and each manufacturer carried out his own research to try to find a resin which was suitable. They had limited success and eventually, in the case of Redland, resorted to a surface coating using the resin of the Monoguard system, whilst Johnstons incorporated this same resin in the outer skin. Price was high because of the special formulation and site ground conditions precluding the use of the standard length 10 m. pipe. Backfilling is critical on this type of flexible pipe to ensure uniform pipe loading, which again increases costs.

Appendix 2 - Location Plan

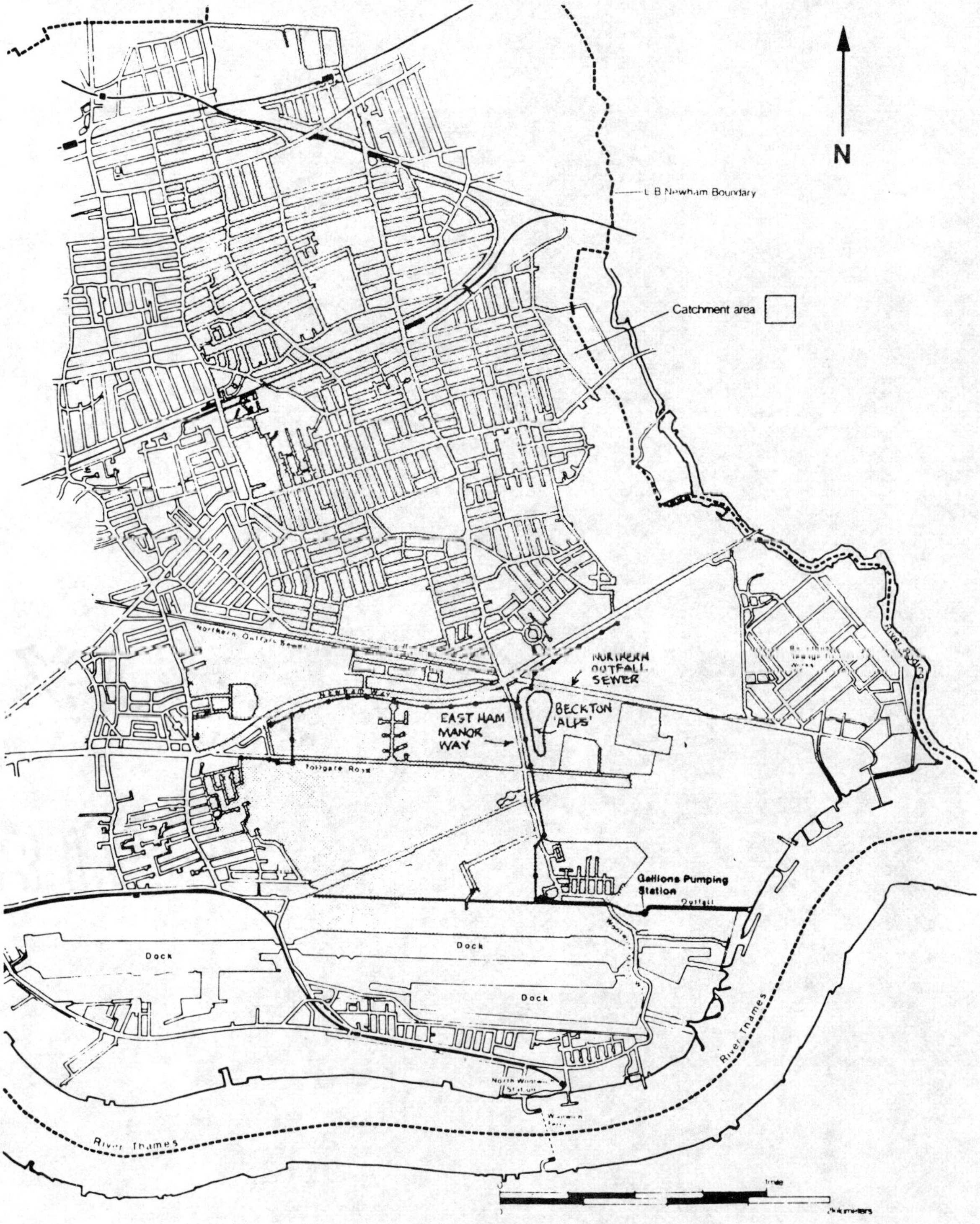

Piling in Contaminated Reclaimed Land

A. S. West

Federation of Piling Specialists

A pile is a structural member transferring foundation load from a structure through weak ground to a stratum adequately strong to support it.

The first piles were made from trees trimmed and driven into the ground. Timber is reasonably durable so long as it is always wet or always dry and have excavated piles driven by Romans still in good condition. But timber rots when it is alternately wet and dry - between wind and water.

Timber was used until the turn of the century when piles made of reinforced concrete were introduced. They were preformed above ground and driven to the required depth. Concrete is more durable than timber but nature has varied the level of the earth's strata and having estimated the length of pile required and cast them, they may prove to be too long or too short involving additional cost. High quality concrete can be used and compacted.

The next development was cast in place piles when a hole is made in the ground either by boring or by driving in a temporary casing with a shoe or plug to the required depth or set and then casting the pile in the hole and withdrawing any temporary casing. The concrete for these piles must be very workable and self-compacting.

Preformed steel piles are also used and these may be steel tubes or H-sections.

The latest development is sectional precast piles with mechanical joints designed to be as strong as the concrete section.

All types of pile can be protected from damage by contaminated soils but with preformed piles, a protective coating may be damaged during driving.

Before deciding what type of pile to use, site investigation information is required regarding strata, type of contamination, details of sub-soil water particularly if moving horizontally, and the position and depth of all services.

For use in contaminated ground, steel or precast reinforced concrete piles are probably best though very much a case of horses for courses. Bored piles involve the disposal of contaminated spoil and handling of this material.

Piling in Contaminated Reclaimed Land

R.S. Veak

Federation of Piling Specialists

A pile is a structural member transferring foundation loads from a structure through weak ground to a stratum adequately strong to support it.

The first piles were made from trees driven into soft ground. Timber is reasonably durable so long as it is always wet or always dry and there are several piles driven in Europe still in good condition. But timber rots when it is alternately wet and dry - between wind and water.

Timber was dominant until the turn of the century when piles made of reinforced concrete were introduced. They were preformed above ground and driven to the required depth. Concrete is more durable than timber but nature has denied the level of the earth's strata and having estimated the length of pile required and cast them, they may prove to be too long or too short involving additional cost. High quality concrete can be used and compacted.

The next development was cast in situ piles where holes made in the ground either by boring or by driving in a temporary casing with a shoe or plug to the required depth or set and then casting the pile in the hole and withdrawing any temporary casing. The concrete for these piles must be very workable and self-compacting.

Preformed steel piles are also used and these may be steel tubes or [illegible].

The latest developments are sectional precast piles with mechanical joints designed to be as strong as the concrete section.

All types of pile can be protected from damage by contaminated soils but with preformed piles a protective coating may be damaged during driving.

Before deciding what type of pile to use a site investigation information is required covering strata, type of contamination, behaviour of the soil water particularly if moving horizontally, and the position and depth of all services.

For use in contaminated ground, steel or plastic, or reinforced concrete piles are probably best although very much a case of horses for courses. Bored piles involve the disposal of contaminated spoil and handling of this material.

DURABILITY AND PROTECTION OF BUILDING MATERIALS IN CONTAMINATED SOILS

by P J Nixon, K W J Treadaway and W H Harrison

SUMMARY

This paper outlines present information on the performance of building materials which are commonly buried in use. It considers the effects, in general terms, of ground contamination on the durability of such building materials and offers suggestions for alleviating durability problems and for methods of protection.

The increasing scarcity of building land with the resulting pressure to re-use land which has been contaminated in some way necessitates a review of building techniques appropriate to this purpose. In addition to the problems posed by the uncertain engineering characteristics of such land (which are being dealt with in other papers to this Conference) the long-term durability of building materials in a potentially aggressive environment is of great importance. This paper therefore is concerned with the resistance to attack by aggressive chemicals of the principal building materials which might be used in construction on contaminated land.

The materials and components which might be expected to be at risk are essentially those which are buried in the ground and the principal ones are listed below:

Concrete - occurring in	- piles
	- strip/pad foundations
	- ground floor slabs (and slabs external to the building)
	- drainage pipes
	- tunnel linings
	- bricks/blocks
	- retaining walls
Metals	- steel reinforcement in concrete
	- service pipes and conduits
	- driven steel piles
	- tunnel linings
Fired clay products	- pipes
	- clay bricks in footings and in linings to manholes etc
Calcium silicate products	- bricks in footings, etc
Plastics	- pipes
	- protective membranes
Rubber	- sealing joints to pipes
Bituminous and asphalt materials	- water proofing and tanking uses.

The durability of building materials depends on the environment in which they are placed and in the present context this means to a large extent the particular chemicals occurring in the contaminated ground. The range of potential contaminants is clearly large but it is possible to identify certain classes of chemical as being potentially aggressive towards particular materials. It is self evident that the degree of seriousness with which a particular contaminant is regarded may be different when its possible action on building materials is being considered from when its effect on human beings is being assessed. Thus the majority of heavy metals are very toxic to humans but are of relatively little consequence to building materials which are most affected by acids, strong alkalis, organics and sulphate and chloride solutions.

In the following sections the individual types of building material are discussed in terms of possible aggressive contaminants.

CONCRETE

A considerable amount of the concrete made in the UK is used below ground in civil and structural engineering. It has been estimated from a sample of over 360 buildings covering different types of construction that approximately 60% of the concrete was used in foundations. Much has been written about the need to protect concrete placed in the ground from the effects of acids and salts, particularly sulphates, but most of the data on precautions to be taken is based on theoretical and laboratory studies with some limited field tests. These are supported by case studies of failures throughout the world which are generally characterised by one or more of the following factors: an unusually high salt or acid content in the soil or fill, thin section concrete, very poor quality concrete or an unusual construction technique. However it is also apparent that a very large proportion of the concrete placed below ground level suffers little deterioration. Thus for most normal soil conditions, certainly in the UK, the present Codes of Practice would appear to specify concrete of adequate durability. However, in sites contaminated with high concentrations of chemicals aggressive towards concrete the extent of deterioration could become much more serious.

The mechanisms of chemical attack on concrete have been classified into three types by O'Brien(1), viz:

- The leaching out of soluble material
- Degradation of the binding quality of the cement by chemical change
- Disruption caused by expansion of reaction products

A further mode of attack of chemical origin which could be added to these is the degradation of concrete by the crystallisation of salts within the concrete pores.

The leaching mechanism is epitomised by the removal of free lime by acids or soft groundwaters. This leaves the concrete in a more porous and weakened condition and also more susceptible to ingress of other chemicals.

Degradation of the cement binder is one of the principal forms of attack by acids but strong alkalis and some salts will also attack the cement hydrates. The severity of attack depends both on the concentration and availability of the attacking chemical and on the nature of the reaction product. If the reaction product is of low solubility it may inhibit further attack but if soluble there will be no barrier to fresh reactant reaching the cement compounds.

When the reaction product occupies a greater volume than the reactant the expansive force produced can cause severe degradation of the concrete.

The attack by sulphate solutions is a classical example of this mechanism while alkali aggregate reaction which could conceivably be induced by alkaline solutions in the contaminated ground has a related mechanism(2). Similarly salt crystallisation, which is particularly a problem where concrete is exposed to wetting and drying cycles, can damage concrete by a physical expansive mechanism.

A number of authors have described in some detail the effects of various chemicals on concrete(1,3,4,5) and we shall not attempt to reproduce all this information here. Of the chemicals which seem most likely to be found in contaminated sites the mineral acids pose a significant problem, particularly sulphuric acid where both the sulphate and hydronium ions are aggressive towards the concrete. Coal tar residues are said to be aggressive because of the expansive crystallisation of calcium phenate from reaction between phenol in the residue and calcium in the cement hydrates(1). Wastes from phenol production are strongly alkaline and could degrade concrete either by attack on the calcium aluminate in the cement paste or possibly by inducing alkali aggregate reaction.

Protective measures

When specifying concrete for use in a potentially aggressive environment the factors which can influence the durability of the concrete are its permeability, the cement type (and to a lesser extent the aggregate type) and the use of any protective membrane or coating.

The best way of making a concrete resistant is to make it with a low permeability. Ways of doing this are generally by imposing a maximum water/cement ratio and/or a minimum cement content. Impregnation of more porous concretes with a resin or with sulphur has been shown to be effective but is too expensive except for specialist purposes.

Use of a sulphate resisting Portland cement will provide a high resistance to sulphate solutions while ordinary Portland cements incorporating appropriate quantities of blastfurnace slag or pozzolanas such as fly ash offer better resistance to sulphates and acids than OPC alone and enable the permeability of the concrete to be reduced. Supersulphated cement which has to be imported into the UK has a good resistance to sulphate except in high magnesium concentrations and also a better resistance to acids than Portland cements. However none of these cements offer good resistance to strongly acid solutions. The effect of the aggregate is relatively secondary. In acid solutions the use of a limestone aggregate, which will corrode along with the cement paste rather than being left standing proud to be broken out of the matrix by any mechanical force, offers some advantage and in strongly alkaline solutions the use of a silicous aggregate which may be susceptible to attack by the alkali should be avoided because in this case the product of attack is expansive.

In the case of sulphate in natural soils the procedure for deciding on the appropriate protective measures recommended in BRE Digest 174 is to classify the site into one of five classes depending on the level of SO_3. Increasing quality requirements for concrete are recommended as the SO_3 levels increase from class 1 to class 4 and finally class 5 conditions call for added surface protection. Sometimes quite large variations in SO_3 values arise between samples from different parts of the site or from different levels and a procedure is suggested to arrive at a single site classification. However the variation in concentration of chemicals in a site used for waste disposal could easily be much greater than any natural variations for which the BRE Digest recommendations have proved adequate and for this reason the presence of any material reported to be detrimental to concrete should be regarded as indicating the need for a high quality concrete. Such concrete will certainly require a low free water/cement ratio not exceeding 0.45 and in most cases will also need a high cement content. The cement content requirements may be relaxed a little in certain specific cases where, for instance, specialised techniques have been developed to produce dense concrete with very low water/cement ratios such as used for certain types of concrete pile and concrete pipe manufacture. Alternatively whilst retaining the overall cementitious content, mixtures of blastfurnace slag or selected pulverised-fuel ash with Portland cement have been found to produce dense impervious concrete with good chemical resistance.

The greatest difficulty will lie in deciding whether any additional surface protection or sacrificial concrete is required. This can be expensive either in the materials used or in the workmanship required to get the protection effectively installed. One of the principal causes of concern is the effect of acid waters. It is generally acknowledged and has been emphasised in a recent paper by Hughes and Guest(6) that if there is a flow of acid over the face of the concrete then the rate of deterioration can be considerable especially with poor quality concrete. However in a 1975 review(7) Eglington found that a 'literature review provided no information on instances of deterioration of concrete in foundations placed in acidic soils or ground waters' and in the paper presented at this Conference by Bartholomew(8) it is reported that 'corrosion of concrete piles is not a serious problem world-wide except in certain Middle East countries'. This general picture suggests that either the quality of structural concrete placed below ground is generally fairly high or that the rate of chemical action in most situations in the field is fairly slow. In fact both explanations are probably correct. A good quality concrete will be specified since the cost of repair work would be high and any deterioration would go undetected until a serious problem had arisen. The rate of chemical action will tend to be considerably slower when the soil or fill is packed against the face of the concrete than when small specimens with vulnerable edges and corners are stored in solutions in the laboratory. Some long-term experiments to compare the results of laboratory tests with the behaviour of concrete on a site where it is in contact with a clay containing aggressive chemicals are being conducted by BRE although in this case site conditions are designed to test the sulphate resistance of the concrete rather than the effect of acids. An interim report of the results of these tests after 5 years will be issued shortly. Where there is ground water at or above the level of the proposed concrete foundations and the pH is lower than 3.5 it would be advisable to use some sacrificial concrete or surface protection. At pH levels above 3.5 the need to protect the surface of good dense concrete would depend to a large extent on the permeability of the soil. The rates at which water sampling standpipes refill after being pumped out or better still the rate at which water has to be added to the standpipe to maintain a head will give some indication of the permeability of the soil.

Thin wall constructions, especially where evaporation can occur from one face such as tunnels or retaining walls, will be more vulnerable to chemical attack and special consideration should be given to their protection.

DURABILITY OF STEEL IN CONCRETE

The durability of buried concrete has already been discussed. This section of the paper deals with the durability of reinforced concrete but as will be seen the performance of the concrete cover has an integral influence on reinforcement durability.

Steel is thermodynamically unstable and when left unprotected in the atmosphere will quickly form a brown oxide (rust) which will continue to grow until scale flakes from its surface. This corrosion process will continue unless some external means is applied to prevent it. The most obvious techniques for prevention of the corrosion of steel are associated with protection of the steel surface from contact with oxygen and water, the two most common requirements for the completion of electrochemical corrosion cells. However alternative systems are available for protection and these include maintenance of the steel surface in an alkali environment of pH ranging between 9.5 and 13. Within this pH range the steel forms a passive film on its surface which stifles further corrosion. Hydrated cement provides an environment in this pH range, the usual values quoted being between 12.6 and 13.0. Thus hydrated cement in concrete provides a chemical environment in which the steel remains in a passive state. The cover also provides physical protection in terms of controlling the rate of diffusion of oxygen and moisture to the steel surface and it is within this context that the physical properties of the concrete controls the porosity and permeability which play a very important role in determining long-term durability of the reinforcement. Porosity and permeability also have other influences on the durability of the concrete in determining the ingress of acidic atmospheric gases including carbon dioxide and sulphur dioxide. These react with and neutralise the alkalis produced during hydration so lowering the pH of the concrete and decreasing the environmental protection which it gives to any embedded steel.

The passivity provided by the alkaline conditions can also be destroyed by the presence of aggressive anions in the concrete even though a high level of alkalinity remains. The chloride ion is particularly successful in promoting de-passivation of steel in concrete although sulphate ions could likewise be expected to promote some de-passivation. Chlorides react with calcium aluminate and calcium alumino-ferrite in the cement to form calcium chloro-aluminates and calcium chloro-ferrites, compounds with low solubility which reduce the concentration of reactive chloride ion. However they only remain stable when in chemical equilibrium with a small amount of chloride in the aqueous phase of the concrete and it is this low concentration of free chloride within the aqueous phase which promotes de-passivation and corrosion of the embedded reinforcement. At low levels of chloride ion in the aqueous phase corrosion will be minimal, but as the concentration increases the risks of corrosion will be significantly greater. Thus the amount of chloride in the concrete and in turn the amount of free chloride in the aqueous phase (which is partly a function of cement content and also of cement type) will influence the risk of corrosion. While the concrete remains in an uncarbonated highly alkaline state the level of free chloride in the aqueous phase will remain low (perhaps 10% of the total chloride content of the concrete). However the effect of carbonation is to break down the hydrated cement phases and in the case of chloro-aluminates to release chloride. Thus more free chloride is available in carbonated concrete than in uncarbonated materials. In a similar manner to chlorides, sulphates form solid calcium sulpho-aluminates so reducing the amount of free sulphate in the aqueous phase. However, in comparison to chlorides, sulphates are less aggressive to steel because of their reduced ionic mobility and ionic charge density.

Durability of buried reinforced concrete

The protection of steel in concrete in buried conditions is similar to that required in atmospheric conditions. In particular the physical properties of the concrete play an important role in determining durability and in this respect their influence on the diffusion of oxygen and water through the concrete are most important. As with atmospherically exposed concrete the main line of chemical protection is through the development of alkaline conditions within the concrete so, providing these alkaline conditions remain and the steel is not de-passivated by the presence of aggressive anions, corrosion is unlikely to ensue. In the case of buried structures, permanently immersed in water it is likely that the concrete will be saturated which will significantly restrict the diffusion of oxygen through the concrete to the steel's surface. Where water immersion is not permanent but the unit is below ground level, diffusion of oxygen will still be restricted because of the lower bulk concentration of oxygen in the ground in relation to that available in the atmosphere, but here, pore saturation is less likely to occur, so the physical properties of the concrete become more critical. Where pore saturation is likely (in permanently immersed structures) chloride diffusion into the concrete could still occur, but, the restriction of diffusion of the limited amount of oxygen available will limit any corrosion reactions resulting from the presence of chloride. However, where only partial pore saturation occurs oxygen diffusion is likely to be less restricted and the presence of chloride in sufficient concentration could lead to early corrosion. This is most likely at the saturated/unsaturated concrete interface or perhaps at the interface between the soil and the atmosphere, and this would be expected to be the most vulnerable area of the component.

The foregoing has been concerned with the physical protection provided by the concrete in controlling oxygen diffusion and hence the electro -chemical reactions of corrosion of steel under aggressive conditions. Long-term deterioration of the concrete can also influence the likelihood and rate of corrosion of reinforcement. The role of sulphate in causing the physical disruption of the concrete is of considerable importance. Under such conditions disintegration of the concrete would lead to complete loss of protection to the reinforcement. A similar situation may occur where the concrete is degraded by acid soil conditions. Here acid-base reactions between the environment and the concrete leaching the alkaline concrete constituents will produce a porous cover so reducing protection to the reinforcement. Even before this occurs the alkaline conditions within the concrete can more easily be neutralised in acid soil conditions especially where the concrete itself is permeable and porous.

The concrete cover therefore has a vital role in the protection of reinforcement in buried structures. This role is twofold, firstly a physical barrier to the reinforcement is provided which restricts the ingress of aggressive anions such as sulphates and chlorides and also restricts oxygen migration to the reinforcement so controlling the corrosion reaction occurring. The second line of defence is provided by the alkaline environment produced by the cement hydrates. This maintains the steel in a passive condition which provides long-term protection assuming that no aggressive anions are present.

METALS

Three main metallic materials are of interest for use underground. Lead, copper and steel are used for services, while steel can also have structural applications such as for piling.

LEAD

The corrosion of lead can be initiated by its burial in adjacent soils of different character particularly where the degree of aeration, the presence of chemicals, bacterial activity or water content can vary. Lead performs well in freely aerated soils but where oxygen access is restricted oxygen concentration cells may develop. Thus closely textured clay soils, especially those containing cinders or other materials containing carbon can create aggressive conditions. Aggressive conditions can also be established by the presence of nitrates in acidic soils, and by alkaline soils. The presence of sulphates assist in the formation of passivating films, while carbonates and silicates can also act as passivators although uneven distribution of these anions may lead to localised protection and the development of galvanic cells which cause corrosion in the unprotected areas.

Underground corrosion of lead can be prevented by the application of bitumen or resins of various types according to the soil conditions. Cathodic protection has also been successfully applied to prevent the corrosion of bare or coated (to protect pinholes in the covering) lead.

COPPER

Generally bare copper can be safely buried in soils without fear of serious corrosion. In some circumstances corrosion may however occur, for example, in acid peaty and clay soils, and those containing cinders where sulphides are present. Made-up ground containing cinders and other carbon bearing materials may also cause problems. Where corrosion of copper is likely to occur protection by bitumen impregnated wrapping tapes or plastics coatings is necessary.

GALVANISED STEEL

The action of the zinc coatings on buried steel is sacrificial which results in a life for the component depending on coating weight and soil conditions. Certain soils which are aggressive to steel, including those containing carbonaceous ashes, acidic peats and some made-up grounds, are even more aggressive to galvanised steel and in these conditions additional protection (bitumen coating or tape wrapping for example) is desirable, although in alkaline soils the corrosion resistance of galvanised steel is most marked.

STEEL

Coarsely grained and well aerated soils with high electrical resistance are the least corrosive to buried steel. On the other hand heavy clays and highly saline soils are the most corrosive. The depth of water table is also important in controlling corrosion, especially in relation to wetting and drying cycles. Bacterial activity in waterlogged clays and similar soils can also influence the corrosion rate significantly. In anaerobic soils containing sulphates, sulphate reducing bacteria can thrive and accelerate the corrosion of steel. In these and other conditions where accelerated corrosion may be predicted additional protection provided by overlapping tape or bitumen coatings will be required.

BRICKWORK

Clay bricks are among the most durable of building materials and are likely to be relatively unaffected by chemicals in the soil. The combined effects of salt crystallisation and frost action have sometimes been found to cause degradation of bricks placed above ground but below dpc and this could be a problem in ground with a high salt content if frost susceptible bricks are used. BRE Digests 164 and 165 recommend the use of special quality bricks in exposed brickwork below dpc. These bricks should satisfy one of three criteria; either evidence of satisfactory service supplied by the manufacturer, satisfactory behaviour over three years in an experimental panel or being of engineering grade. Such 'special quality' bricks should be durable in contaminated ground.

The mortar in the brickwork is likely to be much more susceptible to attack than the bricks and reference should be made to the concrete section of this paper. Use of sulphate resisting Portland cement can give protection against sulphates in the ground while use of a cement/sand mortar containing a plasticiser which is recommended by BRE Digest 160 to give good frost resistance may help to resist other aggressive chemicals. As with concrete the use of an increased cement content can help to increase the chemical resistance.

Calcium silicate bricks are likely to be less durable to frost action and salt crystallisation in contaminated ground than clay bricks. BRE Digest 157 states:

'The (calcium silicate) bricks may be attacked by high concentrations of magnesium sulphate or ammonium sulphate sometimes present in industrial wastes used for filling sites. If contaminated by chlorides eg calcium chloride, common salt or sea water, all classes are liable to suffer if attacked by severe frost'.

PLASTICS

In recent years plastics have been increasingly used in building components. The plastics components which could be at risk in contaminated ground are mainly service pipes, tanks and protective membranes. The durability of a component will depend to a considerable extent on the particular plastics from which it is manufactured but unfortunately it is often difficult to anticipate which plastic is likely to be used or without expert knowledge, to identify the plastics type when the component is received. It is therefore important that a user, who anticipates using plastics in a potentially aggressive environment, should discuss the durability of the particular material with the supplier.

Plastics are much used in general industry for purposes of resisting attack by chemicals and so, although there is only limited knowledge of the behaviour of plastics building components in contaminated ground, little trouble is to be expected provided the appropriate plastics type is used. In general plastics are least resistant to other organic chemicals. Phenols are particularly aggressive towards plastics. Strong acids and strong alkalis may also attack depending on the particular plastics but on the whole alkalis are more aggressive. When attacked by acids, alkalis or other organics, plastics tend to fail by stress-corrosion cracking.

The ground/air interface is a potential problem area because of the deleterious effects of ultra-violet light on plastics which could exacerbate damage due to contaminants in the ground. However plastics components close to ground level are likely to be sheathed in order to protect them from mechanical damage. Some of the building components most commonly made in plastics are given below with comments on the likely plastics type and its resistance to chemicals.

Service and sewage pipes

- gas pipes in high density polyethylene) both resistant
- water pipes in pvc or low density polyethylene) to acids/alkalis

Septic tanks - often in grp. Susceptible to attack by benzenes, hydrocarbons, petrol etc and also by phenols, acids and alkalis.

Dpc/sheet membranes often in low density polyethylene. Resistant to mineral acids and alkalis.

RUBBER

Rubber is most likely to be found in contaminated ground in the form of pipe sealing rings though it is also a component of some mastics and sealants which could be found underground. Both natural rubber and numerous specialised synthetic rubbers are available and the chemical resistance of the various types differs considerably. It is therefore important, as with plastics, that components are purchased from a reputable manufacturer with whom particular problems can be discussed. Information is also available from the Malaysian Rubber Producers Association and the Rubber and Plastic Research Association. Both organisations publish useful literature on the chemical resistance of rubbers.

In general terms strong acids and alkalis would not be expected to have a great effect on rubber except for oxidising acids. Oxidation is the principal cause of degradation in rubber, leading to hardening and cracking, and catalytic oxidation can be induced by some salt solutions, particularly salts of copper, iron and manganese. Oils and organic solvents which cause swelling of the rubber may necessitate the use of an oil resistant rubber.

Much will depend on whether the contact with the aggressive solution is likely to take place over a long period. Occasional contact may not be detrimental, especially as the proportion of the area of a sealing ring which is exposed to the contaminated ground is probably small.

BITUMINOUS AND ASPHALT MATERIALS

Bitumens and asphalts may be used below ground for waterproofing and tanking as well as a binder in surface layers. They are regarded as being fairly resistant to attack by inorganic chemicals such as mineral acids and alkalis though organic contaminants could cause them to soften.

Bitumens may also be used as protective coatings for metal pipes. In this use they are often reinforced by a fibrous mat of, for example, hessian, glass fibre or plastic, in order to protect the coating from mechanical damage. Such fibres may be vulnerable to attack, so weakening the coating.

DISCUSSION

Much of the information outlined in this paper has been obtained in laboratory or small-scale field examinations and while these data are of significant value, extrapolation to the full-size structure may prove complex. In this context monitoring of durability of structures in made-up or contaminated ground, in conjunction with assessment of performance of test specimens in conditions identical to those of the full-size structure, could help in alleviating the problems associated with scaling-up laboratory tests. The localised nature of the distribution of some contaminants in ground can also pose considerable problems in an overall assessment of ground conditions. In this respect the development of risk assessment techniques using the necessarily limited data available from ground examination may prove valuable. Finally it should be emphasised that practical experience of the durability of building components in contaminated soils is very limited and therefore if ground analysis indicates the presence of potentially aggressive contaminants the specifying authority would be well advised to err on the side of caution in choice of materials and in the protection given to them.

ACKNOWLEDGEMENT

The work described has been carried out as part of the research programme of the Building Research Establishment of the Department of the Environment and this paper is published by permission of the Director.

REFERENCES

1 T P O'Brien, Resistance of Concrete to Chemicals, Corrosion and Coatings South Africa, Vol 4 No 6 1977 pp 19-23
2 W Gutt and P J Nixon, Alkali aggregate reactions in concrete in the UK, Concrete, Vol 13, No 5, 1979, pp 19-21
3 I Biczok, Concrete Corrosion and Concrete Protection, English edition, Publishing House of the Hungarian Academy of Sciences, Budapest, 1967
4 Guide for the Protection of Concrete against Chemical attack by means of Coatings and other Corrosion Resistant Materials. Report by ACI Committee 515, Chairman W H Kumming, J of ACI Dec 1966 pp 1305-1392
5 W H Gutt and W H Harrison, Chemical Resistance of Concrete, BRE Current Paper 23/77
6 B P Hughes and J E Guest, Limestone and siliceous aggregate concretes subjected to sulphuric acid attack, Mag Con Res Vol 30, No 102, 1978, pp 11-18
7 M S Eglington. Review of concrete, behaviour in acidic soils and groundwaters. CIRIA Technical Note 69, 1976, (available to members of CIRIA only)
8 R F Bartholomew. The protection of concrete piles in aggressive ground conditions: an international appreciation. Paper presented to Reclan 79

BIBLIOGRAPHY

BRE Digest 174, Concrete in sulphate bearing soils and groundwaters. BRE Feb 1975
BRE Digests 164 and 165. Clay brickwork, Parts 1 and 2, BRE 1974
BRE Digest 160, Mortars for brick laying, BRE Dec 1973
BRE Digest 157, Calcium silicate (sand lime, flintstone) brickwork, BRE Sept 1973
S Buchan, Rubber in Chemical Engineering, The Malaysian Rubber Producers Research Association, 1965
Data sheets on the chemical resistance of rubber published by the Rubber and Plastic Research Association
'Corrosion', L L Shrier, Vols 1 and 2, Newnes

UNDERGROUND PIPEWORK IN RECLAIMED GROUND

Trevor R. Smith
C.Eng., Ph.D., B.Sc., M.I.Gas.E., M.I.W.E.S.

The conference theme is the Reclamation of Contaminated Land and many of the other papers presented deal with specific aspects on the form of contamination. Since there is some doubt as to what can be described as contamination - contamination to one person may be acceptable for a particular use, whereas it may be totally unacceptable to another person for a different use - it is not the purpose of this paper to dwell on details of contamination. It is the intention to highlight some of the problems of installing pipelines in reclaimed land and it will be for the reader to decide about those aspects which pertain to his given situation.

Reclaimed land can be from the sea, industrial redevelopment, from waste or spoil tips or maybe what is known as "made-up" ground where it has been necessary to import materials so that the site can be utilised. In nearly all cases, it is essential to consider the installation of the primary services of water, sewerage and gas in order to meet the requirements of the facilities to be constructed on the reclaimed ground. This applies whether the site is to be for industrial purposes, new housing or even an amenity centre.

The laying of pipes in reclaimed land can be straightforward, but on the other hand can, in certain circumstances, give rise to several installation and operational problems. This paper serves to highlight particular problems of ground movement, heat build-up, corrosion and infiltration of toxic substances so that preparation and precautions can be taken at the planning stage to minimise or eliminate these hazards. In this way, the community as a whole should be able to benefit from the use of reclaimed land with the full security of the basic services that have been provided.

BASIC SERVICES

This paper will be concentrating on sewers, water mains and gas pipes. Pipes and conduits for other applications, such as district heating, steam and compressed air pipelines may also be installed but more particularly in industrial complexes. For these instances, similar considerations have to be given.

It is at this stage worthwhile examining the materials which are used for pipelines for the different applications, since this can have a significant bearing on the selection of materials for a given set of conditions. The use of materials has been summarised in Table 1 which is meant to be a general guide rather than gathering all options.

Ductile Iron Pipe

Ductile iron is probably the most widely used material for water and gas mains. It is available in the U.K. in sizes 80 mm - 1200 mm with push-in, mechanical or flanged joints. Towards the end of 1980 the size range will be extended up to 1600 mm. Pipes are 5.5 m in length, although longer pipes up to 8 m will be available when the size range is extended. Ductile iron pipes are made in accordance with B.S. 4772 and in the case of pipes for gas, British Gas Corporation Specifications BGC/PS/C4 and BGC/PS/C6 covering fittings. Ductile iron pipe has a working pressure, dependent on the size, in the range 25 - 40 bar.

Grey Iron Pipe

Like its derivative ductile iron, grey iron is also widely used for water mains and sewers and there is a long tradition of its use in these applications. Pipes are available in the size range 50 mm - 500 mm and are suitable for working pressures up to 10 or 16 bar dependent on class of pipe. Grey iron pipes are made to conform with B.S.4622.

Concrete Pipe

Concrete pipes are made by several techniques, for use in surface drainage or sewers. Pipes are manufactured in the size range 150 mm - 2550 mm and to the requirements of B.S. 556. It should be pointed out that there can be a wide variation in performance of concrete pipes and therefore serious consideration should be given to ensure that pipes are, at the very least, subject to the British Standard Institution's Kite Marking scheme to ensure that the pipe quality is in compliance with B.S. 556.

Prestressed Concrete Pipe

Prestressed concrete pipe is made in the diameter range 600 mm - 3200 mm by one of two methods. The cylinder pipe which has been available in the U.K. for over 20 years, incorporates a steel cylinder inside which a concrete core has been spun and over which prestressing wires are wound and protected by mortar coating. In the non cylinder version, as its name implies, there is no steel cylinder and it is necessary to prestress the pipe in both longitudinal and circumferential directions to obtain the necessary performance. Prestressed concrete pipes are suitable for working pressures up to 12 bar and are manufactured in accordance with B.S. 4625.

RPM/GRP Pipe

Glass reinforced plastics are new pipe materials to the water and sewerage industry. There are several techniques for making GRP pipe and all methods result in a lightweight flexible pipe. The use of sand and resin to increase the thickness and hence the stiffness results in a pipe which is described as reinforced plastic matrix (RPM). Whilst RPM pipes are available as large diameter pressure pipes, most GRP systems are available only for non or low pressure applications. Pipes are manufactured in accordance with B.S. 5460.

Steel Pipe

Steel pipes are used for water and gas mains. In the case of gas distribution mains, the steel pipe is preferred for pressures in the range 2-7 bar. Steel pipe is used for water trunk mains but the similarity of mechanical performance between cheaper ductile iron and steel has led to a diminishing market share for steel.

PVC Pipe

PVC pipe has been widely used for water mains and sewers. It is available in the size range 25 mm - 600 mm and for pressures up to 12 bar. PVC pipes used to be jointed using solvent cements but the majority of applications now utilise push-in joints. Care should be taken in designing PVC pipe systems where there are surge or pressure fluctuations, since there have been several reported failures. PVC pipes are made to conform with B.S. 3505.

Polyethylene Pipe

The British Gas Corporation is extensively using polyethylene in gas distribution pipe systems. Unlike water pipe which is available to B.S. 1972 (low density polyethylene) and B.S. 3284 (high density polyethylene), the British Gas Specification BGC/PS/PL2 is written for medium density grades of polyethylene which can be jointed using heat fusion techniques. Whilst polyethylene is generally inert to many chemicals, there are specific environments where it is not suitable to install polyethylene pipe. Polyethylene, like PVC pipe, is more sensitive to damage by third parties excavating around the mains and services, when compared with other pipeline systems.

Asbestos Cement

Asbestos cement pipeshave been used for water and sewers for some 50 years. The pipes are available in the size range 50 mm - 800 mm for pressures up to 12½ bar. Pipes are made to B.S. 486.

Clay Pipe

Clay pipe in one form or another, has been used since Roman times and is renowned for its inertness to a wide variety of chemicals. Clay pipes are available up to 600 mm diameter and lengths up to 2.5 m. All joints now rely on rubber gaskets to ensure flexibility of the pipe network. Clay pipes are made to B.S. 65 and 540.

In the U.K., water pipes have to meet the requirements of water regulations and bye-laws. This applies equally for mains laid by the authorities or for pipes laid within the curtilage of private property. Similarly, sewerage systems have to comply with building regulations and local bye-laws. In both cases, pipes should be laid in accordance with B.S. Code of Practice 2010. B.S. Code of Practice for building drainage and sewerage is currently being revised.

Gas pipes should be laid in accordance with British Gas Corporation/Code of Practice or in compliance with the Institution of Gas Engineers Code IGE/TD/3.

INSTALLATION

From the pipeline engineers point of view, reclaimed land can be full of problems. The land may still be subject to settlement, lack of drainage, fluctuating water table, decaying organic material, corrosive soil or toxic waste. It is therefore essential, before any pipes are laid in this ground, that full comprehensive planning of the project to examine all these aspects is undertaken.

Ground Movement

If there is a problem with settlement or ground movement in the reclaimed land, then it is necessary to have pipes which are capable of withstanding the high forces associated with disturbance. Furthermore, it is necessary that the pipeline can flex to relieve the strains along the length of the main. Flexible joints are preferred for this situation. In the case of pressure pipes, special attention has to be given to anchoring the flexible joints. This is important, not only to maintain integrity of the pipeline from pipe to pipe but also to ensure that fittings, where higher thrusts are generated are not blown off.

Where there is a possibility of ground movement or poor structural strength in the soil itself, it can be very difficult to design and construct suitable anchor blocks, and therefore the anchored joint comes into its own. In the case of ductile iron pipes, there have been several recent innovations which allow a standard joint to be easily modified at the production stage or in the field so that anchorage within the joint can be provided. The standard joint for water and sewerage applications is the Tyton push-in joint which consists of a simple specially designed gasket which is located in the socket and allows very simple jointing. For the size range up to 300 mm, it is now possible to achieve anchorage by replacing the standard gasket with a Tyton Anchor gasket which has stainless steel teeth embedded in the rubber. These teeth bite into the spigot end when there is an attempt for the spigot to withdraw, and thus the teeth provide the necessary anchorage. For larger sizes, a new version of the Tyton joint is being introduced which obtains anchorage from a small flange cast on the socket via circlip and gland onto a welded bead close to the spigot end. For gas mains, the Stanlock joint is available in an unanchored form and is easily made anchorable by the provision of a groove in the spigot to engage a locking circlip which is located in a recess in the socket chamber. Sections of these joints can be seen in Figure 1. These give an impression of the jointing mechanisms.

If the ground is liable to excessive ground movement there may be need to consider stabilisation of the soil through the use of polymeric fabrics and the importation of suitable backfill. An alternative is to consider placing the pipeline on piles but here careful attention has to be given to the design of the support and saddle, in order to ensure that the pile is not forced through the pipe. Tolerances on pipe lengths have to be noted if a pile is to be placed in the same position on every pipe.

Extra care should be taken if flexible pipe materials are being installed in unstable ground since any ground movement can cause disturbance to the backfill and pipe surround. This is a vital factor since much of the load on a flexible pipe system is transferred to the pipe embedment and should this be disturbed, it will obviously interfere with the pipe-soil transfer mechanism.

Lack of Drainage

Lack of drainage does not cause any problems at the design stage of a new pipeline, it is merely necessary to consider it at the installation stage. It may be necessary to contemplate the use of dewatering equipment so that the pipes can be safely and correctly installed.

Whilst undertaking this work, consideration should be given to providing some form of drainage, as it is feasible that the pipe trench itself even when backfilled, will become a drainage channel in its own right. There have been cases where this has occurred and the flow of water has eroded the backfill material and caused subsequent operational difficulties for the pipeline.

Fluctuating Water Table

In a similar way to lack of drainage, fluctuating water table has to be considered principally as an installation problem but care should be taken in this particular case to consider the possibilities of pipe flotation. In many cases, fluctuating water table can lead to the recommendation that anchored joints are preferred for pressure pipelines.

Decaying Material

Where decaying organic material is present, this can be considered to be aggressive to ferrous metals and consideration should also be given to the type of cement used in cementitious products. This will be discussed in the subsequent paragraph on corrosion.

Decaying material can lead to heat build-up and this can be a particular problem for plastic materials since they are temperature sensitive. Plastic pipes perform less satisfactorily at elevated temperature and are liable to soften significantly at quite modest temperatures.

Corrosion

Generally speaking, all man-made ground can be regarded as being corrosive and therefore corrosion protection should be considered for ferrous and cementitious products. In natural reclaimed land it is expected that a large proportion of such sites again will be corrosive, but it is possible to undertake soil surveys to verify the corrosive nature of the soil.

Ductile iron pipe is very easily protected against corrosive conditions by the use of site applied polyethylene sleeving. Tubular polyethylene film is pulled over the barrel of the pipe and fixed in position and after jointing, the socket and spigot are also protected by polyethylene or by the use of a thermo-plastic or petrolatum impregnated wrapping tape. This method is extremely successful and, being site applied, is not subject to damage in transit. The sleeving is also capable of withstanding a degree of damage and still providing protection. Polyethylene sleeving provides a corrosion barrier since it isolates the pipe from the soil. It is possible for water to get inside the polyethylene sleeving but generally speaking, this is not a problem since the corrosion mechanism is self-stifling within the sleeve.

Where acidic waste products are likely to be present, it is strongly recommended that the pipe itself be fully wrapped with petrolatum impregnated or thermo-plastic tapes. In the case of steel pipes, it is necessary of course to provide a full cathodic protection system and for this to be successful, a high grade holiday free coating must be specified. The coating should also be capable of withstanding the chemicals in the soil.

For some sites the use of sulphate resisting cement should be used in cementitious products but the sulphate content of the soil should be checked beforehand. In many cases it may be found that the sulphate content goes above the BRE recommendations for sulphate resisting cement and therefore it is necessary to consider wrapping the concrete pipe or surrounding the pipeline with a limestone backfill. If neither of these systems are economic or feasible, then an alternative material such as RPM should be considered.

Toxicity

Toxic waste matters present two problems, firstly from the point of view of installation it is necessary to ensure that all operatives are suitably protected and that all care and supervision is taken to prevent accidents. The second aspect can be more serious and that is the problem of infiltration of toxic materials. In the case of many pressure pipe joints such as Tyton, the joint is capable of withstanding full vacuum and therefore will resist infiltration.

Generally speaking, because of their design, pressure pipes do not have any problems with infiltration unless excessively high external pressures are generated around the pipe and joint. However, in the case of non-pressure pipes, it is well established that there can be problems of infiltration and should the fluid be toxic, this can be conveyed to water courses elsewhere with serious implications for the community. Some joints and pipe systems have better resistance than others to infiltration particularly bearing in mind the possibility of ground movement which can cause the joints to be at the extremes of withdrawal and deflection. The advice is to check with the manufacturers about the suitability of their joints in these circumstances and should there be any doubt, the guideline should be that the larger the volume of rubber and the larger the compression, the greater should be the resistance to infiltration.

Conclusion

All of the foregoing relates to specific problems associated with the use of underground pipework in reclaimed ground.

One of course should not overlook the normal pipeline, technical, legal and financial requirements. Provided that these aspects together with those discussed in the paper are considered at all stages, then there should be minimal problems of installing and operating the service for the normal anticipated life. By minimising or eliminating the hazards, it is possible for the engineer to install pipelines within reclaimed land areas and ensure full security of the basic services which have to be provided if the community is to benefit from the use of this reclaimed land.

TABLE 1

	Ductile	Grey	Concrete	P.S.C.	RPM/GRP	Steel	P.V.C.	P.E.	A.C.	Clay Ware
Water Mains	X	X		X	X	X	X		X	
Sewers	X	X	X	X	X		X		X	X
Gas Mains	X					X		X		

FIGURE 1

JOINTS USED ON DUCTILE IRON PIPES

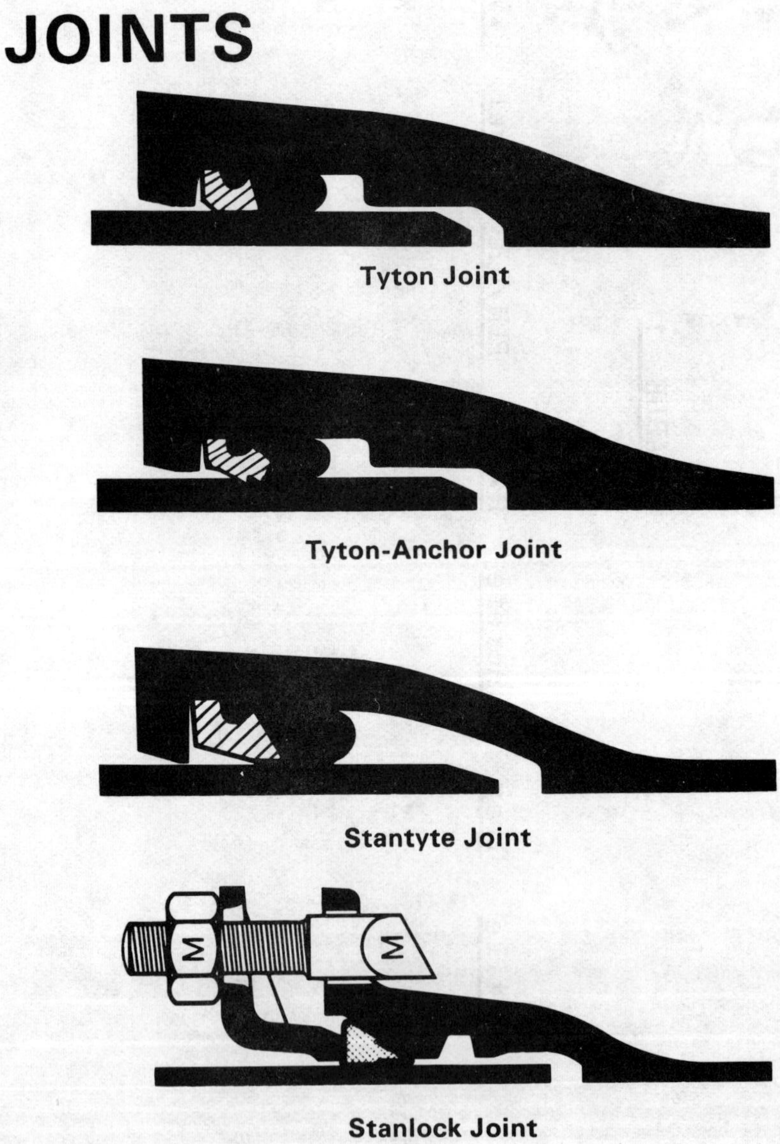

DISCUSSION

Paper E1

The ground treatment techniques described in the paper "Geotechnical problems of building on deep fill" are essentially methods of increasing the density of loose existing fills that had been placed without compaction. In the situation where filling can be carried out under controlled conditions, placement of the fill in thin layers with adequate compaction is the obvious way to substantially reduce subsequent settlement. Where the fill contains a substantial percentage of organic matter which will be subject to decay and decomposition, large movements will occur although controlled filling with compaction may reduce the rate of settlement. In this latter situation ground treatment soon after the completion of filling would not eliminate such settlements and I am not aware of the use of ground treatment methods on such sites very soon after the completion of filling.

ASSESSMENT OF FERROUS SLAGS FOR FILL APPLICATIONS

J.J. EMERY

Civil Engineering Department, McMaster University, Hamilton, Ontario

SYNOPSIS

The engineering, environmental and economic assessment of ferrous slags (blast furnace, steel and ferroalloy) for fill and embankment applications is described with examples from several projects. With the exception of the potential expansive behaviour of steel slags, ferrous slags have very favourable engineering properties for bulk applications. While leachates from ferrous slags do not contain significant levels of toxic constituents, care must be taken with stagnant water conditions in the case of blast furnace (sulphur species) and steel slags (calcite). Steel and ferroalloy slags do have high bulk densities that can influence potential transportation distances. Environmental enhancement usually results when slag replaces natural aggregates in bulk applications.

INTRODUCTION

The ferrous industries generate large quantities of byproduct blast furnace, steel, foundry and ferroalloy slags that are either recycled, utilized as materials, or with the exception of blast furnace slag, still open dumped. While the materials and energy conservation aspects of recycling steel slag to burden, metallics recovery, and uses involving replaceable energy or special features (slag cements and steel slag asphaltic concrete for instance [1]) are recognized and encouraged, this paper is concerned with the performance of ferrous slags in fills and embankments. Reclamation of old slag dumps for construction sites is also included. Such bulk applications often compete with natural aggregates so that a low financial return is involved unless environmental enhancement factors such as decreased dumping and reduced quarry requirements are recognized. Short transportation distances are generally the rule.

Recently, the writer has been involved with major industrial sites and fill applications where the overall assessment of ferrous slag suitability was required. The main concerns of project engineers were stability (steel slag), leachates (blast furnace, steel and ferroalloy slags), and high densities (steel and ferroalloy slags) in terms of transportation costs. It was found that leachates are not a problem except in some cases of stagnant water contacting large masses of blast furnace or steel slags. However, the expansion of steel slag can be a significant problem that calls for caution in confined or structural fill applications. It should be noted that some waste steel foundry sands containing slag may also have a potential expansive behaviour [2].

FERROUS SLAGS

Before describing the overall assessment procedure and its application on several large projects, some general information on ferrous slags and their characteristics in terms of fill applications will be summarized.

Air-Cooled Blast Furnace Slag

By selective cooling of the relatively uniform molten slag that is a coproduct (≈0.23 kg slag/kg iron) of iron production, four distinct types of blast furnace slag are produced: air-cooled (aggregate applications); expanded (lightweight applications); granulated (cementitious applications); and pelletized (lightweight and cementitious applications). While often fully used in conventional aggregate applications, air-cooled blast furnace slag is sometimes available, particularly from old dumps, for base, fill and embankment construction. General features of air-cooled blast furnace slag that make it attractive for such applications include: low compacted bulk density (typically 1200 to 1450 kg/m^3) that reduces dead load, lateral pressures, and transportation costs on a volumetric basis; high stability (CBR > 100) and friction angle (≈45°); stabilizes wet, soft underlying soils at early construction stages; placeable in almost any weather; very durable with good resistance to weathering and erosion; free draining and non-frost susceptible; almost complete absence of settlement after compaction; and non-corrosive to steel and concrete, with the possible exception of concrete-slag interfaces in stagnant water [3-6]. These positive features have resulted in the wide use of blast furnace slag in construction such as steel works structural fill and marine facilities, and embankments over soft ground.

Even though blast furnace slag may exhibit a sulphur odour and early contacting water may take on a greenish tinge, an alkaline condition is maintained (pH 8 to 10, i.e. steel protected), and the leachate does not contain significant concentrations of toxic constituents [7,8 and details given later]. However, since sulphur is present in a potentially leachable and reactive form (mainly calcium sulphate), care must be taken where contacting water can collect under stagnant conditions (concrete protection and isolation from water supplies required). It is undoubtedly the well-known colour and odour of contacting water that has caused concern with the leachate from slag fills rather than toxic constituent considerations.

With simple precautions in stagnant water, there are no construction problems with blast furnace slag. Some processing, such as size reduction from dumps may be required and large vibratory compactors have proven particularly effective during placement.

Steel Slag

Steel slag is a coproduct (≈0.18 kg slag/kg steel) of various batch steelmaking processes (open hearth, basic oxygen and electric arc) that introduce important usage restrictions compared to blast furnace slags. Steel slags are fairly variable in composition and characteristics, have a high bulk density (typically 1900 kg/m^3) and a potential expansive nature (volume changes of up to 10% or greater attributable to hydration of free calcium and magnesium oxides), while blast furnace slag is relatively light and completely stable. Large quantities of steel slag are recycled to blast furnace burden. Since serious damage, and even structural failures, may result from indiscriminate use of steel slags in confined applications, potential long-term volume changes must be checked before such usage [2]. (Obviously, steel slags should not be used in portland cement concretes.) However, there are still many applications for processed steel slag where expansion can be tolerated, has been controlled by suitable aging or treatment (spent acids for instance), or where sealing of the steel slag with an asphaltic coating eliminates any problems.

There is also a large amount of steel slag used for applications such as bases and many old steel slag dumps or fills are being built on, or used for construction purposes. A simple, economical and rapid test procedure for evaluating the potential expansion of steel slags has been developed [1]. It should be noted that aging for long periods in large dumps does not guarantee the elimination of expansive behaviour, particularly if the steel slag is unprocessed so large lumps are involved. Contamination with steel works rubbish should also be watched for. Leachates from steel slags do not contain significant concentration of toxic constituents, but in stagnant water conditions deposits of calcite have been noted. Construction procedures with suitable steel slags are similar to those for air-cooled blast furnace slag.

Ferroalloy Slags

The ferroalloy industry (principally ferrosilicon, ferromanganese, ferrochrome and ferronickel) is the smallest of the ferrous industries and the available slags tend to be localized. Silicomanganese slag (≈1.1 kg slag/kg of silicomanganese) is glassy textured, typically in large pieces, and used for bases or open dumped. Ferrochrome slag (≈1.5 to 2.0 kg slag/kg ferrochrome) occurs in dense 10 to 15 cm pieces with a vesicular surface and is used in road construction or open dumped. Ferronickel slag (up to 31 kg slag/kg ferronickel) is often granulated and used in road bases or open dumped. For silicomanganese and ferrochrome slags the engineering properties are similar to air-cooled blast furnace slag except that heavy slags are involved (1800 to 2000 kg/m^3). Leachates from ferroalloy slags do not contain significant concentrations of toxic constituents.

ASSESSMENT

The assessment of ferrous slags for bulk applications involves a number of interacting steps that are usually required when any alternative materials utilization is proposed [5].

Materials Selection Stage (Feasibility)

- material types and general properties required for site (noting any special requirements)
- inventory of available conventional and alternative materials with desired properties
- comparative cost analysis for available materials (costs at production point, processing costs, transportation costs, placement and compaction costs, extra costs associated with factors such as road strengthening, dust control, corrosion control and special testing to control variability, any credit for decreased disposal costs)
- environmental factors and significance when using alternative materials (positive - reclamation of dumps, decreased disposal, borrow not required;
  negative - environmental toxicity;
  comparative - public nuisance due to transportation dust, noise and vibrations)
- selection of suitable alternative materials for design stage.

At this feasibility stage, transportation charges will often be the governing factor as they make up 30 to 50% of the bulk materials cost. Typical transportation charges and limiting distances are summarized in Table 1. It is clear that truck transportation, which is commonly used, is limited to the 30 to 50 km range.

TABLE 1 TYPICAL BULK MATERIALS TRANSPORTATION CHARGES
Can.$/tonne-km, 1976 [9]

Mode	Charge	Comments
belt	0.018 to 0.054	suitable for relatively short distances
truck	0.045 to 0.072	used for fairly short distances, < about 150 km
train	0.021 to 0.034	used for long distances, up to 800 km
marine (barge, ship)	0.0055 to 0.0083	used for long distances, up to 1200 km
unit train	0.0036 to 0.0081	typically > 650 km, 4000 tonne shipments
slurry pipeline	0.0022 to 0.011	typically > 80 km, > 1 million tonne/yr

Materials Characterization Stage

- geotechnical characteristics of selected alternative materials (noting the inherent variability of large dumps and possible property changes with time)
- solubility testing if necessary (there are many procedures and tolerance levels to choose from, see companion paper by Emery and Matchett)
- rejection of alternative materials that lack required geotechnical properties or involve significant leachate problems
- selection of material design parameters
- development of any special provisions such as avoiding stagnant water conditions or requiring vibratory compaction
- check on cost analysis if significant engineering design changes required to incorporate alternative material
- input data to project engineer for use in overall design.

For ferrous slags, typical U.S. EPA slag solubility test data given in Table 2 have been summarized to show the slags listed are non-hazardous (i.e. limited solubility) [7,8]. For the solubility testing reported in following sections, a modified OECD shaking and decantation procedure has been adopted to determine solubility rates [5, companion paper by Emery and Matchett].

Design and Construction Stage

- in addition to normal procedures, additional quality control may be required due to the inherent variability of many alternative materials.

The key aspects of the assessment are transportation costs, geotechnical characteristics, solubilities and interpretation, and variability control. For ferrous slags, it has been found that potential steel slag expansion and solubility testing to confirm the lack of significant toxic constituents in leachates are the major factors.

APPLICATIONS

Potentially Expansive Steel Slag

Detection of potentially expansive steelmaking slag used in confined or

TABLE 2 TYPICAL FERROUS SLAG SOLUBILITY TEST DATA
Calspan Corporation for Environmental Protection Agency [7,8][a,b]

Slag	Analysis of Filtrate, mg/ℓ Cr	Cu	F	Mn	Pb	Zn	pH
blast furnace (iron)	<0.01	<0.03	1.9	<0.01	<0.20	<0.01	10.6
open hearth (steel)	0.01	0.04	3.1	<0.01	0.30	<0.01	12.5
basic oxygen (steel)	0.03	<0.03	4.0	<0.01	0.20	<0.01	12.5
electric arc (steel)	0.27	<0.03	1.5	<0.01	0.44	<0.01	12.4
iron and steel foundry	0.05	0.25	-[c]	0.06	<0.20	0.12	10.6
ferrochrome	0.02	0.02	-	0.30	0.40	0.20	9.9
ferronickel	0.01	0.74	-	0.07	1.00	2.00	8.5
silicomanganese	<0.01	0.17	-	0.10	<0.20	0.05	6.8
ferromanganese	0.02	0.04	-	2.10	<0.20	0.03	5.9

Notes: a. One solubility test for each slag. Two parts distilled water (pH 5.5) to one part slag gently agitated for 72 hours. Mixture filtered through 0.45 μm micropore and filtrates analyzed. Selected samples probably represent the more comminuted fraction of slag. Ni <0.05 mg/ℓ for all slags and Co <0.02 mg/ℓ for ferronickel slag.

b. All ferrous slags listed considered non-hazardous. (Toxic materials not leached at greater than 1 mg/ℓ. Greater tolerances for F, Mn and Zn.)

c. Not determined.

structural fill application, or in old dumps and fills being built on, is very important. A laboratory procedure has been developed that involves preparing steel slag specimens and a non-expansive control using the standard Proctor method and molds with perforated base plates to allow for moisture movement during the immersion period [1,10]. The specimens are then totally immersed in a water bath at 82±1°C and the vertical expansion with time monitored. Most geotechnical laboratories have the required equipment shown in Figure 1.

The 82°C test temperature was selected on the basis of the initial expansion test series shown in Figure 2. Since the expansion levels of 5 to 9% observed at 82°C (about three times those at 60°C) are similar to the levels of long-term expansion often observed in the field, this test temperature was adopted [2]. A short monitoring period of 1 to 7 days appears adequate to detect potential expansive behaviour. The actual steel slag gradation involved for a project should be tested where possible as the expansion behaviour is clearly a function of gradation, tending to decrease as size increases. Also, surcharge weights can be used to simulate overburden conditions which limit expansion. Support for the laboratory accelerated expansion tests at 82°C were given by the series of long-term expansion tests at 20±1°C shown in Figure 3, that resulted in expansions at 475 days of about half those observed in 7 days for tests at 82°C.

The test series given in Figures 2 and 3, and other tests such as the influence of spent acid (≈12% H_2SO_4) treatment on expansion shown in Figure 4, have indicated that aging in stockpiles (preferably after processing and in small quantities), spent acid treatments, and the use of coarser sizes all tend to limit the potential expansion of steel slags. These results are in qualitative agreement with field observations.

FIGURE 1 EQUIPMENT FOR EXPANSION POTENTIAL TESTING

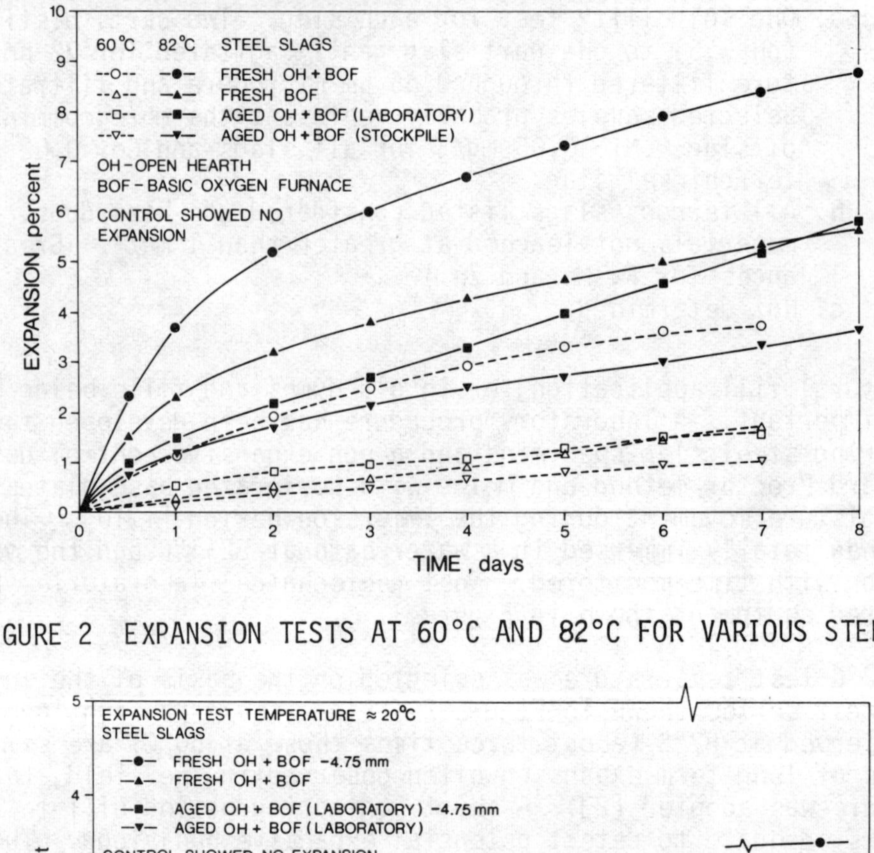

FIGURE 2 EXPANSION TESTS AT 60°C AND 82°C FOR VARIOUS STEEL SLAGS

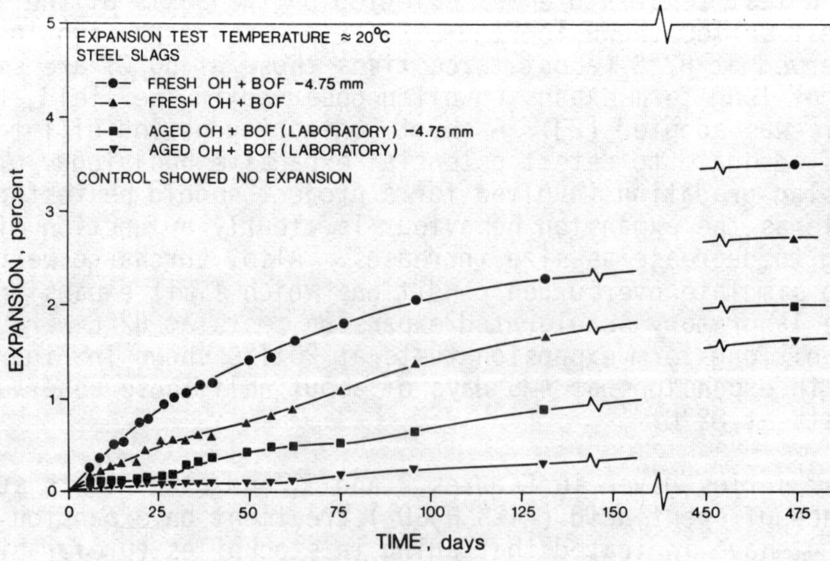

FIGURE 3 LONG-TERM EXPANSION TESTS ON VARIOUS STEEL SLAGS AT 20°C

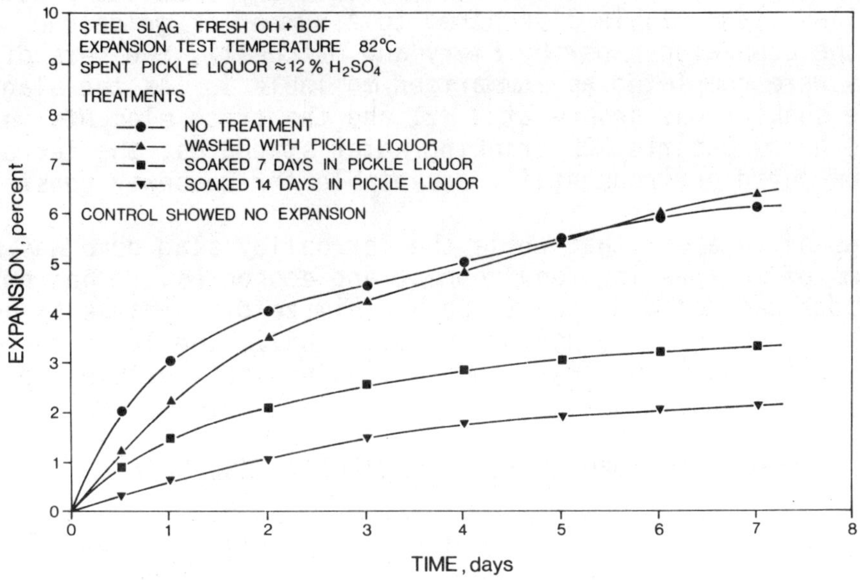

FIGURE 4 INFLUENCE OF SPENT ACID TREATMENT ON STEEL SLAG EXPANSION

This laboratory procedure, in conjunction with test pits to delineate steel slag zones, has been used successfully on several large construction projects involving steel plant additions over fills containing steel slags. A small magnet provides a rapid method of differentiating steel from blast furnace slag as test pits are excavated with a backhoe; all of the steel slag tends to be attracted by a magnet, while only a few pieces of iron in blast furnace slag are attracted. The test pits (or borings if required by depth) are also helpful in locating any contaminants or large pieces of slag and ladle skulls that may cause construction problems.

Two important factors should be noted here again:

1. aging steel slag in large dumps and/or pieces is not very effective in reducing potential expansion; and
2. blast furnace and ferroalloy slags are not expansive.

Ferroalloy Slag Use in Causeway

The proposed use of ferroalloy slag from a dump, to construct an adjacent causeway crossing an old canal, required a review of both the general engineering and environmental suitability. This slag was essentially a heterogeneous mixture of silicomanganese slag (60 to 65%; green, glassy appearing, dense slag in pieces up to a metre in size), ferrochrome slag (20 to 25%; white to dark brown, vesicular appearing, dense slag in pieces somewhat smaller than the silicomanganese slag) and coke and plant rubble (up to 20%; pieces up to 0.3 metre) in a large dump. While no direct geotechnical testing was completed, it was evident from the dump construction, previous use in large berms, and experience with other heavy slags (steel and nickel) that the material was satisfactory from an engineering viewpoint. A friction angle of at least 40° could be assumed and side slopes of up to 45° would result from standard dumping practices. With the large size of the slag and its high bulk density (1800 to 2000 kg/m^3) no wave erosion was anticipated. It was recommended that outer layers of dumped material be selected to avoid coke and plant rubble as much as possible.

While the technical literature did not indicate any leachate problems for such ferroalloy slags (Table 2 compared to a number of guidelines in Table 1 of the companion paper by Emery and Matchett), two sets of solubility tests were completed as summarized in Table 3. As the slag was crushed, the shaking was severe at 3 Hz, and the trace elements in the leachate are below Ontario MOE drinking water standards, the ferroalloy dump was considered environmentally acceptable for causeway construction.

It is perhaps of interest that while the ferroalloy slag dump was acceptable in terms of engineering, environment and economics, it has not been used: a bridge was decided upon to both minimize disturbance of bottom sediments and permit boating use of the old canal. The large ferroalloy slag dump is still available for construction projects within economic transportation distances.

TABLE 3 FERROALLOY SLAG SOLUBILITY TEST DATA[a]

Sample	Analysis of Filtrate, mg/ℓ						
	Cr	Cu	Fe	Mn	Pb	Zn	pH
1. 12 hours shaking	<0.03	<0.03	<0.03	<0.03	0.005	<0.005	9.7
2. 24 hours shaking	<0.03	<0.03	<0.03	<0.03	0.009	0.008	10.2
3. 24 hours shaking	<0.03	<0.03	<0.03	<0.03	<0.001	<0.005	-[b]
4. 36 hours shaking	<0.03	<0.03	<0.03	<0.03	<0.001	<0.005	-

Notes: a. Combination of 60% silicomanganese slag (crushed to minus 19mm), 20% ferrochrome slag (crushed to minus 19mm) and 20% coke material shaken with distilled water (pH 6.4) in a solid to distilled water ratio of 1 g to 0.1 ℓ for the periods indicated. Samples 1 and 2, and Samples 3 and 4 from separate bulk samples.
b. Not determined.

Solubility Testing of Blast Furnace Slag

Before using a large volume of air-cooled blast furnace slag in a large embankment, the designers wanted to confirm its engineering and environmental acceptability. The general engineering qualities of blast furnace slag for bulk applications have been indicated previously (friction angle 42°, compacted dry bulk density 1350 kg/m^3, and vibratory compaction recommended in this case), so that only the solubility testing will be described here.

Solubility tests were completed using the modified OECD shaking and decantation procedure to determine solubility rates [5, companion paper by Emery and Matchett]. This procedure tends to be very severe in comparison to the field situation as the slag pieces are crushed first, and the shaking of crushed slag with distilled water at 3 Hz is then very vigorous and long-term compared to the free-draining state in a large embankment. Both fresh and two-year old, pit-run, air-cooled blast furnace slag were used after crushing to the -13 mm and +300 μm size. Batch kinetics and desorption isotherms on this slag and others indicated a 1 g slag to 0.05 ℓ distilled water ratio and 24 hour shaking to be appropriate. This was repeated 5 times on a 60 g crushed slag sample with decanting and filtering, and the use of fresh distilled water for each 24 hour shaking period. The main concern was sulphates, but a wide range of analyses were completed with the results summarized in Table 4. While this data can be used to determine rates of solubility for a more complete study, it is clear that the slags do not contain significant

TABLE 4 BLAST FURNACE SLAG SOLUBILITY TEST DATA[a]

Slag[b]	Analysis of Filtrate, mg/ℓ									
	Ca	Cu	Fe	Hg	Mg	Ni	Pb	$C\ell^-$	$SO_4^=$	pH
F-1	13.5	0.02	0.02	0.00013	0.60	<0.01	<0.01	13.2	17.0	7.8
F-3	7.2	<0.01	0.04	0.00006	0.84	<0.01	0.02	1.6	3.3	7.8
F-5	6.6	<0.01	0.02	0.00004	1.08	0.03	<0.01	1.4	3.5	7.9
A-1	15.9	0.05	0.03	0.0006	0.60	<0.01	<0.01	12.0	26.0	7.7
A-3	7.5	<0.01	0.07	0.00003	0.76	<0.01	<0.01	1.2	4.2	7.7
A-5	8.0	0.04	0.05	0.00004	0.87	<0.01	<0.01	1.1	3.6	7.6

Notes: a. Using OECD test procedure adopted in Construction Materials Laboratory [5]. See companion paper by Emery for further details. Solid (60 g) to distilled water (pH) ratio of 1 g to 0.05 ℓ. Cr, Mn and Zn <0.01 mg/ℓ in all cases.

b. Blast furnace slag F - sampled shortly after production. F-1, F-3 and F-5 filtrates from first, third and fifth 24 hour shaking, respectively.
Blast furnace slag A - sampled from small pile two years after production. A-1, A-3 and A-5 filtrates from first, third and fifth 24 hour shaking, respectively.

quantities of any toxic constituents (compare severe laboratory condition data with acceptable levels in Table 1 of companion paper by Emery and Matchett). Care will be taken to avoid stagnant water conditions at the site of course.

This type of bulk use will ease the slag disposal problems for a major steel works in an area of Ontario where conventional aggregate outlets are limited.

CONCLUSIONS

It is hoped that the suggested methodology for the engineering, environmental and economic assessment of slags for fill application, and examples of use, will assist others working on similar ferrous slag projects. For steel slags, the main concern is the potential expansive behaviour that can cause significant damage in confined or structural fill applications. With this exception, ferrous slags have very favourable general engineering properties for bulk applications in fills and embankments. While the leachates from ferrous slags do not contain significant concentrations of toxic constituents, care must be taken with stagnant water conditions in the case of blast furnace (sulphur species) and steel slags (calcite). The high bulk densities of steel and ferroalloy slags may be an important limiting factor on bulk use due to transportation costs. The methods outlined can also be used during the assessment of non-ferrous and incinerator slags.

ACKNOWLEDGEMENTS

Many of the studies described here were completed in conjunction with the Trow Group Limited, Consulting Engineers. The assistance of Dominion Foundries and Steel Limited has also been particularly important.

Any opinions expressed in this paper are those of the writer, and not necessarily those of any associated organizations.

REFERENCES

1. Emery, J.J., 1978, Slags as industrial minerals, Proceedings, Third Industrial Congress, Paris, pp. 127-142.

2. Crawford, C.B. and Burn, K.N., 1969, Building damage from expansive steel slag backfill, Journal of the Soil Mechanics and Foundations Division, ASCE, Vol. 95, pp. 1325-1334 (also discussions and closure).

3. NSA, 1971, Processed Blast Furnace Slag, The All-purpose Construction Aggregate, National Slag Association, Alexandria, Virginia.

4. Bauman, E.W., 1966, Blast Furnace Slag, Ideal Backfill Material For Steel Sheet Piling, National Slag Association, Alexandria, Virginia.

5. OECD, 1977, Use of Waste Materials and By-products in Road Construction, Organization for Economic Co-operation and Development, Road Research, Paris.

6. Lake, J.R., 1978, Waste materials and industrial by-products in highway engineering in Great Britain, International Conference on the Use of By-products and Waste in Civil Engineering, Paris, Vol. 1, pp. 43-51.

7. Calspan, 1977, Alternatives for Hazardous Waste Management in the Metals Smelting and Refining Industries, EPA Contract 68-01-4312, Calspan Corporation, Buffalo.

8. Environmental Protection Agency, 1977, (I) Executive summary, (III) Ferrous smelting and refining, (IV) Appendices, Assessment of Industrial Hazardous Waste Practices in the Metal Smelting and Refining Industry, NTIS, Springfield.

9. Emery, J.J., 1978, Utilization of wastes and byproducts as construction materials in Canada, Conservation and Recycling, Vol. 2, pp. 31-41.

10. Emery, J.J., 1974, A simple test procedure for evaluating the potential expansion of steel slag, Selected Workshop Papers, Roads and Transportation Association of Canada, Ottawa, pp. 19-40.

USE OF STABILIZED INDUSTRIAL SLUDGE FOR LAND RECLAMATION

J.J. EMERY AND D. MATCHETT*

Civil Engineering Department, McMaster University, Hamilton, Ontario

SYNOPSIS

A stabilization process for an industrial sludge has been developed that meets both environmental and engineering requirements for use as fill in the reclaimed area. Various combinations and proportions of sludge with fly ash, waste kiln dusts and cements are used in a combined drying/pozzolanic/hydraulic process, depending on the desired strength development. Laboratory solidification and solubility testing are described along with details on preliminary field trial procedures and results.

INTRODUCTION

The expansion of an integrated iron and steel plant in Hamilton requires both the removal of a large quantity of very soft, high water content, industrial sludge from an old harbour disposal area, and replacement with suitable fills to provide a construction site. Some treatment of the sludge to meet landfill disposal requirements is required [1], and various stabilization methods have been evaluated in the laboratory and in small field trials. It has been found that the sludge can be satisfactorily stabilized/solidified to meet both environmental (solubility of toxic constituents) and engineering (bearing capacity) requirements for industrial site fill applications. Thus, the proposed disposal area reclamation process which is at the final scale-up stage will consist of removing the sludge by stone fill displacement below the water surface and dredging (in progress), stabilization, and then return to the site above the water table to act as fill in the reclaimed area.

The stabilization process involves a number of possible combinations and proportions of sludge with stabilizing agents - fly ash, waste cement kiln dusts, waste lime kiln dusts, steel slag fines, portland cement, and slag cement - to give various strengths and rates of strength gain for the highly variable sludge. Industrial byproducts have been used where possible, with low level, or no cement addition, to minimize stabilization materials' costs. A wide range of these byproducts has also been evaluated to ensure supply continuity. Construction soil stabilization concepts have essentially been adopted involving initial drying by high surface area materials (fly ash, dusts) and then pozzolanic and/or hydraulic reactions as required. For disposal to landfill, just some drying and fairly low strengths (shear strength ≈0.01 to 0.02 MPa, equivalent to soft clay) are desirable, while for fill, initial handling strength and then high strength development (shear strength >0.1 MPa, equivalent to very stiff clay) is required.

Details of the laboratory solidification and solubility testing, and field trials, are given in this technical note and slides of the field

---

*Now with Ontario Ministry of the Environment, Cambridge, Ontario.

aspects will be shown during the conference. As the various combinations and proportions of materials used are largely a function of the specific industrial sludge involved, masses of test data will not be given, but rather the methodology, and typical combinations and strengths. For the reader interested in a more complete review of stabilization/solidification technology than can be incorporated here, the volumes edited by Pojasek [2] are excellent.

INDUSTRIAL SLUDGE

The industrial sludge involved (approximately 300,000 m^3) has been deposited over a number of years under lake level in an old harbour slip area adjacent to the plant. Geotechnical studies indicate the sludge is fairly consistent in appearance along the slip - very loose, oily, black, organic metallic (mainly iron oxide) waste - but highly variable in moisture content (29 to 75%, weight of water to total weight used throughout), insitu bulk density (1300 to 1650 kg/m^3), and loss on ignition (7 to 31%). While mainly of industrial origin from rolling mills, at early stages the slip received municipal sewage. In summary, a rather unpleasant and variable industrial sludge is involved.

LABORATORY STABILIZATION TRIALS

A simple laboratory procedure for preparing and monitoring stabilized sludge was developed so that a wide range of stabilizing agents and addition rates could be considered to provide information for field trials and costing. This involved mixing the sludge with the selected quantities of stabilizing agents in a small Hobart mixer (or by hand) until consistent, placing this mix in 10 cm diameter by 10 cm high cans, sealing the can with both a twist lid and double plastic bags to maintain moisture conditions, and then taking penetrometer* readings to monitor unconfined compressive strength development with time at 20°C [3]. Supplemental tests included pH, moisture content, temperature and bulk density monitoring.

This approach is considered more representative of conditions in a large mass of solidifying sludge than open curing where significant strength development may be due to simple drying, rather than cementitious reactions. Other conditions can be readily simulated in the laboratory such as lowered or elevated temperatures and curing under water.

Some typical laboratory stabilization trials for a 46% moisture content sludge (combinations and proportions by weight, with 1, 7, 14 and 21 day penetrometer strengths in brackets) would include:

80% sludge/15% lime kiln dust (quick)/5% portland cement
(0.29, 0.43, >0.43, >0.43 MPa)
80% sludge/15% lime kiln dust (quick)/5% slag cement
(0.25, 0.31, 0.41, 0.43 MPa)
80% sludge/15% cement kiln dust (bypass)/5% portland cement
(0.04, 0.42, 0.43, >0.43 MPa)

---

*Soiltest Pocket Penetrometer CL-700 which gives a direct measure of unconfined compressive strengths up to 0.43 MPa (4.5 TSF). For a safety factor of 3, these readings are very close to the safe bearing capacity for a footing at the surface of a cohesive soil. However, care should be taken in extrapolating to the field as correlations have not been checked.

80% sludge/15% lime kiln dust/5% portland cement
(0.12, 0.20, 0.22, 0.24 MPa)
80% sludge/15% fly ash (≈8% carbon)/5% lime kiln dust (quick)
(0.07, 0.12, 0.17, 0.18 MPa)
75% sludge/15% lime kiln dust/10% fly ash (≈8% carbon)
(0.16, 0.27, 0.30, 0.34 MPa)
75% sludge/20% cement kiln dust/5% portland cement
(0.02, 0.20, 0.43, >0.43 MPa)
75% sludge/15% fly ash (≈8% carbon)/5% lime kiln dust/5% portland cement
(0.12, 0.24, 0.43, >0.43 MPa)
75% sludge/15% fly ash (≈8% carbon)/5% lime kiln dust/5% slag cement
(0.04, 0.12, 0.15, 0.31 MPa)

These results for stabilization trials meeting strength requirements for landfill and/or fill indicate that the high organic content is not stopping pozzolanic and hydraulic reactions, but probably inhibiting them to some extent. Quick (i.e. unslaked) lime kiln dust is much more efficient than ordinary calcitic or dolomitic lime kiln dust and slag cements require a fair degree of alkaline activators. The additive costs are in the Can.$1.50 to $4.00 range per tonne of final stabilized material. The combinations and proportions for satisfactory stabilization of a wide range of sludge moisture contents have been established.

To check that miniaturization of the laboratory strength gain tests is valid, some large, 90 kg samples were prepared with various stabilizing agents and mixed in an Eirich R-7 mixer using only the drum rotation to simulate a large stabilization plant pug-mill. Both sealed curing and curing under a 0.8 m head of water were considered at a 7 to 13°C temperature range to simulate cooler field conditions. Strength development monitored with a penetrometer and probe rod indicated similar results to parallel small samples. These results, particularly for the submerged curing, supported the laboratory test methodology adopted. While bearing capacity is the major geotechnical factor of concern, it is planned to complete laboratory triaxial, consolidation and permeability testing on specimens recovered from the final field trials and early full-scale stabilization work [4].

SOLUBILITY TESTING

The major environmental concern with any industrial sludge is the potential level of soluble toxic constituents and their significance to the fill design. However, there are a number of solubility testing procedures and acceptable toxic constituent levels to choose from [5-9]. The writers have adopted and modified an OECD solubility test procedure [7, see Appendix for a full description] as it gives the required solubility rate for each constituent of interest. Selection of the maximum allowable concentrations of toxic constituents is not easy as indicated by the range in Table 1.

The results for typical solubility tests on the raw and stabilized industrial sludge are given in Table 2. Batch kinetics and desorption isotherms indicated a 1 g sludge to 0.05 ℓ distilled water ratio and 24 hour shaking to be appropriate (i.e. similar to slags). Unfortunately, the shaking at 3 Hz is rather aggressive, even on stabilized sludge, so there was continuous turbidity in the shaker flasks. After 24 hours, the shaker flask contents were allowed to settle, the supernatant drawn off and filtered, and the filtrate stored in a refrigerator until chemical analysis. This was repeated four more times although only the first,

TABLE 1 TYPICAL MAXIMUM ALLOWABLE CONCENTRATIONS OF SUBSTANCES IN WATER

	Maximum Level in mg/ℓ			
Substance[a]	EPA Drinking Water Standards[b]	EPA Extract[c]	Ontario Drinking Water [9]	Ontario Industrial Effluent Discharges [8]
	Interim Primary			
Arsenic	0.05	0.50	0.05	
Barium	1.0	10.0	1.0	
Cadmium	0.01	0.10	0.01	0.001
Chromium (VI)	0.05	0.50	0.05	1.0
Flouride	1.4-2.4		1.2-2.4	
Lead	0.05	0.50	0.05	1.0
Mercury	0.002	0.02	0.0002 water quality	0.001
Nitrate (as N)	10			
Selenium	0.01	0.10	0.01	
Silver	0.05	0.50	0.05	
	Secondary			
Chloride	250			
Copper	1			1.0
Iron	0.3			
Manganese	0.05			
Nickel				1.0
Sulphate	250			
Tin				1.0
Zinc	5			
pH	6.5-8.5			5.5-9.5

Notes: a. Only ions and pH listed here. For pesticides, radioactivity, etc. see references cited.

b. EPA interim primary and proposed secondary drinking water standards. For full listing see Reference 5.

c. Extract based on EPA extraction procedure. Levels specified are ten times the EPA national interim primary drinking water standards. For full details and listing see Reference 5.

third and fifth shaking results are given in Table 2. While not shown here, it is also possible to draw solubility curves from the data for each constituent as cumulative solubility in mg/g against cumulative filtrate in ℓ/g.

The solubility data summarized in Table 2 indicate that the stabilization has reduced trace element constituent levels in all cases except copper and lead, but these are still well below the maximum allowable (EPA extract or Ontario industrial effluent in Table 1 for instance). Indeed, the leachate from the raw sludge also probably meets these standards. While the alkalinity that results from the stabilization materials is important and very effective in limiting the solubility of many trace metals, this is probably not the case for copper and lead. Since the stabilized sludge was broken up before testing and then continually reduced during shaking (constant rates of solubility indicate this), the solubility test is far too severe to represent field conditions. This is now being checked by using a much gentler shaking action and column tests [6]. Cadmium will also be considered in future solubility testing.

TABLE 2 COMPARISON OF SOLUBILITY TEST DATA FOR RAW AND STABILIZED INDUSTRIAL SLUDGE

	Analysis of Filtrate, mg/ℓ											
Sample	Ca	Cu	Fe	Hg	Mn	Mg	Ni	Pb	Zn	$C\ell^-$	$SO_4^=$	pH
R-1	8.3	0.01	0.16	0.00032	0.08	2.5	0.02	<0.01	0.07	5.8	10.9	6.6
R-3	11.1	0.02	0.07	0.00011	0.57	1.5	0.03	<0.01	0.10	0.8	19.6	7.6
R-5	7.2	0.04	0.18	0.00012	0.71	2.2	<0.01	<0.01	0.37	1.3	24.7	7.4
S-1	160	0.04	0.06	0.00008	0.01	0.02	0.02	0.04	<0.01	9.0	2.9	11.9
S-3	81	0.02	0.06	0.00018	<0.01	0.06	<0.01	0.03	<0.01	7.0	1.4	11.5
S-5	111	0.04	0.06	0.00005	<0.01	0.03	<0.01	0.02	<0.01	9.1	3.4	11.7

Notes: a. Using OECD test procedure adopted in Construction Materials Laboratory [7]. Solid (70 g) to distilled water (pH 6.4) ratio of 1 g to 0.05 ℓ. Cr <0.01 mg/ℓ in all cases.
b. Raw industrial sludge R. R-1, R-3 and R-5 filtrates from first, third and fifth 24 hour shaking respectively.
c. Stabilized industrial sludge S - 75% wet sludge/15% slag cement/10% lime kiln dust (quick), cured 77 days and crushed. S-1, S-3 and S-5 filtrates from first, third and fifth 24 hour shaking respectively.

Regardless, it is clear that the stabilized sludge is fully acceptable from an environmental viewpoint.

FIELD TRIALS

Field trials to check the possibility of completing the stabilization in-situ (generally under water) without removing the sludge were completed early in the study. This involved the use of a large drag-line to mix cement kiln dust into the sludge. Where adequate mixing was achieved there was a noticeable stiffening of the sludge. Unfortunately, this process required too much machine time and there were substantial pockets of unstabilized sludge. It was decided at this stage that the sludge would have to be put through some sort of stabilization plant to produce the desired strength properties, and stabilized material would probably only be used for fill above the water table. A few simple trials using a large grader to mix sludge with waste kiln dusts confirmed that "proper" mixing resulted in the required stabilization. At this stage, the laboratory testing was emphasized to develop a range of suitable combinations and proportions of stabilizing agents.

Recently, field trials have been initiated to prepare for the proposed stabilization process. A rather efficient method has been developed for large field trials that allows a number of batches to be tested each day: a chute has been constructed that allows excavated sludge to be fed into a standard ready-mix truck; then the stabilizing agents are added using a small belt conveyor with the most reactive agent going in last to allow for visual control of initial stiffening; and after mixing, the batch is discharged at a location where strength monitoring can be completed on the stabilized sludge "lump". Samples are also taken for laboratory testing. As the summer weather in Hamilton can be very hot, the stabilized sludge is excavated to below any surface influences before using the penetrometer. In several cases the stiffening has been so rapid that water additions have been necessary to get the stabilized sludge out of the truck. After several days, it is possible to drive vehicles onto the stabilized sludge. This ready-mix truck operation is proving so efficient that it appears to be a viable alternative to a large base stabilization plant for the production rates required.

While testing is still in progress, some typical results can be given:

72.8% sludge/13.7% fly ash/8.1% lime kiln dust/5.4% portland cement (0.41 MPa at 6 days)
83.6% sludge/12.9% lime kiln dust/3.5% portland cement (0.28 MPa at 5 days)
75.7% sludge/11.5% fly ash/9.9% lime kiln dust/2.9% portland cement (0.22 MPa at 5 days)

Based on these types of field results, final plant designs and stabilization agent selection(s) (i.e. economics) are being completed. It should be noted that conditioned fly ash (≈13% moisture) is being used as this eliminates the need for more costly pneumatic transport trucks and silos. It is estimated that the total cost for removal, stabilization and placement as fill will be about Can.$5.00 per tonne of stabilized material. This is most favourable compared to sludge disposal and fill material costs.

CONCLUSIONS

It is hoped that by the time of the conference it will be possible to report on the actual final stabilization process and agents adopted. The concept of industrial sludge stabilization using industrial byproducts and low cement contents, and construction stabilization technology, appears most attractive. Hopefully other sludge applications will develop.

ACKNOWLEDGEMENTS

The authors wish to acknowledge the continuing technical and financial support provided by Dominion Foundries and Steel Limited for the sludge stabilization study. W. Rombough, M. Greenfield and G. Ertel of Dofasco have provided much assistance during the stabilization/solidification process development.

Any opinions expressed in this paper are those of the writers, and not necessarily those of any associated organizations.

REFERENCES

1. MOE, 1978, New Reporting System and Guidelines for Hauled Liquid Industrial Wastes, Ontario Ministry of the Environment, Toronto, December.

2. Pojasek, R.B., 1979, Editor, Toxic and Hazardous Waste Disposal, Volume 1 - Processes for Stabilization/Solidification, Volume 2 - Options for Stabilization/Solidification, (other volumes in press), Ann Arbor Science, Ann Arbor, Michigan.

3. Snyder, G.A., Zuhl, F.A. and Burch, E.F., 1977, Solidification of fine coal refuse, Mining Congress Journal, December, pp. 43-46.

4. ASTM, 1978, Recommended Practice for Use of Process Waste in Structural Fill, draft, American Society for Testing and Materials Subcommittee E38.06.05, personal communication, June.

5. EPA, 1978, Hazardous waste - proposed guidelines and regulations and proposal on identification and listing, Federal Register, Vol. 43, No. 243, December 18.

6. MOE, 1978, Determination of Leaching Potential, Ontario Ministry of the Environment, Toronto, July 26.

7. OECD, 1977, Use of Waste Materials and By-products in Road Construction, Organization for Economic Cooperation and Development, Paris.

8. MOE, 1978, Objectives for the Control of Industrial Wastes Discharges in Ontario, Ontario Ministry of the Environment, Toronto.

9. MOE, 1978, Water Management - Goals, Policies, Objectives, and Implementation Procedures of the Ministry of the Environment, Ontario Ministry of the Environment, Toronto, November.

APPENDIX
SOLUBILITY TESTS AND INTERPRETATION [7]

Purpose

Solubility testing is used to determine if materials contain substances which may be leached out by water acting over a long period, and the concentrations resulting from such action. Since the identification and quantification of dissolved substances is a matter of analytical chemistry, this aspect will not be given here. It should be emphasized that the suggested solubility testing is not concerned with determining what quantity of the substance may dissolve in say one litre of water, or what the concentration of the saturated solution will be. Such information is of little practical use during monitoring and facility design since leaching is most unlikely to produce a saturated solution, and it is the solubility rate that is of real interest. What a solubility test should provide is an indication of the types and quantities of substances liable to dissolve in a volume of water comparable to that likely to be available during the life of the facility (landfill, fill, etc.) in which the material is incorporated. Use of the solubility test data in practice also requires information on the long-term environmental conditions (climatic and hydrogeologic) at the site, the storage capacity of the strata involved, and a clear indication of the nature of anticipated problems, i.e. potential toxicity, sulphate attack of concrete, corrosion, settlement, etc.

While a complete chemical analysis of the material under consideration may be useful for the orientation of solubility testing, it will not generally supply the required information since the solubility of a substance may depend on factors such as the presence of other dissolved substances and the pH of the solution. Further, relating potential problems to chemistry can be extremely misleading since the leaching action may never release the particular toxic constituent to the environment.

Methodology

In general, a known weight of a dry sample (care being taken during drying to not modify the material) is submitted to the leaching action, by percolation or shaking, of a known quantity of distilled water or solution representative of the site conditions (with buffering if necessary). The undissolved solids are then separated, by filtration, decantation, or centrifuging, from the liquor containing the dissolved substances. This liquor is then analyzed for the substances leached out (type and quantity), and since the first extraction is probably not complete, the leaching of the solid residue, separation and analysis procedure is repeated until no significant amount of solute is found in the liquor. The composition of the dissolved substances may vary from one liquor to the next, and must be carefully monitored, with pH often an indicator of changes (phases of the leaching process).

Various extraction procedures can be used - filtered percolation, shaking and decantation, Soxhlet extraction - but it should be noted that while the Soxhlet method can provide a full extraction of soluble substances and dilution ratios information, it does not provide a quantitative accounting of the different phases of the leaching process. Some other comments on general procedures would include: test temperatures should be comparable to site conditions; the suitable contact period and solid to water ratio can be determined readily by batch kinetics and desorption isotherms; while some reduction in size may speed up testing, caution must be taken to ensure that this does not change the nature of the material or potential leaching process; and test miniaturization must be compatible with the heterogeneity of the material involved and precision of instruments used.

The development of full procedures for measuring the solubility of materials has been considered by OECD Road Research Group C12 [7] and is currently an activity of ASTM Committees D19 and E38, and the U.S. EPA [5]. These efforts should ultimately result in procedure(s) compatible with the requirements of both producers and monitoring agencies, recognizing the growing concern for characteristics of hazardous wastes such as toxicity. In the Construction Materials Laboratory, the use of the OECD shaking and decantation method, in conjunction with standard analytical chemistry procedures (including trace elements) has proven most effective for measuring the solubility of wastes and byproducts. Columns for comparative filtered percolation studies (saturated and unsaturated) are also being established [6].

Interpretation

The results of solubility testing are interpreted to evaluate the possible effects of leaching by comparing the laboratory solubility rates with the actual amount of water that can be expected to act at the site on a unit mass of the material. From typical solubility rate data (in mg/g per ℓ/g) for each constituent of interest based on the suggested method for measuring solubility, the quantities of dissolved substances can be determined for any time period.

Acknowledgements

The writers would like to acknowledge the background for this Appendix developed from participation on, and the report of, OECD Road Research Committee C12 Use of Waste Materials and By-products in Road Construction (OECD, Paris, 1977). Discussions with D. Gorle and R. VanGanse of the Belgian Centre de Recherches Routières have been most helpful in developing the writers' approach to industrial waste solubility problems and engineering solutions.

INVESTIGATION AND TREATMENT OF RAVENFIELD TIP

A. Q. KHAN

B.Sc., M.Sc., M.I.S.W.M. (SOUTH YORKSHIRE COUNTY COUNCIL)

SYNOPSIS

Ravenfield Tip became a problem site due to past indiscriminate, unscientific and uncontrolled deposition of highly toxic and polluting waste in a series of sandstone quarries. Complaints are reported to have been made during the operation of the site and afterwards by local residents with apparently no success. Soon after re-organisation the existence of this completed tip was realised by the officers of the County Council and it was regularly monitored. During late 1975 and early 1976 it was noted that sludges and liquids (acid tar) started coming to the surface. Subsequent analysis showed these to be highly acidic and, therefore, a potential hazard to the public and particularly to children.

Despite the fact that the land was in private ownership and the County Council had no legal responsibility to deal with the situation, immediate steps were taken by the County Council to safeguard the public by completely fencing off the affected area and distributing warning notices to householders, schools and press.

The Council further decided, after considering all the environmental implications of the situation, that it would accept the moral responsibility to make the area safe and to restore the land for beneficial use.

After buying the land (by Compulsory Purchase Order) the County Council initiated an investigation to assess the magnitude of the problem and after consideration of all options treated the materials in situ.

Ravenfield Tip, an area of 6.5 acres, is situated south-east of Rotherham. The site was originally a series of sandstone quarries. Planning permission was granted on 2nd December 1958 by West Riding County Council for the deposit of industrial waste. From 1958 to 1971 an oil reclaimer used this site for the deposit of acid tars and fuller earths. In early 1971 a private waste disposal contractor bought the site and deposited various toxic and non-toxic industrial wastes. Planning enforcement was taken against the waste disposal contractor in May 1973 and was found to be in breach of planning conditions and was fined £10. He appeared in court on two subsequent occasions for non-compliance with enforcement proceedings and at the court hearing in July 1974 was given conditional discharge on having stated his intention to seek fresh planning permission to formalise the over-tipping.

RAVENFIELD TIP SCHEMATIC DIAGRAM OF TREATMENT PROCEDURE

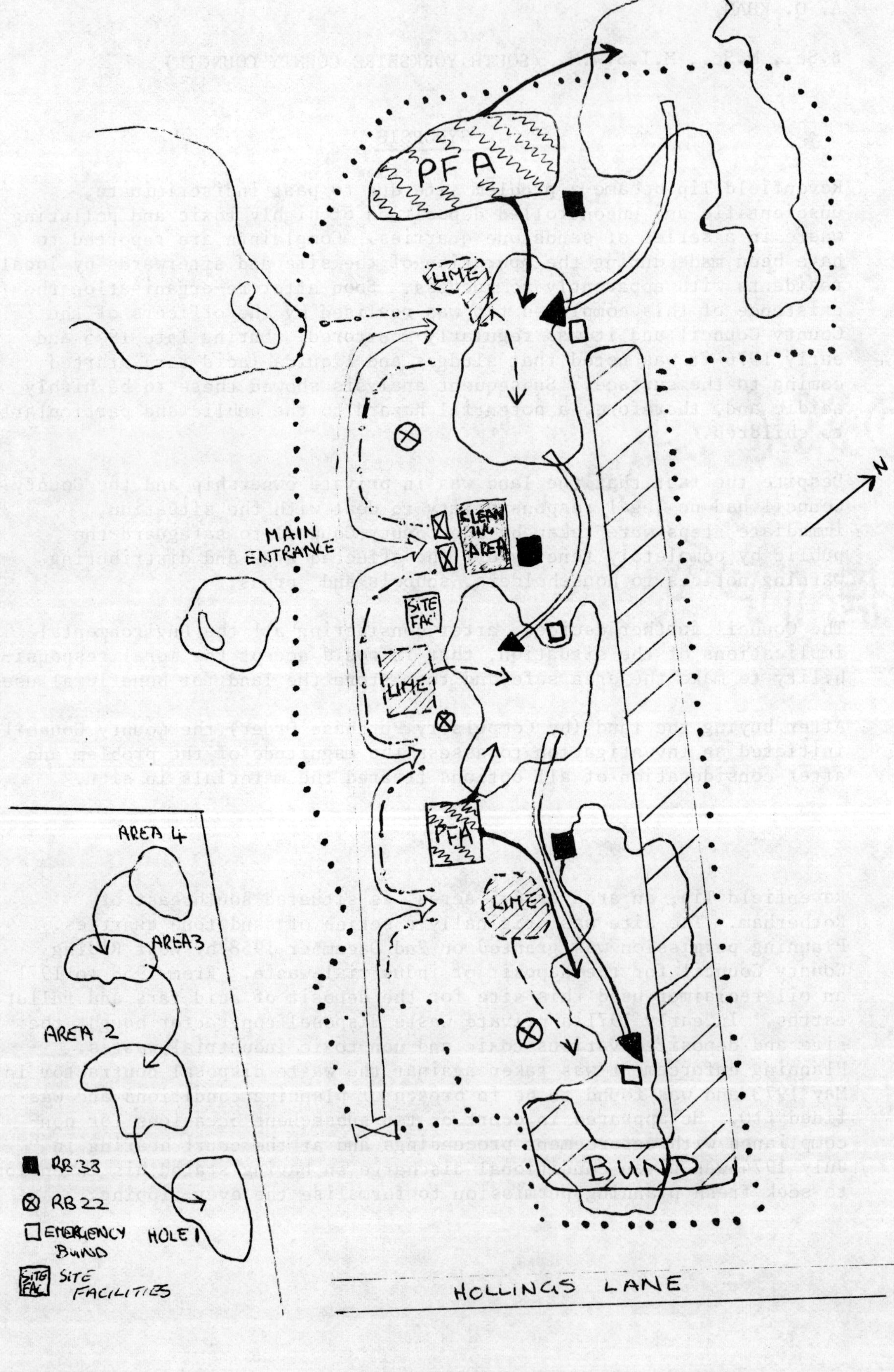

Ravenfield Tip - RB22 depositing waste mixed with lime. The steam generated can be seen on the right-hand side of the photograph. Note the nearness of the housing estate.

Ravenfield Tip - Treated materials being deposited in the quarry. The steam and lime dust can be seen rising from the heap.

A planning application was submitted for extension of the tip to contain seepage of "oily" waste but was found not valid as it involved land which was not in his ownership but included two other properties.

Further enforcement action was instituted against all the three owners of the contaminated land. However, appeal was kept in abeyance when it was discovered that there was a greater potential risk in compliance of the enforcement action as it would have exposed pools of acid tars and other potentially hazardous chemical wastes.

INVESTIGATION

As the site was in use pre 1972, very little information was available of the quantities and qualities of the waste that had been deposited in the quarries. There was no factual information available even as to the accurate dimensions of the quarries.

It was considered useful to get the information, although approximate, on the dimensions of the quarries and some indication (if possible) on the physical nature of the waste and whatever other information was possible by the use of aerial photographs which were taken during 1948, 1959 and 1967. A consultant was engaged and the information thus obtained was quite useful.

A full scale investigation was carried out on the site during October 1977 in order to determine the types and characteristics of tipped materials, the occurrence of water bearing layers, the geology of the site and the degree of contamination into undisturbed rocks and other inert materials.

The basic principle of the investigation was to excavate holes and trenches using a Hymac type of machine, on a programme designed to produce data from as much of the site area as possible within a limited period.

Forty-two holes/trenches were excavated covering 30 different areas of the site, the work of investigation being carried out by a team of officers from the Environment Department of South Yorkshire County Council under the team leadership of the Author. Dr. P. G. Oliver, Principal Hydrogeologist of West Midlands County Council provided extremely useful scientific and technical assistance by full involvement during investigation and by personally directing all the excavation work. His involvement in this project was under the "Exchange of Expertise Scheme" which exists between the two authorities.

Excavations were made up to depths of 7 metres, samples of liquids and solids were taken and where liquid was found in significant quantities, slotted pipes were inserted for monitoring purposes. 15 such pipes were inserted and the excavations were immediately backfilled in order to make them safe.

The contents of drums which were found during excavations were examined and samples taken where necessary. Detailed notes were taken and photographic records made of the entire operation.

The south quarry was considered to be potentially more hazardous and, therefore, in view of its proximity to houses, when excavation of this area was carried out temporary evacuation of residents was undertaken in the interest of their safety.

A total of approximately 100 samples of solids and liquids were taken .

RESULTS OF INVESTIGATIONS

(a) Materials which had been deposited on the site mainly consisted of acid tars, oil contaminated fullers earth and general industrial waste and slags, most of which appear to have arisen from the Rotherham area.

(b) The major problem was the acid content of the tars. This was up to 32% of concentrated sulphuric acid.

(c) No cyanide or dioxine (TCDD) was found.

(d) Analysis of samples from the monitoring pipes showed that no acid was leaching from the wastes.

(e) The pipes inserted in the north-east quarry where oil contaminated fullers earth was deposited have shown that as a result of disturbance during excavation, oil has been released from the waste.

(f) There was no recognisable pattern of tipping. Highly acidic and alkaline waste were found within inches of each other.

(g) A comprehensive report by Dr. Oliver recommended that waste materials are not removed but are treated on site primarily by neutralisation.

OPTIONS CONSIDERED

1. Disposal at County Sites

Consideration was given to the possibility of taking quantities of this waste, either treated or untreated, and, using the principle of dilute and disperse, to distribute it in smaller quantities at some of the County Council sites. Consultation with the Water Authority of the area was very helpful and their attitude to the problem was exceptionally co-operative. However, this option was rejected as it would have serious public relation problems and modification of planning permission and the Section 11 resolution (Control of Pollution Act 1974) would be required.

2. Disposal within the Region

Discussion with the neighbouring County also indicated similar problems for private as well as public sites.

3. Disposal at other County Council Areas Including West Midlands and South of England Regions

This was not considered practicable as either the planning permission or site licence issued by the Authorities did not allow or the authority of the area considered this to be unnecessary transport with associated problems if the disposal was for landfilling. The Council agreed with this view.

4. Disposal at Sea by Seaborne Incineration

A company which intended to operate a specially equipped incinerator vessel had offered its services. We did not think it was possible to incinerate this waste for it was contaminated with various other materials such as slags. This was rejected as unsuitable.

5. Building of a Concrete Raft over the whole Area

It was not thought to be a practical method.

6. Installation of Permanent Containment/Collection Sumps

One or more underground collection sumps could be installed at suitable places in association with a series of collection drains. The advantage with this system was that there was no need to carry out extensive excavation work or large scale removal of contaminants. Waste draining into sumps could be collected for disposal after treatment to other suitable sites. The surface of the land could be restored for any purpose. The disadvantage with the system was that the sumps and associated works needed to be maintained, waste collected and disposal provided for an indefinite period. Members of the public would not have agreed to this treatment.

7. Removal of a Calculated Quantity of Contaminated Materials

A calculated quantity of the contaminated materials removed to another quarry on the site or some other suitable sites. The volume this created to be filled with a mixture of lime and selected domestic or commercial waste which would absorb (mop up) the remaining quantity. The method appeared to be quite reasonable but it was not possible to say that it could be a total success.

8. In Situ Incineration

This was not considered to be a practicable method for this contaminated material.

9. Encapsulation Method

This was not considered a suitable method either at specialised plants or by in situ installation of special plants.

10. Treatment and Deposit at the Site

The excavation of all waste materials then mixing with quicklime and deposit at the site was considered to be the most suitable method.

TREATMENT

As previously outlined three quarries on the site had been used for the deposit of acid tars, oil contaminated fuller's earth and other industrial wastes. The actual extent and depth of the quarries was unknown as was the amount of waste deposited within and above them, although the amount of contaminated material to be treated was estimated to be of the order of 10,000 tonnes.

In order to facilitate deposit of treated material, a fourth empty quarry, adjacent to the most southerly of the affected quarries, was purchased by the County Council and included within the site. This quarry was enlarged to provide further capacity.

Treatment consisted of excavating the contaminated material by means of a RB38 drag line and, using a RB22 crane and grab, mixing it with a calculated quantity of quick lime.

The proportion of lime was varied between 1:4 and 1:10 according to the degree of acidity of the excavated material, but at all times "overkill" was ensured.

Mixing was carried out in batches and each batch was then transferred by the grab to the deposition area, where the chemical reaction was allowed to proceed. A considerable amount of heat was generated during the reaction, with the evolution of steam.

When reaction was complete the treated material, which was then friable and innocuous, was tipped in layers, compacted and covered with subsoil within the deposit area, using a CAT 951.

Treated waste from the second quarry was deposited in the now empty first quarry and from the third quarry into the second one when excavation had been completed. The final empty quarry was utilised for mainly the disposal of uncontaminated wastes and any surplus subsoil which has been used to create protective bunds.

The choice of machinery was designed to ensure that the operators were kept as far away as possible from the contaminated and reacting materials.

Supervision and control of operations was carried out by three scientists who work on a weekly rota basis from within the team of thirteen. Three was the minimum number necessary to comply with Health and Safety at Work Act requirements and working for only one week at a time enables their sense of smell, which was of vital importance under the conditions prevailing on the site, to recover. Furthermore, since duty at Ravenfield was additional to the normal work load of the officers, the rota system allowed for other essential duties of the Department to be maintained.

Protective clothing, including dust mask, respirator, safety helmet and goggles was provided for all operators working on the site. Breathing apparatus and other safety equipment were also provided on site and all members of the team were fully trained in its use.

Monitoring of the site for toxic gases was carried out by the duty teams in co-operation with Environmental Health Officers from Rotherham District Council.

In the early days of the operation some difficulty was experienced, and complaints received, regarding emissions of dust and smell from the site. This was largely due to a period of exceptionally strong winds blowing across the site towards the nearby housing estate, the nearest house being only 27 feet from the treatment area.

However, improvements were made in the methods of handling and depositing the treated material and with subsequent lighter winds from more favourable directions no further complaints were received.

The entire site was enclosed by a 2 metres high chain link fence surmounted by an angled top with 3 strands of barbed wire and permanent guard duty was provided by an independent security firm.

The deliberate policy of using the minimum number of operatives and the safety precautions which have been established had ensured that no incidents or accidents to staff or plant operatives had occurred.

Comprehensive records of all operations, including still photography, were maintained.

THE PRESENT STATUS OF THE SITE

At the time of writing this paper the site is being landscaped by creating a flat area for a football pitch and additional landforms for other recreational purposes.

LIAISON WITH OTHER AUTHORITIES AND ORGANISATIONS

Before investigation and treatment started very detailed discussions were undertaken with various organisations.

The Department of the Environment at London and the Regional Office at Leeds provided very useful assistance and advice. The aerial photographs were also provided by them. The Hazardous Material Sciences at Harwell were involved in various discussions and their advice has been beneficial. The Water Research Centre also provided assistance in expressing their views on the lime treatment of acid tars. As mentioned earlier the Waste Disposal Department of West Midlands County Council gave invaluable assistance by allowing their Principal Hydrogeologist to work with us during the investigational work and in submitting his detailed report at that stage. Various departments of the County Council were actively involved throughout the operation, the most important being the Chief Executive's Department, the County Treasurer's Department and the Planning Department. The first two departments had arranged approval to bypass standing orders where requirement of any equipment or material was urgently necessary and there was not time to go through the normal procedure.

The Water Authority, National Gas Board, Central Electricity Generating Board and the Post Office assisted us in giving details of their services in and around the site so that no interference would be caused to them by this operation. The Water Authority also gave their view on the presence of the materials in the site and their view on the suggested method of treatment.

The Health and Safety Executive gave the benefit of their expertise on the choice of safety equipment and protective clothing. They also monitored the operation regularly and gave advice where appropriate.

The Fire Service provided a tender at the site during the investigational stage. They also gave training to all the persons involved in the use of breathing apparatus. Similarly, the continual interest of the Chief Constable and his senior officers has been useful throughout the operation and in particular when the residents of the area were evacuated for two days. They provided for cordonning off the area and dealt with associated problems. They also provided a control room for communicating with the standby officers throughout the operation.

The Environmental Health Department, Education Department, Social Services Department and Planning Department of Rotherham MBC and various other departments have been very actively involved. They made all the arrangements, in liaison with County Emergency Planning Department, for informing people about the evacuation procedure and for carrying out evacuation and provision of all necessary arrangements for evacuated families. The Environmental Health Department also carried out independent monitoring of air pollutants during the exercise and provided constant advice.

Ravenfield Parish Council were involved in keeping abreast with the activities and took an active part in a series of public meetings which were held in the area. The meetings were organised by the Chairman and Vice-Chairman (Councillor Sturrock and Councillor Willers respectively) of the County's Environment Committee, in order to keep the public fully informed of progress.

The National Coal Board and NUM officials were also closely involved similarly in the progress at the site as the Silverwood Quarry Colliery is adjacent to the site.

The Area Health Authority provided useful assistance by their constant involvement in all the public and other meetings and by agreeing to provide information on the unusual symptoms (if any) in any of the residents of Ravenfield. The Ambulance Service were also on standby during the operation and had assured us of an ambulance at the site within five minutes of contact.

In my experience I have never come across such a nice and effective co-operation between so many departments, authorities and organisations. We got our reward when the leader column of the local newspaper commended the work.

FUTURE MONITORING OF THE SITE

There is no need to monitor the site as all the materials have been treated. However, the Planning Authority, Central Electricity Generating Board, Yorkshire Electricity Board, Gas Board, Water Authority and Post Office have been advised to consult us if there is any reason to disturb the area.

COSTS AND TIME SCALE

The total cost of treatment and reclamation was approximately £200,000.

The problem became known during 1976 but active investigation started in late 1977. The treatment finished in April 1979. The final landscaping should finish this year.

CONCLUSION

The lessons of general interest that can be derived from this experience are as follows:

1. There must be many other sites throughout this Country where indiscriminate and unscientific tipping has taken place in the past. Pollution Control officers of the County Councils and District Councils should try to collect this information on such sites and make the information available to Planning Officers of the area.

2. Wherever such sites are known the appropriate authorities should attempt to make these sites safe by their direct involvement. Delay in treating such sites will make the problems more difficult.

3. Monitoring of all sites within the authority should be such that the sites which are being used nowadays do not cause similar problems. The calibre of the Pollution Control Officer should be such that they must be able to anticipate problems and take immediate action.

RECOMMENDATION

To minimise the risk from the waste disposal site to the later uses of the site I would make the following recommendations. These are my personal views.

1. Any waste which can cause environmental problems in the after-use of the site should be prohibited from the deposition on the last layer of waste material and the penultimate layer only to be allowed when there is no other suitable sites available, i.e. attempts should be made to keep toxic substances at least 8 feet below the final surface.

2. Discourage disposal of drummed waste for landfilling, particularly liquid waste in drums/containers.

3. Consideration should be given to either inclusion in the site licence a condition requiring the licence holder to obtain a suitable "Environment Impairment Liability Insurance", which should be such that it could pay for any remedial work even after (say 25 years after) the tipping finished, or for the local authority to have a blanket policy to cover for any such sites found in its area. This recommendation is particularly important as once the site licence has been handed over no provisions under the Control of Pollution Act could be used to make the last licence holder take remedial action if the situation arises.

4. All local authorities (Waste Disposal Authorities) should prepare a list of all old and new known waste disposal sites or industrial sites where toxic chemicals may have contaminated the land. It should be the statutory duty (modification of legislation may be needed) of the Planning Authority to consult the Waste Disposal Authority when any planning permission is made for such land or surrounding lands which could be adversely affected by such a site.

5. Similar information should be available through Land Registry Authorities by including an appropriate question in the standard forms so that during transaction of such property information becomes available to new owners. Better still this information should be included in the deeds of the land.

6. Consideration should be given for discouraging building of any houses near potential landfill sites until the site has been filled and properly reclaimed.

CONTACT FOR FURTHER INFORMATION

A. Q. Khan, Principal Toxic Waste Officer, Environment Department, South Yorkshire County Council, Barnsley. Telephone Barnsley 5131 Ext 27.

ACKNOWLEDGEMENTS

I would like to thank County Councillor Sturrock, County Councillor Willers, Chairman and Vice-Chairman of the Environment Committee, Mr. G. A. Thomas, Chief Environment Officer, Mr. I. R. Wingfield and Mr. C. I. Tunaley for the permission to present this paper and for giving continuous encouragement and guidance during the operation. My thanks also to the 12 members of the team, Messrs. Bratley, Hook, Hackney, Whittles, Street, Timmington, Brennan, Hardy, Stanbra, Taylor, Cook and Mrs. Connolly and Mrs. Redgrave for their invaluable and devoted involvement in reclamation of the site. Thanks to Dr. P. Oliver for his invaluable assistance. Last but not least to all those people who contributed towards the success of this project.

2. Discharge of [illegible] of [illegible] water [illegible] intermittently [illegible] in [illegible] containers.

3. Consideration should be given to [illegible] the [illegible] public liability insurance [illegible] should be [illegible] paid for any remedial [illegible] the [illegible] policy [illegible] such sites [illegible] the situation [illegible]

4. [illegible]

5. Further information about the availability [illegible] should be [illegible] in the [illegible]

6. Consideration should be given for [illegible] until the [illegible] satisfactorily [illegible]

CONTACT FOR FURTHER INFORMATION

[illegible]

ACKNOWLEDGEMENTS

I would like to thank [illegible] Council [illegible] for [illegible] the permission [illegible] the success of [illegible]

A NEW APPROACH TO LAND RECLAMATION

Prof. Dr. rer. nat. Friedrich Bölsing

Institut für Organische Chemie der Universität Hannover

The Principle

There are numerous simple chemical reactions in which reaction products are formed, which, compared with the input materials possess a considerably greater surface area. A significant example is the reaction of aluminum alcoholates with water.
For example, if a liquid alcoholate is exposed to air, it gradually turns into a fine-grained white powder under the influence of atmospheric humidity.
Reactions of this kind can, to a certain extent, be used as auxiliary reactions in order to disperse homogeneously other materials.
In the given example a material dissolved beforehand in the alcoholate, e.g. an oily material, would be transferred during the auxiliary reaction into the newly-developing structure of the reaction product, i.e. of the pulverulent solid. The reaction product, now charged with the oily material, retains its powdery qualitics within a wide range of conditions; the oily material can no longer be recognized as such because it is now homogeneously enclosed and adsorbed in every single small particle.

The aluminum alcoholate reaction described is a particularly demonstrative example; for practical application, however, there are many more materials or mixtures of materials available which altogether satisfy the condition laid at the beginning with regard to an increase in surface area by chemical reaction. They are summarized in the designation 'ARC' (Auxiliary Reaction Compound).

The Application

The process has general applicability; it can, as a matter of fact, be used also in the field of environmental protection as far as chemical processes figure here. This applies both to simple dispersion process as well as to subsequent secondary reactions on the basis of an increased chemical reactivity as a result of the high degree of dispersion. The individual process depends solely on the kind of problem.

In the case of the conversion of oils and oil-like substances or mixtures of substances which contain oils and oil-like substances one has to take into consideration some general guidelines.
As was shown in the example of the model reaction, each material which is to be dispersed by an auxiliary chemical reaction must first be completely pre-dispersed in the ARC.

Since almost all auxiliary reactions that can be used in this range of application are based on a reaction with water, it must be distinguished very precisely, whether the oils and oil-like substances mentioned are on their own or do occur together with water.
In the first case, that is, when anhydrous or largely anhydrous materials, e.g. used bleaching earth, waste oils and acid sludges are to be converted; this can be done by anology with the model reaction. If, however, there are oils and oil-like substances in aqueous systems, special precautions must be taken to prevent the applied ARC reacting with water before a pre-dispersive complete adsorption of the oleiferous phases has taken place.
For this purpose, ARCs are used which, on the one hand, are greatly reduced in their chemical reactivity to water but which, on the other hand, can very easily adsorb the substances to be dispersed.
A further special case occurs when, for certain reasons, a further use of the converted materials is not expected, but a harmless disposal is required. Problems of this kind can also be solved generally using the dispersion process. This is for example the case with so-called oilhazards or in the treating of polluted water or polluted soil in general. The reaction procedure of the operation is now illustrated and explained by some practical examples and with reference to the specific circumstances described above.

Starting from the fact that oils in soil are - to the most part - biodegradeable it is obvious that this biodegradation will be effected much faster, if the oil does not occur in form of a closed phase but is most finely dispersed in the soil.
In order to achieve the desired high degree of dispersion the contaminated soil is treated with ARC which contains water-soluble non hydrophobic reaction retarders.
The treatment follows according to the situation given either in a suitable mixing equipment or on the spot, whereby the ARC is worked into the soil homogeneously by means of a simple rotary mixer. The particles of the ARC which because of the reduced chemical reactivity do not at first react with the soil humidity, are thus brought into close contact with the oilphase already present in the soil. The oil is adsorbed and pre-dispersed in the ARC.
By this procedure the blocking of the chemical reactivity is released with the effect that the reaction with the humidity can take place. The semi-liquid homogeneous paste of ARC and oil is hereby transformed into a finely dispersed, bone-dry powder which can be readily distributed into the surrounding soil. It was established in model experiments that 55 % of an oil which had been finely dispersed by chemical reaction after its distribution in farm ground was decomposed in a trial period of 33 days.
For the test a crude oil fraction was used. It was one which could be distributed mechanically and readily in a soil sample; although the other conditions were the same, nevertheless only 27,5 % were decomposed.
Experiments with oils which cannot be distributed mechanically and which, therefore, exist in form of a closed

phase in the soil could be neglected since the biological decomposition here is practically impossible anyway.
Thus, for example, no biodegradation at all can be recognized on a piece of hardened asphalt. After transformation according to the process into a bone-dry fine powder a biological decomposition however starts without more ado; the speed and extent of this biodegradation - under otherwise identical conditions - are only dependent on the composition of the sample.

In the case of oil spills on water the treatment is carried out in the same way as described for the so called hydrophilic treatment of oil in soil.

In case of a thick oil layer on the water surface the reagent is applied in some suitable manner, i.e. by sprinkling or spraying etc.
The reagent, charged with oil, sinks at first to the bottom; this is, however, of no importance, since, after some time a violent chemical reaction starts, whereby the finely dispersed powder which is formed is distributed homogeneously in the water. The oil as a phase can no longer be recognized. Because of the extremely fine dispersion the oil incorporated in the reaction product is subjected to a fast biological decomposition.
It should be emphasized that the finely dispersed powder existent in the water can under no circumstances neither accumulate nor be laid together with greater aggregates, as can always be observed when oils are emulsified with unsuitable emulsifiers, with the result that the emulsion which develops at first breaks again later, e.g. when contacting sand.

It goes without saying that in the two cases mentioned above,'oil in soil' and 'oil on water', only reagents can be used which cannot have negative effects in these systems and which, in particular, do not bring about any concentration of salt.

To what extent some so-called waste materials can be reused through the use of the dispersion process is best illustrated by the fact that it is possible to exploit even those oils in soil for exacting technical purposes.

In a concrete example there were 3.000 - 4.000 cubic metres of oily substances in 9.000 - 10.000 cubic metres of soil on a factory site. An assembly area with a very high bearing capacity was to be set up here. The subsoil consisted mainly of parts of soil in which cohesive parts appeared alongside sandy structures. It is well known that sandy soils are stabilized with bitumen or cement and that cohesive soils are stabilized with lime.
In the given circumstances bitumen did not need to be conveyed from outside to the spot since the soil contained ample (up to 80 % !) oily substances. In order to construct a subsoil with a very high bearing capacity the reagent therefore was combined with the soil-stabilizers mentioned but without bitumen. Layers with a thickness of 30 cm could be worked on. After the soil containing oily

materials had been brought out, the reagent, now a hydrophobic one, was brought up onto this broad layer and hoed in immediately.

Owing to the intensive intermixing the oily substances penetrated into the hydrophobic reagent within a few minutes and after 10 - 30 minutes a chemical reaction began in the whole area in the course of which the oily substances were turned into a very fine bone-dry powder of a brownish colour.

Through the chemical process just described, the effectivity of the components stabilizing the soil is not impaired. On the contrary it is shown that in the whole subsoil a much higher mechanical compressibility can be attained. In this way a supporting layer, 90 cm in height and with exceptionally good technical qualities, especially extremely water repelling, could finally be constructed.

It goes without saying that this material is as well suitable for road construction, especially for the construction of high ways and the like.

Among the recently performed measures some data of two are added for further information.

Bodenuntersuchungen
Versuchsstrecke Marchegg

Ergebnisse der Laborversuche

Bodenbezeichnung		hydrophobes Reaktionsprodukt eines ölverunreinigten Erdreiches (schluffig sandiger Kies)	
Korndichte (g/cm³)		2,45	
Korn-größen-verteilung (%)	Kies	51	
	Sand	34	Abb. 1
	Schluff	)	
	Ton	) 15 +)	
lockerste Lagerung (g/cm³)		1,40	
Porenanteil hierzu		0,43	
dichteste Lagerung (g/cm³)		2,01	
Porenanteil hierzu		0,18	
Wiener Routine-Scherversuch		35,5° 33,5°	
Abb.:		2	
Kompressionsdurchlässigkeitsversuch bei lockerer Lagerung		Abb. 3	
Durchlässigkeit (cm/s)		10^{-5}	
Kompressionsdurchlässigkeitsversuch bei dichter Lagerung		Abb. 4	
Durchlässigkeit (cm/s)		10^{-7}	
Glühverlust (%)		21,3	

+) Schlämmanalyse nicht durchführbar, da Korn wasserabweisend

Wien, 1979-07-30

gez. H. Borowicka

Prof.Dr.H.Borowicka
Vorstand des Institutes für
Grundbau und Bodenmechanik
Technische Universität Wien

Bodenuntersuchungen

Versuchsstrecke Bahnhof Marchegg

Tragschichte aus hydrophobem Reaktionsprodukt eines ölverunreinigten Erdreiches
Schichtstärke 60 cm auf weichem Humus

Ergebnisse der Feldversuche vom 1979-06-22

Punkt	Natürl. Wasser-geh. (%)	Dichte feucht (g/cm3)	Dichte trocken (g/cm3)	Poren-anteil (%)	Lage-rungs-dichte	Modul bei Lastplattenversuch nach SNV 70 317			Modul bei Lastplattenversuch nach RVS Kapitel XXII		
						Laststufe M_E 15,0-25,0	Laststufe M_E 25,0-35,0	Laststufe M_E 35,0-45,0	Laststufe E_{V1} 15,0-25,0	Laststufe E_{V1} 25,0-35,0	Laststufe E_{V1} 35,0-45,0 +)
1	5,9	1,90	1,80	0,265	0,66	4650	5750	5750	3500	4300	4300
1	1,1	1,92	1,90	0,225	0,82	5450	6500	7100	4100	4900	5350

+) (N/cm^2)

Lastplattenversuch des Untergrundes

Modul nach SNV 70 317
Laststufe M_E

15,0-25,0 (N/cm^2) Versuchsplatte bei Spannung von 30 N/cm^2 eingesunken

195

Wien, 1979-07-30

gez. H.Borowicka

Prof.Dr.H.Borowicka
Vorstand des Institutes für
Grundbau und Bodenmechanik
Technische Universität Wien

HOCHSCHULE für TECHNIK BREMEN
Prof.Dr.Ing. Jagau
Laboratorium für Bodenmechanik,
Erd- und Grundbau

Bremen, 28.1.1977
Langemarckstraße 116
Ruf (0421) 5905-330

REFINERY SITE DOLLBERGEN (ARAL)

Chemical treatment of oil-containing and oil-like substances

2. Results

2.1 Plate loading tests (see Apps. 1 - 3)

After the application of the reaction products was largely finished three plate loading tests were carried out, with a plate diameter of 30 cm, in order to estimate the bearing capacity of the resultant soil body.

The test sections were labled as follows:

Test 1 : Surface - approximately 24 hrs after compaction

Test 2 : Surface - straight after re-compaction

Test 3 : 0.6 m below surface - approximately 5 days after compaction

The bearing values, E (see Apps. 1 - 3), evaluated by these plate loading tests are shown in the following table:

test	bearing values (kp/cm^2)			E_2/E_1
	E_1	E_2	E_3	
1	83	265	275	3.1
2	28	145	152	5.1
3	136	422	482	3.1

These results show that the time factor is of great importance when filling in the reaction products. After compaction the chemical coagulation process continues and the consistency of the soil becomes much better due to the continuing water loss.

Moreover, as shown by the use of the vibratory sheepsfoot roller, it will be noted that the optimum compaction is achieved at a depth of about 20 - 40 cm. Consequently, test 3 shows the true bearing values (of the compacted and filled in soil), which rise with the depth and reaction time.

2.3 Density measurement

Undistributed samples were taken from the three test sections for the determination of the density. The results are listed in the following table:

sample	test section	depth (m)	density (Mp/m^3)
1	1	0.15	1.59
2		0.15	1.71
3	2	0.15	1.33
4		0.15	1.35
5	3	0.60	1.42

Because of the oil present the normal determination method of the water content used in soil mechanics is not possible. The densities are of the soil composition found in situ after the filling in.

A mixed sample was heated to remove the volatile components. These were found to have a value of:

$$G_V = 29.1 \%$$

It is in this fraction that the volatile hydrocarbons and the pore water of the oil-containing soil are to be found.

To compare the density established on the project site the material of the mixed sample was tamped in a Proctor mould in the laboratory. In this standardized test (in which the impact numbers and the layers were chosen in the simple Proctor test), the following density was obtained:

$$\gamma_p = 1.42 \ Mp/m^3$$

This value lies near the mean value found on the project site, and, hence, an optimum density was obtained on the site.

2.4 Water absorption test

The potential water absorption of the compacted, filled in soil was evaluated for sample 2. In order to do this, a 50 cm high water column in a vertical cylinder was put on the sample. After a period of 14 days no measurable water absorption was detected for the sample. Only the surface was slightly wet to a depth of about 0.5 mm. Beneath this the sample was completely dry.
These test results show that the soil is practically water resistant.

2.5 Freezing and thawing test

To assess the susceptibility to frost action a cylindrical sample (10 cm diameter x 12 cm length) was frozen to -30°C and then thawed to 20°C, several times. It was seen that there were no changes in the form or dimensions of the sample.

After the freezing and thawing test the sample was put in a vessel containing some water so that it stood about 3 cm deep in the water, and then frozen with the water. Here, also, no changes in the form or dimensions of the sample were detected.

After this, the sample was completely immersed in water, frozen, and thawed again. Again, no changes in its shape and dimensions were observed.

The sample was then broken up. No changes in its structure, e.g. by the formation of ice lenses during the freezing, resulted. Due to the hydrophobic nature of the material water is not absorbed. The compacted soil is therefore frost-resistant.

When the sample was again frozen in its broken state, after the addition of water, distinct ice lenses were formed. It is therefore essential to have sufficient compaction of the filled in material for frost-resistance, as was achieved at the project site.

4. Conclusion

The built in reaction product obtained by the treatment of oil-containing and oil-like materials was well-compacted and well-stabilized by the stabilization steps. Its former pasty to fluid consistency was transformed, by the mixing in of reactants and subsequent compaction, to a semi-solid to solid consistency. After this stabilization and compaction, the soil absorbs no water and is frost-resistant due to its own high, natural strength. Should the soil be used directly for road surfaces, it should, as a precaution, be sealed by bitumen surfacing.

In the plate loading tests, when a period of more than 24 hours after compaction had elapsed, a bearing capacity,

E_2 = 150 kp/cm^2, was calculated.

When a load is applied on the stabilized soil, the subgrade reaction may be expected to have a value of = 2.0 - 3.0 kp/cm^2. The foundation settling values which can be expected are generally in the range of about s = 1.0 - 2.0 cm. As settling depends on the values and proportions of the load and the dimensions of the foundations, these settlings values should be tested more carefully, taking the required supporting material from here as a basis.

(The full expert's report contains further details as well as a description of the tests carried out).

gez. Dr. Jagau
(Prof. Dr. Ing. Jagau)

RE-DEVELOPMENT OF DERELICT LAND AS PART OF THE OPERATION OF THE CENTRAL ELECTRICITY GENERATING BOARD

MR P A SNELL, DIRECTOR, BRITISH ELECTRICITY INTERNATIONAL LIMITED

DR J BROWN, HEAD OF CHEMISTRY, MIDLANDS REGION, CEGB SCIENTIFIC SERVICES

INTRODUCTION

The CEGB burns about 70 million tonnes of coal per year, resulting in the production of 12 million tonnes of ash. Most of this ash is fine material known as Pulverised Fuel Ash (PFA) and has the characteristics of a light silty soil. About 4 million tonnes of this fine ash finds a use in the building and construction industry and the remainder, because of its particular properties, is used for land reclamation. Several hundred sites of varying sizes have been reclaimed using this material, often for agricultural purposes and in this paper some examples are described showing their diversity.

On occasions, the Generating Board has used derelict land and industrial waste tips for new operational installations and an example is mentioned for completeness.

TYPES OF SITE RECLAIMED

Probably, the largest number of available sites for ash disposal are the worked-out sand and gravel quarries lying, for the most part, along river valleys. These generally tend to be comparatively shallow and because they often form part of the river washlands, filling is usually restricted to original ground level to prevent obstruction of the flood plain. When this consideration has not been of importance and where the local authority and the owner have been in agreement, embankments have been raised using clay materials from the overburden or from the pit bottom to increase the usable capacity of the quarries, and filling has continued to well above previous ground level.

Clay workings associated with the brick and earthenware products industries, have provided other important opportunities for reclamation. These are frequently quite deep and can cause their own particular problems which will be touched upon later.

Other types of re-development have involved: (i) The construction of carefully contoured mounds of PFA above the original ground level in Yorkshire and Staffordshire; (ii) The reclamation of salt marsh, again with PFA; and (iii) The restitution for operational purposes of land blighted by other industrial use, eg the dumping of chemical waste.

METHODS OF TRANSPORTATION OF ASH

In general, the most economical system for transporting PFA is by mixing the ash with water to form a slurry containing about 20% ash and pumping the mixture through a pipeline to its destination. A detailed account of the Generating Board's use of hydraulic transport systems is already available (Wright and Brown - 1979).

Although the initial cost of this type of scheme can be high, the running cost is comparatively low. Delivery by pumping has been used for distances up to about 13 km, and could no doubt be carried out successfully at substantially greater distances, but at some stage repeater stations would be required to boost pressure. When the slurry reaches the particular area being filled, the ash is allowed to settle out and the water is decanted and then either allowed to discharge to waste or is pumped back for re-use. If the water is discharged to a natural watercourse, an additional settlement lagoon is interposed between the main area and the discharge point to ensure the high quality of the decanted water with respect to suspended solids.

Sometimes, an entirely closed system of water for ash transportation is installed with the same water being used time and time again, although a small make-up has to be available because some of the water is retained in the ash. The problems arising from this type of disposal are described later.

Where disposal areas are at a considerable distance from the Power Station, it may become cost effective to transport the ash dry in pressurised rail tankers similar to those used for transporting bulk Portland Cement. The wagons are emptied by compressed air at the unloading station at the disposal site and as the ash is discharged, it is mixed with water and pumped to the particular part of the area being reclaimed. It should be emphasised that an additional advantage from both of these disposal methods, is the considerable reduction in lorry traffic on the public highway.

At many of the smaller Power Stations, transportation of ash by lorry has been the most convenient method of moving the material. The ash is usually "conditioned" with up to 20% by volume of water being added to the dry ash for ease of handling and to prevent the ash from blowing during its journey. At the disposal point, the ash is tipped and spread and compacted by tracked vehicle. Much of the ash sold for structural fill - in roadworks and bridge embankments for example - is in this conditioned state with the moisture content carefully adjusted to the optimum for maximum compaction.

Other transportation methods have been investigated and these have included moving conditioned ash by conveyor where only short distances are involved, and the pumping of high density ash water mixes. The former method has been adopted at Barlow for the disposal of PFA from Drax Power Station. The latter method has looked very promising but would require further development.

EXAMPLES OF RECLAMATION SCHEMES

The following examples illustrate the types of site, including transportation and water management, already outlined above:-

Gravel Workings

Worked out gravel pits at Besthorpe take pumped ash from High Marnham Power Station a distance of about 5 km. The pits, which are in the river washlands, are being filled to the original ground level and the lagoons are bunded by the gravel company for the CEGB to a height of about 3 metres to prevent overtopping by flood water and the possibility of ash being washed in to the river during the filling process. The decanted water is allowed to discharge to the river. To date, about 80 hectares have been returned to agricultural use, the crops including cereals in addition to grass.

Ash from Cottam Power Station is pumped some 13 km to Sutton and Lound. Some of the sand and gravel workings on this site are in the river washlands and are therefore being filled to ground level, and as before are protected by bunding during the filling process. Other pits outside the flood plain have been raised to 10 metres above original ground level. The largest of these has a capacity of 1.3 M tonnes. The decanted water in both cases is passed through a secondary settlement pond incorporating a gravel filter prior to discharging to the river.

Worked Out Brick Clay Pits

An important site in the Midlands is the worked out London Brick Company clay pits at Peterborough. Here is an example in which ash from Ratcliffe and West Burton Power Stations is loaded into pressurised rail wagons and taken some 130 km to the site which is being reclaimed to the original ground levels. On arrival the ash is discharged by compressed air, mixed with water to form a slurry and pumped to the particular area being filled. The water for the slurry is supplied from a central reservoir holding 2,000 million litres (450 million gallons), and located within the rail loop.

When the ash has settled out from the slurry, the water is decanted and returned to the reservoir via an intermediate collecting pond. There is no purge but a small amount of make up water is required because the deposited ash retains some water. The make up is obtained from a local "flash" stream. However, to all intents and purposes, this constitutes a closed system. The operation is on a large scale, with about 1 million tonnes of ash deposited each year. To date, about 40 hectares have been returned to agricultural use and ultimately this figure is expected to reach about 400 hectares.

Substantial embankments are raised to contain the disposal lagoons and these in turn become reservoirs as defined by the Reservoir Act and have, of course, to be treated as such.

Geological Cleft

Two steeply sloping valleys at the Devils Dingle in Shropshire, provide a natural disposal site for ash from Ironbridge Power Station. Conditioned PFA is brought by lorry for use in the construction of embankments across the mouths of these valleys, the valley streams being connected and culverted in the bed of one of them. Slurried ash from the Power Station is then pumped in to the lagoons formed by the embankments and the sides of the valley.

In this case, the decanted water is passed through a tower in to the culverted stream previously mentioned and then to two settlement tanks before discharge to the River Severn. The embankments and decanting towers are raised in stages and will give the scheme a total capacity of 3.7 M tonnes over a period of about 25 years.

Landscaped Mounds

In South Yorkshire, at Gale Common, a carefully contoured mound is being constructed with ash from Ferrybridge 'C' and Eggborough Power Stations. The site was chosen because it was low lying and difficult to use agriculturally and was liable to further subsidence when coal was mined from beneath it.

The ash is transported to the site by pumping as a slurry - the maximum distance being about 7 km. When it reaches the disposal area, some of this slurry is diverted to a vacuum filter plant where its moisture content is reduced to 20%. This material, together with waste shale delivered by lorry by the NCB from a nearby colliery, is then used to form the lagoon banks.

The balance of the slurried ash reaching site, is handled in the usual way - pumped to the lagoons, allowed to settle and the decanted water returned to the River Aire via secondary settlement ponds to reduce the concentration of suspended solids in the final effluent to acceptable standards.

The final height of the landscaped mound will be 50 metres above original ground level and will, when the full scheme is completed, ultimately cover an area of some 300 hectares.

The restored land will be partly planted as woodland and the remainder will be returned to agricultural use.

Nearby, at Drax Power Station, the ash is being disposed of at an adjacent derelict munitions dump at Barlow.

The ash is conditioned to approximately 20% moisture content and as already noted, taken by covered belt conveyor to form an artificial mound - again carefully contoured. The material is spread, levelled and lightly compacted as it arrives and as soon as final level is reached, is topsoiled and seeded.

At this site, we have sheep grazing within yards of the disposal point.

The first phase of the operation will cover an area of 48 hectares and the maximum height of the mound will be 29 metres above original ground level. Since 1974, when the first ash was placed, 3 million tonnes have been used in this reclamation.

Reclaimed Salt Marsh

At some coastal or estuary sites, ash has been used to reclaim saltings - the usual method in this case has been by end tipping conditioned ash from lorries. Connahs Quay in the Wirral is a typical example and the land has been returned to agriculture.

Reclaimed Chemical Tip

In the early sixties, the CEGB was faced with the need to build a large switching station in North West Birmingham where suitable sites were both scarce and expensive. Finally, an 11 hectare site was selected which consisted for the most part of two 28 metre deep marl holes half full of grossly polluted water. Nearby, was a 30 metre high mound containing 600,000 cubic metres of chemical waste which formed a notable landmark and eyesore known as Blue Billy. This was moved bodily into the marl holes and topped off with waste from a nearby quarry. The remainder of Blue Billy was used to build landscaped contoured banks around the site which were topsoiled, seeded and tree planted, The switching station covers only part of the site and the remaining area has been made available to the public.

AGRICULTURAL CHARACTERISTICS OF PFA FILLED AREAS

As already mentioned, PFA is similar to a fine silt. It contains the inorganic materials present in the plants from which the coal was formed, but the bulk of the material is extraneous mineral matter resembling the soil in which the coal measure plants grew. Consequently, PFA is similar in general composition to many present day mineral soils.

Before combustion, the coal is ground to the consistency of face powder and is burnt in suspension in air. As a result, the ash produced is largely composed of spherical glassy particles with a particle size distribution similar to that of a light silty soil. About 60% of the particles are within the "fine sand" category (0.2 - 0.02 mm) and about 40% "silt" (0.02 - 0.002 mm) fraction. The material is therefore promising from an agricultural point of view, because the available water capacity of the soil is related to the fine sand fraction. (Salter et al - 1966). On the other hand, its detailed chemical composition, particularly in respect of trace elements, is such that initially it requires careful management. As a result of the combustion process by which it is produced, PFA is completely sterile and virtually devoid of nitrogen. Other plant nutrients are present in acceptable quantities but so are small amounts of elements to which some plants are sensitive (eg boron).

The nitrogen deficiency can be remedied by suitable fertiliser programmes. As the ash contains no colloidal matter or exchange properties, the fertiliser is not easily held and the maxim is "little and often". Alternatively, a slow release fertiliser may be used. As the ash is cultivated over several years, it acquires its own organic content which improves fertiliser retention. Plants which fix their own nitrogen may be successful without fertiliser applications but because of the initial sterility of the ash they should, preferably, be inoculated before sowing with an appropriate strain of rhizobium bacteria.

The most important of the trace elements governing the range of crops that can be successfully grown on new deposited PFA, is boron. However, after about two years, the available boron is reduced to acceptable levels as a result of weathering. Until this stage is reached, the reclamation site should be used for cultivation of boron tolerant plants, for example grass/clover ley. If the ash is laid down from a water based slurry, then by the time the material is firm enough to work it is capable of growing a good grass sward.

Other trace elements are not significant from the point of view of restricting the pattern of use, and extensive studies both within CEGB and elsewhere have, to date, revealed no major problems associated with the uptake of trace elements into plants grown on the ash. (Townsend and Gillham - 1973). These studies have included the health of animals grazing on lands newly reclaimed with PFA.

Although crops can be grown on pulverised fuel ash without topsoil, in general addition of topsoil improves yield and at the same time limits wind erosion which would otherwise affect the ash in much the same way it would a sandy soil. The optimum amount of topsoil is likely to be small. For example, experiments at Besthorpe already reported (Townsend and Gillham - 1973) at a site with a high water table, have shown that in a dry year the best yields of Winter wheat are obtained with only 100 mm of topsoil, well mixed into the ash so that the water retaining properties of the ash still extend to the surface. By contrast, 300 mm of topsoil unmixed, gave results which were no better than those obtained on bare ash. These findings have since been confirmed with other crops.

With careful planning, it is not difficult to restore ash filled areas to full agricultural use with minimum application of topsoil, a material frequently in short supply. On sites which are free draining, cultivation can commence very soon after final filling. When the lagoons are not free draining as at Peterborough, it may take some time before the surface is sufficiently stable to be worked by agricultural machinery. The subject of agricultural restoration is dealt with in more detail in the paper previously mentioned.

ENVIRONMENTAL CONSIDERATIONS

About 1% of the ash consists of floaters, which as the name suggests, are particles lighter than water which consist of small hollow glassy spheroids. These tend to collect at the lagoon outlet and can be blown by the prevailing wind to form a crust along the windward embankment. The floaters can become airborne and create a dust nuisance, particularly in windy weather in a filled lagoon which has been left to dry out. Spraying with water gives a temporary cure but a more permanent solution is to spray with resin emulsions. Soiling and planting provide the final cure.

PFA contains about 2% soluble material including trace amounts of some metals. Consequently, each disposal scheme must meet the requirements of the appropriate Water Authority who will consider, inter alia, proximity to and possible effect upon potable water sources and the arrangements for the return of decanted water to the local river system. A detailed discussion of the disposal of PFA in water supply catchment areas, is already available in the literature. (Brown, Ray and Ball - 1976).

CONCLUSION

The burning of coal for the production of electricity, results in a large quantity of solid waste which has proved to be a valuable filling material for reclaiming derelict areas left by other industries.

Although not large, viewed against the annual consumption of often prime agricultural land by housing and motorway development in this country, we have contributed, so far, 37 square kilometres of reclaimed land to help redress the balance.

ACKNOWLEDGEMENTS

The authors wish to thank the Central Electricity Generating Board for permission to present this paper and colleages in the CEGB for their assistance in its preparation.

REFERENCES

1 WRIGHT, D AND BROWN, J (1979) - Hydraulic Disposal of PFA from CEGB Midlands Region Power Stations -

Paper to the 6th International Conference on Hydraulic Transport of Solids in Pipes, Canterbury, England: 26 - 28 September 1979.

2 SALTER, P J, BERRY, G AND WILLIAMS, J B (1966) - The Influence of Texture on the Moisture Characteristics of Soils III - Quantitive Relationship between Particle Size Composition and Available Water Capacity.

Journal of Soil Science: Volume 17: Pages 93 - 98.

3 TOWNSEND, W N AND GILLHAM, E W F (1973) - Pulverised Fuel Ash as a Medium for Plant Growth -

15th Symposium of British Ecological Society: 10 - 12 July 1973. (The Ecology of Resource Degradations and Renewal).

4 BROWN, J, RAY, N J, BALL, M (1976) - The Disposal of PFA in Water Supply Catchment Areas -

Water Research, Volume 10: Pages 1115 - 1121.

The Rehabilitation and Reclamation of Scottish Colliery and Oil Shale Waste. The problems and potential.

M.J.S. Edwards, Dip LA (Manc) ALI.
Winston Churchill Fellow in Environmental Reclamation 1977.

The scale of the task

The opportunity for the rehabilitation of Colliery and Oil shale waste in Scotland is afforded by successive Governments political commitment to environmental improvement and the consequent opportunities for renewed industrial vigour and regeneration of the economic viability of depressed areas.

Derelict land is the broad definition that allows for the inclusion of Colliery and Oil shale waste in environmental improvement programmes. Ostensibly, this is land that lies in an unproductive and derelict state after the abandonment of some 'industrial' activity.

The Lothian Region, comprises 700 sq. miles (1126 sq. kilometres) centred on the City of Edinburgh and includes three District Councils, Mid, East and West Lothian. The 6,075 acres (2,459 ha.) of derelict land in the Region represents 15.6% of the Scottish derelict land problem, estimated by Central Government to total 39,000 acres (15,783 ha.). Hence 1.4% of the Regions total area is thus defined as derelict land, 61% of which lies in West Lothian, 28.5% in Midlothian, 6.6% in East Lothian, and 3.5% in the City of Edinburgh.

Shortly after Local Government reorganisation in Scotland, 1975, and the inception of the Scottish Development Agency (December 1975), a programme of over 50 sites was drawn up as the basis of a ten year rehabilitation programme. In 1976 the Region submitted an initial five year programme to the S.D.A. comprising 1,098 acres (444 ha.) at an estimated cost of nearly £4.95 million, at 1976 prices, amounting to an average annual expenditure of £1 million. The overall task would be shared between the Region and the Districts, with the Region concentrating on the largest projects and the District smaller, local environmental improvement schemes. As the entire funding of the programme was to come from the S.D.A. the Region requested the granting of an Agency Agreement, under Section 8(4) of the Scottish Development Agency Act 1975, to enable the first five years rolling programme for the rehabilitation of derelict land in Lothian Region to proceed. On the granting of the Agency Agreement the Regional Council set up a multi disciplinary team of officers with the required skill and expertise to carry out the annual programme. Today the Unit comprises seventeen members; Engineers, Landscape Architects, Ecologist, Soil Scientist, Geologist, Researchers, Technicians, Estates Surveyors and Administrative Staff. (Derelict Land Rehabilitation Unit, D.L.R.U.).

The nature of the task

Over half of the area of derelict land in the first five year programme comprises Colliery waste slag heaps. The remaining area is covered by spent Oil shale. Both Colliery waste sites and Oil shale sites are attended by abandoned mine buildings, foundations, and disused railway lines. Much of the Colliery waste is burning or burnt, consequently there are general similarities in the chemical and physical/

characteristics of spent Oil shale and burnt Colliery waste.

The Scottish Oil Shale industry started in 1851 extracting oil by destructive distillation of coals and shale mined in the Bathgate area, West Lothian. The South East Scotland or Lothian Oil shales are found in the upper Calciferous sandstones of Carboniferous age. A larger and more extensive reserve of Oil shale is also known to extend in a broad band from East Yorkshire to Dorset in the Kimmeridge clay formation of Jurassic age.

In 1865 there were 120 works in the Lothians producing burning oil (paraffin). The majority of the Oil shale was won by underground mining using the 'pillar and stall method', with access from inclined shafts. At the peak of production in 1910, over 3 million imp. tons of shale were mined producing:

Gasolines	600,000	imp gals.
Other Distillate	4,400,000	" "
Kerosene	20,000,000	" "
Gas/Fuel Oil	10,000,000	" "
Wax	25,000	" "
Sulphate of Ammonia	54,000	tons

The average yield was 20.4 gallons (90 litres) per ton of oil product per ton of shale. Needless to say, the majority of the material mined remained after processing and this spent shale or 'Blaes' was piled into slag heaps or as they are known in Scotland, 'Bings'. More than 100 million tons of blaes remains today dominating the landscape, particularly in West Lothian.

By 1950 only a handful of mines remained active and in 1962 the Government withdrew the reduced exise duty privilages offered by the Finance Act 1918, and Oil shale mining was no longer considered profitable.

Rehabilitation - Potential and problems

Lothian Regions Derelict Land Rehabilitation Team has built up a close working relationship with the Land Renewal Section of the S.D.A., which administers the programme and approves the design parameters and budget for each scheme. Every scheme passes through numerous stages from concept, design briefing, land acquisition, sketch designs, cost estimations, through to detailed design and contract preparation, culminating with implementation, maintenance and management. Each sketch scheme goes through detailed consultations with both District Authorities and Statutory Bodies.

The Districts are invariably the eventual owners of the land.

Much can be said regarding the importance of various design philosophies, depending on land use requirements, land take, the quality and cost effectiveness of newly created agricultural land, etc./

but for the purpose of this paper I will concentrate on the following points:-

(a) The chemical and physical problems associated with the establishment of vegetation on Colliery and Oil shale waste,

(b) the various options to solving some of the problems encompassed by (a), and the research involved,

(c) the maintenance, management and ecological stability of the vegetation solutions.

<u>The chemical and physical problems</u>

Although much of the Colliery waste is burnt or burning a few sites exhibit the well known plant growth limiting factors associated with many of the coal wastes from English coal fields. Toxicity caused by the oxidation of iron pyrite on these sites is a major problem. Neutralising carbonate or clay minerals are not present in any quantity and the sulphates resulting from pyrite oxidation can be easily detected in both ferruginous mine drainages, and extensive areas of barren unvegetated wastes. Spoil pH varies considerably on each site and little can be done to utilise any weatherhead reserves of a less acid nature due to the practical problems of stripping such thin and irregularly dispersed deposits.

On one site the mean pH is 2.7 with a mean pyrite (FeS_2) content of 1.5%. High potential acidity, measured and assessed on the basis of both pH and pyrite content, is, as elsewhere, artificially neutralised by the application and incorporation with the spoil, of ground limestone ($CaCo_3$). In some cases as much as 70-100 tonnes per ha. are applied. The problems associated with high lime rates are not easy to overcome, and grass and particularly clovers often show symptoms of lime induced iron deficiency chlorosis. However, grass swards do survive and develop slowly and satisfactory cover on bare spoil can be achieved.

The importance of clovers within selected grass seed mixes can not be over emphasised especially in relation to nitrogen and phosphate status and spoil fixation capacity.

In general toxicity problems associated with colliery spoil are well known, and correction practices are well tried and are relatively successful. As a result a greater emphasis is now being given to the physical problems of Colliery waste 'soils'. Particular attention is being given to soil structure, compaction characteristics, soil moisture characteristics, organic matter content, and the roles of microflora and fauna.

Spent Oil shale, blaes, and burnt Colliery waste are invariably alkaline, sometimes saline and always low in available nutrients. Mean pH lies around 7.5. As many of the bings have been weathering for a number of years, conductance lies around 2.5 (mm hos/cm at 25°C) where growth reduction effects are only seen in susceptible plants.

As/

As with Colliery waste, blaes, due to high pH, suffers from immobilisation of phosphates through the formation of insoluable calcium phosphates. In addition trace element deficiencies can also occur and iron deficiency chlorosis symptoms can often be seen in the grass sward. However, the degree of alkalinity does not cause serious plant establishment problems per se, the real problems lie in nutrient availability, low organic matter content and poor 'soil' structure.

Even though a high degree of natural vegetation cover evolves on many shale bings, often it is necessary to remodel a bing to achieve the desirable landform for redevelopment or aesthetic purposes and the cover must disappear. In this event unweathered, 'fresh' spoil; on Colliery waste sites this material is often unburnt, is exposed, presenting a new actively oxidising surface manifesting all the initial pH problems of freshly produced mine spoil. In the case of blaes salinity can once again become a major plant growth inhibitor. In addition, it now appears that much of the engineering works involved in the process of producing a remodelled landform, can seriously hinder the development of a satisfactory 'soil' structure. Work on soil structure and the performance of grasses has shown that not only yield, but root growth and distribution are dramatically affected by increased degrees of spoil compaction. Shallower rooting will inevitably affect grass establishment and survival and several completed schemes already show reduced grass yield. The most serious problems seem to be on Colliery waste, particularly those sites with a high proportion of unburnt spoil.

Oil shale blaes and burnt Colliery waste, however, have different weathering characteristics to unburnt Colliery waste. In general, a non engineered blaes is free draining with an open structure, but once engineered it tends to compact too. In time, freeze thaw action and general particle size degredation aid the process of compaction. Grass root penetration, confined to a surface horizon as a result of compaction is far more susceptable to moisture stress.

Despite extensive pre-seeding cultivation including, different ripping treatments, little benefit is seen in sward density or general condition as soon as six months after seeding. Nevertheless there maybe some real advantages during the establishment stage particularly on unburnt Colliery waste.

Although, several researchers now advocate the use of bulky organic ammendments to resolve some of the soil structure problems of Colliery waste, most of the research work has been carried out on unburnt Colliery waste using topsoil. It is hoped that the principles of organic ammendment transferred to Scottish Oil shales will exhibit similar results.

In July, 1979 the D.L.R.U. was granted a Science Research Council, C.A.S.E. (Co-operative Awards in Science and Engineering) project grant. The aims of the project are:

(a) To examine the use of amendment additions, such as lime and organic matter in the reclamation of acidic colliery spoil from the Central Scotland Coalfields, for amenity and recreation purposes.

(b) To examine the use of selected annual and perennial plant species in the reclamation of colliery spoils as a low cost alternative to high fertiliser and organic matter usage. Particular attention to be given to the examination of the creation of 'soil conditions' for the development of minimal maintenance vegetation establishment.

In the course of this three year research project it is hoped that some of the reasons for poor grass sward establishment and survival and slow build up of 'soil conditions' will be highlighted, and new or corrective site preparation techniques recommended.

Already, in contract, individual site trials, using various organic ammendments are in hand, utilising sewage sludge, chicken litter, weathered pig manure, canal dredgings and peat. Along with the required soil structure, moisture retention etc., improvements, in association with the Institute of Terrestrial Ecology, detailed studies on the interacting roles of microflora and fauna in decomposition process and nutrient cycling are being considered. Root-microbe interactions have been well studied in the U.S.A. with particular reference to the inoculation of devastated soil with endomycorrhizal fungi and coniferous tree seedling inoculation with ectomycorrhizal fungus. It is intended that work in this field will aid in the assessment of the relative values of various organic ammendments in the creation of an active soil microflora. In addition to the known use of leguminous herbaceous and woody species, it is thought that once soil structure conditions are good, soil toxicity problems will take on a reduced significance as a result of a developing soil harmony.

Maintenance and management of vegetation

Great care is taken in the selection and matching of both herbaceous and woody plant species to site conditions, and plants tolerant of the adverse geophysical and geochemical substrate are used extensively. Lists of plants tolerant to both low nutrient status, adverse toxicity, moisture and physical conditions, are readily available in the Reclamation Press and selection of tolerant species is a relatively straightforward exercise.

Ecology and habitat surveys undertaken by the D.L.R.U. have reinforced and added to the available information on the components of tolerant ecotypes. However, aside from the relative success of establishing certain pioneer grassland communities, trees still present many establishment problems.

The majority of tree planting relies on a large number of 'pioneer' species, particularly those with nitrogen fixing capabilities (Alnus spp.) Planted as bare rooted forestry transplants, they generally establish well and quickly provide shelter and nutrients for other species. However, results can be variable, and attention to planting technique, aspect, associated herbaceous layer, and maintenance is extremely important. Often both Foresters and Landscape Architects have been tempted to plant trees which are too large and of a species that is out of place in the pioneer community. Trying to force ecological maturity by planting of species out of the natural ecological succession has worked rarely and this does not really substitute for the passage of time and the natural development of diversity. On the other hand introduction of long life species can be successfully achieved as an enrichment process after say 10 - 15 years of growth, when pioneer trees have reached a conopy closure stage where shelter and a build up of 'soil' and nutrients can enhance establishment.

Often little attention is given to the origin of selected tree species and research work in hand, with the Institute of Terrestrial Ecology, is seeking to establish whether clonal differences between species, especially those species that invade or naturally colonise unreclaimed waste sites (e.g. Salix, spp. Betula, spp.) can be exploited.

The problems of tree establishment, exposure, moisture stress, smothering by long grass, fires and consequent high failure rates, results in costly and repetative replantings. Bare rooted transplants on our sites, in general, do not achieve the higher rates of those planted on less harsh 'soils'. At present the Unit is undertaking trials with containerised transplants using Lodgepole Pine (Pinus contorta) of South Coast origin. Work in both the U.S.A., Scandinavia and the U.K. on commercial containerised planting reveal considerable advantages in establishment rates and general performance. The advantages of providing an immediately available supply of nutrients and water through an undisturbed root system in an individual pocket of planting medium appear to be central in achieveing a high establishment rate and vigorous development. Field trials have shown that with careful planting into established grass swards, (Fescue - Bent - Clover mixtures) of both bare root and containerised transplants, that establishment rates are considerably higher with containerised stock. However, bare root stock can, and does, achieve similar growth rates, but most often it is the second year growing season that shows the containerised stock as the best performer. Longer term differences are yet to be observed, but the establishment advantages do, to a high degree, make containerisation a cost effective alternative to repetative and costly replantings.

An additional opportunity in the pursuance of containerised plantings is seen to be in the use of mycorrhizal fungi. I have already mentioned American work on spoil inoculation and its role in the creation of soil organic matter and the accumulation of nitrogen for plant growth, but the problems and practibilities of introducing mycorrhizal fungi in association with trees planted on a near sterile substrate, have not been adequately investigated in the U.K. The possibilities of emulating successful American work, that has shown that coniferous transplants innoculated with specific ectomycorrhizae have produced root systems with a higher degree of persistance and a greater ability to adapt in adverse substrates, is open to us, and working in association with the Institute of Terrestrial Ecology and others eminant in the Mycorrhizal field, the D.L.R.U. is helping to prepare a research programme.

Recent Government 'cutbacks' have changed the direction of the D.L.R.U's programme towards an even greater emphasis on derelict land that can be rehabilitated for Industrial Development. However, this has not reduced the necessity or potential for vegetation establishment. Although the extent of tree planting may be reduced slightly, the chemical and physical problems of both Colliery and Oil shale waste still present a considerable challenge.

Depite ongoing research in the fields highlighted in this paper and the ad hoc in contract experimental work, solutions to our problems of vegetation establishment will continue to arrive in an intermittent and ponderous manner, not only because of the organic nature of the material under examination but due to the relatively small research effort, and lack of research funds.

The extent of the rehabilitation task is considerable and although successive governments have changed the emphasis and funding of derelict land rehabilitation the necessity for the utilisation of derelict land is likely to continue, especially when one takes into account the fast diminishing reserve of development land. Those rural sites that continue to produce a polluting influence in the environment, whether in terms of toxic drainages, public safety, to severe visual and sociological detrement, are also likely to retain a prominent position in rehabilitation programmes.

However, in order to secure maximum advantage from the rehabilitation work being carried out today, and the techniques and solutions used in the past, an even greater effort should be afforded towards the recording, evaluation and dissemination of rehabilitation practises.

A most useful instrument in pursuance of this objective is the Standard System of Monitoring devised and promoted by the Standing Local Authority Officers Panel on Land Reclamation - Vegetation and After Management Sub Group, in association with the Department of the Environment and the Agricultural Development and Advisory Services. A far greater emphasis should be placed on evaluation of techniques, so that the continuing political commitment to Derelict Land Rehabilitation and the considerable funds made available to us can be utilised to the full.

* The views expressed in this paper are those of the author and do not necessarily reflect those of the Lothian Regional Council, or the Scottish Development Agency.

References

Cameron, I.B., McAdam, A.D., The Oil-shales of the Lothians, Scotland: present resources and former workings. Institute of Geological Sciences, Report 78/28 1978.

Chadwick, M.J., Nutrient problems in relation to vegetation establishment and maintenance on colliery spoil. University of York, 1978.

Daft, M.J. and Nicolson, T.H., Arbuscular mycorrhizas in plants colonizing coal waste in Scotland. New Phytol., 73, 1129-1138, 1974.

Gemmell, R.P., Colonisation of Industrial Wasteland, Inst. of Biology's Studies in Biology No. 80, 1977.

Lothian Structure Plan, Report of Survey, Department of Physical Planning, Lothian Regional Council, 1977.

Lothian Region Structure Plan, Department of Physical Planning, Lothian Regional Council, 1978.

Macleod Mathews, A.F., O.B.E., J.P., U.K. Oil Shales, Past and possible future exploitation. Report to Secretary of State for Energy (Department of Energy), H.M.S.O., 1975.

Marx, D.H. and Barnett, J.P., Mycorrhizae and containerised forest tree seedlings. Great Plains Agricultural Council. Pub. no. 68. 1974.

Marx, D.H., Mycorrhizae and establishment of trees on strip-mined land. Ohio J. Sci 288-297, 1975.

Parkinson, D., The Restoration of soil productivity. The Breakdown and Restoration of Ecosystems, Plenum Pub. Corp., 1978.

Rimmer, D.L., and Colbourn, P., Problems in the Management of soils forming on Colliery spoils, University of Newcastle Upon Tyne, 1978.

Schramm, J.E., Plant colonisation studies on black wastes from anthracite mining in Pennsylvania. Amer. Philoso. Soc. 56: 1 - 194, 1966.

Coal Related Wastes; Present and Future Problems

L. Rosborough

Institute of Planning Studies, University of Nottingham

1. Introduction

The quadrupling of oil prices in 1974 with the ensuing uncertainty over the future availability of adequate supplies appears to have halted the decline of the British Coal Industry and even to have encouraged its regrowth. If coal continues to consolidate its position as a major energy source there will be profound social and environmental consequences for Britain, including the potential problem of increased land contamination.

2. Current Problems Relating to the disposal of Colliery Spoil

The country already has a significant backlog of coal-related dereliction; about 15,000 hectares are occupied by colliery spoil tips (McKechnie Thomson, 1972) and the National Coal Board alone owns 1,500 disused tips. The significance of derelict land in the surrounding community is being increasingly recognised, in particular the economic depression and air of neglect often found in the older mining areas (Barnsley Metropolitan Borough Council 1979).

The National Coal Board is still the largest producer of mineral waste in the country, in spite of its reduced coal output over the last fifteen years. It currently tips about 55 m tonnes of coal per annum whereas the only other large waste producer, the china clay, industry disposes of only 27 m tommes per annum. Total waste production from all other mines is about 3 m tonnes per annum. The land take of the coal industry is also very significant in comparison to other industries. Annual land utilization by the National Coal Board until recently exceeded that of any other extractive industry and is now second only to sand and gravel. Colliery spoil tipping requires 200 hectares of land per year. (McKechnie Thomson, 1972)

The activities of the coal industry present a significant problem to planners in the coalfields and one of the major issues is that of waste disposal. The visual impact of a tip extends far beyond its site boundaries and so a county containing a high proportion of tipped land may have a serious problem. Nottinghamshire, for example estimates that it has about 15 square kilometres of land occupied by tips or committed for tipping with the prospect of this area being increased by the turn of the century.

	1920	1930	1950	1954	1970	1974
waste production /m tonnes	—	7[2]	20[3]	25[4]	60[2]	55[5]
coal production /m tonnes	—	200[2]	200[3]	210[6]	140[2]	122[5]
% waste	1-2[1]	3	9	11	30	31

1 Harding 1970.
2 McKechnie-Thomson 1972.
3 Taylor 1973.
4 Tanfield 1972.
5 Stevens 1976.
6 Energy Policy Green Paper 1978.

Table 1.

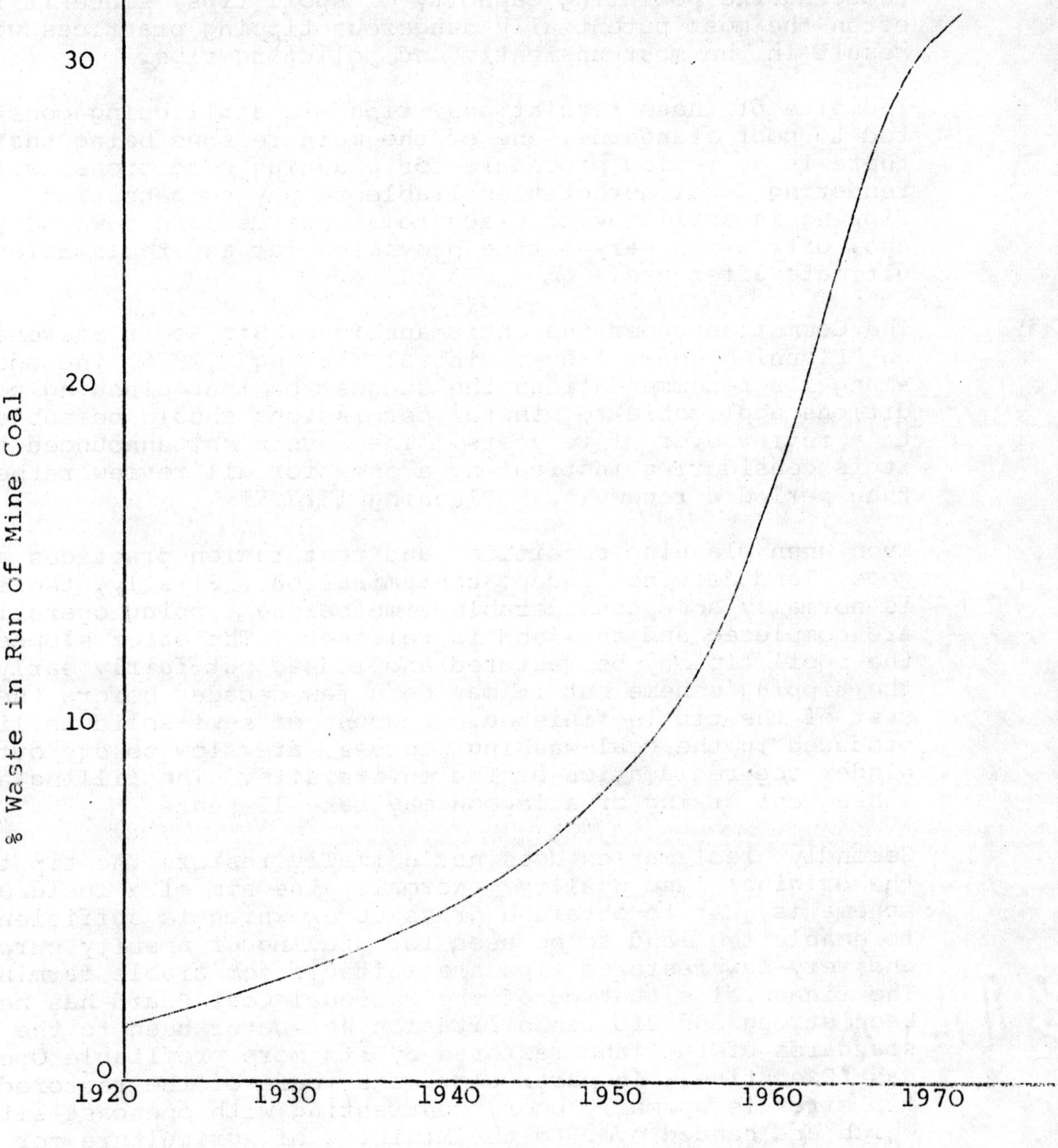

Figure 1. The Changing Composition of Run of Mine Coal.

Data Table 1.

The impact of colliery spoil tips is controlled by two main types of legislation, that relating to tip safety which was drafted as a result of the Aberfan disaster and conventional planning legislation. Safety regulations along with the planning process, are important in improving the appearance and reducing the polluting capacity of spoil tips, since it is often the most potentially dangerous tipping practices which result in the most unsightly and polluting tips.

In spite of these regulations, tips are still being constructed to poor standards, one of the main reasons being that there is no review procedure for planning conditions, without rendering local authorities liable to pay compensation. Tipping is still taking place to standards laid down 30 years ago, often with very little provision for any reclamation or ultimate after-use.

The Committee under the Chairmanship of Sir Roger Stevens on "Planning Control over Mineral Working" (1976) included among its recommendations the suggestion that planning conditions applicable to mineral permissions should be subject to a review every five years. The Government announced that it is considering instituting a once for all review rather than periodic renewals. (Planning 13.7.79)

Even when planning conditions and restoration practices are good, land does not escape contamination. Firstly, there is normally some considerable time before tipping operations are completed and the land is released. The outer slopes of the spoil tip may be restored and leased out fairly early in the tipping scheme but it may be a few decades before the rest of the tip is finished. Lagoons of semi-solid tailings, produced in the coal-washing process, are slow to dry out and hinder the reclamation of the entire site. The filling and subsequent drying of a lagoon may take 12 years.

Secondly, reclamation does not normally restore the tip to the original land quality. Normally the aim of a reclamation scheme is just to obtain a grass cover which is sufficient to enable the land to be used for grazing or amenity purposes and very few restored tips are suitable for arable farming. The financial situation of the National Coal Board has never been strong and tip rehabilitation has never been to the standards of the land restored by its more profitable Opencast Executive. In particular after-care of the restored tip sites is normally poor, contrasting with opencast sites which are handed over to the Ministry of Agriculture for a five-year management period once the over burden has been replaced. Spoil tips are often leased to a local farmer after the top soil replacement and as a result the standard of management in the critical first few years is very variable, depending on the tenant's long-term interest in the land.

Figure 2. The Distribution of Collieries with an Output in excess of 1m tonnes per year.

Data: NCB 1977/1978

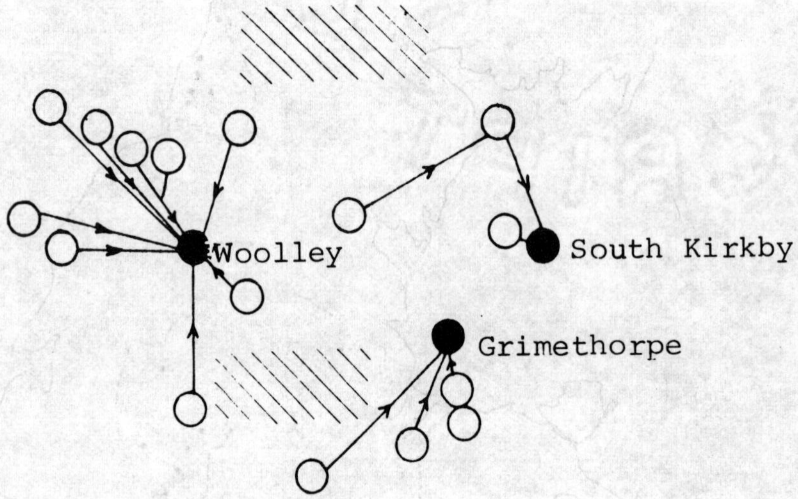

Figure 3. Planned reorganization of the 19 Barnsley Area's Pits, concentrating output at 3 collieries.

Source: Adapted from Mining Engineer January 1979.

3. Future Problems of Colliery Waste Disposal

The NCB's policy documents "Plan for Coal" and "Coal for the Future", (NCB, 1974) and (Department of Energy, 1977) conceived in the wake of the oil crisis, envisage a production of 135m tonnes by 2000 (compare 1978 figure of 117m tonnes). This would mean an annual production of 65m tonnes and 85m tonnes of colliery spoil respectively, in comparison to today's 55m tonnes, assuming a constant ratio of coal to waste production. The latter is not necessarily so. Figure 1 indicates that the ratio of waste to coal has been increasing for many years and particularly since 1950. This is due both to increased mechanization and more efficient washing techniques and with continued investment in mechanization the ratio could well continue to increase. The NCB report "Mining beyond 2000 A.D." (NCB 1977) which considers mining technology for the beginning of the next century, says about mined coal, "Those properties of run of mine coal which affect the ease, accuracy and efficiency by which it can be prepared", (including increased waste content), "have been declining steadily for several decades, and there is no shred of information to suggest that this trend will or can be reversed." Therefore it seems likely that, provided the NCB can meet its targets for coal production, the quantities of colliery waste produced, over the next 25 years, will exceed those stated above.

It also seems likely that over the next few years, coal production will become increasingly concentrated in just a few very productive counties. (North and Spooner, 1978) Already the peripheral Welsh, Kent and Scottish fields are declining at the expense of the more profitable Yorkshire and Midlands coalfield and the NCB has indicated that it intends to direct the investment earmarked for existing pits to those fields where production costs are significantly lower than the national average, again primarily in Yorkshire and the Midlands, (Plan for Coal 1974). Large new developments are planned for Selby and North East Leicestershire - both extensions of the Nottinghamshire - Derbyshire - Yorkshire coalfield. In all, in order to meet the NCB's proposed production figures for 2000 about 60m tonnes of entirely new capacity must be created, this is equivalent to about 60 of the current day largest mines, (Department of Energy, 1978)

In summary the Coal Board's plans for the future appear to involve:-

a. a progressive run down of the peripheral mining areas
b. continued investment in the productive Yorkshire and Midlands coalfields. The current proposals for the Barnsley Coalfield are an example of the type of development to be expected. The NCB proposes to bring the entire output from 19 collieries to the surface at just three handling points (see Figure 3).
c. a new generation of mines, many in areas with no history of mining development.

The implications of this in terms of land contamination are as follows.

a. There is a danger that a progressive run-down of the mining industry in old mining may further increase their already serious problem of dereliction, as mines become exhausted and tips are abandoned.
b. Tipping sites are likely to become scarce in the coal-producing counties, particularly where collieries are reorganised so that output is concentrated at just one pit-head.
c. New mining areas face the problem of losing large areas of useful land for tipping and the danger that if restoration practices prove inadequate the new tips may become tomorrow's derelict land.

In addition to the future problem of colliery spoil disposal there is a danger that as coal becomes an attractive fuel to industrial energy consumers, disposal of coal ash may become a problem. At the moment the CEGB is the main coal burner and it sells most of the ash produced through an efficient marketing board, using the rest for very successful landfill projects. If large quantities of ash are produced by many relatively small diffuse consumers, the ash may not continue to be disposed of as effectively.

4. Reducing Contamination Caused by Colliery Spoil

There is no doubt that the disposal of large quantities of colliery spoil is, and is likely to continue to be, a serious planning problem. What methods can be used to reduce the land contamination caused by spoil disposal?

a. Local Tipping

This is currently the normal method of colliery spoil disposal because of the lower cost but as has been shown, it is not always performed to a high standard. The activities of the Opencast Executive have demonstrated that provided there is plenty of top-soil available, and reclamation and maintenance procedures are adequate, restoration can be very good. The National Coal Board is aware of its poor record in this field and has indicated in its proposals for the North East Leicestershire Coalfield that restoration practices will be different to those utilized at other tips. It hopes to use a system of continuous reclamation so that the land is only out of agricultural production for a short time and secondly to dispense with the need for tailings lagoons by drying fires in filter presses so that the material can be safely mixed with coarse spoil and placed on the tip.

b. Remote Disposal etc

Spoil can be tipped at a distance from the pithead either at a site which is more suitable, or into an existing excavation. In the latter two potential areas of dereliction are removed in one operation. It has been suggested that if the North East Leicestershire project does go ahead, the colliery spoil which would be produced should be transported to Befordshire to be used for reclaiming brick-clay excavations. It has seen calculated that on the basis of 50 years tipping, with no allowance for inflation, remote tipping in Bedfordshire would add about £1.24 to the cost of a tonne of coal produced from the coalfield. (Local authorities - NCB working party, 1978) The total extra cost imposed by remote tipping would be of the order of £8 million per year.

If the proposed development does take place, it is likely that the local authorities would encourage the NCB to dispose of the waste in Bedfordshire but the Board will be unwilling to commit itself to such a large extra expenditure, especially since this would establish a precedent for future developments. It is arguable that government funding should be available for this type of scheme.

c. Waste Utilisation

Colliery spoil can be used as a fill material for earthworks and for highway construction and about 6% of the annual colliery spoil production is utilized in this way. The Coal Research Establishment is investigating other uses including light-weight aggregate, road dressing, mineral wool and concrete blocks. Unfortunately spoil is produced at a rate far exceeding any possibility of using it. (Stevens 1976)

d. Underground Stowage

Underground stowage of coal wastes has been carried out on a large scale in the past and has the advantage of reducing subsidence damage at the same time as limiting the spoil disposal problem. The NCB consider it unecononmic, hazardous and incompatible with modern mining methods. Attention is often drawn to the fact that in Germany waste is still deposited underground and the NCB has been urged to emulate this example. However, the practice is dwindling in Germany for similar reasons to Britain and is retained at a few mines, chiefly in an attempt to reduce what would otherwise be severe surface subsidence damage.

e. Underground Gasification or Liquefaction

In the longer term future and outside the scope of this particular paper new technologies may ensure that waste need no longer be extracted from the coal-face - solving the problem of land contamination by coal wastes, although other environmental problems would still remain, in particular the problem of subsidence.

References

1. Barnsley Metropolitan Borough Council, Coal Study, 1979.

2. Department of Energy, Coal for the future, progress with plan for coal and prospects to the year 2000, London, HMSO, 1977.

3. Department of the Environment, Planning Control over Mineral Working, Report of the Committee under the chairmanship of Sir Roger Stevens, London, HMSO, 1976.

4. Department of the Environment, Energy Policy. A Consultative document, London, HMSO, Cmnd.7101, 1978.

5. Harding, C.P., Plant available nutrients in colliery spoil, Thesis. University of York, 1970.

6. Local Authorities - National Coal Board Working Party, Interim Report, 1978.

7. McKechnie Thomson, G. and Rodin, S., Colliery spoil tips after Aberfan, London, Institute of Civil Engineers, 1972.

8. National Coal Board, Plan for Coal, London, NCB, 1974.

9. National Coal Board, Mining beyond 2000 A.D., NCB, MRDE, 1977.

10. North, J. and Spooner, D.J., The geography of the coal industry in the 1970s: Changing directions?, Geojournal 2.3 255-272, 1978.

11. Planning, Overhaul for Minerals Regs. 13 July, 1979.

12. Tanfield, D.A., Utilization of coal mining residuals in the UK, Katowice, Poland, WHO Training course on solid waste disposal, 1972.

13. Taylor, A.R. Coal mining refuse disposal today and tomorrow, Colliery Guardian Annual Review, 91-96, 1973.

Reclamation of site of Herbicide Factory

Lars-Gunnar Lindfors

Swedish Water and Air Pollution Research Institute

In October 1977 the County Council of Malmöhus commissioned the Swedish Water and Air Pollution Research Institute (IVL) to investigate and present alternative solutions for the cleaning up of the BT-Kemi area in Teckomatorp, Svalövs Community, in the south of Sweden.

BT-Kemi had before the factory was closed in 1977 produced and formulated herbicides and insecticides for 10 years. Waste products and solid wastes contaminated with phenoxy acids and chlorinated phenols had been disposed of within the area. This affected the water quality of the small nearby river Braån.

Crops in market gardens located downstreams had been damaged and the smell, particularly from chlorinated phenols and cresols from the factory was causing inconveniences to the local population.

The contaminated area consisted of the plant itself with buildings and installations originally constructed for suger manufactoring and surroundings in total 100.000 m^2. Drums containing waste products and solid wastes had been buried mainly in an area north of the plant close to the river Braån. The plant itself was in bad shape and heavily contaminated.

The company had before it was closed taken some action to protect the river Braån. Drainage pipes had been installed around the disposal area. The drainage water was pumped to lagoons within the area wherefrom it was pumped to an activated carbon adsorber before beeing discharged. The adsorber was however frequently out of order. Great quantities of drainage water were stored in the lagoons (Fig 1).

Leakage from the contaminated area due to bad performance of the drainage system increased the concentration of phenoxy acids and chlorinated phenols in the river from in average 1 µg/l up-streams to in average 5 µg/l down-streams. Effluents from a municipal sewagetreatment plant situated next door to the factory caused a further increase in concentration to in average 7,5 µg/l due to leakage of contaminated drainage water into the sewers.

One of the first steps was therefore to stop contaminated drainage water to reach the Braån. A bentonite shield (a mixture of bentonite, sand and water) was built down to firm clay at 6 meters level between the river and the disposal area, a distance of 500 meters. The concentration of contaminants in the Braån outside the area was then decreased from 5 µg/l to 1,1 µg/l. The discharges of contaminants from the municipal sewers were then the main problem. In order to solve this a comprehensive investigation of the complicated sewer system was carried out. The leaks could then be traced.

The drainage water from the now sealed area was stored in four lagoons. About 10.000 m^3 of contaminated water were stored at the end of 1977 with a concentration of chlorophenols and phenoxyacids of 150-200 mg/l. Several treatment methods were tested but activated carbon adsorption was found to be the only one giving a satisfying result. The costs were however rather high. Activated carbon could reduce the concentration from 200 mg/l to less than 0,005 mg/l with a carbon consumption of 1 kg/m^3.

A full scale unit with a capacity of 20 m^3/h was installed. Between December 1977 and April 1978 about 35000 m^3 of water were treated from which 2,7 tons of contaminants were captured on 18 tons of activated carbon. The cost was 11 $Sw.Crs./m^3$. The treated water had a BOD_7 content of about 200 mg/l caused by unadsorbed but biodegradable organics. It was therefore pumped to the nearby municipal sewage treatment plant (activated sludge) before being discharged to the river.

A comprehensive program was set up for the control of ground water movements in different layers within the contaminated area. Several hundreds of soil samples were taken and analysed for contaminants and physical properties.

The contaminated soil could be divided into three parts; about 2-3000 m^3 with a higher concentration of contaminants than 500 mg/kg, about 15.000 m^3 with a concentration between 50 to 500 mg/kg and about 150.000 m^3 with a concentration less than 50 mg/kg. The most contaminated soil had a rather low permeability in contrast to the less contaminated one. Several samples were also analysed for dioxines especially TCDD, a highly toxic compound well known from the Seveso case. This compound was however never detected.

It was considered necessary to destruct the contaminated soil and several methods were tested. Incineration of the most contaminated soil and waste in a cement kiln where the formation of dioxines could be controlled was considered This was tested on liquid wastes with the same contaminants with an excellent result. The plant needed however to be rebuilt in order to accept solid wastes. No conventional incineration plant that was able to treat this type of waste at acceptable costs was found.

Tests with composting of the soil together with bark and sewage sludge showed that 90 percent of the phenoxy acids and chlorophenols could be degraded within a month. The quantities of bark and sludge that were necessary to add to the soil would however increase the quantity to be treated more than twice.

The adsorption characteristics of the soil and the ground water conditions did however show that it was possible to treat most of the soil in situ by leaching with water, which was considered as the best solution. The process is slower than the alternatives but cheaper and safer.

The investigations resulted in the following plan for the restoration of the area:

The buildings and process equipment is cleaned up so far that it can be deassemled without any special security arrangements.

Drums are excavated and their liquid content will be incinerated together with other wastes already stored.

The contaminated soil is leached with water. A ditch is excavated in the middle of the area and a perforated pipe is layed down. Water is then pumped through the pipe and enters the soil. The water is then collected by the drainage pipes and treated in the activated carbon plant. (Fig. 2)

The system will have a capacity of 7,5 m^3/h controlled by the permeability of the different soil layers. The low contaminated soil will be clean within 1-2 years. After 5-6 years the concentration of contaminants in the drainage water will be less than 0,5 mg/l. The drainage water will then be pumped directly to a sewage treatment plant without the activated carbon step.

The costs for this program are estimated to 10 million Sw.Crs. The total costs for the project will then be about 20 million Sw.Crs.

Fig. 1

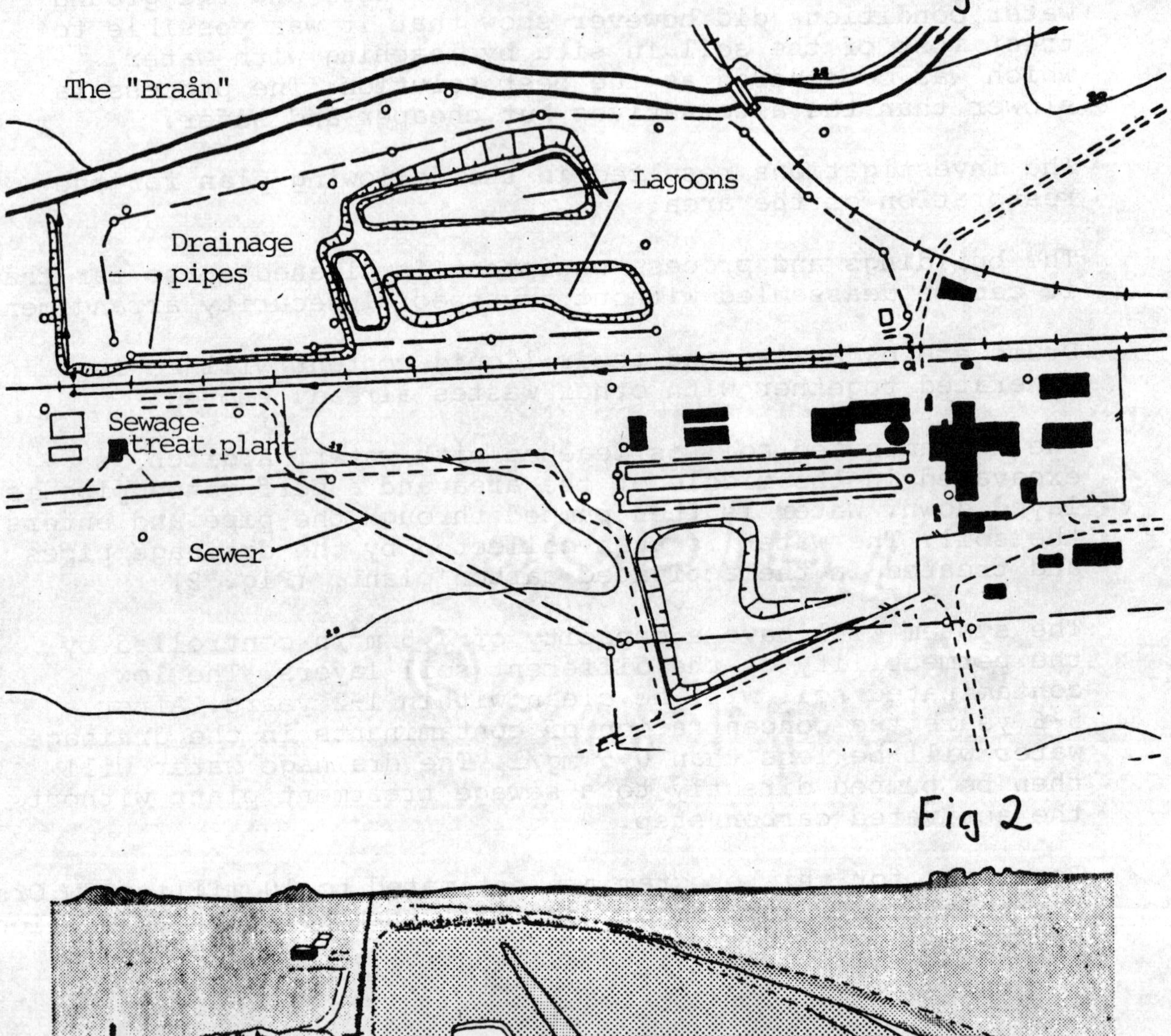

Fig 2

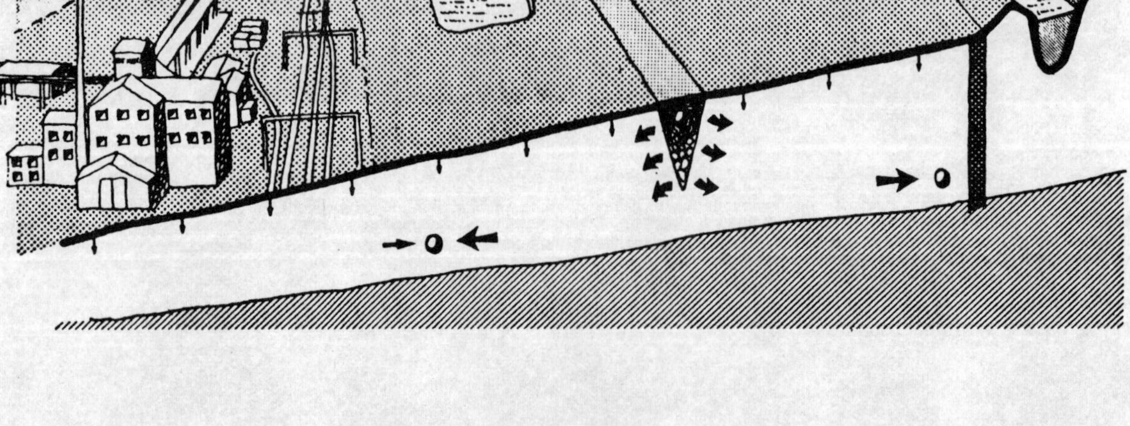

SUCCESSFUL REDEVELOPMENT OF GASWORKS SITES

by

F. E. Dean, B.Sc., C.Eng., F.I.Gas E., F.Inst.E., A.Inst.Pet.
and B. B. Goalby, D. Phil., M.A., C.Chem., F.R.I.C. C.Eng., M.I.Gas E.

British Gas Corporation Headquarters, Marble Arch, London

Synopsis

This paper outlines 18 outstanding examples of successful development of gasworks sites. The examples chosen are typical in that contamination was confined to small specific parts of the site and, within contaminated parts, the distribution of contaminants was not uniform.

None of the examples quoted have exposed any unusual environmental problem regarding either safety or health hazards to workers on the sites or construction and permanence of buildings and structures. These examples represent only a small proportion of the total number of gasmaking sites successfully redeveloped by British Gas Corporation or other developers.

INTRODUCTION

Some idea of the gasworks situation can be formed by recalling that at vesting date, May 1st, 1949 when the gas industry in Great Britain was nationalised, there were 1037 separate undertakings, all making coal gas and some chemical by-products by coal carbonisation on a more or less continuous basis. These works occupied areas ranging from about one acre to hundreds of acres each. Since creation of the British Gas Corporation all gasworks and sites have been consolidated into the 12 Regions listed later, formerly known as Area Gas Boards.

This paper consists of one or more examples of successful development of disused gasworks from each Region. These cases represent only a small fraction of Corporation sites which have been successfully developed.

To show the independence of successful development as regards location, size, end-use and date of site four principles have been adopted in selecting the examples:-

(1) Choice on as wide a geographical basis as possible

(2) Sites of widely different areas

(3) Widest spectrum of type of development

(4) Sites of different ages

Cases are presented of sites (a) developed by the Corporation for its own purposes (b) sold to other developers.

We can say that all of the sites will have contained the following contaminants in varying degrees:-

Tar liquors and residues
Spent oxides
Cyanides
Phenols
Coal and Coke residues
Catalysts
Hydrocarbons
Acids and Alkalies

These materials will have arisen from the gasmaking plant including the following:-

Coal Stores
Retort House
Coke conveyors and stores
Tar separators, tanks and wells
Sulphate plant and stores
Naphthalene and ammonia scrubbers
Gas purifiers
Compressors and governors
Gasholders

However, it should be emphasised that, contrary to popular belief, none of the sites have been contaminated uniformly overall. In all cases contamination is confined to small, specific areas of a site, representing on average no more than 10% of the individual site area.

As an illustration of this Beckton gasworks in North Thames Gas, the largest gasworks in the world, (about 500 acres) was founded in 1868 and since that time has had not only numerous houses on the site but virtually a complete village including a school, church and administrative buildings. The gasworks plant was laid out on a grand scale but despite the amount of gas making activity no report has ever been made of a problem due to contaminating materials. The same absence of problems is true in the cases detailed below and this is remarkable when one considers the wide range of end-users involved. It should be noted, however, that the British Gas Corporation is not thereby lulled into a sense of false security or carelessness regarding the future.

Nevertheless a situation can exist in which the number and magnitude of problems can be proportional to the level of knowledge about contamination. Generally speaking, contemporary thought is that the more one knows about contamination the more one suspects that a hazard problem is present or will arise. Two aspects aggravate this thought and easily lead to exaggerated statements about the problems of developing contaminated land:-

(1) The increasing tendency of analytical techniques to detect smaller and smaller quantities of contaminants and to be more readily available.

(2) An instinctive reaction to consider, as real hazards, trace substances whose occurrence was not even recognised in the days of less sensitive or less accessible analytical techniques.

At the time of the developments described chemical analysis was not as sophisticated as it is now. The consequences of this were two-fold:-

(a) Few difficulties of technical interpretation of analytical information arose

and (b) There was less opportunity for "imaginary" problems to arise.

Over the past few years we seem to have rapidly arrived at a situation where difficulties of statistical sampling and interpretation of analytical data have caused exaggeration of problems to occur and perhaps overmuch attention to be given to matters of academic but no practical significance in the industrial field. The authors, in fact, detect an excessive tendency for some enthusiasts to climb onto the environmental bandwagon and to claim importance for particular courses of study or action, often stimulated by commercial motives, when their significance has not been adequately assessed.

One objective of this paper is to correct exaggerated statements about the problems caused by contaminants on gasworks sites. Short case histories enable this to be done and are provided below. Restrictions on the length of this paper mean that the case histories have to be very brief but the authors hope that there will be a better opportunity at the Conference to illustrate the underlying theme of this paper, both at the presentation of the paper and in the permanent Conference exhibits.

	Gas Corporation Region	Site	End-Use
1.	Scottish	a. Keith b. Broughty Ferry c. Inverness	a. Landscaped Garden b. Award winning housing development c. Car park and light industrial development
2.	North-West	Altrincham	Regional Headquarters Offices
3.	North-East	a. New Wortley b. Harrogate	a. Gas Training Centre b. Wharehousing and NEGAS developments
4.	Northern	Minton Lane North Shields	80 Houses and School on part/part Northern Gas Depot/old gasworks offices now used by Local Authority as Social Welfare Centre
5.	West Midlands	Dudley	Regional West Central Area Office
6.	East Midlands	a. Gainsborough b. Oakham	a. Petrol Station and Supermarket b. Local Authority development of 60 dwellings
7.	Eastern	Albert Road Barnet	Regional Area Headquarters Offices
8.	North Thames	Bromley	Regional Central Stores

9.	South-East	a. Eastbourne	a. Food processing plant
		b. Guildford	b. Multi-storey carpark, flats, Police Head-quarters, Law Courts.
10.	South-West	Torpoint	Small formal park
11.	Southern	a. Southampton	a. Warehouse/Industrial Estate
		b. Brockenhurst	b. Housing Estate
12.	Wales	Grangetown	Regional Control Office, Training School, etc.

SPECIFIC EXAMPLES

Keith site in Banff (Scottish Region) dates from 1905 and was in continuous operation as a gasworks until 1968 when the plant, apart from the gasholder, was demolished. In 1973 part was sold to Keith Town Council, who developed it as a landscaped garden, the construction being carried out by the then Borough Surveyor. The latter found no difficulty in establishing the garden. It is now in the charge of Moray District Council who have not reported any problems, although the garden is open to the public.

Broughty Ferry was first developed in 1842 and became a gas holder station after 1913. The one gasholder left on site at nationalisation was demolished in the 1950's. The site was sold to Dundee Corporation in 1970 and re-sold to the Isabella Agnes Smith Housing Association in 1973. Messrs. James Parr & Partners, Architects were employed by this Trust to design houses for single ladies in the area. The architects reported no problems due to contamination. The main problem was dealing with the gasholder tank foundations and the large diameter gas mains. The development was highly successful as is shown by the grant of a SALTIRE Society Award in 1977.

The interest of Inverness lies partly in the age of the site, which dates from 1826. In 1969 a section of it was sold to the Town Council, who partly developed it for carparking and sold the rest to A.I. Welders Ltd., for extension to their premises. Messrs. Thomas Munro & Co., Architects, Inverness carried out the development. No serious problems due to contamination were reported, although an old tar and liquor well occupied a major portion of the site. Liquor was removed from the tank but the tar deposit was left and the tank filled in with rubble. Piling was adopted in the tank area, the remainder of the building being on normal pad foundation. There is no record of any settlement of the building, either generally or differentially between piled and pad foundations.

North-West Gas now use Altrincham site as the Regional Headquarters (Welman House). It started as a gas-making works in 1847 and closed in 1957. In the early 1960's the site was selected for the new office complex, cleared, levelled and piled. All foundation work was in sulphate-resistant cement. Welman House, which is a very large building, was occupied in 1964. It can be stated that the residues of 110 years of gasmaking did not give rise to problems during building or after.

The Altrincham development was given the North West Regional Civic Award for Architecture by the RIBA.

New Wortley is now the home of the new North-East Gas Training Centre and Customer Service Depot. The original gasworks occupied some 16 acres. The whole site was the subject of major road works involving land exchange and construction of a large roundabout, placing half of the land on an island site. No problems due to contamination (affecting buildings or health of individuals) were encountered.

At Harrogate the original gasworks occupied about 16 acres and some 6 acres were sold for warehousing after demolition of the gasworks plant and two gasholders. The purchaser was Commercial Development Projects Ltd., Elland, Yorks. There have been no reports of contamination problems during or after redevelopment. The remaining 10 acres have been developed by North-East Gas for gas distribution purposes and for a new Service and Distribution Centre.

An example from Northern Region is Minton Lane, North Shields. The gasworks there was on a 14 acre site. After closure some of the land was retained by Northern Gas and redeveloped as a low pressure gas storage and governor complex with a Distribution and Customer Service Depot. The latter was completed in 1974. 10 of the 14 acres were sold to the Local Authority and 80 houses and a school have been built. The old gasworks offices were taken over by the County Borough of Tynemouth, redecorated and are now in use as a Social Welfare Centre. British Gas Corporation is unaware of any particular problem encountered during or after development, which can be attributed to contamination.

West Midlands Gas has its West Central Area office located on Dudley site, where 250 staff are employed, together with 100 employees in the adjacent Distribution Depot. Gas manufacture began on the site in 1820 and continued until 1963, when the gasmaking plant was shut down. Demolition commenced some 3 years later. Redevelopment started in August 1971 and the new offices were occupied in June 1973. The site area is some 5 acres. No very specific information on contamination was available at the time of the development but nothing was revealed which led to any modification of the intended end-use or layout.

Tar and liquor wells were reported to have been emptied at the time of works closure, but filling of one well was found to be incomplete because of the arched structure of its roof, which partly inhibited entry of rubble. Some 25,000 gallons of tar were taken away by a waste disposal contractor. About 20,000 gallons of supernatant contaminated water were pumped to foul sewer under existing Water Authority consent. The liquid tar was disposed of by an authorised waste distribution contractor who also buried the very viscous sludge in a marl hole. All disposals were in accordance with the Deposit of Poisonous Wastes Act 1972.

No specific remedial measures were employed and no unusual construction methods were involved. The major part of the office block had been built when the tar well was found partly unfilled in May 1972. Large foundations of the old gasworks buildings were not taken up but were used advantageously as bases for the new building. No problems manifested themselves during or after development. Some small quantities of spent oxides remain buried on the site.

A lesson for the future from this development is that tar wells may be discovered unknown or unfilled. Additionally, existing large in-ground obstacles, e.g. foundations, can be turned to the developer's advantage.

The Gainsborough works in East Midlands Region has been developed as a petrol station and supermarket. It dates from 1856. Gas making ceased in 1962 and the holders and buildings were demolished by 1968. In 1972 the major part of the site was sold off, approximately 2000 sq. yards to the Lincolnshire Co-operative Society Ltd., and approximately 5000 sq. yards to Graycourt Properties Ltd. According to the records there have been no contamination problems.

Gasworks were established at Oakham in 1840 and gasmaking continued until 1955. Redundant land was sold off at intervals, the major part in 1972, to the Urban District Council for housing development. Sixty dwellings built for the Local Authority are now on the site. The Council paid, in addition to the purchase price, a sum of money to divert all mains out of the land and to relocate the governing equipment. They were also required to fill in all holes left by removal of the gasholders. Again no problems due to contamination have been reported, either during or after development.

The example from Eastern Region is Albert Road, Barnet, where gasmaking was carried out for 111 years until the works shut in 1965. It is now the site of the Outer Metropolitan Area Headquarters of Eastern Gas and of the Transport Depot. After all these years of gasmaking the ground would have been contaminated fairly strongly and irregularly. There is little specific information on contamination available but examination for sulphate was carried out and the levels were thought to be too high even for sulphate resisting cement, although the old concrete foundations were found to be in good condition. In view of this finding one might perhaps speculate that normal building methods could have been used but in the event an unusual piling technique was used with complete success and no problems have subsequently arisen. This method employed a technique developed by G.K.N. Ltd., - Vibrated Aggregate Piles. Briefly a hole is sunk using a heavy metal vibrator column, this is withdrawn and stone aggregate is poured in. This is then vibrated to compact it against the surrounding soil. The process is repeated as necessary.

Bromley Works is a site selected from the North Thames Region. Development involved demolition of a gasmaking plant and the construction of the present Gas Central Stores. During demolition, tar on the site was used to fire works boilers and tar emulsions with more than 50% water were transferred to another site. Tar sludge which was difficult to pump was taken away by a demolition contractor. Spent oxide was sold to a continental firm. The development took place before today's legislation on disposal of toxic materials had been enacted. It is noteworthy that no problem due to contaminants has arisen.

Eastbourne site, where gasmaking began in 1845 and stopped in 1965, is located in South-Eastern Region, and is a particularly interesting example in that a food processing plant owned by Birds Eye Limited has been built on the site of the former gasworks. During demolition and development between 1969 and 1971 some underground storage tanks with liquid sludge in them were found and were removed to a tip in Essex by a waste disposal contractor. Apart from this this building contractors were told not to remove any material from the site and were instructed to carry out all underground concrete work in sulphate-resistant cement laid on polystyrene sheeting. A food processing plant might be anticipated to be unusually sensitive to contamination, but no problem attributable to this has been reported.

South-Eastern Region also includes Guildford, where a multi-storey car park with flats over it was constructed on the contaminated part of the former gasworks. Gasmaking began there in 1860 and finished in 1963. Redevelopment began in 1973 and the major part of it was completed in 1978. Prior to construction of the car park and flats the contaminated material and soil were excavated and removed to a tip by a waste disposal contractor and there sealed in a clay lining. The Police Headquarters and the Law Courts were also built on the old gasworks site on non-contaminated ground. No problems due to contamination have been reported with the Guildford site.

Torpoint in South-West Region has now been developed as a small formal park, the land having been leased by the gas company from the Antony Estate Company in 1929 on a 60 year lease. The site was handed back by British Gas Corporation in 1970. No particular problems due to contamination were experienced during the making of the park and nothing untoward has been reported since the park was completed and used. Torpoint gasworks was only a small undertaking carbonising about 1400 tons of coal a year and serving 820 domestic customers. It generated about 13,000 gallons of tar per year and 1,300 tons per year of coal and breeze.

Southern region has two examples typifying successful development. Southampton, site of a gasworks for many years, was leased to a firm of builders in 1972 and a warehouse/industrial estate has been erected on the site. The estate is now thriving and no problems have been reported. The Region included their standard disclaimer clause in the leasing documents. The Brockenhurst site was sold in 1962 and there is now a small distinctive housing estate on it. The Corporation knows of no problems with this site.

Grangetown Gasworks near Cardiff in Wales Region started gasmaking in 1821. Increased demand led to the building of a larger works in 1865. Various major additions of plant were made until gasmaking ceased in 1972. At this point an appraisal of the situation regarding waste disposal was made. About 3,000 tons of spent oxide were removed and tipped in a reclaimed area in the City boundary. Tar emulsions were incinerated. About 100 tons of catalysts were re-sold to the manufacturers. Liquor from the Stretford Process for hydrogen sulphide removal was mixed with the tar emulsions and incinerated.

The site now consists of a South Area Control Office for Wales Gas, together with a large Training School, car parks, pipe storage areas, Distribution training areas and expansion space.

DISCUSSION

In the above examples the authors have tried to summarise the main emphasis of the presentation and the status of the sites. Basic tests, such as soil acidity, presence of sulphates and building strength tests will have been carried out for the sites and in some cases systematic drilling for core samples and analysis will have been conducted by the developers. In all cases of land sale, prospective purchasers were told that the site had been a gasworks and that it would therefore contain some or all of the contaminants listed at the beginning of this paper. They were also warned of the presence of underground tankage and pipework, the exact layout of which was unknown, in some instances, even to British Gas Corporation (due to age of a site or inadequancy of its site plan). In most cases the Region of British Gas responsible for sale of a site would have included a disclaimer clause declining responsibility in the event of a problem, attributable to contamination, arising during or after development work.

In developments of this type contributions and involvement by practitioners in different disciplines are clearly indispensable in arriving at no problem situations. The British Gas Corporation is very pleased to have the opportunity of publicly thanking all those in the public and private sectors who enabled these developments to be carried out so successfully.

Due to restrictions on the length of this paper it has not been possible to give a detailed account of what measures were adopted to reduce, if necessary, contaminant occurrences. Remedial measure have to be tailored to the requirements of specific sites. Only one of the examples, Albert Road, Barnet in Eastern Region, has involved any unusual building method.

We do wish to emphasise that no gasworks site is uniformly contaminated. Where contamination is present in soil it is usually clearly visible, either in terms of unusual colours or textures or lack of vegetation. In the great majority of cases site plans of gasworks enable one to identify those areas in which contaminants would be present. The main problem in gasworks site development arises, in fact, not from contaminants as such, but from those hidden tar wells and pipework, the existence of which is not evident from the site plans of very old works or the memories of former employees.

Generally speaking contamination considerations would point to the desirability of industrial/commercial development on a former gasworks site, rather than residential development, although, as indicated above, there is no evidence from the specified sites to show contamination problems in either development category.

Experience from the developments specified above shows that attention to the following points will be of assistance in the future:-

1. In cases where it is vital to remove toxic materials this should be done as early as possible because costs of removal have risen disproportionately with time because of shortage of reliable landfill sites. It should be noted, however, that in the opinion of the British Gas Corporation many toxic materials are best left in sites undisturbed and built over.

2. It is essential to be fully acquainted with the consent conditions established by Local Authority Environmental Officers, Water Authority Pollution Officers and the like.

3. As much information as possible about the history of a site should be collected, prior to redevelopment, from primary sources such as company records and secondary sources, such as old employees.

4. During site levelling prior to building, spreading of uncertain material over a wide area should be avoided. Back-filling of part-empty tanks should also be avoided until all liquid has been removed.

5. It is advantageous to know, if possible, future site development plans before demolition and waste disposal are started.

The authors wish to thank all those in British Gas Corporation who have supplied information and helpful criticism during the preparation of this paper.

SPENT OXIDE POLLUTION ON FORMER GAS WORKS LAND

David C. Wilson and Brinley C. Hudson

Environmental Safety Group, Harwell Laboratory, Didcot, Oxon OX11 ORA

ABSTRACT

Much of the contamination on gas works sites has its origin in the "spent oxide" produced during gas purification. This material contains high concentrations of free sulphur, sulphates, cyanides and acidity: levels of sulphide and phenols are generally low. An experimental investigation of its leaching characteristics, gas generation and microbiological activity was undertaken, primarily to assess the acceptability of any wastes removed during redevelopment for landfill disposal. It is concluded that the co-disposal of spent oxide wastes with domestic refuse ie environmentally acceptable, provided that certain simple conditions can be met.

INTRODUCTION

The redevelopment of redundant gas-works land is an attractive proposition as most sites are located near the centre of urban areas. However, before redevelopment may proceed some remedial work will usually be required due to pollution of the land. From a review of gas making practices and of observed pollution at a number of sites, the major sources of contamination have been identified as "spent oxide" from the gas purification process and by-product tarry residues (1). In some cases identifiable deposits of waste material may be present and should be removed for off-site disposal. More generally the problem will be one of contaminated soil, when each case must be treated on its merits.

This study of spent oxide wastes was commissioned by the Toxic Waste Management Section of the Department of the Environment. Its purpose was to assess the acceptability of any wastes which require removal for disposal at a landfill site, but the results are also relevant to the behaviour of contaminated soil in-situ.

HISTORICAL BACKGROUND

The carbonisation of coal to produce gas, initially used primarily as an illuminant, is generally attributed to William Murdoch during the late 18th century. The process was carried on until the 1960s when it was superceded first by hydrocarbon reforming processes producing gas from various petroleum distillates, and ultimately by natural gas.

The carbonisation was accomplished by heating the coal at about 1000°C in sealed retorts or coke-ovens. The heat was provided by burning producer gas made by passing air through red-hot coke; the manufacture of water gas concurrently with carbonisation by steaming the charge was also practised. A ready supply of coke was available in the retorts after carbonisation. The crude gas was purified in a complete series of operations, depending on the by-products which were required. Tars and ammonia were removed in a series of water condensers, precipitators and scrubbers; naphthalene, benzole and organic sulphur could be removed simultaneously in an oil scrubber, or, as economics dictated, could be removed separately.

The final purification step aimed to reduce the hydrogen sulphide concentration in the finished gas below the statutory limit of 0.7 ppm. Early processes used slaked lime, the primary reaction being

$$Ca(OH)_2 + 2\ H_2S \longrightarrow Ca(HS)_2 + H_2O$$

This process was irreversible; large quantities of "foul lime" being generated for disposal. Weathering caused the evolution of hydrogen sulphide gas and thus a smell nuisance.

This and related problems led to the replacement of slaked lime with an alternative absorbent, hydrated iron oxide, towards the end of the nineteenth century. The oxides first used were chemically prepared, but these were soon superceded by naturally occurring "bog iron". This is a formation of α-ferric hydroxide monohydrate in a matrix of organic, peaty material: it combined chemical scavaging efficiency with a porous open structure which caused little pressure drop when used in thick layers. It had the particular advantage that the gas purification reaction

$$\alpha\text{-}Fe_2O_3.\ H_2O + 3\ H_2S \longrightarrow Fe_2S_3.\ H_2O + 3H_2O$$

could be reversed by aerial oxidation

$$Fe_2S_3.\ H_2O + {}^3/_2\ O_2 \longrightarrow \alpha\text{-}Fe_2O_3.\ H_2O + 3S.$$

Regeneration was frequently accomplished in-situ by an air bleed into the crude gas but this did not allow total regeneration. The oxide was removed from the purifier boxes and laid out in the open air in thin layers, or with some degree of turning, before being reused. On some larger works, extensive oxide handling plants were used to water, crush and mix different batches in order to achieve maximum utilisation. The oxide was considered "spent" when it could no longer absorb hydrogen sulphide. The free sulphur content of spent oxide varied between 30-60%, depending on the quality of the original oxide and the efficiency of oxide regeneration.

In favourable circumstances, the spent oxide was a saleable by-product. It was roasted in pyrites burners, the sulphur being oxidised for use in the Chambers process to manufacture sulphuric acid. The roasted oxide could be reused, although it was physically and chemically inferior to bog iron. To improve its performance as a purifier it was activated by weathering or adding a mixture of lime and ferrous sulphate. The organic support for this roasted oxide was frequently sawdust of some kind.

The pollution potential of spent oxide is associated largely with two groups of side reactions. Any hydrogen cyanide or cyanogen in the crude gas also reacted with the iron oxides and iron sulphides to form principally ferric ferrocyanide or Prussian Blue, and ferric thiocyanate, which is blood red in colour. The product $Fe_2S_3.\ H_2O$ also tended to decompose, particularly if the temperature rose above 35°C and/or acid conditions prevailed, to give Fe_8S_9, which on oxidation yielded ferrous sulphate.

Spent oxide contamination on gas works sites arose from a number of practices:

- aerial oxidation by spreading in thin layers on the ground
- spillage
- eventual disposal of inert or unsaleable spent oxides, which were used for example as hardcore in road and car parks
- disposal of oxides in process when the works closed. These may be present as deposits of "neat" spent or part-spent oxide, or may have been spread over the site.

Various processes using scrubbing towers and liquid absorbents were developed as alternatives to iron oxide for hydrogen sulphide removal from gas. Although commercial plants were built, the predominant position of iron oxide was never seriously challenged.

THE POLLUTION POTENTIAL OF SPENT OXIDE WASTES

Selection of samples

Samples of spent oxide were obtained from the Beckton and Uxbridge works of North Thames Gas. These samples were all neat oxides left in process when the works closed. Analytical data for selected samples are shown in Table 1. For comparison, data are also shown for bog-iron and for selected samples of contaminated soil from gas works sites. Contamination levels in the spent oxide samples were about 36-60% free sulphur, 2-3% sulphate and about 3-6% total cyanide, mainly as thiocyanate and complex cyanides. The cyanide concentrations are not wholly consistent; both these and various organic analyses were hindered by interference effects. The soil samples show lower levels of free sulphur and cyanides, but the sulphate concentrations range up to 20%. These high sulphate concentrations are presumably due to oxidation of free sulphur, perhaps microbiologically, in the soil. Most of the samples studied were acidic, with pH values in the range 2-7. Heavy metal concentrations in the spent oxide samples are moderate, and can be accounted for solely in terms of impurities in the original bog iron. Levels of zinc, nickel, chromium, lead, and copper in the soil samples are rather higher: the first three possibly have their origin in the various catalysts used in gas works and the latter two in the piping. The soil samples also showed evidence of contamination by tar residues, with concentrations of coal tar derivatives of up to 2%.

From these results, it was concluded that the most significant pollutants from spent oxides are sulphate ion, which makes building work difficult through corrosion, particularly of concrete; low pH; various cyanides; and possible generation of hydrogen sulphide or hydrogen cyanide gases. Experimental work was confined to three samples of neat spent oxide from Beckton, with free sulphur concentrations of 36, 45 and 56.5% (Table 1); the pollutant levels, with the exception of sulphate, are generally higher than for contaminated soil, so that the results should represent the worst case.

Laboratory tests were performed to assess the leaching of pollutants from the spent oxide samples, and to measure the resultant gas generation. A field test to provide some confirmation of the leaching experiments was carried out at Southampton gas works. A dump of some 4,000 tonnes of oxide was accumulated at Southampton from various Southern Gas sites. A trench through this tip and three feet down into the underlying clay was dug using a mechanical excavator. Five samples were taken in a vertical line, three in the tip and two in the sub-soil.

Leaching tests

The experimental procedure used to examine the leaching behaviour of the spent oxide under landfill conditions has been developed by the Waste Research Unit. Three leaching fluids were used: distilled water, real leachate from a domestic refuse landfill and 5000 ppm acetic acid. Repetitive extractions using samples of the three selected oxides were carried out by shaking 50g of oxide with a single bed volume (approximately 50 ml in this case) of the leaching fluid, until solution equilibrium was attained; the resulting leachate was filtered and analysed, and the extraction repeated with a fresh aliquot of leaching fluid.

The leaching results were broadly similar for all three samples of oxide and all three leaching fluids, and are illustrated in Fig. 1 for just one case, that of a "typical" spent oxide, containing 45% free sulphur, with real domestic waste leachate. The only significant differences between the tests were for pH, where the buffering capacity of the real leachate resulted in a pH > 6, higher than for either distilled water or acetic acid. The results for the more acidic high sulphur oxide are also shown in Fig. 1; even in this case the pH had risen to 5 after 5 consecutive bed volumes had been leached.

Sulphate ion concentrations in the leachate show a maximum about 7000 ppm, while the maximum for total cyanide is about 7-800 ppm, accounted for almost entirely by soluble thiocyanate; after 5 bed volumes have been leached these concentrations have fallen to about 4% of the maximum level. Sulphide concentrations were all insignificant, < 1 ppm.

Larger scale tests using spent oxide from the Beckton works are currently in progress; the Waste Research Unit are using a 15 cm diameter column of pulverised domestic refuse overlying a layer of oxide while the Water Research Centre have incorporated about 40 tonnes of oxide into a simulated domestic waste landfill experiment at Edmonton in London. While these results are awaited, the results of our laboratory scale experiments may be compared to those from the simple field test at Southampton.

Samples taken from Southampton were each centrifuged at 4,500 r.p.m. in an MSE Minstral 6L for one hour to remove interstitial fluids. The centrifuged solids were then subjected to repetitive extraction by distilled water to remove the soluble species. Analytical results for the interstitial fluids and for the initial water extraction (which gave maximum pollutant concentrations) are shown in Fig. 2. Sulphate concentrations below the tip are still significant, with a total water soluble concentration 3 feet down of 3,600 µg/g of soil. The pH values rise sharply at the base of the tip, but are still around 3-4 in the soil. The cyanide concentrations are both low and fairly constant with depth;

however, in the acid conditions of the samples both cyanide and thiocyanate degrade rapidly, so these values are very unreliable. Sulphide concentrations were not measured; phenol levels were low, particularly below ground level.

It may be concluded that the most significant species leached from spent oxide are sulphate, thiocyanate and hydrogen ions. The results of further work currently in progress will allow confirmation of this conclusion, and will allow quantification of the attenuation of these species by domestic refuse.

Permeability tests

A sample of Beckton spent oxide was tested in a laboratory permeability cell, the permeability being measured as between 0.06 and 0.11 mm/second. This is well in excess of the average daily rainfall rate in the UK of 2 mm/day, but is low enough to allow relatively high concentrations of pollutants to be dissolved from thickly deposited layers of oxide: it is estimated that equilibrium concentrations of pollutants could be achieved within 1 metre depth of oxide.

Gas generation

Gases evolved during the laboratory shaker tests were passed, using a nitrogen purge, through a Draeger gas-detector tube for hydrogen sulphide detection and then bubbled through 50 ml of 2N sodium hydroxide solution for hydrogen cyanide collection.

No hydrogen sulphide was detected, even from the more acidic shaker experiments. This test does not, however, take into account any possible gas generation by bacteriological processes on the sulphur or sulphate content of the spent oxide; anaerobic bacteria commonly found within domestic refuse landfills reduce sulphate to sulphide.

Small quantities of hydrogen cyanide gas were evolved. The maximum collected was equivalent to 0.3% of the free cyanide in the solid sample, or to a volume of 2 litres per tonne of spent oxide, evolved over a period of several months. The risk to landfill personnel would be minimal.

Microbiological tests

A sample of spent oxide from Beckton and all the samples from the Southampton tip were examined bacteriologically. Small quantities of each of the spent oxide samples were introduced aseptically to flasks of nutrient medium and incubated at 25°C for 7 days, after which stained slides were prepared for microscopic examination. To confirm these observations, samples of the inoculated medium were incubated in flasks of fresh nutrient medium for 24 hours and slides were again prepared. A further sample of Beckton inoculated medium was incubated in a flask of fresh nutrient medium under anaerobic conditions at 25°C for three days.

In the aerobic cultures, microscopic analysis indicated that growth occurred both in the Southampton soil culture at 8 ft. down and in the Beckton spent oxide culture. The organisms were identified as yeast cells, gram-negative cocci and positive rods: the Beckton sample exhibited a greater density of organisms than the Southampton samples.

The samples of cultured medium which were cultured further exhibited turbidity due to bacterial growth in all cases, except the spent oxide sample from the base of the tip at Southampton.

The cultured Beckton medium grown on under anaerobic conditions turned turbid after 3 days incubation. The organisms identified were as for the aerobic culture but the population density was very much reduced. This is indicative of a facultative response by the aerobes in the change to anaerobic mode.

DISCUSSION AND CONCLUSIONS

This study was concerned principally with contamination on gas works sites which has its origin in the past use of iron oxide for gas purification. Its purpose was to assess the acceptability of any identifiable deposits of spent oxide wastes, or of soil contaminated with spent oxide, for disposal in a landfill site. The major contaminants in spent oxide are free sulphur up to 60% w/w, sulphate ion up to 3% and cyanides, mainly thiocyanate and complex cyanides, up to 6%. Soil contaminated with spent oxide generally contains less free sulphur and cyanides, but may contain more sulphate, heavy metals and coal tar derivatives. Our experimental work was confined largely to neat oxide samples.

There are broadly three ways in which landfill disposal of toxic waste may have an adverse effect on the environment:

(i) pollution of water supplies. Laboratory leaching tests show that the major water-borne pollutant from spent oxide is sulphate ion, with a maximum equilibrium concentration about 7,000 ppm. Total cyanide, mainly as thiocyanate, is also present at levels up to 800 ppm. These results are broadly confirmed by field tests at a spent oxide tip in Southampton, although measured sulphate levels were higher and cyanide levels were lower: both species were still present in clay soil a metre down. The leachate from spent oxide may also be acidic. Soils contaminated with spent oxide may produce leachate with higher sulphate concentrations. Domestic waste has a considerable attenuation capacity for both acids and cyanides, while sulphate would be reduced under anaerobic conditions to sulphide: the sulphide would probably be precipitated as an insoluble metal salt. Permeability tests show that spent oxide will allow passage of rainwater, at a rate such that a thick layer of a metre or more would be necessary to allow the attainment of equilibrium concentrations. The deposition of spent oxide in thin layers (< 50 cm) and in dry conditions should help to limit leachate concentrations.

(ii) gas generation. Measurements made during the laboratory leaching tests showed no evolution of hydrogen sulphide and only limited generation of hydrogen cyanide. However, in a landfill site anaerobic conditions favour the reduction of sulphate to sulphide: co-disposal of acid wastes with spent oxide must thus be avoided to reduce the risk of hydrogen sulphide generation.

(iii) effect on landfill microbes. The stabilisation of a landfill site, and thus also its future use, is dependent on an active microbial population. The disposal of wastes toxic to these microbes should thus be avoided in a domestic waste landfill. Simple laboratory tests showed that even neat oxide itself contains microbes, so there should be little problem in this respect.

(iv) future land use. It is possible that landfilling spent oxides could simply move the problem of contamination from one site to another. Burial of large quantities of spent oxides in contained sites with impermeable cover is simply a form of storage which would sterilise the land. Large quantities of spent oxide have been landfilled in the past and experience shows that revegetation is difficult or impossible unless the oxide is at least 3 metres below the finished surface.

In conclusion, spent oxide is a waste for which controlled landfill is an acceptable disposal route. Co-disposal of thin layers (< 50 cm) of oxide between layers (> 2m) of domestic refuse is recommended. The site should not be allowed to become water-logged, and co-disposal with acid wastes should be avoided. The oxide layer should be at least 3 metres below the finished surface of the site.

ACKNOWLEDGEMENTS

This work was carried out by the Waste Research Unit on behalf of the Department of the Environment. The authors would like to express their thanks to North Thames Gas and Southern Gas, to the Analytical Services Section headed by Mr. J. Stevens at Harwell, and to Mr. J. King and Mr. K.W. Pearce of the Environmental Safety Group. The paper is published by kind permission of the Department of the Environment and the Director of the Harwell Laboratory, U.K.A.E.A.

REFERENCE

(1) K.W. Pearce (1979). Problems posed by the need to reclaim land formerly used for coal carbonisation plants. This meeting.

TABLE 1 - ANALYSES OF SPENT OXIDES, CONTAMINATED SOILS AND BOG IRON

Sample \ Species	Sulphur			Cyanides				Heavy Metals								pH
	Free S	$S^=$ (ppm)	$SO_4^=$	Total CN	SCN^-	Free CN^- (ppm)	Complex Cyanides	Fe	Mn (ppm)	Zn (ppm)	Cr (ppm)	Cu (ppm)	Ni (ppm)	Pb (ppm)	Cd (ppm)	
Bog Iron(1)	<0.1%	11	2.1%	280 ppm	54 ppm	3	750 ppm	25.2%	2,580	31	49	18	56	28	<2	3.5
Spent Oxides																
Beckton - weathered material(2,5)	45%	122	2.4%	3%	4%	800	4.6%	7.3%	2,310	39±4	23±4	21±4	23	20±15	2±2	5.6
- weathered material(2)	54.7%	118	2.1%	3.4%	4.3%	780	5.3%	3%	2,050	22	29	13	36	20	<2	5.1
- weathered nodule(3,5)	56.5%	42	2.8%	4%	1.9%	670	6.1%	7.9%	868	32	29	17	57	38	4	2.5
- part-spent oxide(4,5)	36.3%	63	1.4%	2.5%	4.9%	640	4.1%	5.7%	2,180	50	43	21	72	48	2	5.8
- nodule(3,4)	56.5%	195	2.2%	4.5%	5.3%	720	6.3%	3.4%	3,650	52	30	64	78	<15	2	5.5
Uxbridge	36.2%	68	2.2%	2.4%	5.3%	500	3.4%	9.4%	3,060	42	29	53	81	32	3	5.1
Contaminated Soil																
Site A - sample 1	2.1%	4	19.4%	7560 ppm	n.d.	24	7480 ppm	n.a.	n.a.	188	46	128	110	158	6	5.2
sample 2	14.2%	n.d.	5.5%	2280 ppm	450 ppm	10	2250 ppm	n.a.	n.a.	245	36	20	48	82	4	7.2
Site B	←	27%	→	n.a.	n.d.	n.d.	n.a.	n.a.	500	300	200	69	57	100	n.a.	2.4
Site C	0.6%	40	0.6%	900 ppm	n.d.	n.a.	1600 ppm	n.a.	n.a.	n.a.	n.a.	n.a.	n.a.	n.a.	n.a.	7.2

Notes:
(1) material remaining at Beckton gas works
(2) powdery material dumped in the open about 5 years ago
(3) consolidated material, fused into a refractory solid
(4) material remaining in the purifer shed, exposed to air but under cover
(5) sample used in laboratory leaching tests
n.a. not available
n.d. not detected

FIG. 1. LEACHING FROM BECKTON SPENT OXIDE.

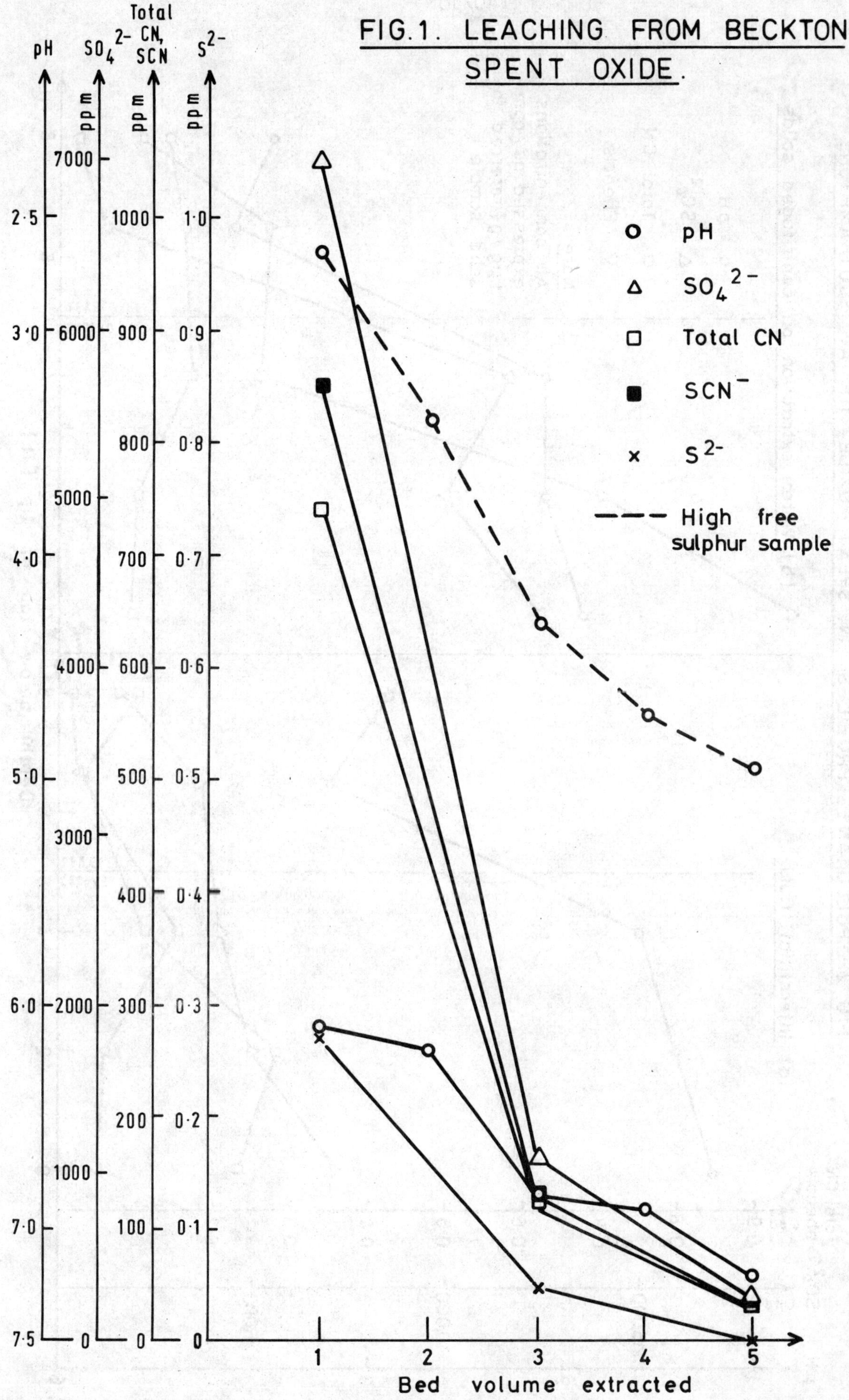

FIG. 2. POLLUTANT PROFILES IN SPENT OXIDE TIP AT SOUTHAMPTON.

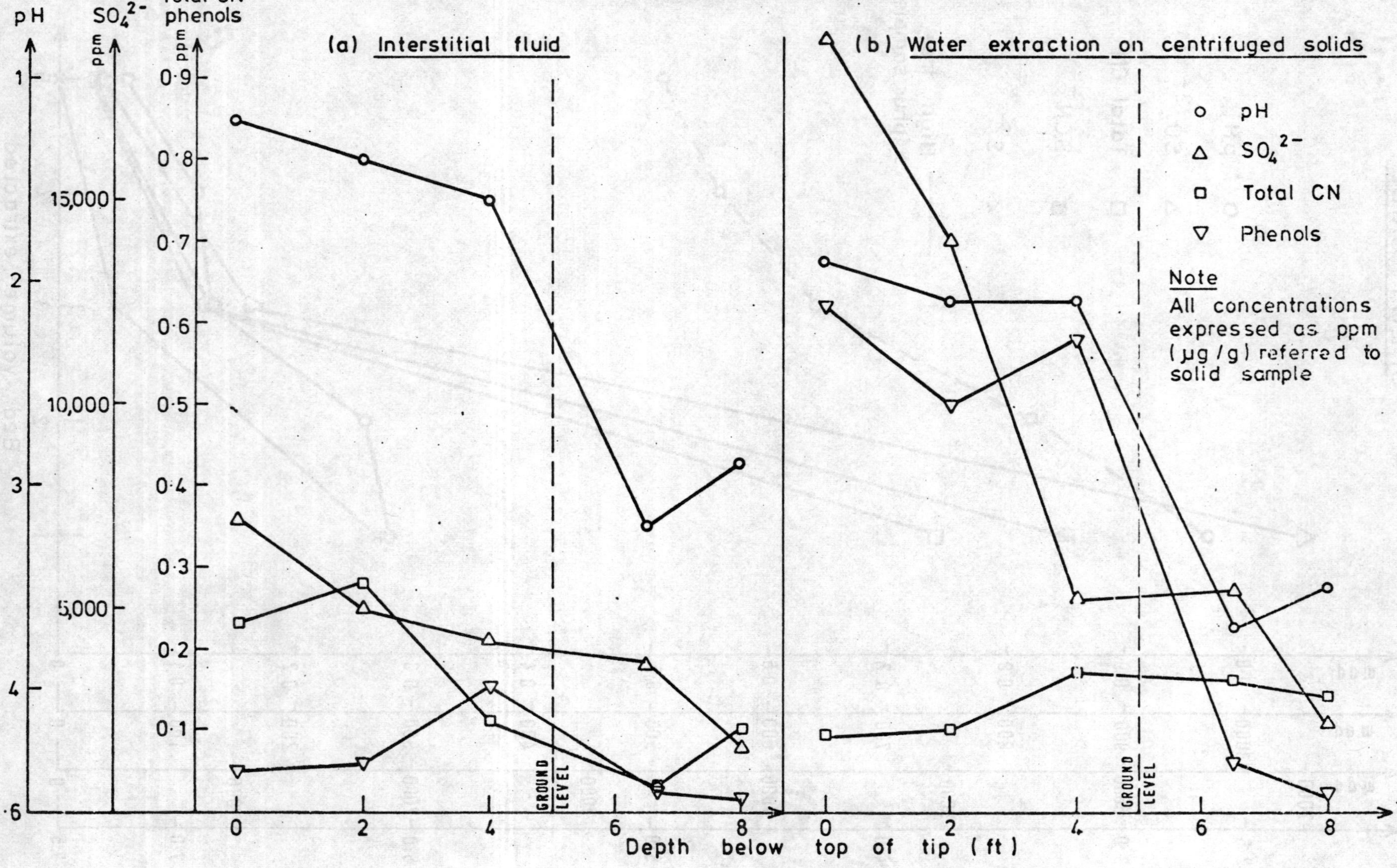

ESTABLISHMENT OF VEGETATION ON GAS WORKS WASTE

T.M. Roberts and R.P. Gemmell, Central Electricity
Research Laboratories and Greater Manchester Council

INTRODUCTION

Gas purification waste or spent oxide may cause quite serious problems during the reclamation of disused gas works sites and on other land where this waste has been tipped. In particular it is extremely toxic to plant growth, it may cause water pollution through acid run-off, and it often emits noxious gases such as hydrogen sulphide and various hydrocarbons. Large amounts of complex cyanides may be present and free cyanides may occur under certain conditions.

No published information is available relating to establishment of vegetation on the waste and control of pollution. In this investigation the chemical nature of spent oxide and its effects on plant growth were studied. Various treatments to control toxicity were evaluated in glasshouse and field experiments.

ORIGIN OF SPENT OXIDE

Hydrogen sulphide was removed from coal-gas by passage through layers of moist ferric oxide, ferric sulphide or ferrous sulphide and free sulphur being formed:

$$Fe_2O_3\ H_2O + 3H_2S \longrightarrow Fe_2S_3 + 4H_2O$$

$$Fe_2O_3\ H_2O + 3H_2S \longrightarrow 2FeS + S + 4H_2O$$

The material was then exposed to air and moisture which caused oxidation to ferric oxide and sulphur:

$$2Fe_2S_3 + 3O_2 \longrightarrow 2Fe_2O_3 + 6S$$

$$4FeS + 3O_2 \longrightarrow 2Fe_2O_3 + 4S$$

After recycling several times the waste accumulated up to 50% of free sulphur. In some cases the material was reprocessed for its sulphur content and ferrocyanides were occasionally extracted, the latter originating from the fixation of hydrocyanic acid in coal-gas to form ferric ferrocyanide or Prussian blue.

CHEMICAL ANALYSIS

Acidity

Samples of spent oxide were obtained from a tip at Blackley, approximately 4 km north-east of Manchester city centre. At four sampling points waste was obtained from depths of 0-10 cm, 30-40 cm and 60-70 cm, air-dried at room temperature and passed through 2 mm mesh. pH values were determined in distilled water (1:2.5 ratio waste:water) and lime requirements were determined using the method of Woodruff (1948).

The waste is intensely acidic (Table 1) with pH values as low as 1.5 at 60-70 cm depth. This is presumably due to oxidation of free sulphur. The amounts of calcium carbonate required to raise the pH to 6.5 for 15 cm depth are extremely high, being at least 100 tonnes/ha and therefore an order of magnitude above the highest levels normally applied to acidic soils. These values, however, do not allow for neutralisation of further acid which may be released from sulphur oxidation.

Table 1 pH and lime requirement values of spent oxide ($\bar{x} \pm$ S.E.)

Depth (cm)	pH	Lime requirement (tonnes/ha $CaCO_3$)
0-10	2.08 ± 0.03	99.5 ± 2.2
30-40	2.05 ± 0.04	124.8 ± 6.6
60-70	1.54 ± 0.02	124.0 ± 9.4

Values are means of 4 replicates but only 3 replicates were taken at 60-70 cm depth due to presence of iron pan

Toxic Elements

To determine whether the waste contains toxic levels of specific elements, two samples of material obtained from depths of 0-10 cm and 30-40 cm were examined by neutron activation analysis. The waste at both depths contained particularly high concentrations of calcium, iron and substantial amounts of aluminium (Table 2). Concentrations of these three elements were higher in the unweathered waste at 30-40 cm depth than at the surface. Other elements were present at relatively low concentrations.

Table 2 Total levels of some elements (ppm) in spent oxide

Element	0-10 cm depth	30-40 cm depth
Al	3130	5750
Ca	97740	127000
Cu	138	140
Fe	11000	54200
Mn	380	200
Zn	< 10	56

Analyses were carried out by Dr. L. Salmon at the Environmental and Medical Sciences Division, A.E.R.E. Harwell.

Some idea of the amount of iron available to plant roots was obtained by extraction of samples in 1 m ammonium acetate at pH 3.0 for 1 hour, using a waste-extractant ratio of 1:10. In the 0-10 cm layer, the concentration of iron was 775 ppm (± 144 S.E.) but increased to 30500 ppm (± 4790 S.E.) at 30-40 cm depth and 37330 ppm (± 13640 S.E.) at 60-70 cm depth. These high concentrations are probably toxic to plants and may cause immobilisation of phosphate fertilisers.

Plant-available calcium was extracted similarly but at pH 7.0 using a waste-extractant ratio of 1:17. At 0-10 cm depth 5550 ppm (± 155 S.E.) was present with a small increase to 6220 ppm (± 510 S.E.) at 60-70 cm depth.

Hydrocarbon Gases

Samples of unweathered waste were enclosed in 30 ml glass vials fitted with silicone rubber septa. Samples of gas were withdrawn at intervals and analysed by Dr. A.W. Davison (Department of Botany, University of Newcastle) for ethylene using gas chromatography. The ethylene levels ranged from 0.1 to 6.0 ppm with evidence of the presence of other hydrocarbons at relatively high concentrations.

EFFECTS ON PLANT GROWTH

Treatment with Lime and Fertiliser

To raise the pH of the waste, calcium carbonate was incorporated at 25, 50, 75 and 100 tonnes/ha. Phosphorus was applied at 100, 250 and 500 kilograms/ha P_2O_5 as superphosphate to overcome the deficiency of phosphorus and its fixation by iron. Nitrogen was applied at 125 kilograms/ha N as sodium nitrate and the pots seeded to *Lolium perenne* at 30 kilograms/ha.

At 12 weeks from sowing growth was extremely stunted, even at the high limestone and phosphorus fertiliser applications. Nevertheless, there were considerable growth responses to limestone and phosphorus with a positive interaction between 100 tonnes/ha $CaCO_3$ and 500 kilograms/ha P_2O_5 (Table 3).

Regeneration of Acidity

Monitoring of pH during the preceding experiment showed that pH values were favourable for growth at the beginning of the experiment (Table 4). After 12 weeks, however, significant acidification has occurred with the pH falling from 6.2 to 2.4 after liming at 25 tonnes/ha.

Table 3 Effects of calcium carbonate and superphosphate on growth of Lolium perenne (mg fresh weight/plant) in spent oxide

$CaCO_3$ (tonnes/ha)	Phosphorus (kilograms/ha P_2O_5) 100	250	500	Means
25	0	0	0	0
50	10	15	15	13
75	13	21	22	19
100	14	15	33	21
Means	9	13	18	

Values are means of 6 replicates (L.S.D. = 5) completely randomised

Table 4 Changes in pH of spent oxide treated with calcium carbonate

$CaCO_3$ (tonnes/ha)	Time from treatment (weeks) 0	8	12
25	6.2	4.1	2.4
50	7.0	6.5	4.6
75	7.1	6.8	5.5
100	7.1	6.9	6.1

Values are means of 6 replicates

Addition of Organic Matter

In view of the stunted growth in the preceding experiment being caused by factors other than pH, the use of organic matter in alleviating the residual toxicity was investigated. Peat and sewage sludge have been used previously to remove metallic cationic toxicities by chelation (Street & Goodman 1967) and anionic toxicity by ion association (Gemmell 1972).

Unweathered waste was treated with calcium carbonate at 100 tonnes/ha and diluted with peat to give waste concentrations from 0 to 50%. The peat was also limed at 10 tonnes/ha. A control series using quartz sand was set up. Fertilisers were applied at 100 kilograms/ha N, 50 kilograms/ha P_2O_5 and 50 kilograms/ha K_2O prior to seeding of Lolium perenne at 30 kilograms/ha.

At 38 days from seeding, growth in the peat series was vastly superior to that in waste diluted with sand (Table 5). 10% waste reduced growth in sand by 67% but reduced growth in peat by only 19%.

Table 5 Growth of Lolium perenne (g dry weight) in spent oxide diluted with peat

% waste	Control (sand)	Peat
0	1.2	5.4
1	0.9	4.7
5	0.5	4.4
10	0.4	4.4
20	0.4	2.1
50	$<$0.1	0.3

Values are means of 3 replicates

ACIDIFICATION AND ITS CONTROL

Continuous acidification is likely to pose the greatest problem for establishing permanent vegetation on spent oxide. It is also responsible for the water polluting effects of the waste and may mobilise large quantities of soluble iron and possibly free cyanides.

Rates of Acidification

Site experiments were established following the removal of the surface 10 cm layer of weathered waste. Ground calcium carbonate was applied at rates of 0, 50, 100 and 200 tonnes/ha and incorporated to a depth of 20 cm. Further treatments examined were 5 cm sewage sludge (from the activated sludge process) incorporated to 15 cm depth and soiling with a sandy-clay subsoil to give a surface covering 10 cm deep which was limed at 10 tonnes/ha. Prior to soiling, all the plots were treated with superphosphate giving 1000 kilograms/ha P_2O_5 incorporated to 10 cm depth. NPK fertiliser (10:10:10) was applied to the seed-bed at 1000 kilograms/ha followed by seeding to Lolium perenne at 200 kilograms/ha. The design was a randomised block, replicated once, with 1 m x 1 m plots.

Satisfactory growth occurred only on waste amended with 200 tonnes/ha ground limestone plus 5 cm sewage sludge but, even on this treatment, regression occurred after grass cutting. All the soiled plots failed to support acceptable grass cover and complete dieback occurred after several months.

The problem is clearly acidity as demonstrated by the dramatic falls in pH (Table 6). At 330 days from treatment there was no residual pH effect of 100 tonnes/ha ground limestone. Throughout the treatments the pH was lower at 10-15 cm depth than at the surface.

Table 6 Acidification of spent oxide treated with calcium carbonate

$CaCO_3$ (tonnes/ha)	10	40	75	330
	pH values at 0-5 cm depth; Time from treatment (days)			
0	3.2	2.1	1.8	1.0
50	4.4	2.2	1.9	0.9
100	5.5	2.7	2.4	1.1
200	5.8	4.2	3.4	3.0
	pH values at 10-15 cm depth			
0	2.4	1.8	1.7	1.0
50	3.6	1.9	1.8	0.9
100	4.0	2.7	1.7	1.1
200	5.0	3.3	1.7	2.1

Values are means of 8 determinations

Sewage sludge raised the pH of the waste initially but increased the rate of acidification (Table 7). This effect was most pronounced after 75 days following 100 tonnes/ha ground limestone application and after 330 days following 200 tonnes/ha limestone addition.

Table 7 Effect of sewage sludge on acidification of spent oxide

$CaCO_3$ (tonnes/ha)	Sewage sludge	pH values at 0-5 cm depth; Time from treatment (days) 10	75	330
50	Nil	3.8	2.0	1.0
	5 cm	5.0	1.9	0.8
100	Nil	5.6	3.0	1.3
	5 cm	5.5	1.7	0.8
200	Nil	6.0	3.9	4.1
	5 cm	5.6	3.0	1.9

Values are means of 4 determinations

Acidification of Soil Cover

The dieback of grass on waste amended with a 10 cm layer of soil was also caused by acidification (Table 8), upward movement of acid occurring from the underlying waste. Limestone incorporation into the waste at 200 tonnes/ha prior to soiling failed to prevent this. Incorporation of sewage sludge prior to soiling caused an increase in the rate of soil acidification.

Table 8 Acidification of soil cover over spent oxide

$CaCO_3$ (tonnes/ha)	Sewage sludge	pH values at 0-5 cm depth: Time from treatment (days) 10	330
0	Nil	5.1	3.0
	5 cm	6.1	1.6
50	Nil	5.8	1.8
	5 cm	5.9	1.7
100	Nil	5.8	2.3
	5 cm	4.1	1.8
200	Nil	5.8	4.9
	5 cm	6.0	2.0

Values are means of 2 determinations

To investigate the control of upward movement of acid through soil cover, waste was amended with 10 cm layers of various covering materials having different physical and chemical properties. A further series involved 5 cm layers of these materials overlaid with 5 cm of topsoil. Both series were repeated usung waste treated with calcium carbonate at 200 tonnes/ha as controls. The experiment was conducted in a glasshouse using 15 cm diameter plastic pots. All the treatments received NPK (10:10:10) fertiliser at 1000 kilograms/ha and were seeded to Lolium perenne at 200 kilograms/ha

Determination of pH values at 12 months from treatment revealed considerable acidification of most of the covering materials, even in the surface 0-2 cm layer (Table 9).

Table 9 Acidification of cover materials over spent oxide

Cover and initial pH		pH values after 12 months: Depth (cm) 0-2	2-4	6-8	8-10	Underlying waste
Sandy clay	(4.4)	3.6	3.0	2.5	2.1	1.6
Sand	(5.3)	3.2	2.6	2.2	1.7	1.2
Power station ash	(7.3)	7.4	6.7	4.0	2.4	1.5
Clay	(4.2)	3.4	2.9	2.6	2.4	1.9
3 mm grit	(6.0)	5.5	5.1	4.1	3.2	2.0
Boiler ash	(5.3)	4.3	3.5	2.7	2.2	1.4
Sewage sludge	(5.6)	5.4	4.1	2.8	2.5	1.7
Peat	(4.0)	2.4	2.4	2.1	2.0	1.5
Lime waste	(7.7)	7.8	7.8	7.6	7.6	1.7

Values are means of 2 replicates

Mobilisation of acid was greatest through sandy-clay, sand, clay and peat. No acid movement occurred through lime waste due to its high calcium carbonate content (80%) and the lime content of power station ash (3% calcium carbonate equivalent) probably neutralised most of the mobile acid during the course of the experiment. The restricted movement through 3 mm grit was probably caused by its low capillary properties.

Sampling of growth at 12 months from treatment confirmed the efficiency of lime waste in controlling acidity (Table 10). Again, power station ash was more effective than the other materials and the restrictive effect of 3 mm grit on acid movement reduced the depressing effect of the underlying waste on growth.

Table 10 Growth of *Lolium perenne* (g dry weight/pot) on cover materials over limed and untreated spent oxide

Cover material	$CaCO_3$ at 200 tonnes/ha	Nil $CaCO_3$	Effect on growth
Sandy clay	5.6	3.7	- 34%
Sand	9.8	4.8	- 51%
Power station ash	9.7	8.9	- 8%
Clay	5.9	3.3	- 44%
3 mm grit	3.7	2.9	- 22%
Boiler ash	8.1	2.6	- 68%
Sewage sludge	10.5	6.7	- 36%
Peat	5.0	2.3	- 54%
Lime waste	5.2	7.8	+ 50%

Values are means of 4 replicates

Acidification of topsoil placed over 5 cm layers of the covering materials was influenced by the type of cover. As expected, no significant acidification occurred with lime waste (Table 11). The most suitable material for restricting acid movement was 3 mm grit, the pH of soil over this falling only slightly from 5.7 to 4.8. Over all the other materials, the soil pH fell to below 4.0 which is too acidic for growth.

DISCUSSION AND CONCLUSIONS

Extreme acidity is the principal growth limiting factor in spent oxide. Other contributory factors are high concentrations of soluble iron, complex cyanides, aluminium, hydrocarbon gases and extreme nutrient deficiencies, particularly phosphorus. These secondary toxins and deficiencies can be controlled by liming, organic matter incorporation and fertiliser additions but the continuous generation of acid makes permanent neutralisation costly and difficult.

Table 11 Acidification of soil overlying spent oxide treated with 5 cm of different covering materials

Cover and initial pH		$CaCO_3$ at 200 tonnes/ha	Nil $CaCO_3$
Sandy clay	(4.4)	5.2	3.7
Sand	(5.3)	5.6	3.8
Power station ash	(7.3)	6.4	3.8
Clay	(4.2)	5.5	3.2
3 mm grit	(6.0)	5.7	4.8
Boiler ash	(5.3)	5.8	3.9
Sewage sludge	(5.6)	5.7	2.6
Peat	(4.0)	5.5	3.0
Lime waste	(7.7)	6.3	5.6

Values are means of 2 replicates

pH monitoring of site trials indicates that the waste consumes well over 200 tonnes/ha of calcium carbonate per annum if limed to 20 cm depth. It would be uneconomic to lime the waste annually at rates around this level, quite apart from the costs of limestone incorporation and reseeding. It would also be uneconomic to apply sufficient calcium carbonate initially to neutralise all the potential acidity from sulphur oxidation, particularly in the case of waste containing up to 50% sulphur.

These arguments suggest that reclamation should involve isolating the toxicity of the waste from plant roots by covering with soil or other material. However, upward movement of acid causes contamination of the cover, as reported by Gemmell (1974) for chromate smelter waste. The use of lime waste appears to control this but this method is only practicable where lime waste deposits are situated close to spent oxide sites. Power station ash is unsuitable, its high capillary properties and low lime content resulting in control of acidity for a limited period only.

Evidence from the behaviour of 3 mm grit suggests that control of acid movement can be achieved by using cover material which prevents capillary contact of the underlying waste with the plant rooting medium. Possible alternative primary cover materials with low capillary properties include gravel, coarse boiler ash, coarse quarry waste and demolition rubble. However, a problem may arise if fine particles from the secondary or soil cover move down into the primary cover, effecting capillary contact. Some kind of membrane or intermediate clay seal may be required to control this.

One adverse effect of the preceding method is that leaching of the waste through rainfall percolation will continue, leading to possible pollution problems. There was no evidence from the experiments that acid generation was reduced under covering material, as might be expected in view of the oxygen dependence of sulphur oxidation reactions. However, oxidation may well be reduced under substantial depths of cover, particularly if layers greater than 30 cm depth are employed. This requires further research.

The behaviour of clay in the experiments suggests that this material is permeable to acid although certain types may behave differently. If clay could be used as primary cover as practised in the reclamation of some municipal refuse tips, the construction of a field drainage system beneath the secondary cover should prevent acid leachate production.

ACKNOWLEDGMENTS

We thank Manchester City Corporation for financing the initial part of this study, Liverpool University Department of Botany for providing facilities, and Mr. D. Holmes for research assistance.

REFERENCES

Gemmell, R.P. (1972). Use of waste materials for revegetation of chromate smelter waste. Nature, Lond. 240, 569-71.

Gemmell, R.P. (1974). Revegetation of derelict land polluted by a chromate smelter. II. Techniques of revegetation of chromate smelter waste. Environ. Pollut. 6, 31-37.

Street, H.E. & Goodman, G.T. (1967). Techniques of revegetation in the Lower Swansea Valley. The Lower Swansea Valley Project (Ed. by K.J. Hilton), pp. 71-110. Longmans Green, London.

Woodruff, C.M. (1948). Testing soils for lime requirement by means of a buffered solution and the glass electrode. Soil Sci. 66, 53-63.

Problems posed by the need to reclaim land formerly used for coal carbonisation plants

K.W. Pearce
Environmental Safety Group, Harwell Laboratory

ABSTRACT

Coal is the starting material for many products used both for domestic and industrial purposes. Processes for its conversion to other useful products are or have been used by various industries including the gas, steel, smokeless fuel and chemical industries.

In the past many by-products from these industries have been discarded, generally due to the lack of a suitable market for them. In addition, catalysts and other materials used in conversion processes will have been stockpiled or discarded at the end of their useful life.

The presence of such materials in land required for other purposes will necessitate an extensive on-site evaluation, including chemical analysis of selected samples. Any remedial measures needed will be dictated by such results assessed in conjunction with the proposed site development plan.

INTRODUCTION

When man first built protection cover for himself and his possessions demolition was soon to become a reality. Most structures have a useful life and eventually fall down or are demolished. Methods of demolition are twofold - firstly complete destruction of all materials or secondly, careful dismantling of valuable items followed by ruthless spoilation. The sequence is therefore dismantle or knock down, load up and cart away. These operations are usually carried out at great speed in order to replace the once useful building and plant with a more up-to-date one.

HISTORICAL

In the case of coal carbonisation, the industry had its origins in the first half of the eighteenth century. (The first reported use of coal distillation was in 1739). Initially, coal gas was the primary product and its use as an illuminent started in 1792. In 1808 limited areas of London were first lit by coal gas. Because the gas needed to be distributed through a network of pipes, once established, it generally followed that the site selected would continue to be used for coal carbonisation. It therefore follows that because of the timescale involved and inevitable process developments many coal carbonisation sites featured more than one type of production plant and, hence, today will almost inevitably present a complex picture to those involved in their reclamation for other uses. (Coke and town gas are of course still produced in connection with the manufacture of iron and some coal-tar and chemical works are still in operation).

PRESENT SITUATION

On many sites the coal carbonisation plants are still standing, on others the plant and buildings have been demolished and present a barren unattractive landscape. As a consequence of urban development many sites which were once situated at the edge of a community now find themselves in an area ripe for commercial or housing development and as such are attractive to local authorities and property developers.

Until fairly recently little heed was taken of the previous activities carried out on any site prior to redevelopment. Any site investigations undertaken were generally directed towards engineering considerations for foundation design purposes. Chemical considerations were usually limited to soil pH, sulphate and sometimes chloride concentrations. Problems encountered on site were normally dealt with on an ad hoc basis by those directly involved with the problem.

For a complex variety of reasons the 'environment' and 'quality of life' are now high on the list of concern to the human race. In consequence, potentially harmful features of most activities need to be evaluated to answer questions which will inevitably be posed. The development of land formerly used for coal carbonisation purposes is one which needs careful evaluation if the environment is to be protected.

PROBLEM AREAS

The chemical contamination of sites which have been used for coal carbonisation processes has arisen from a combination of accidental spillage, leakage and the deliberate dumping of residues. In some cases the situation has been compounded by the manner in which demolition and clearance of the site has already been undertaken. Soil contaminants found in the land included heavy metals, sulphides, sulphur, sulphates, cyanides (simple and complex), cyanates, thiocyanates, phenols, coal tars, organic sulphides, etc.

Depending on the geological and hydrogeological features of the site such contamination may have moved to considerable depth and horizontal distances, especially along sub surface soil/clay interfaces, where it has been carried along the water table.

Many by products from coal carbonisation have been recognised to be toxic, skin irritants, carcinogenic or phytotoxic in their own right. Deliberate or accidental lighting of fires on sites have caused problems in generating noxious fumes from the combustion of contaminants present there.

It should be recognised in developing a site that sulphate, magnesium and acid attack on concrete must be dealt with in an appropriate manner. The possibility of creating a leak path for contaminants to underground strata when piling is required for foundation purposes which could for example contaminate an aquifer used for drinking water supplies must be considered. Plastic pipes may be porous to organic materials present in the site. Rubber gaskets and joints may be swollen and degraded. Disturbance of the site may produce odour problems.

SITE INVESTIGATIONS

Experience has shown that a significant amount of useful background information can be obtained from a variety of sources before setting foot on a site to be investigated. Some of the sources of information are as follows:- The previous owners and/or operators of the site may have plans and drawings. The Local Authority Planning Department may hold useful records. Various editions of Ordnance Survey maps should be obtained and differences in building layouts noted. Local inhabitants or ex-employees may provide much useful information. Seek out geological information - the Water Authority and/or the Institute of Geological Sciences will be able to help. Visit the site several times to compare the existing situation with information obtained from others. Mark out significant features on the site and identify where plant processes were carried out. Armed with this information it will be possible to decide where soil samples should be taken and to what depth.

Opinions differ as to whether trial holes should be drilled or trial pits excavated using a mechanical digger. The author always opts for trial pits since one can generally see much more, and a pit can easily be extended if interesting features are observed. However at depths greater than 3-4 metres one may have to resort to drilling.

If liquids are encountered in underground tanks, wells, sumps, catch pits, etc care should be exercised in sampling at various depths to ensure that samples are taken from each of the various phases almost inevitably present.

Having obtained samples which are considered to be representative of the site typical chemical and physical analyses required will include:-

Inorganic cations:	Arsenic, Antimony, Barium, Cadmium, Chromium, Cobalt, Copper, Iron, Mercury, Molybdenum, Nickel, Lead, Vanadium, Zinc.
Inorganic anions:	Ammonia, Chloride, Cyanate, Cyanide, Thiocyanate, Ferricyanide, Ferrocyanide, Nitrate, Nitrite, Phosphate, Sulphate, Sulphide.
Organic compounds:	Solvent extracts, Oils, Phenols, Coal tars, Sulphides, Total Organic Carbon, Chemical Oxygen Demand.
Others:	Asbestos, Sulphur. pH. Plant available Copper, Nickel and Zinc.

ASSESSMENTS

This is probably the most difficult task to undertake since the decisions taken will almost certainly affect a significant proportion of the cost of reclamation. In arriving at decisions as to whether contaminated land in a particular area should be removed from the site for disposal elsewhere, be buried in a suitable area on the site, remain where it is and covered or dispersed over a suitable area of the site, consideration must be given to the ultimate use of the site and the concentration of contaminants it contains. Deriving acceptable residual concentrations in the soil also needs careful consideration.

If covering with an inert material is an option, it needs to be demonstrated that the material selected is in fact inert and its load bearing and other properties will be suitable for the proposed development. A decision on the depth of cover also has to be reached. Various cover materials which have been considered suitable include clay, fly ash, limestone dust, building rubble, etc depending on availability. Clay or similar material has been used as a seal covered with suitable top soil. Up to 2 metres total depth may be required.

REMEDIAL OPERATIONS

The analytical results obtained from samples taken from the site will dictate the protective measures required to safeguard the health of operatives involved in the reclamation. The exposure of workers will normally be of limited duration and little difficulty should be experienced in affording them proper protection. Direct contact with chemicals contaminating the soil and the presence of toxic gases particularly in confined spaces are probably the two major problems to be guarded against.

Protection of the population at large from harmful concentrations of vapours and gases should not present significant problems. However, since many of the contaminants present in the soil may be highly odorous, from a public relations not withstanding a toxic hazard viewpoint, it may be necessary to limit the area under excavation at any one time to minimise the dispersion of odorous or toxic materials.

In this paper I have tried to outline the general problems one may encounter and need to resolve in reclaiming land which has been used for coal carbonisation plant processes. Similar considerations will be required in reclaiming any contaminated land but usually to a lesser extent. In investigating any site it is necessary to draw up a checklist of possible problems which may be encountered. Unfortunately the list is not exhaustive and can always be added to in any particular situation.

DISCUSSION

Paper F1-F4

Dr Emery's first paper refers to the need to take into account not only the total amount of potentially leachable substances in a material such as ferrous slags but also probably leaching rates under site conditions. He also pointed out the dangers of using steel slags as fill because of possible expansive hydration of lime present in the slag, even when the slag has been stockpiled for decades. Support for the former view came from Dr J Brown of the CEGB. He said that there is too much emphasis on simple indicators such as total or available trace elements. He referred to work carried out by the Water Research Centre on pulverised fuel ash (fly ash). The physical characteristics on "hydrology" of the ash when placed reduces leaching rates to a very low level. The porosity of underlying and surrounding materials were also important. (Water Research Vol. 10, p.1115 - 1121) Dr Emery in reply said that in contrast to fly ash, slag when placed tended to be highly permeable but as leaching could only occur at the surface of the slag, which tends to be dense and crystalline, very little leaching of toxic materials could take place. Dr Nixon of the Building Research Station spoke in support of the need for great care in the use of steel making slags particularly if little is known about the particular slag in question. BRS heard of far too many failures of ground floor slabs etc due to the use of materials described simply as "slags" without any proper characterisation having been made. Replying to a question Dr Emery said that where heave is already taking place due to the expansion of steel slag it is difficult to find effective remedial measures, apart from the engineering solution of excavation and underpinning. One possibility is to encapsulate by pressure grouting with hot asphalt (cementitous materials will not work) to prevent ingress of further moisture and to provide, where appropriate an insulating blanket between the slag and any source of heat - this had been done in the case of a coke oven. If no expansion has yet taken place some assessment of the potential for expansion could be made by laboratory tests and steps should be taken to prevent any changes in ground water conditions.

There were no questions on Dr Emery's second paper or the paper presented by Mr Kahn.

Prof. Bolsing was questioned about the cost of his process, the proportion of special reagent required in terms of percentage of dry solids, health and safety precautions during storage, transport and placement, and whether it could be applied to domestic sewage sludges which frequently have 20% ether extractable matter in its dry solid content. Costs are clearly proportional to the amount of reagent which has to be used and this in turn will depend on the nature of the material to be treated. The process has been used now in a number of countries to treat a variety of materials, and costs found to be in the range 60 - 100DM (£16 - 26) per tonne of material to be moved. Light oil requires an approximately 1:1 mix with reagent and although consequently expensive to treat costs are still within the quoted range. The amount of reagent required can be reduced by the pre-mixing of fly ash or other fine residues (Rapporteurs Note - The Ravenfield Tip (Mr Kahn's paper) was treated using liming and fly ash and reference was made in discussion of gas works sites of the use of fly ash to bind tar-like materials). Heavily polluted soil requires approximately 20% special ingredients. In one particular case the cost of removal of 40-50000m^3 of oil to another site would have been

12M DM(£3.16M) compared to the actual cost of 2.7M DM(£0.71M) using the process described.

Some work has just been completed on sewage sludge for the Federal Ministry of Agriculture, Food and Forestry. German peat reserves are expected to run out in about thirty years and alternative materials are being sought. A suitable material has been produced on a pilot scale using special ingredients, selected sludges with low contents of trace metals, and chemically pretreated bark. Two continuous process plants are now under construction at sewage works in Germany.

Rapporteurs Note: There was no reply to the question on health and safety precautions.

Papers F5-F8

The session Chairman, Dr J N M Firth, in his introductory remarks said that the main topic for discussion would be the wastes arising from the exploitation of coal and oil shales.

Pulverised Fuel Ash

Mr Snell remarked that paper F5 was in two parts, the first dealing with the reclamation of land using pulverised fuel ash and the second considering the chemical behaviour of the material. The CEGB produces approximately 12 million tonnes of ash per annum from burning 70 million tonnes of coal. Wherever possible PFA is disposed of in quarries and gravel pits along river valleys in which filling is restricted to the original ground level. For some power stations embankments have been constructed and the ash retained above the general ground level. Ash is also discharged into abandoned brick pit workings, which are much deeper than river valley gravel pits, and also on reclaimed marshes.

The usual treatment before disposal is to slurry the ash (20% ash) and then pump through pipelines (up to 30km). The discharge into the lagoon is allowed to settle out and the water is either recycled or alternatively passed through settling ponds to a watercourse. Rail and road tank wagons are also used and at some power stations the dry ash is transported by conveyor. 4 million tonnes per annum are sold by the CEGB for use as structural fill.

Mr Snell illustrated his paper with examples of PFA disposal from Cottam, Ironbridge, Ferrybridge, Drax and Connah's Quay; he also described the reclamation of the clay pits at Peterborough. Floaters can be a problem in the lagoons.

PFA, with a grading of 60% fine sand and 40% silt, has a similar particle size distribution to many agricultural soils, but when cultivated it needs careful management as it is devoid of nitrogen and trace elements. Boron is high initially but reduces after two years or so.

A speaker described how the problem of floaters at Gale Common had been solved naturally where reedmace (Typha angustifolia) has invaded and floats on the surface of the floaters. He asked whether this, and other species, could not be sown on other lagoons. This would be cheaper and more effective than an emulsion spray.

Oil Shale Waste

Mr Edwards in introducing Paper F6 said he would confine his remarks to the spoil arising from the oil shale industry of the Central Coalfield of Scotland, concentrated particularly around Bathgate in West Lothian. Extraction of the shale from seams in the Upper Carboniferous Series took place from the 1850's up to 1962. The Scottish Development Authority is responsible for the reclamation of derelict land in the country and in this part of Scotland has delegated powers to the Regional Council. There are approximately 2500 ha. of derelict land in the Lothians Region of which 61% is in West Lothian.

The shale was heated in retorts to produce marketable light and heavy fractions. The waste, of which there is some 100 million tonnes was stacked in spoil heaps (bings) in the form of burnt shale (red blaes).

In his paper, Mr Edwards had concentrated on the chemical and physical problems, the options open to solve them, and the importance of maintaining and conserving the ecology of habitat reserves found on derelict sites.

A contributor asked for further information on the use of sewage sludge on the bleas with respect to the effect on soil texture. Mr Edwards replied that experiments were in progress using sewage sludge and the results of these will be published. The work is taking into account the effects of metals on the sludge.

Colliery Waste

In introducing Miss Rosborough, the Chairman said that he was looking forward to hearing views from outside the industry. Presenting Paper F7 Miss Rosborough commented that she had studied the future of coal mining following the quadrupling of oil prices in 1974 together with the general acceptance that the coal industry now had a long-term future. As production of coal expanded again there would be more waste for disposal per tonne of coal produced than in the past. There were two main reasons for this, firstly the use of mechanical mining which was less selective at the coal face, and secondly the use of modern coal washing plant. The latter produced tailings as well as dry waste.

The peripheral coal fields in Scotland, Wales and to a lesser Kent were contracting and current investment was in the expansion of the central English coalfields of Yorkshire, Nottinghamshire, Derbyshire and Leicestershire, future investment would probably be in new fields in the Vale of Belvoir, Oxfordshire and Warwickshire.

She projected an increased annual production of 60 million tonnes of coal per annum by the year 2000 equivalent to 60 new collieries.

It was most important to consider now the best methods of waste disposal and the alternatives which Miss Rosborough referred to were discussed in her paper.

Wastes From Herbicide Manufacture

Mr Lindfors (Paper F8) described the problems which had arisen during the decontamination of an area around a factory at Teckomatorp in South Sweden. Phenoxy acids and chlorinated phenols, as the waste products from the manufacture of herbicides and insecticides, had been disposed of around the factory and had contaminated the nearby river. The contamination was exacerbated by the effluent from the municipal sewage treatment plant adjacent to the factory, as the effleunt was affected by leakage of contaminated water into the sewers.

The first stage in the reclamation was to construct a cut-off trench filled with bentonite slurry to limit the infiltration into the river. Water was drained from the sealed area and stored in four lagoons from which it was pumped for treatment in an activated carbon plant. The treated water was passed to the sewage works as the remaining contamination was bio-degradable.

A speaker suggested that it would have been feasible to heat the waste to approximately 900°C at which temperature the organic compounds would decompose and could be discharged to the atmosphere. Hydrochloric acid was likely to be the most noxious of the end products. Questioned on how it had been determined that it was safe to demolish the factory, Mr Lindfors answered that samples had been taken and tested. It transpired during the discussion that the contaminated activated carbon had been exported to England for treatment.

Papers F9-F12

The Chairman, Mr G S Cribb introduced the session by saying that the discussion would cover the four papers dealing with the problems relating to the redevelopment of gas works sites.

Disused Gas Works

Mr Dean introduced Paper F9 and commented that British Gas had inherited a century-old problem and since 1949 had made a positive contribution to land reclamation. At the time of nationalisation there were some 1000 separate undertakings and 1100 plants making gas by the destructive distillation of coal. Some of these works had been rebuilt for hydrocarbon reforming producing gas from petroleum distillates and these plants have now been closed down following conversion to natural gas. Some sites are still used for gas storage.

Mr Dean described the redevelopment of 14 sites throughout the United Kingdom. He said that there had been few real problems and these had sometimes been caused by the finding of unsuspected tar wells. Some heavily contaminated areas had been used as car parks.

Mr Lord said that there were often high sulphide levels up to 4000 mg/kg in the wastes at gas works, and on disturbance hydrogen sulphide was given off. Dr Pankhurst replied on behalf of British Gas and said that nothing could be done to prevent the microbiological reduction of sulphide. She said that the degree of reduction and sometimes oxidation depended on the organisms present. The problem was a very old one and occurred when iron oxide boxes were used; iron oxide was not sterile as it was often thought to be.

British Gas were questioned on the thoroughness of their investigation into potential problems. A speaker considered that in most cases investigations would have been limited and in the absence of data on what exists at a site it is unsafe to say there are no problems. He remarked that British Gas offered no data on either site conditions or the extent of remedial measures carried out.

One speaker remarked on the bad contamination found at the Norwich Gas Works during the construction of the new Crown Court on the site. Another example critical of British Gas policy was the possible disposal of the site of the Redheugh Gas Works at Gateshead, where there was wide contamination of free sulphur, cyanides and spent oxide.

It was suggested that records of former gas works should be held at the Land Registry.

Spent Oxide Pollution

Mr Wilson introduced Paper F10 by saying that much of the contamination on gas works sites had its origin in the spent oxide produced during gas purification. The bog iron process yielded spent oxide as waste which could be burnt to produce sulphuric acid.

He discussed the disposal problem and future land use as described in the paper.

The first questionner asked whether the maximum use had been made of manufacturing sulphuric acid from spent oxide. Mr Wilson replied that when the coal carbonisation plants were shut down in the early 1960's, there was no shortage of cheap sulphur from alternative sources. Spent oxide, therefore, had been difficult to sell although it was high in sulphur.

Disposal of spent oxide by burning was also discussed. Where this had been tried it had proved difficult to control the fire, and there had been excessive release of sulphur dioxide.

Establishment of Vegetation On Gas Works Waste

Dr Gemmell stated that Paper F11 considered the establishment of vegetation on gas works waste on a tip some 4km north of Manchester. This was a pilot investigation involving the treatment of the waste with calcium carbonate between 25 and 100 tonnes/ha and above. The pH at the surface was originally 2.1 and this decreased at depth to pH 1.5. The effect of liming at a rate of 25 tonnes/ha was to increase the pH to 6.2 falling after 12 weeks to pH 2.4. An application rate of 100 tonnes/ha increased the initial pH after treatment to 7.1 but this fell only to pH 6.1 after 12 weeks. In the paper the use of cover materials was discussed, including pulverised fuel ash, grit and lime waste. In the case of PFA its efficiency in this context depended on the source.

The plant used in the experiments was perennial rye-grass (Lolium perenne) and satisfactory growth occurred only on waste additionally treated with sewage sludge. Regression occurred after grass cutting and complete die-back after several months. Continuous acidification is the greatest problem in establishing permanent vegetation on spent oxide.

In the discussion it was pointed out that it was important to distinguish between the PFA resulting from the combustion of bituminous coal, as burnt in the United Kingdom, and that derived from brown coals (lignite). The calcium carbonate equivalent in PFA from bituminous coal is about 3% whereas from brown coal it is up to 20%. The brown coal ash is thus much more effective in neutralising spent oxide. The same speaker was critical in that in seeking an absolute minimum cost solution the possible health hazards had been ignored.

Dr Brown commented that there was little point in discussing at length the minor differences in the neutralising capacity of PFA from different sources. The basic point is that no alkaline material can provide a permanent solution unless it is mixed with the spent oxide in sufficient quantities to neutralise the whole of the acid that might be formed from the sulphur content with time. Otherwise there is little option but to incorporate a capillary break layer in the overlying material. The importance of keeping adequate and permanent records of contamination was stressed.

There was criticism of the choice of species in that Lolium is not tolerant of acid soils. Surely, it was argued, it is unreasonable to draw general conclusions about the revegetation of spent oxide from experiments with a single species. The authors agreed that wavy hair grass

(Deschampsia) and Yorkshire fog (Holcus) generally have a greater tolerance to low pH than Lolium perenne, but said that no grasses were tolerant of acid soils with pH less than 3, and there was no other option than to neutralise.

Reclamation of The Sites of Coal Carbonisation Plants

Mr Pearce (Paper F12) started from the premise that most structures have a useful life and eventually fall down or are demolished. Until fairly recently, little attention was paid to the previous activities carried out on a site prior to redevelopment. The development of land formerly used for coal carbonisation processes merits more careful consideration than in the past.

The principal areas of contamination in a gas works are the process zones and the sites of the gas holders. Mr Pearce considered the cost effectiveness of reclamation and described the methods of site investigation, the analyses required and the assessment of the results.

In the discussion, one speaker felt that ignorance of the hazards was a principal reason why gas works sites had not received the attention they merited. He also suggested that occasionally known facts had been withheld for financial gain.

In correspondence received subsequent to the discussion, it has been suggested that geophysical methods could be used effectively to trace the positions of underground tanks and tar and liquor wells, which have frequently been capped with concrete and concealed.

It has also been questioned, particularly in the case of gasworks sites sold for residential development, that British Gas have not been sufficiently diligent in disclosing known information. Consequently no allowance was made by the purchaser for expenses which might be incurred in dealing with the contamination. British Gas point out that most problems have arisen related to undiscovered tar wells or tanks at works demolished before nationalisation. Examination of site values realised by the British Gas has shown that, as the value is made up of several factors, mainly market demand, planning restrictions and ground conditions, the cost of decontamination is of little consequence in relation to the overall construction cost. However, there have been some contentious disposals, generally after sale to a Local Authority for residential development.

THE ROLE OF THE LANDSCAPE ARCHITECT IN LANDSCAPE RECLAMATION

Brian Clouston dip.L LO NOH DH Edin FILA

R.D. McLean C.Eng. BSc. MICE GI Env.Sc.

Synopsis

Landscape architectural philopsophy is the total integration of restored land into the pattern of the adjoining landscape and the creation of a viable land use which allows the land to be managed in perpetuity as part of a total landscape. This paper develops this theme and cites several examples where it has been adopted on contaminated land.

The Nature of Derelict Land

Derelict land is a result of industries need for materials extracted from the ground, such as coal, iron ore, slate, stone, sand gravel and clay and of changing economic and commercial cycles which lead to the decay of the obsolete machinery and buildings associated with the extraction and the wastes generated from use of the materials.

Light industries may cause little disturbance below ground, but leave a rotting superstructure on site and a compacted ground surface. Heavy industries such as smelting, iron and steel production, and manufacture of machinery may leave holes in the ground, or massive foundation and sub-ground such as ducts, pipes, etc. Waste tips liable to spontaneous combustion and composed of toxic materials may pose special reclamation problems.

The biggest dereliction in the UK is associated with coal mining. Twenty million tons of coal have been mined in the last 100 years, leaving 2,000 spoil heaps, each containing between 1 and 4 million tons of waste on 10,530 hectates of land. Large tips generally have the greatest impact on the landscape because of their height, unnatural, uncompromising shape and barrenness.

Sand and gravel extraction creates the largest amount of land disturbance forming water filled or dry pits on a large scale. Often these are filled with town refuse or the water filled pits used for sailing and boating. The heaps of toxic waste from chemical industries and energy producing plants pose particular problems regarding restoration, planting and after use of the sites. These problems are likely to become more complex with the development of nuclear power.

This then, is the nature of the problem which faces countries which possess the natural resources required by industrialized societies. Derelict land is useless land, often ugly, often dangerous. It can be reclaimed for future use only at a cost, which once accepted can turn a liability into an asset. An important member of the restoration team is the landscape architect, who with his special skills can restore the land as part of its original landscape through earth shaping and planting, or turn it into land suitable for new uses. The role of the landscape architect will be discussed at some length later in this paper but first it is pertinent to consider the motivation behind reclamation schemes in the UK and elsewhere since this is a formative influence on our theme.

The United Kingdom was one of the first countries to promote programmes for the reclamation of derelict land, although until 1948 action was purely voluntary. The present position where large sums of money have been provided in the form of government grants can be attributed to a gradual realisation of the extent and nature of the problem over the past twenty years. Increasing pressures on the land created by a growing population with limited space to expand, underlines the need to make the best use of the land we have at our disposal.

Efforts by local authorities and enlightened mineral operators to reclaim land provided examples of what could be achieved where the will to tackle the problem existed. Technological advances provided the means in the form of heavy earthmoving equipment developed for civil engineering and mining. Scientific study showed how wastes and sterile soil could once more bear plant life. But these efforts were not co-ordinated at the national level. There was no clear plan of action. Initiative and responsibility were confined tothose enlightened people in positions of authority or influence who could no longer accept the degradation of their environment and determined to do something about it.

The tragedy in the village of Aberfan, was a glaring example of the inadequacy of our measures to deal with derelict land. On 21st October 1966, 28 teachers and 116 children were buried in the coal sludge of the tip which engulfed the village school. The disaster provoked a national outcry. Gradually, however, through the 1960's and early 1970's dereliction has been recognised as a land use problem rather than the negative view which emphasises the danger to the public from subsidence and unstable spoil heaps which was previously more popular. The government has actively encouraged reclamation since April 1967 when the whole country was covered by a substantial grants scheme for this purpose.

Despite the improvement in attitudes towards reclamation, the progress so far gives us no grounds for complacency. It was estimated in 1974 that the net area of official derelict land is increasing at a rate of about 1500 hectares per year, and that in England alone there are still about 48,600 hectares.

Thus, despite the impact of the Aberfan disaster, the main motivation behind the reclamation programme in Britain is the immediate and long term creation of jobs. Footloose industries are encouraged to move to run down areas by government incentives and the improvement to the attractiveness of the environment in these areas.

The most commendable efforts are those which stem from local initiative and attempt comprehensive organisation of reclamation on a large scale. Some have been motivated by University research teams bringing a wide variety of specialist skills to focus on the problem, as at Swansea and in Newcastele. Others have been motivated by co-operation at local government level, transcending, for once, local government boundaries, as in the Lea Valley and Hillingdon. If schemes like these can be successfully operated, others may follow their example, and land reclamation may be seen throughout the country as landscape reclamation, something far more positive and enduring.

It can be seen from the examples in this paper that reclamation programmes bring obvious changes to the appearance of the landscape, in some cases land use changes, and in others the abatement of air and water pollution. But the political motivation behind each reclamation programme may be quite different. These differences are important. Important because until we gather together forceful, well argued cases for land reclamation, based on the premise that it is right to use land, but wrong to use it up, we will constantly be falling back on inadequate legislative devices. While they allow reclamation to take place, they fall short of the ideal that there should be a planned and viable after use of the land which should ensure its total integration into the pattern of local landscape. It is only when this basic philosophy is universally accepted that reclamation teams comprising the minerals operators, local planners, landscape architects, land agents, chemists, ecologists and land economists will be formed to preplan the restoration of mineral workings and the reclamation of derelict land waste tips. Only when we plan for a viable after use can we hope to achieve land management in perpetuity.

Reclamation and the Landscape

> "The real tragedy of England, as I see it, is the tragedy of ugliness. The country is so lovely; the man-made England is so vile ...."

Here D.H. Lawrence was speaking of Nottingham and the coal industry of the Midlands but the scene he no doubt visualised in writing these words would equally have applied to hundreds of other similar industrially blighted areas. Since the 1930's attitudes have changed. The UK is now, as a nation, well advanced in a programme of reclamation of the inherited derelict land left over from the first industrial age. But too often the aim is simply to tidy up the mess, covering it with smooth greenfields, totally devoid of landscape features such as trees, knolls and streams. Thus restoration work can, at its most unimaginative, mean the loss of those characteristics which distinguish one landscape from another, and hence the loss of that richness, variety and close texture which forms the traditional English landscape. We must understand our own landscape, the history of its development, and its unique character, if we are to attempt to restore to it some of the beauty it has lost to man's exploitation of minerals.

The landscape of lowland Britain, Lawrence's admired 'lovely country' is man made. It is acknowledged throughout the world to be a special kind of landscape. Why does the agricultural landscape of Cape Province, New England States and parts of Natal resemble in so many ways the English landscape but contrast so profoundly with the landscape of the much later developed areas.

Perhaps the answer can be summed up quite simply by saying that the differences are due to differences in attitude - differences in attitude developed over a comparitively short period of time and influenced by the seventeenth and eighteenth century landscape architects working in England, commissioned by the great landowners of the time. Their influence extended far beyond the boudaries of the estates upon which they worked. Landowners with less means became their own landscape gardeners and the Commissioners responsible for the sub-division of the landscape during the Enclosures imitated what they saw so successfully that a complete and similar pattern of landscape was developed in a matter of a hundred and fifty years and thereafter carefully husbanded.

The English landscape was thus created and managed under the stewardship of landowners and farmers to become a viable economic landscape, a lovely landscape. Just as the pattern of the agricultural industry was influenced, changed and improved by the landscape gardeners of the past, so too can the pattern of the extractive industry today be influenced by the landscape architect.

As Kenneth Wallwork has written:

> There is much more to successful long term reclamation of derelict land than unleashing mighty machines in order to devour spoil heaps and fill abandoned quarries.

A major part of the despoiled landscape of the UK was created because the simple planning logic of the 18th Century landscape movement had been ignored and quick profits meant more than beautiful and functional usable surroundings. The role of the landscape architect contributing to the planning of mineral workings is the most simple and practical to ensure that a straightforward logic is applied to the planning, working, restoration, after use, and after management of the mineral operation. A great deal of money being allocated to the reclamation programmes in Europe, South Africa and North America is being badly spent because on completion of the reclamation works no viable after use for the land has been created which is capable of management in perpetuity as an integrated part of the landscape.

The earliest work in the UK took place during the 1930's; it was started initially as a means of relieving unemployment in the old industrial areas rather than reclaiming land for a positive after use. Attempts were made to integrate spoil heaps into the landscape by planting them with tree. This method of rehabilitation did not wholly succeed because the unnatural form of the pit heap remained obvious and planting techniques were insufficiently developed to ensure success on every occasion. It did, however, show that trees and vegetation could be established on coal and other waste material and provided valuable guidance for future works in this field. Following the Second World War more planting took place, particularly in Lancashire (north west England) Durham and Northumberland (north east England), in further attempts to disguise and camouflage, despoiled land. Grassing experiments followed, but it was soon realised that unless full scale reshaping of the land took place and a viable after use was created, there was little hope of a reclamation programme being anything more than an expensive cosmetic exercise and a continuing maintenance burden. During the 1960's, and following the injection of capital after the Aberfan disaster both the rate at which derelict land has been reclaimed and the quality of the reclamation work increased dramatically. Today there are many examples throughout the country which show that the application of grading, new vegetation and drainage, working towards a planned and viable after use of the land, allows the reclamation site to be fully absorbed visually into the surrounding landscape. It is now possible, for example, to plant trees direct into shale wastes without the added expense of importing topsoil, or to seed colliery spoil and pulverised fuel ash. Many areas of dereliction cannot now be seen as anything more than useful farmland, recreational open space, or urban development except to the expert eye.

A case in point is Bowden Close colliery in County Durham. This pit closed between the wars leaving ruined buildings, the foundations of the mine headgear and its associated coal and tar works. Waste had been dumped across the mouth of the valley, obscuring the landform. Land previously underlying the tip is now used for agriculture and an extension to the local golf course. Volunteer wood and scrubland areas have been conserved as public open space, forestry areas and a nature reserve.

Silksworth Colliery on the edge of the Sunderland conurbation is a much bigger area of dereliction undergoing transformation. This project illustrates the difficulties which can be encountered by the landscape architect in reclamation work, and the effort required to turn despoiled land into a useful resource. Two-thirds of the sixty four hectare site lies derelict, dominated by a coal shale heap tipped along the valleyside, clearly visible from the town below. The tip presented the most difficult reclamation problem as the greater part of it is either burning or potentially combustible, and in certain parts salinity is at a level toxic to plant growth. These problems have influenced the form of the earthworks, surface treatment and drainage provided. Sunderland now has a much needed new recreational complex, visually integrated into the surrounding landscape by the creation of a sympathetic landform and planting. An athletics stadium, ski slope, sports centre, playing fields and two lakes are among the facilities proposed.

In Scotland, reclamation work is being tackled to improve the coastal environment and provide recreation facilities capable of withstanding the predicted pressures created by the development of Irvine New Town, near Glasgow. The scheme has been designed as a Beach park providing a wide variety of activities on a site formerly used for manufacturing industry, the making of explosives, fly tipping and sand winning. A 20 metre high, steep sided tip of chemical waste has been regraded to provide a ski slope and outdoor amphitheatre and a large area of sand winning has been further excavated to form a lake. The edges of the lake are varied with concrete, larch posts and several shingle beaches for paddling. Special precautions were necessary to counteract wave erosion.

The problems created by the need to plant over toxic wastes have been considerable, and on-site trials were established to test the efficiency of different types of blinding in inhibiting the upward movement of chromates. As a result the exposed or deposited chromates, were sprayed with liquid sewage sludge which reacted with and neutralised the harmful chromates, then blinded with 350mm of lime waste material from the site, followed by a further metre of sand and topsoil. Grass is now growing satisfactorily over all areas underlain with toxic material. There were special problems concerning the sand dunes which necessitated trials to observe the rate and nature of sand accretion and experimentation with the establishment of dune vegetation. Careful management should ensure that the regraded and planted dunes are soon restored.

Gateshead Riverside Park, now the venue for some of Brendon Foster's International athletic events was formerly a combination of colliery and chemical industry dereliction. Reclaimed in the early 1960's, the toxic waste heaps have been remodelled, capped with protective soil and drained. Vegetation is now well established, restoring green banks to the revitalised waters of the River Tyne.

Common to every reclamation project is the need for adequate and effective drainage. Although topsoils and subsoils can protect vegetation and the public from contaminated wastes, rainfall will percolate through these layers, and leach out potentially harmful compounds, possibly to reappear in the drainage system at a lower level. Leachates and their influence on the feasibility of any water feature and existing waterways requires particular attention. At Hillingdon's Willow Tree Lane Project contaminated leachates were collected and passed into the existing sewerage syste. Upward movement by capillary action of saline water and water contaminated by heavy metals was prevented by incorporating a granular capillary break layer.

Brenda Road Slag Banks in Hartlepool resulted from the iron and steel industry and associated coal carbonisation plant. Currently the site is being reclaimed for industry and recreation. The ferous slags with their potential for expansive chemical reactions are being excavated, relaid and compacted to provide industrial plots. A two year monitoing programme by the British Steel Corporation's Slag Utilisation Department will provide background data for research into some of the problems of utilising older ferous slags. Potentially toxic wastes from the carbonisation plant were tipped in lagoons around the site. By incorporating these in deep fill within the recreational landscape zone sufficient cover was placed over these difficult wastes to render them innocuous.

These examples give some idea of the range of work being undertaken on land which is already derelict. But it is not sufficient to reclaim land it has been despoiled. There is much that can be done to aid reclamation while extraction is in progress, as is illustrated by the work of the Opencast Executive of the National Coal Board. Unlike the deep mining section of the NCB, this group has earned itself an admirable reputation for reclamation work.

Other extractive industries are not so well organised. Deep coal mining still results annually in around 405 hectares (1000 acres) of land being buried by spoil, and little reclamation work has been done. Sand and gravel extraction is the biggest disturber of land. About half the new pits are restored by tipping after use. Less reclamation takes place after limestone clay and shale, chalk and china clay workings.

The degree to which dereliction affects the landscape depends on the type of material being worked and the method and extent of extraction. The skills of the landscape architect can be utilized from the outset in order to ensure that the work has the minimum effect on the landscape both during and after the operation. Work in progress can be concealed by shelter planting; unnatural cuts into the hill skyline avoided, and material extracted in such a way as to make the job of reclamation for a specified use easier. If the landscape architect is employed at this stage, the operators are more likely to ahcieve success in planning applications to extend their works and the local authorities concerned will, in all probability, be better satisfied with the after use of the land than they would be otherwise. In the long term the skills of the landscape architect can help ensure that new areas of dereliction are not formed while work is in progress, and that restoration takes place as soon as possible. But no scheme will be successful if the reclaimed land is not properly managed and maintained after initial works, keeping a careful watch on soil levels and plant growth.

An increasing amount of mineral extraction is taking place in south east England and in other intensively developed areas where industry communications, housing, quarrying, agriculture and amenity compete for the same limited area of land. The different land-use relationships which emerge from the struggle are frequently incongrous and incompatible.

The landscape suffers as a result. The landscape architect can help determine the impact of mineral workings on the local environment, and suitable after uses in the form of a comprehensive plan which aims to avoid conflict between the workings and other competing land-uses. This involves the integration of some of these other land uses into the development of the mineral working and its restoration plans. Multiple use of this kind offers a flexible and economical solution to the worsening problems of land shortage.

The types of after use which can be considered will depend upon the needs of the locality and the planning proposals for the area. Some sites can be re-used for industry or housing, thus leaving other open land undisturbed. Often derelict land in the UK is located in the heart of urban areas where it may form the only open space left. In such cases re-use for recreation and parkland may be appropriate, providing much needed local facilities. Hard level surfaces in particular can often be formed in urban areas for organized sport such as tennis, squash, running, etc. It was estimated in 1074 that about one fifth of all land reclaimed in the UK was used for open space and recreation and that the proportion was increasing. The increasingpopularity of water based sports can provide an ideal after-use for sand and gravel pits; often the presence of such facilities create their own demand where none apparently existed previously.

Conclusion

In Britain we have a fine record in the field of derelict land reclamation. Many of the planning and design skills and the revegetation techniques learned during the implementation of the restoration programme can usefully be applied to restoration work in current mineral operations and in the planning of new works. At the"Countryside in 1970"conference in 1965 it was argued that comprehensively planned reclamation policies could have an effect on our landscape comparable to that of the creation of parks andcountry estates in the eighteenth and nineteenth centuries. We certainly have the design and technical skills to transform our landscapes for the better. But it has yet to be universally accepted that the high price we must pay to restore our degraded land is a necessary cost, as surely as it is necessary to our economy to exploit our mineral resources. It has been said that 'the prevention of future dereliction depends on ensuring that mineral working is more strictly controlled and that restoration is funded as a normal production cost,rather than a state financed afterthought. Often neglecting to spend money restoring land at the earliest possible time simply increases the cost that will eventually have to be paid.

Beauty is priceless. Although it cannot adequately be translated into language of profits and losses, the effect on our lives of a beautiful environment can be measured in terms of human well being. The landscape architect plays a key role in improving the quality of the landscape. His responsibility extends across the whole field of land use planning for a landscape must be capable of management during its most vulnerable years of growth if its future quality and suitability for afteruse is to be ensured. There are more tangible benefits to this. Reclamation particularly in areas close to large centres of population can provide much needed land for housing, industrial development, and precious open space where it is most required. A change of heart on the part of less enlightened mineral operators should be accompanied by a willingness on the part of the government to accept responsibility for organising reclamation on a continuing basis, and strengthening the legislative backing for such an operation. Then those responsible for exploiting mineral resources can work hand in hand with those whose job it is to restore and create landscapes of quality and utility.

THE VALUE OF DERELICT LAND FOR ENVIRONMENTAL EDUCATION

CHRIS BAINES, B.Sc., Dip L.A., M.I. Biol. Lecturer Birmingham Polytechnic and Partner in The Landscape Design Group, Kidderminster.

ABSTRACT

This paper discusses the need to conserve rich areas or urban dereliction, highlighting the value of such areas as gravel quarries, pit mounds and railway lines both for wildlife and for creative childrens' play.

The paper also stresses the need to restore derelict sites at a much more human scale, in order to re-create those qualities which make intimate landscape attractive.

Emphasis is given to the need for a more professional approach to the techniques of applied ecology being used in land reclamation, and after siting specific technical examples, the paper advocates the need for a much stronger management input, and identifies professional managers as having a dual role to play both as creators of the rich detailed landscape and as the wardens and interpreters needed to develop a new environmental awareness amongst children, and amongst the public in general.

Children need to grow up in a landscape which is rich in opportunities for adventure, discovery and mild destruction. The need to build dens, light fires, dig holes, catch tadpoles and pick buttercups is absolutely fundamental.

In the past twenty years, Britain has spent thousands of millions of pounds sterilising its open spaces and as children have become more and more deprived of rich environmental experience, so the boredom and aggression of youth has increased. For the most part, our public parks cater only for formal ball games and the acceptable passive recreation of the middle-aged. The only excitement a child of eight or nine is likely to have is the thrill of being chased out of the bushes by the "parkie".

The space around schools is surely the most obvious place for children to study the natural environment, but here too the "gang-mower rules - O.K." and more often than not the entire area is manicured to cricket pitch standard. Many of the sites proposed for new schools possess such valuable assets as hedgerows, ditches and marshy areas, but by the time the building receives its first intake of eager young minds, the hedges have been grubbed, the ditches culverted and the wet areas drained - all in the interest of grounds maintenance simplicity, and with little thought for the education of the children.

Whilst planners, engineers and parks superintendents have been ironing out the rich wrinkles in the towns, farmers have been doing a similar job in the country. Here again the power of the machine, the increase in economic pressure, and the "tidy mind syndrome" have combined forces to produce a clinical countryside - beautifully managed to feed the body but not the soul.

A farmer with an area of thorn scrub on his land, an un-drained wet field or un-fertilized pasture is regarded by his neighbours at best as an eccentric to be humoured, and at worst as a disgrace to the farming community, unable to cope with modern methods and likely to generate problems of pest and disease for adjacent farms. All this pressure for uniformity, simplicity, mechanisation and tidiness combines to produce an environment with very little of the richness so essential for well-balanced wildlife and well-balanced people.

Shafts of sunlight are at last beginning to filter through this big black cloud. Dramatic cuts in public spending have produced such unexpected bonuses as flowery road verges and un-pollarded street trees. Schools are devoting more and more time to Environmental Education and some local authorities are making efforts to enrich school grounds through changes in management and through positive action such as tree planting, pond excavation and construction of artifical nesting sites.

Our own consultancy work includes the creation of a community butterfly garden in the middle of a high-rise housing estate in Deptford, wedged between Milwall Football Ground and the rich ecological corridor formed by the New Cross railway embankment.

We are also streamlining the grounds maintenance of a large psychiatric hospital at Epsom, to make the management more effective, whilst at the same time creating rich habitats for wildlife and developing the opportunities for psychiatric therapy.

The real silver lining to the big black cloud lies in the recognition of our unreclaimed 'derelict land' as a valuable environmental resource. W.G.Teagle in his recent book, "The Endless Village", highlights the value of urban wasteland for wildlife, and certainly anyone who enjoyed a childhood on the urban fringe as I did will have vivid memories of fists full of coltsfoot flowers presented to Mum "courtesy of British Railways", or jam jars crammed with frogspawn or sticklebacks from the quarry pond. Back in the fifties the opportunities for adventure and discovery were endless, but now shallow ponds and patches of rosebay willow herb are desperately scarce, and those that do survive are eroded by the desperate patter of tiny feet, and coveted by the Parks or Environmental Health Dept. as prospective candidates for "reclamation". These magical places are drained, levelled, cleared of scrub, and transformed into features known as C.P.A.'s (Childrens' Play Areas?)

For the most part derelict land has so far enjoyed only two standards. BEFORE, when the site is totally disregarded by The Authorities, abused by flytipping, damaged by fires which get out of control and is full of very real dangers which do little for the quality of either the wildlife or the childrens' play; - and AFTER, when all the spirit has been beaten out of the landscape, no slope is steeper than 1 : 4, the bramble and hawthorn have been replaced by skeletons of well staked standard Sycamores, and the kids are labelled vandals for continuing to do all the things they did without question before the land was "improved".

If we are to reclaim our derelict land more positively, then we must find a compromise between the BEFORE and AFTER of the 60's and 70's. We must find ways of making sites safe without removing all scope for adventure. We must find ways of revegetating slopes without producing a uniform green sward or a stand of colonally-selected standard trees - fine on plan but hopeless in section.

I believe the biggest advance will come with the recognition that any landscape is something to be enjoyed from within. Our open spaces are for people, but they are all too often designed at a scale no larger than 1.2500, on a flat piece of paper in an office remote from the site and the client. The designer, be he engineer or landscape architect, must recognise the need to conserve old wrinkles and - even more difficult - create new ones. As human beings, we know how we use land, but as professionals we seem all too often to forget that we are human beings too. In recognising that which is attractive before reclamation - the pond, the flowery meadow, the three foot 'cliff' and the hollow-centred elderberry bush, we should recognize the key to successful restoration. Landscape works for people at a very detailed level. Obviously there is pleasure in standing on an exposed hill top, looking out across rooftops to the open countryside or the sea, but this pleasure is soon exhausted. It is the intimate spaces which stimulate the senses - narrow valleys, woodland glades, lakeside bays, but most important of all it is the variety of space; the change in character from sunlit to shady, from damp to dry, from enclosed to open. Good picnics take place in spaces as small as a living room, but it is impossible to get the "Letraset" into a space that small on a 1 : 2500 plan, so the result is picnic spaces the size of football pitches, sprinkled with British Standard picnic tables and peopled by families exposed, embarrassed, uncomfortable, but with nowhere else to go.

Cynicism is all very well. Criticism is easy with hindsight but I do recognise the need to be practical. The economics of much-shifting demand that the landscape be re-shaped by huge insensitive machines, and yet those designers who spend more time waving their arms about on site, and actually talking to the machine driver, and who keep working drawings to a minimum, manage to achieve remarkably sensitive results with those same brutal tools.

Assuming a reasonably sympathetic scale at the engineering stage, a great deal more can be done through sensitive re-vegetation. Not only can trees and shrubs create small, soft spaces; if chosen properly they can very quickly add a fourth dimension by bringing the site to life. To sit amongst trees is enjoyable, but to sit amongst trees filled with birdsong is magical.

It is possible to re-create rich habitat, but the techniques of creative ecology demand every bit as much professional skill as that needed to establish a bowling green, or a stand of high productivity commercial forestry. By planting native species, by fostering both a tree and a shrub canopy of pioneer vegetation, by supplementing these with a seed mix of native herbs and grasses, and by avoiding the lavish over-use of rich top soil it is possible to give the new landscape a substantial boost in the direction of a rich balanced community, but the real success will only come through sensitive professional management. The rigid, regular maintenance regime of a playing field is simple compared with the creative management needed to develop a rich habitat from a basic planting matrix. The decision to stop suppressing weed growth, and to begin encouraging those same invasive competitors to form a rich herb layer; the decision to coppice certain trees and remove others in order to promote variety in the canopy and renewed vigour in the stems; the decision to introduce climax species such as Oak or Beech when they fail to appear naturally after ten or fifteen years; the constant battle to remove Sycamore seedlings, and prevent these and other non-native invaders from dominating the copse - all these actions can be anticipated to some extent at the design stage, but the precise timing is often critical, and this can only be determined by a manager who understands the design objectives which underly the scheme, the general principles of ecology which form the basis of this kind of approach, and the idiosyncracies of the particular site in question. In other words, the success of a piece of rich landscape design will depend upon the skill of the manager. This is vitally important, and fortunately it is the key to the success of these same schemes in influencing the quality of life of the people who use them. The local knowledge of the manager can bring the landscape to life. Areas of rich unreclaimed urban dereliction, and hopefully the wave of newly created reclamation sites which are to come, have the enormous attraction of accessibility. They are much more available to people than the country park or the heritage coast, but familiarity breeds contempt, and these areas are abused through lack of education. They are not recognised as being of any importance, and this is because for the past twenty years urban man has been losing touch with nature. His affinity with the countryside needs to be re-kindled, and this can best be done at familiar local level by those same managers whose detailed knowledge is so essential to the development of the site. By changing the emphasis from maintenance to management, by employing sensitive communicators to act as wardens, rather than in-sensitive machine drivers and uniformed patrolmen, and by spending money on low key techniques closely akin to nature, instead of clever feats of horticulatural technology destined to survive no longer than the defects liability period, our derelict areas can become the foundation of a new environmental awareness, and our most deprived communities can become our most privileged.

In recent years mechanisation has become synonymous with economy and efficiency. Suddenly the oil crisis is bringing that philosophy into question, and in this field of natural environmental improvement the need for a labour-intensive approach can be of double benefit. Firstly, it can allow a much finer degree of detail in the new landscape. Individual plants can be conserved, irreplaceable features such as ant hills and bat roosts can be saved and rich surface soils and leaflitter can be kept undisturbed. Secondly, there is the enormous social benefit of real public participation. The gradual development of rich habitat through sensitive management simply will not fit into the straight-jacket of the competitive tender system. It does not lend itself to time-and-motion studies or incentive schemes, but it is ideal for groups of enthusiastic 'helpers' keen to get muddy, light bonfires, dig holes and just be out in the fresh air for a day. Generally speaking, our derelict sites are tough. They can cope with over-enthusiasm very well, and through the continuity of the management process the people involved will develop an intimate understanding of plants, animals and the environment in general.

This opportunity for direct, active involvement with nature should be available to all sectors of the community, but it is perhaps especially important for young children. There seems little doubt that an active, close involvement with plants and animals at junior school age promotes a greater respect for the environment in the teens. There are several examples of housing areas close to derelict, unimproved wasteland where the level of vandalism in the adjacent housing is negligible compared with those estates whose children have only the swing, slide and see-saw for amusement. There is a site in Birmingham too, where the 'vandals' who were destroying the landscape have been inspired by the enthusiasm of a dedicated teacher and been persuaded to join the management group. After a summer of hard work these same "vandals," are proudly showing off their work to the Establishment.

There is one vital factor which must be born in mind when involving the community, and children in particular. Enthusiasm is not enough. It is essential that the work is handled in a professional way. Involving children in tree planting, for example, is a marvellous way of promoting an enjoyment of their environment, but if the trees die then the disappointment can be soul destroying. Sad to say, a great deal of the well meaning enthusiastic landscape management carried out by schools, youth groups and the like does fail, and often for lack of the most fundamental knowledge. Simple improvements in technique can make a dramatic difference. Planting trees only in the dormant season; planting native pioneer species; choosing small plants with a healthy root system, and avoiding standard trees which are unlikely to survive the move from sheltered nursery to slag-heap; pruning, and in many cases cutting back to the ground to compensate for root loss and encourage vigorous basal growth; soaking-in to prevent soil settlement- using polyuronide root dip to increase early water up-take; protecting against rabbit and vole damage - all of these techniques and many more are simple to implement, some represent a cost saving, and they can all make a great deal of difference to the success of new planting, but they seem, perhaps not surprisingly, to be almost unheard of amongst school teachers, youth leaders and others who initiate this kind of work.

So our areas of dereliction represent an enormously rich resource. If we are to exploit it positively, then we must be much more delicate in our conservation of that which is rich, and much more skilful in adapting that which is poor. We must promote landscape with detailed, intimate character. We must echo nature, and create through gradual, sensitive management rather than dramatic capital expenditure, and most importantly we must involve people. The richness of the landscape experience lies in the changing seasons, the dramatic growth of vegetation, the detailed observation of plants and animals and the intimacy of a close involvement with natural surroundings. Our remaining derelict land represents a fantastic opportunity for people to renew their acquaintance with nature, and we have a responsibility to change from creating smooth, simplistic, sterile scenery to developing complex, intimate landscapes, rich in opportunities for fun, adventure, discovery and understanding.

Bibliography

RUFF, A.R. Holland and the Ecological Landscapes
c/o Manchester University.

WARD, C. The Child in the City
Architectural Press 1977.

TEAGLE, W.G. The Endless Village
Nature Conservancy Council.

EDUCATION AND COMMUNITY INVOLVEMENT IN RECLAMATION
THE LOWER SWANSEA VALLEY

S. J. Lavender, B.Sc., M.I.Biol.
Conservator of the Lower Swansea Valley.
University College of Swansea, West Glamorgan.

Abstract

The reclamation of the Lower Swansea Valley is still in progress after over twenty years. Major works are carried out by the City Council and contractors. Smaller but vitally important projects were, and still are, carried out by local youngsters. This paper explains the background to the Project and the educational value of community involvement in reclamation work.

Introduction

Two hundred years of copper, zinc and tinplate industry changed this once attractive South Wales Valley into a vast area of dereliction and polluted land. Most of the 475 hectares of the Valley were covered with tip material from the smelters, disused abandoned buildings, or badly eroded boulder clays. Following the second World War much of the City's finance was put into rebuilding and redeveloping the City centre after the terrible bombing raids. Consequently there were no funds available to reclaim the badly polluted Swansea Valley. As the 1950's passed by, so Swansea people turned their backs on the Valley's problems. Together with the lack of finance both at local and National level, a situation developed where it was likely that nothing would ever be done about 'the most concentrated and uninterrupted area of industrial dereliction in Britain'. (Barr 1967).

The Project

In 1961, through the initiative of members of the University College, Swansea, together with financial assistance from the City and County Council, the Welsh Office and in particular, the Nuffield Science Foundation, a four-year study was established to investigate the problems of the Valley and possible methods of bringing about its redevelopment.

The thirteen study reports, Conclusions and Recommendations were published in the Lower Swansea Valley Project Report. (ed. Hilton).

The major recommendations included the fact that the first step to redeveloping an area of derelict land is to bring about its visual improvement. An area is much more acceptable to a developer if it looks attractive rather than if it looks ugly.

The botanical studies of the Project attempted to provide cheap, simple methods of covering tips and eroded boulder clays by grassing and tree planting. The grass trails and experiments succeeded in obtaining an acceptable grass cover on all types of tip material. By using cheap sources of fertilizer and particular grass mixtures, a surprising variety of plants, fungi, mosses and liverworts were observed to grow on tip material.

The cover of vegetation obtained was suitable only as a cosmetic measure as the vegetation would not withstand any recreational pressure. Whilst the grass cover would not be suitable for sports pitches and so on, there are two major reasons for establishing a cover of vegetation on tips.

The first reason has already been discussed. If an area looks attractive (i.e. green and not black), it is much more likely to be taken up by developers. When the ultimate land use of a derelict site has not been finalized it is sensible to cover the tip directly with a tolerant specie of grass rather than carry out expensive landscaping work.

Secondly, if a tip is covered by vegetation, the amount of wind-blown material from tips can be reduced. In the case of tips with unusually high levels of heavy metal, any reduction of wind-blown fall-out would be of great benefit to surrounding vegetation. If grass cover could be established on steep sided tips the action of the roots holding the material would also prevent further erosion by rainfall washing the material down the slope.

Trees would not grow directly into tip material. Experiments did however, establish species which would grow on the badly polluted and eroded boulder clays. Of the 26 species tried on the Valley soils, four species were selected and planted on specific sites in the Valley. Lodgepole Pine, Japanese Larch, Corsican Pine and Silver Birch were planted on an agency basis by the Forestry Commission. A total of ten thousand trees were planted in 1962-3.

Within a very short time most of these trees were killed, not by the conditions, but by the weeds, vole, rabbits, fire and by vandalism.

It was now realized that something had to be done to combat the serious problem of fire and vandalism. This had to be done by showing the local people the value of the forest areas. If these problems could not be overcome the forest areas would be destroyed by the very people for whom they were intended.

Suggestions that fencing and patrolling should be used were turned down because in the first place they would be too costly, and secondly, these ideas were not compatible with plans for this area.

The Conservator and the Schools

It was eventually decided that visits to local schools by a member of the Project team might help the local young people realize the value of the planted areas.

Permission was granted to visit schools by the Director of Education and so the Projects Forest Officer (later to be called the Conservator), visited schools and talked about the History of the Valley to the children.

Early in the Project, the Forest Officer would encourage school children to join in with treeplanting and in some cases school forest plots were established. It was hoped that these young people might regard the forests as assets rather than as a target for vandalism. Since this part of the Project began young people have been involved in treeplanting every year and many thousands of the Valley's trees have been planted by school-children. Teachers gradually realized the value of children being involved with reclamation Projects. Consequently a demand developed for more information about the Project area. Visits were planned, Nature Trails laid out and a duplicated trail leaflet was produced.

In 1966 the funds from the Project providing the employment of the Forest Officer were starting to run out. It was essential that the involvement of the local children was continued in order to ensure the future security of the forest plantations.

By continuing the visits to schools it would be possible to help each generation understand the value of the trees and the value of protecting them. In late 1967, the City Council agreed to provide funds for the employment of a Conservator who would be based at the University College, Swansea, and have a special responsibility towards the reclaimed areas.

The City Council provided yearly grants to the University and each Conservator stayed one or two years and used the position as a stepping stone to a more permanent appointment. In 1978, the Council gave the Conservator a three-year appointment. This enabled him to plan more long term projects and also to obtain assistance through the Manpower Services Commission.

As the years passed, so more schools, colleges, Polytechnics and Universities realized the value of the Lower Swansea Valley as a Study Area. It was now necessary to produce resource material to satisfy the demand for information from these groups.

The duplicated Trail Leaflet evolved into a proper leaflet which was published by the City Council in 1975. This leaflet - 'Discovering the Lower Swansea Valley' - follows the original trail and explains to the visitor different sites in the Project Area. A second leaflet - 'Industrial Archaeology Trail' - was published shortly after, (again by the City) and follows a footpath through another part of the Valley.

Both Trail leaflets are still used by local people and visitors who wish to know more about the changing Valley. Supplementary leaflets were produced by the City Council after groups of children had carried out specific studies in the Project Area. The first leaflet was produced following a study by a group of Ornithologists who investigated the bird life of the Valley during a full year. Two other studies were carried out by 'O' Level schoolchildren who investigated the Fungi and the Wild Flowers of the area.

The pamphlet about the Bird Life of the Valley shows how some birds have been able to live through the time of industrial dereliction and even thrived on the lack of competition, whilst other species have only recently arrived, being able to take advantage of the new sources of food provided by the woodlands. The Kestrel is one example of the type of bird who made its home amongst the derelict buildings, whilst the Coal Tit, is one example of the type of bird exploiting the new ecological niche provided by the coniferous forests.

Following on from the close involvement of schoolchildren and teachers in the Valley, a new-style trail leaflet was designed for young people visiting the Valley. With assistance from teachers and the local Library and Resource Centre a 'Response Trail' entitled 'Nature Reclaimed' was produced.

The leaflet follows the original trail but rather than being completely descriptive, it asks questions and has spaces for drawings and written answers. In this way it allows the child the freedom to make his own decisions on a variety of topics presented by the Trail. The topics in the trail include Reclamation, Transport, Pollution and Tree Recognition.

Once completed the Trail is kept by the child who may carry out further work in school and then take it home.

In 1978 a slide book entitled 'The Reclamation of a Polluted and Derelict Landscape' was published for schools and Colleges. Through the slides and short text, the history of the Valley and the various techniques used to reclaim derelict sites are explained.

The leaflets and books have all been produced in response to an increasing demand from visitors and local teachers for information on a variety of topics about the Valley. In early 1979 a full-scale pictorial exhibition was opened by Lord Sandford of Banbury at Swansea Museum. This is another resource for children and adults who wish to learn more about the history of the Valley.

During 1979 the Conservator obtained assistance in four areas from young people employed through the Special Temporary Employment Programme. A carpenter built the exhibition area in a gallery at Swansea Museum, a Forest Worker/Warden carried out patrols and maintainance work within the planted areas of the Valley; a Research Worker/Technician carried out a variety of investigative projects on new problems developing in the Valley; a fourth worked as an assistant to the Conservator and helped to extend the work done in schools and continued the valuable voluntary work carried out by groups of children and adults during their own time.

Community Involvement

In the early days of the Forest Officer, the Reverend E. Hunt, then the Vicar of the Glantawe Parish in the Valley, involved the young people of the Church Youth Club in small work Projects. This involvement showed that the young people were willing to spend their own time and energy protecting the Valley.

From 1968 to 1976 when the Conservator was employed on a yearly contract, the Reverend Hunt was invaluable in providing continuity, not only by holding the interest of the young people, but also by introducing each new Conservator to the area, and most of all to the young people. In 1976 the Reverend Hunt left the Glantawe parish for a different part of the City. In that year the Conservator did not change and he was able to bring the work of the youngsters to the attention of the new vicar.

The Glantawe Youth Club had been involved in small work tasks right from the start of the Project in the early 60's. By 1970 the value of this work was recognised by the grant of a Prince of Wales Award.

Each year from 1970, the Prince of Wales Committee for the Environment has made Awards to any groups or individuals who they believe have made an outstanding contribution to the physical aspect, or to the better understanding of the Welsh landscape.

As the forest areas grew so the young people were involved in building Nature Trails, Fire Breaks and setting up bird boxes to attract new life back into the Valley.

Following on from the Youth Club Projects, International Work Camps were organised and Overseas students worked around the area during a number of Summers building store shelters, a small dam, and helping with stone picking and fertilizing of trees. It was decided after the 1974 International Camp, that local youngsters would benefit more if such holiday schemes could be arranged for them.

Financial aid for an Adventure Holiday Scheme was obtained from the Urban Aid Programme. The Holiday Scheme was organized for young people who lived around the Valley. Each week the youngsters, aged between eleven and sixteen years would take part in small work projects and also be involved in visits to other parts of South Wales. By mixing work and play during the day and orginizing feature films and discos for the evenings, it was hoped that these young people might take more interest in their surroundings.

The work projects involved building steps on new Nature Trails, spreading fertilizer around the base of the young trees, litter-picking and keeping a constant fire watch. The visits included days out on the Gower beaches, Land Rover safaries through more mature forest areas, ice skating, strawberry picking and even attempting an Army assault course.

The effect of the Holiday Schemes has been remarkable. A great deal of work has been carried out by the local youngsters whose involvement encourages them to protect the forest areas rather than treat them as a target for vandalism.

The success of the Schemes prompted them to continue and such projects have been organized each summer since 1975. The concept of the Adventure Holiday Scheme won a Prince of Wales Award in 1976. Following the publicity of the Scheme a number of other holiday projects have been organized through the Prince of Wales Committee throughout the Principality.

Adult involvement developed through the enthusiasm of the children and from informal gatherings at a local public house. The Halfway Inn at Llansamlet is often used by Urban/Field Study groups after a visit into the Valley. Here students are able to talk informally with the Conservator and chat to local people about the changes taking place. In 1978 a social group within the pub, the 'Hatters' organized money raising events. As their interest in the development of the Valley increased so they wished to be more involved. Members of the group joined in tree planting and firefighting as well as working voluntarily as supervisors on the Adventure Holidays. In 1978 they donated a cheque for £500 to the Conservator to organize a formal Club for the young people who are involved with project work in the Valley.

With support from the local Carnival Committee and local industry the Lower Swansea Valley Rangers was established. Anyone who is concerned about the Valley and all the aspects of its life may join. By 1980 there were over 300 Rangers of all ages, recruited from the Valley schools.

Together with teachers, and the local Library and Resource Service and the Conservator's assistant a magazine was produced for the Rangers during 1979. Distributed through the schools the 'Valley Monthly' contained articles about wild life of the Valley, local history and topical articles. By combining these features with quiz pages and a diary of events, the magazine has been useful to children, teachers and parents by being instructive and amusing.

The Rangers Group has grown faster than anyone might have expected. The size of the group demonstrates the demand for activities and facilities for this age group.

Young people have a tremendous amount of energy. If this energy can be used in a helpful way much can be achieved. One of the most important lessons to be learned from the Lower Swansea Valley Project is the value of the Conservators' close involvement with the local young people, both in school and during their spare time.

Conclusion

Vandalism of trees and wild life is now minimal, and this success had depended considerably on the high level of community involvement. This degree of involvement has been achieved after a long struggle against apathy and required great enthusiasm, energy and patience on behalf of those involved.

The continuing education and constant involvement of young people in reclamation work will ensure that the future for the Valley's forests is secure.

References

Barr J.	Derelict Britain, Penguin 1967.
Hilton, K. J. (Ed)	Lower Swansea Valley Project, Longman 1967.
Lavender S. J.	Swansea Valley Scars are Healing. Geographical Magazine, March 1978.
Lavender S. J. and Wainwright S. J.	Reclamation of a Derelict and Polluted Landscape. Focal Point, Portsmouth 1978.
Swansea City Council	Discovering the Lower Swansea Valley.
Swansea City Council	Industrial Archaeology Trail: Lower Swansea Valley. Swansea, 1975.
Swansea City Council	Birds of the Lower Swansea Valley. Swansea, 1976.
Swansea City Council	Nature Reclaimed. (Produced by the West Glamorgan County Council and the Conservator). Swansea, 1978.

Current Legislation relating to Contaminated Land

R G P Hawkins MA Barrister at Law (Redland Purle Limited)

Abstract

The law and planning practice concerning contaminated land are historically reviewed; the contemporary legislation is summarised. The context of the 1972 environmental hazard offence is recited; licensed contamination of land is contrasted with societally unacceptable contamination; EPA practice is adduced.

UK planning experience is related; reform in development control is suggested. The complementary role of HASAWA 1974 is explained; a brief summary of the case history to be related at the Conference is given.

Synopsis of Contents

A. Historical Backcloth

"I have never felt that the contemplation of the past, with the knowledge that it cannot come again, need be a source of sorrow"

Duff Cooper - in Old Men Forget

1. Ever since England in the early 19th century moved from an arable into an industrial society, its Executive and Legislature have been attempting to control the changing environment by a progressive series of local and national enactments generally as a result of recommendations by government appointed working parties. The Public Health Act 1875 was the direct result of urban development in Metropolitan London following the industrial build up in the Greater London area.

   The 1898 Royal Commission on Sewage Disposal which sat for 10 years at the end of the century set the Biochemical Oxygen Demand (BOD), which still is the standard used by the 1974 constituted Regional Water Authorities.

2. In 1936 the Public Health Act imposed certain duties both on local health authorities and on occupiers of houses and industrial premises adequately to prepare their contaminated materials for further deposit. Again under Sections 91-100 of the 1936 Act which were not repealed by the Control of Pollution Act (CoP) 1974, local authorities are empowered to take action to secure the abatement of "statutory nuisances". Among the definitions of statutory nuisance is "any accumulation or disposal which is prejudicial to health or a nuisance". Thus sites, including land on which toxic or contaminated wastes have been disposed of dangerously, can be considered as statutory nuisances.

However, a nuisance must exist before it can be abated and it clearly cannot be good practice to await a dangerous situation and then to take action. The short Public Health (Recurring Nuisances) Act 1967 gave power to local authorities to deal with statutory nuisances which had abated, but which were likely to occur again on the same premises. These Acts remained in effect the only law for 30 years which could control the surface effect of the deposit of any matter on to land apart from certain Local Acts, the provisions of the Town and Country Planning Acts of 1962 and the 1963 General Development Order together with the Rivers (Prevention of Pollution) Acts of 1951 and 1961.

3. Because of the lack of public interest - for example in the local council chambers - there was in these years an absence of firm central and local government co-ordination. The dumping from council vehicles and the tipping up of contractors' lorries <u>was</u> waste disposal; that is one of the fundamental reasons why in the latter half of the 1970s after the implementation of the first phase of the 1974 CoP site licensing, it is more accurate to use the phrase waste management both in terminology and as a corporate aim rather than the old fashioned, and inherently less disciplined, waste disposal.

B. <u>Contemporary Legislation Summarised</u>

1. The student of societally acceptable management of contaminated land in the late 1970s is therefore faced with a multi-disciplinary subject embracing a wide range of skills and professional and scientific disciplines. In the context of such management, no recognised training forces or adequate professional qualifications exist, and (more important) little or no initiative to provide them is being taken by central or local government or the existing professional associations despite amendments[1] to this effect at the Report Stage in the House of Commons of the Control of Pollution Bill in July 1974.

2. It is appropriate also by way of introduction to recite the plethora of recent legislation affecting the management of contaminated land generally:-

   (i) The four parts of the CoP together with its 1974 to 1979 implementing statutory instrument legislation and descriptive circulars especially in the context of refused material (quaere whether this is synonymous with waste material) e.g. macro and micro prospecting for suitable geological, hydrological and hydrogeological areas suitable for contaminated land - either with reference to historical, present or future deposits - access to such sites, the conditions for licensing such sites and their relationship to existing and future planning permissions under the

(ii) Town and Country Planning Acts and General Regulations under the 1971, 1972 and 1973 legislation.

(iii) Health and Safety at Work etc. Act (HASAWA) 1974 together with its dependent legislation.

(iv) British Standard Institute and other Codes of Practice such as that for Asbestosis or the Waste Management Papers from which can be evidenced the general duty of every employer and employee to ensure, so far as is reasonably practicable, the health, safety and welfare at work of all his fellow employees.

(v) Recent and imminent developments in transport legislation for the adequate description of loads, the vehicle construction suitability for the carriage of homogenous or non-homogenous contaminated materials before or after rain dilution including the unfortunately titled Dumping at Sea Act 1974.

(vi) Factories Act 1961, CoP Part IV, the Alkali Acts, the Highly Inflammable Regulations under the Factories Act 1961 in the context of new techniques in waste destruction.

(vii) Rivers (Prevention of Pollution) Act 1951 and 1961. Water Resources Act 1963 and CoP Part II as far as they are concerned with management of leachate.

(viii) Various legislation concerning the care and control of rural and urban real property e.g. Occupiers' Liability Act 1957, Civic Amenities Act 1967, Farm and Garden Chemicals Act 1967, Dangerous Litter Act 1971, Defective Premises Act 1972.

C. Initial Disciplines

1. Ten years ago and a fortiori previously, the disposal of materials which seemingly had no further use was a localised service throughout the United Kingdom, the responsibility of small organisations with few resources. The deposit of industrial or other wasted substances was either undertaken by the manufacturer himself, often simply dumping the wastes within his factory's curtilage, or by a small local quasi specialist haulier who would be contracted to collect the materials for which no further use had been found and which were probably no more precisely defined than in a tea stained crumpled docket, "chemical wastes", and left to entrepreneurial skills to arrive at a suitable grave. The contractor therefore provided what was essentially a transportation service and the majority had little technical understanding of the nature of the wastes or the environmental danger that could result from their disposal in such an uncontrolled fashion.

2. Many thousands of disposal sites were in use and the wastes received therein in the same disparate form as the consideration invested to obtain their entrance. The total gulf which existed between the owner of the industrial waste and the owner of domestic refuse, the collection and disposal of which was the responsibility of the former 1400 local authorities, meant that there was a marked formal reluctance (but not necessarily reflected in the pragmatic attitudes displayed by some urban or rural district council tip foremen), to allow private sector industrial liquid waste on public sector domestic sites. Sites were slected for their proximity and their

convenience more than their suitability[2], their interrelationship with the considerable national problem of the restoration of derelict land was wholly ignored and many of the sites were extremely small and quite incapable of being administered efficiently or in an environmentally acceptable manner in relation to their contamination potential[3].

3. Early in 1972 an incident involving the concealment of cyanide wastes in the industrial Midlands of England rapidly became a national issue. Although no damage to persons, stock, etc. nor bruising to the environment had occurred - and the statistics show that there has been, albeit fortuitously, very little evidence of harm resulting from these historically crude practices - there were demands for tighter legislative control. The preamble to the Deposit of Poisonous Waste Act 1972 (DPWA) set out its intentions, viz to penalise the depositing on land of poisonous, noxious or polluting wastes so as to give rise to an environmental hazard, and to make offenders liable for any resultant damage; to require the giving of notices in connection with the removal or deposit of waste, and for connected purposes.

By S.1 it became an offence to deposit or abandon poisonous, noxious or polluting waste on land where it might give rise to an environmental hazard. The waste must be regarded as a hazard if it were present in sufficient concentration or quantity to threaten death or injury to persons or animals, or alternatively, if it were to threaten to contaminate a surface or underground water supply. Deposit of the waste in containers was not to be taken as excluding the risk of such a threat; the degree of risk was to be assessed by having regard to any measures which the owner or occuper might have taken to minimise the risk and from preventing children or other people from tampering with the waste. The phrase poisonous, noxious or polluting occurred in the Rivers (Prevention of Pollution) Act 1951 but it has never been the subject of close judicial analysis and the DPWA 1972 provided no further enlightenment; nor was there provided any definition of contamination.

4. A person charged with an offence under the DPWA can broadly plead either he acted under instructions from an employer or that he relied on information supplied by others, provided that he had no reason to believe that the waste was of such a kind that it would be an offence to deposit it. Alternatively, he might show that he had taken all such steps as were reasonably open to him to ensure that he had not permitted an offence.

5. S.3 of the DPWA established a system of notification procedures; thus it became an offence not to notify that land might be contaminated even to a level below causing an environmental hazard unless notices specified

a. the premises from which it was to be removed and the land on which it was to be deposited

b. its nature and chemical composition

c. the quantity to be removed or deposited, together with details of the number, size and description of any containers

d. the name of the person who was to undertake the removal.

A sub-section excluded from this requirement waste of a prescribed description or waste deposited in a prescribed way cf SI No 1017. The authorities to whom notification was to be made included the local government authority and the Regional Water Authority (or River Purification Board in Sc otland) both in the area from which the waste was to be removed and in the area in which it was to be deposited. A person charged with failing to give proper notice could plead that, with no reason to suppose otherwise, he had relied on information that the notices had been given by others. By S.4 it became an offence for the operators of commercial land to fail to give notice within 3 days that the waste had been deposited.

D. <u>The Effect on National and Local Government</u>

The DPWA despite its shortcomings had a number of significant effects viz:-

1. Under the notification procedures if (and a big if) local authorities and/or the RWA properly and assiduously collated the detailed data, then the scale and nature of the potentially hazardous waste problem in England could, to a large extent, be determined for the first time. The fact that the Dolcis Syndrome occurred (put everything into shoe boxes and forget it) has inspired the suggestion that if the notification procedure is to be changed under the implementation of S.17 of the CoP, then there should be an offence of inadequate collation of information supplied. A salutary thought indeed.

2. The elevation of the administration of, and assessment of the industrial waste disposal problem to a regional, and in some instances, to a national level.

3. The direct involvement for the first time by the county and metropolitan councils in the surveying and planning of the problems of regional and industrial refused materials management.

E. <u>The Effect on the Management of the Waste Materials' Industry</u>

1. Increased professional standards and disciplines within the waste management industry.

2. The closure of many unfenced, fly blown, rodent infested, litter scattered sites, which were simultaneously a children's playground, a totter's heaven and also the local public dump, and which could eventually be replaced by fully prospected, engineered and controlled licensed landfill sites.

3. The development of applied techniques for the selection and evaluation of landfill sites suitable for the disposal of all contaminating materials.

4. Technical developments and the commercial assessment of specialised regional plants for the treatment or destruction of all forms of the more intractable materials.

5. Increased movement of contaminated wastes from the point of origin to the point of disposal sometimes involving transportation over very considerable distances which necessitated

6. the enhanced safety standards in the design of equipment for their

safe handling and transportation together with suitable training and integration with the emergency services.

F. The Control of Pollution Act 1974

1. Controlled waste (see S.30) which will arise within the catchment area of a Waste Disposal Authority (WDA) must be surveyed in order to prepare waste disposal plans under S.2. Each WDA must include in its plan the kinds and quantities of materials including soils which the WDA expects to dispose of itself and which the WDA expects to be disposed of by other persons together with the methods of such disposal in association with any reclamation possibilities (and see S.20) together with the estimated costs of the methods of disposal. This does not mean that each WDA must make its own arrangements for waste arising or being imported into its county; its duty is to ensure that arrangements exist for such land or materials which are, or are likelty to be, contaminated in its area. In order to ensure that its fiat runs through its own county, the DPWA offence of environmental hazard is repeated under S.3 together with a second offence under the same section of the depositing of any controlled waste or indeed, the use of any plant or equipment to achieve the same object for the disposition of controlled waste unless the land has been licensed under S.5; the penalty for depositing waste which is poisonous, noxious or polluting and its presence on land which is likely to give rise to an environmental hazard and it can reasonably be assumed to have been abandoned, is marginally more severe.

2. The requisites of a site licence are stated under S.6 and are further amplified in Waste Management Paper No.4 and Circular No.79/77. If the provisions of the Waste Management Papers, the CoP 1974 and its subsidiary legislation are followed, then not only is it possible to contaminate land within the law but, quite properly, to receive the encouragement of the legislature and the executive in so doing.

3. Indeed, such would seem to be the worldwide practice. The Environmental Protection Agency (EPA) in the USA have published an account of the design, construction and evaluation of landfills[4]. It is submitted that reasonable care would have been shown by a prudent site operator if he had reviewed the EPA summary of ideal conditions with his own site which should

: be easily accessible in any kind of weather to all vehicles expected to use it
: have safeguards against water pollution originating from the disposed solid waste
: have safeguards against uncontrolled gas movement originating from the disposed solid waste
: have an adequate quantity of earth cover material that is easily workable, compactible, free of large objects that would hinder compaction, and does not contain organic matter of sufficient quantity and distribution conducive to the harbourage and breeding of vectors (sic)
: conform with land use planning of the area.

In addition the proposed site

: should be chosen with the highest possible regard for the sensitivities of the community residents
: should be the most economic site available commensurate with the ultimate requirements for solid and other materials' deposit.

It should of course be large enough to accommodate the community wastes for a reasonable interval.

4. The standard of care required by recent United Kingdom legislation in the management of waste is essentially a self critical process. Practical realisation of this attitude of mind should be shown by the existence of

   : monitoring on a systematic basis
   : continuous technical audits
   : an efficient weather station
   : lysimeters for soil analysis and surface vegetation
   : a mathematical model for leachate production and quality forecasting
   : adequate cover material - vide the 1971 Sumner Report.

G. Planning for Controlled Contamination

1. Town and Country Planning law began with the Act of 1909 which gave very limited powers to authorities to make schemes so as to control development on land as yet undeveloped and to prevent its contamination. In fact the statutory powers to control or prevent land uses judged to be environmentally harmful were so restricted prior to 1947 that some local authorities sought other means to prevent a proliferation of refuse dumps (sic) within their boundaries. Such a practice had resulted in environmental disasters such as the South Hornchurch (Continuous Ignition) Dumps which after 35 years were some ¾ mile in length, 1/3 mile in breadth, with further layers of refuse added even after the summit had reached 90 feet. Several county councils were given powers by local private Acts of Parliament to prohibit the deposit within their areas of contaminated materials emanating from outside their boundaries. Such powers can be found for example in the Surrey County Council Act 1931 S.94, the Essex County Council Act 1933 S.146 and the Hertfordshire County Council Act 1935 S.26. These Acts all had the same object and gave much the same control. If anyone wished to deposit contaminated materials within the county, they had to seek permission from the authority. The permission, if given, could be subject to conditions regulating the operation of any disposal site. In most cases applicants for permission had a right of appeal either to the Minister or to an arbitration tribunal in case of refusal to grant or decision to withdraw permission.

2. The Town and Country Planning Act 1947 came into force on 1st July 1948. Thereafter planning permission was required for the development of any new contaminated materials site or of any old site by extending its superficial area or height. This was the effect of S.12(3)(b) of the 1947 Act now re-enacted in slightly more economical language as S.23(3)(b) of the 1971 Act; likewise Ss.12 and 13 of the 1947 Act are now S.23 of the 1971 Act. Certain limited types of waste disposal did however fall within the categories of permitted development set out by the General Development Orders. Currently Class VIII(2) of the 1977 Order permits an industrial undertaker to deposit waste from an industrial process on any land comprised in a site used for that purpose on 1st July 1948 whether or not the area or the height of the deposit is extended.

3. The planning system has long since recognised that applications to

change the use of land in connection with contaminated materials tend to arouse controversy and residential opposition above the average. To the cognoscenti all blame cannot be laid on public and private sector past malpractices or on the enthusiasms of local Pullitzer searching sub-editors; larger international manufacturing producers should be reminded that in the very same year that man could tread the moon, it was not on the UK earth considered necessary for example to include a hydrogeological profile of a potential site for acceptably contaminated materials in the necessary papers submitted to a planning committee for their deliberation. Whilst only a disciple of Job would presently apply to develop a dump or to promote a pit, it may be possible in the foreseeable future to obtain conditional permission to engineer a controlled landfill site. Whilst planning applications go before the lower tier district authority and the upper tier authority is the WDA charged with the duty to make adequate arrangements for the disposal of all contaminated materials arising within their area, there will inevitably be a tension between the planners. To-day only the deposit of mineral waste is automatically classified as a county matter; see Schedule 16 para. 32 of the Local Government Act 1972. This reluctance by councillors to grant planning permission for environmentally acceptable contaminated land is understandably partly derived from an alert psephological ear lent to their constituents' reaction to swingeing Fourth Estate hyperbole on toxic tips and poisonous plants. The best informed articles are alas not assimilated by the average parish and district representative and thus is the irony that exaggerated criticisms directed at a local waste management operation, gaining a spurious have-you-heard currency, will almost certainly increase the reluctance or an adjacent planning committee council to grant permission for a local facility, the result of which if granted, would be a reduction in notifiable waste exported to the site which was originally unjustly criticised.[5]

H. The Health and Safety at Work etc. Act 1974

1. The Health and Safety at Work etc. Act 1974 (HASAWA) in S.3 provides that it is the duty of every employer to conduct his undertaking in such a way as to ensure, so far as is reasonably practicable, that persons not in his employment who could be affected by his business are not exposed to risks to their health or safety. Likewise, under S.4(5) of the CoP, where the presence of waste on land gives rise to a statutory environmental hazard even if the waste has been deposited in such a manner or in such a quantity as to subject persons or animals only to material risk of death, injury or impairment of health, actual damage caused to the third party is not a condition precedent of an offence.

Again, S.3(3) of the HASAWA 1974 provides that it is the duty of every employer, and indeed every self-employed person, to give to persons (not being his employee) who may be affected by the way in which the owner conducts his business, information about such aspects of the way in which he conducts his undertaking as might affect their health or safety. Whilst the section is applicable only in circumstances prescribed by a Statutory Instrument, and no such relevant Statutory Instrument has been prescribed so far in the context of waste management, this sub-section does give a clear indication of the future thought processes of the Health and Safety Executive and of a Crown Court.

2. It is clearly prudent therefore for manufacturers whose production processes result in contaminated materials for discard, adequately to monitor, analyse and describe their, for example, industrial effluent even if it is to be deposited within the curtilage of their factory. The CoP has not yet in 1979 wholly been able to arrest the continued use of archaic lagooning systems fuelled by a regurgitating pipe which has been the end of the production process since Mafeking, at any rate S.2(1) of HASAWA 1974 - it shall be the duty of every employer to ensure, so far as is reasonably practicable, the health, safety and welfare at work of all his employees - indicates that such past practices will be unlikely to receive the approbation of a visiting member of the Health and Safety Executive.

3. Indeed, under S.21 HASAWA 1974, if an Inspector is of the opinion that an employer is contravening the provisions of S.2 or S.3 or indeed any of the relevant sections or provisions, or has contravened one or more of those provisions in circumstances that make it likely that contravention will be continued or repeated, the Inspector can serve on the erring employer an Improvement Notice. This will require the employer to remedy matters within a definte period, unless he wishes to follow an appeal procedure.

4. Likewise, a Prohibition Notice bringing all work to an immediate halt can be issued if the Inspector is of the opinion that any activities will involve a risk of serious personal injury and the Inspector will specify the matters which in his opinion give rise to that risk.

All of this underlines the observation that if the implementation of the CoP Act 1974 has been somewhat akin to the progress of a wallowing Moby Dick in the Doldrums, the HASAWA 1974 through its own uncompromising implementation has provided the necessary jaws, and sometimes indeed, a crimson sea.

5. The industrialist whost production process might well be a direct or indirect cause of contaminated land, will therefore be well advised to prepare an adequate waste specification; the minimum criteria should be :-

1. General description
2. Process source
3. Principal components
4. Physical form
5. Quantity and frequency of service
6. Normal (maximum) concentrations for significant constituents
7. Collection point and method of site storage
8. Maximum temperature
9. Maximum density
10. Safety/handling advice (attach standard data sheets if available)
11. Whether any of the following are present indicating level and nature :-

Cyanide	Corrosives
Phenol/cresol/other biocide	Strongly odorous material
Toxic metals	Chlorinated hydrocarbons
Flammables, etc. etc.	

12. Whether material sampled and despatched, in what form and under what conditions
13. Department of Environment waste classification
14. The contact responsible for waste materials (the necessity

to supply a home telephone number concentrates the mind wonderfully).

Appendix I - Case History Outline

In the late 1960s, a partially worked out aggregate pit on the edge of a country village was acquired for the purposes of limited restoration to agricultural land with contaminated materials from various industrial, commercial, trade and household resources. Well sited in its catchment area, radius some 40 miles, the partially controlled landfill was operated for some 3 years for progressive restoration and provided an important service for local business.

A certain amount of biodegradation had taken place when the site was unexpectedly compulsorily purchased for a new motorway, whose line ran directly through the recently infilled part of the quarry. Piling of the motorway was considered but after more test borings, it was resolved to move all the recently deposited contaminated materials and soil down to the very top of the chalk aquifer without disturbing the sealing aquiclude ceiling.

The infilling had occurred before site licensing and the DPWA; recorded descriptions ex factory were therefore as brief as they were inaccurate e.g. had the contents of a drum labelled "toluene" been replaced with the bric à brac from the staff Christmas canteen party? Had every drum to be approached as if it contained the materials stencilled on its outside? To what extent would the River Authority underwrite and assist in the removal and the second restoration? Although before HASAWA, what precautions were taken, e.g. was there a nurse permanently on duty? Did the neighbouring villagers have to close their windows? What were the actual hazards encountered or in the event, were they only potential hazards arising out of the removal of the potentially (and not actual) polluting, noxious, toxic, dangerous, difficult or contaminated land?

Plus some 20 contemporaneous slides.

Appendix II - Sources

1. Amendments to the Control of Pollution Bill at the Report Stage in the House of Commons tabled 25th July 1974; R G P Hawkins.
2. Research into Leachate Generation and Attenuation on Landfill Sites - J D Mather and J Bromley; Land Reclamation Conference 1976.
3. But see for the general environmental effect - or rather lack of - Co-operative Programme of Research on the Behaviour of Hazardous Wastes in Landfill Sites. HMSO 1978.
4. Also see Resource Conservation and Recovery Act of 1967. USA. S.1004. "The term 'open dump' means a site for the disposal of solid waste which is not a sanitary landfill within the meaning of S.4004".
5. And see District /County Functions: Government's overlap tidy up plans to be announced this week. Municipal Engineering 17th July 1979.

LEGISLATION TO CONTROL DEVELOPMENT ON CONTAMINATED LAND

M.J. McCarthy Dip Arch RIBA AMBIM (McCarthy Davies Hines)

THE NEED FOR LEGISLATION

The newly-formed Land Council believe that land is Britain's most precious and irreplaceable natural resource, and one which is being squandered. Certainly there has been a steady and insidiuous errosion of the countryside whilst each year more land degenerates into dereliction. The nation has an acute land use problem. The Civic Trust sets out a case for revitalisation of dormant land most tellingly in its report "Urban Wasteland". There is no doubt that there will be increasing pressure to encourage more appropriate use of waste and derelict urban land, both to regenerate the inner city areas and to prevent further unnecessary loss of agricultural land. Development of derelict land will become essential, and much of that land is contaminated.

Development on contaminated land is a matter affecting public safety. Thus, if the public is to be afforded the protection it has the right to expect, there must be appropriate legislation.

PUBLIC HEALTH ACT 1936

Section 54

Section 54 of the Public Health Act 1936 deals with ground filled up with offensive material. The Section, as amended by the Public Health Act 1961, ss.5 (1), 11 (2) and Sched.1, Pt.III, is such that local authorities (district councils and London boroughs)* are bound to reject a plan submitted under the Building Regulations for the erection of a building on land which has deposited upon it any material impregnated with faecal or offensive animal or vegetable matter, unless it has become or been rendered innocuous. The Section applies only to animal or vegetable matter, and makes no provision for other environmental hazards such as heavy metals, asbestos, phenols or other non-metallic contaminants. Thus, the Section does not afford a means of control of development on contaminated land.

The prime indication that animal or vegetable matter has not become innocuous is a significant rate of methane generation. Thus, the Section places an absolute prohibition on the erection of buildings on sites where methane generation is taking place. However, there are now proven techniques for erecting buildings on such sites. The Section should be amended to empower local authorities to stipulate appropriate precautions to be taken in the design of buildings on such material rather than impose an absolute prohibition.

Rationalisation of the system of building control has led to the Building Regulations becoming wider in scope and more self-contained. The Section

*The meaning of "local authorities" varies between Acts. The appropriate general meaning is given after the first reference to local authorities under each Act. For the precise meaning and the variations in Wales, Scotland and the City of London, reference must be made to the Acts.

amended so as to be wider in scope but not prohibitory, should be brought within the framework of the Building Regulations.

Section 92

Section 92 of the Public Health Act 1936, which must be read together with the Public Health (Recurring Nuisances) Act 1969, supplements the common law of public and private nuisance by creating the concept of a "statutory nuisance". The relevant definitions of "statutory nuisance" are given in Subsection (1) paragraphs (c) and (d) as any accumulation or deposit which is prejudicial to health or a nuisance; and any dust caused by manufacture or a process being prejudicial to the health of, or a nuisance to, the inhabitants of the neighbourhood. Local authorities cannot institute summary proceedings under these paragraphs of the Section without the consent of the Secretary of State for the Environment if proceedings might be instituted under the Alkali, Etc. Works Regulation Act 1906. Furthermore, the Act provides defences which may be used if proceedings are instituted under these paragraphs of the Section. Subsections (4) and (5) of Section 94 of the Act provide a defence if the defendant can prove that the accumulation or deposit was necessary for the effectual carrying on of manufacture, has not been kept longer than was necessary, and that the best practical means have been taken for preventing it from being prejudicial to the health or a nuisance; and that the best practical means have been taken for preventing, or counteracting the effect of the dust. "Prejudicial to health" means injurious, or likely to cause injury to health. Although "best practical means" is defined with some precision in relation to the Noise Abatement Provisions of the Control of Pollution Act 1974, the expression is not defined in the Public Health Act 1936 or the Alkali Etc. Works Regulation Act 1906, and its meaning has not been considered by the superior courts.

In respect of the development on contaminated land, no nuisance will arise until either the operations cause contaminated dustblows, or until occupation of the completed buildings. Although the defence afforded by Section 94 is not relevant to proceedings taken at common law for a public nuisance, it is clear that Section 92 is an unsatisfactory means of control.

COVENANTS

Where contaminated land is in local authority ownership and is to be sold for private development, the possibility of creating covenants arises. By virtue of Section 126 of the Housing Act 1974, local authorities (county and district councils) are empowered to enter into Agreements for the development of land containing covenants in respect of positive obligations such as the carrying out of works. Previously it was not possible to impose an obligation to do a positive act which will bind the land. Both positive and restrictive covenants will run with the land and will be enforceable in equity against successive owners.

The most effective and economical remedial measures to be taken in development on contaminated land will depend upon the nature and details of the proposed development. That being so, conditions would be difficult to formulate in advance of specific design proposals and a covenant would of necessity be framed in generalities rather than with precision. Problems of enforcement would undoubtedly ensue. There is also the possibility that the covenant would make the authority liable for the professional advice implicitly imparted by virtue of its imposition. Clearly, it would be imprudent to make remedial measures for development on contaminated land the subject of a covenant.

TOWN AND COUNTRY PLANNING ACT 1971

Section 52

Section 52 of the Town and Country Planning Act 1971 deals with Agreements regulating the development or use of land. Subsection (1) empowers a local planning authority (county or distict councils as specified in the Local Government Act 1972 ss. 182 and 183, Scheds. 16 and 17) to enter into an Agreement with any person interested in land for the purpose of restricting or regulating its development, and Agreements may contain necessary or expedient incidental and consequential provisions. Positive obligations, prohibitions, and restrictions imposed upon the use of the land by an Agreement under this Section are registrable as a local land charge (Land Charges Act 1925 s. 15 (7) as substituted by the Law of Property (Amendment) Act 1926 Sched.). They will run with the land and will be enforceable in equity against successive owners.

The considerations in respect of difficulty of framing clauses that will be precise, and possible liability for the advice implicitly imparted that applied in the case of positive and restrictive covenants equally apply to an Agreement under Section 52. Thus, such Agreements are inappropriate to control development on contaminated land.

Sections 29, 30, 87, 88 and 90

Section 29 of the Town and Country Planning Act 1971 deals with determination of planning applications. Subsection (1) empowers local planning authorities to grant planning permission, either unconditionally or subject to such conditions as they think fit. Section 30 deals with conditional grant of planning permission. It empowers local planning authorities to impose conditions on the grant of planning permission for regulating the development or use of any land under the control of the applicant. Any conditions subject to which planning permission is granted under this Section are registrable as local land charges (Local Charges Rules 1966).

Although the power to impose conditions is very wide, authorities cannot do just as they wish in exercising their discretion to add conditions to a planning permission. When it is desired to restrict or regulate development in a planning permission in any way, it is necessary to impose a valid condition supported by a good reason. To be valid, a condition must be fit, proper, and appropriate from a planning point of view, and must fairly and reasonably relate to the permitted development. Local planning authorities are required to state in writing their reasons for any grant of permission subject to conditions and are not at liberty to use their powers for an ulterior object, however desirable that object may seem to be in the public interest.

Guidance to local authorities on the application of these Sections is contained in Department of the Environment Circular No.5/68 "The Use of Conditions in Planning Permissions". The advice is based upon the experience of the Department in dealing with applications and appeals and sets out some general principals governing the imposition of conditions. On the question of relevance to planning, the Circular states that there are matters which, though of concern in the exercise of development control, are regulated by other Statutes or by the common law. To seek to control such matters by attaching a condition to a planning permission is usually an undesirable duplication and it invites confusion if the impact of the condition is different from that of the specific control and provides different penalties. On the question of enforceability, it states that a condition should not be imposed if it cannot be made effective, and makes the point that if a condition can only be worded in a positive form, it is likely to be difficult to enforce. On the question of precision, it states that every condition must tell the developer from the outset just what he has to do.

If authorities mistake or misuse their powers the courts can interfere by declaration or injunction. The courts will hold a condition ultra vires if it does not fairly and reasonably relate to the permitted development, and have laid down general limitations to the planning authorities' statutory power to impose conditions. They include that conditions must not be such as to effect a fundemental alteration in the general law, that conditions must not be unreasonable, and that conditions must be sufficiently certain and unambiguous.

A condition requiring that remedial measures be undertaken to contaminated land could be held to be in connection with the proposed development, but there is no direct case on the point and the question is certainly not free from doubt. The doubt stems from the difficult question of whether the existence of contamination is a planning consideration. Furthermore, as in the case of covenants and Agreements under Section 52, the most appropriate precautions for any particular site can only be determined when detailed proposals for its development have been established. It follows that any condition in a planning permission, particularly an outline planning permission, will necessarily be vague as to what the developer has to do. A condition so worded could be held void for uncertainty.

The effect on a planning permission of invalidity of a condition would depend upon the particular circumstances in any case. The courts have ruled that an invalid condition can be severed from the permission, but it has also been held that where a condition was ultra vires, the effect was that the permission itself was null and void.

The Department of the Environment are unable to assist in individual cases because it is precluded from commenting on the framing of conditions which could come before the Secretary of State on appeal. Additionally, as it is for the courts to interpret the law, the Department is unable to confirm whether a condition relating to contamination of land is a planning matter.

At this point it is legitimate to ask whether planning permission to develop contaminated land should be refused on the grounds of public safety because the local authority has no other statutory powers to ensure that adequate precautions are taken. However, the same problems of interpretation which relate to the validity of any condition would apply equally to a planning refusal on the same grounds. This could be tested on appeal but the bringing of derelict land into use should be actively encouraged and not made subject to lengthy and costly proceedings.

Section 82 deals with the power to serve Enforcement Notices. It requires an Enforcement Notice to be served on the owners and the occupiers of the land, and that the alleged breach of planning control, the steps required to be taken to remedy the breach, and the period for compliance be specified. "A breach of planning control" includes non-compliance with any conditions of a planning permission. A Notice cannot be served before duly authorised by the local planning authority and thus the serving could be delayed to the prejudice of public safety. Furthermore, unless the required steps to be taken by the Notice to remedy the breach of planning control are sufficiently certain to enable those affected to know what they are required to do, it may be quashed by the court. A Notice must specify when it shall take effect which shall not be less than 28 clear days after it is served. There is the right of appeal to the Secretary of State if the period for compliance is considered unreasonable.

Section 88 of the Town and Country Planning Act deals with appeals against Enforcement Notices. Appeal may be made to the Secretary of State on the grounds that there has been no breach of planning control, and that the steps required to be taken exceed what is necessary to remedy any breach of planning control. Thus, the doubts regarding the validity of a planning condition can be raised as a defence against an Enforcement Notice, thereby at least ensuring a substantial delay to the effective date.

Section 90, as amended by the Town and Country (Amendment) Act 1977, deals with Stop Notices. The section empowers local planning authorities who have served an Enforcement Notice to serve a further Notice, referred to as a Stop Notice, which prohibits the carrying out of an activity on the land. A Stop Notice must specify when it shall take effect which shall be not less than 3 days after it is served. Compensation may be payable for loss due to a Stop Notice by virtue of Section 177 of the Act. The possible liability for such compensation could be a considerable deterrant to an authority considering the service of a Stop Notice where there was doubt regarding the validity of the planning condition. The comments made regarding the defence against Enforcement Notices equally apply to Stop Notices. Furthermore, the liability of a local planning authority to pay compensation will be decided on a narrow point of law rather than whether the intent of the Notice was in the public interest. It is no consolation that Section 30 does not create a duty to impose any conditions, and no action can lie against an authority in negligence in respect of the imposition or lack of imposition of conditions on the grant of planning permission, as would be the case in respect of application of the Building Regulations.

Town and Country Planning General Development Order 1977

Article 4 of the Town and Country Planning General Development Order 1977 deals with directions restricting permitted development. It empowers local planning authorities to direct that permitted development shall not be carried out unless permission is granted on an application. A direction by an authority under this Article requires the approval of the Secretary of State, and it has been his practice to approve directions restricting development only where it could be established that there was a specific danger or a positive threat to the environment.

The reason for proposing an Article 4 direction is to extend planning control in order to ensure that owner-occupiers are required to carry out remedial measures in connection with the erection or other development within the curtailage of dwelling houses. Excavation of soil in the preparation of development could be hazardous unless measures were taken to cap contaminated material or otherwise dispose of it safely. The permitted development affected would be anything in Class I or Class II (1) and (2) of Schedule 1 of the Order.

DEPOSIT OF POISONOUS WASTE ACT 1972

Most of the provisions of the Deposit of Poisonous Waste Act 1972 have been repealed by Part I of the Control of Pollution Act 1974. However,

Section 3 relating to the duty to notify the local authority (county councils and the Greater London Council), and the river authority in respect of the removal or deposit of waste remains in force. Although the Deposit of Poisonous Waste (Notification of Removal or Deposit) Regulations 1972 exempts certain classes of waste, the exemption does not apply to waste which contains any hazardous quantity or hazardous concentration of any poisonous, noxious or polluting matter. Thus, Section 3 would apply to the removal of earth from a building site on contaminated land. However, the Act contains no provisions to afford control on development of contaminated land.

CONTROL OF POLLUTION ACT 1974

Section 3 of the Control of Pollution Act 1974 prohibits unlicensed disposal of waste. Although the Control of Pollution (Licensing of Waste Disposal) Regulations 1976 excepts certain classes of waste, the exception does not apply tc waste which is poisonous, noxious or polluting and the presence of which on the land is liable to give rise to an environmental health hazard. Thus, Section 3 would apply to the removal of earth from a building site on contaminated land, or relocation of earth within it. However, the Act contains no provisions to control development on contaminated land.

HEALTH AND SAFETY AT WORK ETC. ACT 1974

Section 61

Section 61 of the Health and Safety at Work Etc. Act 1974 deals with amendments of enactments relating to Building Regulations. Insofar as it amends Section 61 of the Public Health Act 1936, it enables the Secretary of State for the Environment to make Regulations in respect of the design and construction of buildings for the purpose of securing the health, safety, welfare and convenience of persons in or about buildings, and of others who may be affected by buildings or matters connected with buildings. It stipulates that Building Regulations may include provision as to the inspection and testing of work and the taking of samples.

Appropriate Regulations made under this Section could effectively deal with development on contaminated land in respect of both buildings and space about buildings. In order to remove all doubt from the application of the Town and Country Planning Act 1971, and in the interests of public health, safety, welfare and convenience, suitable Regulations should be made.

Section 68

Section 68 of the Act deals with power to require or carry out tests for conformity with Building Regulations. It enables local authorities (district councils, the Greater London Council, and London boroughs) to require reasonable tests in connection with building work to be carried out or to carry out the tests themselves, and to take any samples necessary to enable them to carry out such tests. The Section defines that the matters with respect to which tests may be required or carried out, include tests of the soil or subsoil of the site of any building.

It is clear that this Section is important in relation to establishing the extent and nature of contamination on suspect sites. However, it has yet to be brought into force. It can be brought into operation by the Secretary of State by order made by statutory instrument. A commencement order should be made in respect of Subsection (3) paragraph (a) which deals with testing of soil or subsoil of the site of any building, together with Subsections (5) and (6) which deal with the expense of carrying out any tests and the resolution of questions arising under the Section.

Section 62

Section 62 deals with further matters for which the Building Regulations may provide. It enables the Building Regulations to make a prescribed person or class of persons responsible for performing prescribed functions instead of local authorities. It also enables the Building Regulations to require authorities in prescribed circumstances to consult any specified person or body before taking steps in connection with building regulation applications. For example, the power could be used to require consultation by an authority with the fire authority or the Health and Safety Executive before approving plans. Having regard to the expertise necessary to deal with the complexities of contaminated sites and the comparative infrequency with which such sites are developed, it would be wasteful of resources for each authority to employ staff to administer regulations made under Section 61. The Building Regulations should prescribe appropriate persons or class of persons to fulfil this function.

SUMMARY OF RECOMMENDATIONS

Section 54 of the Public Health Act 1935 should be amended so as to be wider in scope but not prohibitory and should be brought within the framework of the Building Regulations.

Until local authorities have specific powers to control development on contaminated land, it would be prudent for local planning authorities to impose conditions in planning permissions notwithstanding the doubts and the difficulties that attach to this course of action. Bearing in mind that permitted development may involve excavations which would disrupt remedial measures, it would be prudent for local planning authorities to seek the Secretary of State's approval to an Article 4 direction withdrawing permitted development rights.

The Secretary of State should make an order by statutory instrument giving effect to paragraphs (3) (a), (5), and (6) of Section 68 of the Health and Safety at Work Etc. Act 1974.

The Secretary of State should make regulations under Section 61 of the Health and Safety at Work Etc. Act 1974 enabling local authorities to control development on contaminated land, and such regulations should prescribe that appropriate persons are either to be consulted or be responsible for the application of such regulations.

"The relevance of the Health and Safety at Work etc Act 1974[1] and Regulations made under the Factories Act 1961".[2]

Graham F Walker HM Deputy Superintending Inspector, Health and Safety Executive.

SYNOPSIS

The paper explores the effect of the above legislation on work being undertaken on reclaimed sites. Two separate eras will be discussed - the situation before the 1974 Act came into force, and the position after it became law. Current legislation made under the Factories Act 1961 includes the Construction (General Provisions) Regulations,[3] the Construction (Lifting Operations) Regulations,[4] the Construction (Health and Welfare) Regulations,[5] the Asbestos Regulations,[6] and other relevant codes. Examples of their scope and application in excavations, earth works, tunnels and demolition will be given. Specific examples of the hazards faced by workers on sites are detailed in a separate paper presented by L C Smith (Health and Safety Executive). Consideration is given to the Joint Standing Committee Report on "Safety in Demolition Work" and to BS94: 1971 "Demolition". For comparison purposes broad references are made to analogous legislation such as the Control of Pollution Act 1974[7] Public Health Act 1936[8] and the Deposit of Poisonous Waste Act 1972.[9]

BACKGROUND

1. Throughout the paper reference is made to criminal and not civil law. The statutory duties created by Section 2 onwards of the 1974 Act do not in themselves give rise to any civil cause of action. Acts or omissions which contravene the requirements of those Sections may, however, coincidentally give rise to a civil action under the common law tort of negligence. Finally, a civil action for breach of statutory duty may be possible for certain of the 'relevant statutory provisions' listed in the 1974 Act Schedule 1, for example, the Factories Act, 1961 and Regulations made under it.

2. There is a variety of legislation covering different aspects of land pollution. Responsibility for its direct application falls to various bodies; eg, Waste Disposal Authorities and the Health and Safety Executive. The legal provisions relating to the collection and disposal of household and trade waste are contained in the Public Health Act 1936 and the Control of Pollution Act 1974; the Deposit of Poisonous Waste Act 1972 introduced notification schemes for some wastes. It is the Control of Pollution Act 1974 which provides the framework for a systematic and co-ordinated approach to waste collection and disposal. The Health and Safety Executive has a statutory consultative role in connection with sites licensing under the Control of Pollution (Licensing Waste Disposal) Regulations 1976[10]. Other papers will give more details of these aspects of the legislation. It is sufficient to say here that the Health and Safety At Work etc Act 1974 is concerned with the protection of people. Waste Disposal Authorities have a wider remit concerned not only with people but also fauna, flora and amenity. It is not easy to define where one aspect stops and another begins. "Grey" areas are under discussion with other Government Departments.

3. One general example can be given to illustrate how the legislation can have overlapping provisions. Road vehicles attract the provisions of the Road Traffic Acts, the Public Health Acts and the Control of Pollution Act. Vehicles must be provided with efficient brakes. "Noise" levels can be monitored by virtue of the Control of Pollution Act. At the same time the cab of a lorry can be a place of work and the lorry is "plant" as defined and therefore the provisions of the Health and Safety at Work etc Act equally apply. In the case of "brakes" there is a specific requirement under the Road Traffic Acts Legislation and since the 1974 Act is "broad" based the provisions of the former legislation will take precedence. In general terms, and referring to the reclamation of land, there will be occasions when in respect of the health and safety of persons employed, the Health and Safety at Work Act may be the most relevant legislation. There will be occasions too when the health and safety of the public will have to be considered under the terms of this Act. The fifth report of the Royal Commission on Environmental Pollution recommended that a central body be formed to take an overall view of pollution problems arising from appropriate industrial processes although water and waste disposal authorities would retain their own controls. No decision has been taken yet on the implementation of this recommendation.

THE PRE-1974 ERA

4. Before 1974, legislation made under the Factories Act 1961 gave statutory protection for the health, safety and welfare of persons employed on site preparation, demolition and rebuilding operations. This legislation remains in force and the 1974 Act imposes additional requirements with wider implications.

5. The earlier requirements applied only to defined operations on land or water and included the construction, structural alteration, demolition repair or maintenance of a building, and the preparation for and laying

the foundation of an intended building. The legislation covered the construction, structural alteration, and repair or demolition of docks, harbours, inland navigations, tunnels, bridges, viaducts, waterworks, sewers, sewage works etc. In 1966 the definition was extended to include roads, airfields, sea defence works and other similar works. The legislation was framed to protect the health and safety of the persons employed only - the protection of the general public was not included. Duties were laid on contractors and employers of workmen.

The Construction (General Provisions) Regulations 1961

6. The legislation affects the methods of working whether or not the land being reclaimed is derelict, contaminated or otherwise. Referring first to the General Provisions Regulations the code can be divided into four grouped headings: (a) Excavations (b) Fumes (c) Demolition (d) Appointment of a Safety Supervisor. It should be said, however, that these groups over simplify the full coverage of the Regulations because other topics eg transport, explosives and a number of miscellaneous items can be included as well.

(a) Excavations, Shafts, Earthworks, Tunnels (Regulations 8-14)

As early as practicable in the course of the work adequate steps must be taken to prevent falls or dislodgements of earth or other materials (eg by timbering). The requirements assume importance when land is being reclaimed because that land may have been back filled or made less stable because of previous contamination. Inspections of excavations by a competent person have to be undertaken daily. Thorough examinations must be undertaken weekly and written reports made. Timbering erection and alterations have to be performed under the direction of a competent person and undertaken by experienced workmen. If there is a danger of rising water means have to be provided to enable persons to reach positions of safety.

(b) Fumes (Regulations 20 & 21)

If, during the manipulation of material injurious dust or fumes can be given off, all reasonably practicable measures must be taken to prevent the employees from inhaling the contaminants. There should be provision for adequate ventilation or, if this is not practicable, suitable respirators must be provided and worn. In given cases, there will be a need to monitor and test the atmosphere depending upon the risk (either established or anticipated). Danger might be encountered in excavations, pits, holes, tunnels, shafts or other confined spaces. Again, there should be adequate ventilation and a need to maintain an atmosphere which is fit for respiration. Further, fumes dust and other impurities should be rendered harmless so far as is reasonably practicable. Finally, persons cannot be allowed to use a working place until it is certain that they cannot be overcome by poisoning or asphyxiation.

(c) Demolition (Regulations 39, 40 and 41)

The special dangers facing demolition workers are recognised and contractors have to appoint competent persons experienced in the operations to supervise the work. There must be consultation with any other contractors and their competent persons too regarding the method of work to be used and the time at which it is to take place. Before demolition is commenced and during the progress of the work all practicable steps have to be taken to prevent danger from risk of fire or explosion through leakage or accumulation of gas or vapour. Again, risk of flooding has to be considered. The requirements imply that contractors ought to establish before-hand what risks can be expected.

(d) The appointment of the Safety Supervisor (Regulation 5)

The Regulation states that contractors who normally employ more than 20 persons should appoint suitably qualified persons to advise them of the requirements for the safety and protection of employed persons. The appointed persons have to exercise a general supervision of the observance of the requirements and promote the safe conduct of the work generally.

The Asbestos Regulations 1969

7. Turning to the Asbestos Regulations; there has been considerable publicity on the subject in recent years. The Regulations apply to work undertaken on building operations and works of engineering construction. Any contractor undertaking a "process" (eg the removal of asbestos lagging) has to take precautions not only in relation to any persons employed by him but also in respect of any person employed there who is liable to be exposed to asbestos dust. The Factory Inspectorate has to be notified when the work involves crocidolite. The employer has to provide exhaust ventilation equipment and protective clothing. All loose asbestos (when not in "use") must be placed in suitable closed receptacles and has to be despatched in the same manner. Finally, receptacles which contain crocidolite should be clearly marked "blue asbestos - do not inhale dust".

Other relevant pre-1974 legislation

8. Apart from the Construction (General Provisions) Regulations and the Asbestos Regulations there are other relevant references in Factories Act legislation. For example, Regulation 27 of the Electricity (Factories Act) Special Regulations 1908 - 1944[11] discusses the need to take special precautions if there are flammable surroundings or explosive atmospheres. Thus it may be known that a "dangerous" atmosphere could be expected in a particular vicinity and it would be

necessary then to consider whether certified flameproof apparatus or intrinsically safe equipment should be provided. Again, precautions in respect of live electricity cables or apparatus are referred to in the Construction (General Provisions) Regulations. Contractors must take all practicable steps to prevent danger eg by rendering overhead lines and underground cables electrically dead.

9. The Construction (Lifting Operations) Regulations 1961 have some relevance to work on reclaimed sites. Precautions have to be taken to ensure the stability of lifting appliances used on soft or uneven

surfaces or on slopes. Next, appliances have to be securely anchored or adequately weighted by suitable ballast. The anchorage or ballast should be examined by a competent person on each occasion before the equipment is erected. The requirements imply that any contractor ought to discover the possible risks first, e.g. he should ask whether there are subterranean passages or water courses which could affect the decision to use a particular appliance in a particular place.

10. Finally, the Construction (Health and Welfare) Regulations 1966 state that first aid boxes have to be provided in reasonably accessible positions. If more than 50 persons are employed the person in charge of the box must be trained in First Aid treatment. If more than 25 persons are employed by the contractor, a suitable stretcher(s) is/are necessary. Further, the Local Health Authority has to be informed of the site location and nature of the operations. If more than 250 persons are employed on a site as a whole, any one contractor who employes more than 40 persons has to provide a first aid room.

11. Finally, the Diving Operations Special Regulations 1960[12] and the Work in Compressed Air Special Regulations 1958[13] could be relevant to work on reclaimed sites. The former code requires that no person shall be employed under water as a diver unless he has had previous experience and a knowledge of diving practice to the depth at which he is to operate. Alternatively he has to be undergoing training under the immediate supervision of an experienced diver. The latter code has similar requirements relating to competent supervision of compressed air operations and to experience of those operations.

THE POST 1974 ERA

12. The Health and Safety at Work etc Act 1974 has altered the situation prevailing previously. Until 1974 only the limited defined operations were covered. The new Act applies to <u>all</u> work situations (apart from limited exceptions which are not relevant here). The Act will apply when surveyors and others arrive at the site before work commences. The Act will apply at the sites of abandoned quarries and mines if work of <u>any</u> nature is being undertaken. A word of caution is required because in certain cases there is interaction between the 1974 Act and Mines and Quarries legislation[14]. The 1974 Act introduced into legislation enforced by the Health and Safety Executive the concept that not only must employees be protected but also other persons (including the general public) who could be affected by the operations.

13. The Act covers the keeping and use of dangerous substances and their unlawful acquisition. Controls can be applied wherever dangerous substances are found, even if they are not being kept or used in connection with the work being performed. This means that a single, integrated set of requirements can be made, imposing controls over dangerous substances in all circumstances.

14. Requirements can be established for control of airborne emissions of noxious or offensive substances from prescribed premises, including those which may not be a danger to health but which could cause a nuisance or damage to the environment. This includes processes covered by the Alkali etc Works Regulation Act[15].

15. Some of the matters employers will need to consider are:-

a. Are systems of work safe? Thorough checks of all operations, especially ones carried out infrequently, should ensure that danger of injury to health is minimised. This may require special safety precautions such as "permits to work".

b. Is the work environment regularly monitored to ensure that, where known toxic contaminants are present, protection conforms to current hygiene standards.

c. Is monitoring also carried out to check the adequacy of control measures?

d. Is safety equipment regularly inspected? All equipment and appliances for safety and health, such as personal protective equipment, dust and fume extraction equipment, monitoring and testing devices, need regular inspection.

e. Are risks to health from the use, storage, or transport of "articles" and "substances" minimised? The term "substance" is defined as "any natural or artifical substance whether in solid or liquid form or in the form of gas or vapour" (Section 53(1)). All storage and transport arrangements should be kept under review.

Duties to Others

16. Employers must, so far as reasonably practicable, have regard for the health and safety of self-employed or other contractors' employees who may be working close to their own employees; and also for the health and safety of the public who may be affected by their activities. This covers, for example, the emission of noxious or offensive gases or dusts into the atmosphere (eg asbestos). Also included is the conveyance of dangerous substances, and the possibility of radiation entering public places. Employers must, so far as reasonably practicable, conduct their activities so as to ensure members of the public are not exposed to risks to health and safety (Section 3(1)).

17. The Safety Representatives and Safety Committees Regulations 1977[16] have changed the picture again. Employers have a duty to consult Trade Unions appointed representatives' with a view to the making and maintaining of arrangements which will enable both the employer and the employees to co-operate effectively in promoting and developing measures to ensure health and safety at work.

OTHER RELEVANT PUBLISHED MATERIAL

18. Where land has been contaminated there is a fair chance that it has become so because of previous industrial activity. Demolition companies clearing old plant and buildings before redevelopment may find, in the broadest sense, that the premises may have been contaminated in numerous ways. The Report of the proceedings of a Sub-Committee of the Joint Advisory Committee on Safety and Health in the Construction Industries "Safety in Demolition Work" published recently made a number of recommendations which have relevance to work on reclaimed sites. These include:-

a. A survey of the physical characteristics of buildings and plant to be demolished should be made mandatory, and a written record of the survey should be kept.

b. A duty should be laid on the client when inviting tenders for a contract to supply the contractor with all the information available to the client, about potential hazards involving the property to be demolished or dismantled.

c. A contractor should obtain certificates from the public utility authorities that all known supplies have been cut off (except those required for the execution of the work).

d. Greater thought should be given to the planning of each phase of demolition work.

e. When "hot work" is to be undertaken where there may be flammable/explosive substances the demolition contractor should invariably institute a permit to work system.

19. BS 94: 1971 "Demolition" offers advice on statutory requirements and safety precautions as well as detailed aspects of demolition work. For example, it recommends a survey and states that enquiries should be made concerning previous uses to determine the nature of any risk of danger (eg from petrol tanks, gaseous material, flammable or explosive material, injurious chemicals and radio active substances). It states "in such cases special provision should be made." It recommends that HM Inspectors of Factories, or the Local Authority Petroleum Officer or the Fire Officer should be consulted. When work is to be performed on old chemical works, gas works and similar establishments the Code recommends that the premises should be examined in conjunction with a chemist so as to determine the nature of any chemical deposits and how they might influence the method of demolition.

RECOMMENDATIONS

20. Many of the recommendations in the "Safety in Demolition Work" report and the BS94 can be considered to have a close relationship with the requirements of Sections2-4 of the 1974 Act. If these reports are studied and followed there is more chance of a safe system of work being evolved.

21. Other specific action to be taken by an employer/contractor includes:

a. Having obtained the necessary information a check should be made of available sources to decide what precautions are necessary.

b. The employer will need to see whether or not the machinery he uses will be affected by the existence of hazardous conditions such as the presence of underground tunnels, shafts, etc.

c. The employer will need to decide whether regular air sampling is necessary and what other allied precautions have to be taken to prevent danger.

d. The employer will need to consult other employers, self-employed and persons in control of the premises to see whether any other activities will be affected as a result of actions taken by him. He will have to consider the effects on the general public.

SUMMARY

1. Before 1974 - the contractor had to ensure the safety of his own employees and only in a limited way did he have to consider other employees.

2. Post 1974 - further requirements were imposed on contractors/ employers not only in respect of their own employees but also in respect of employees of other contractors/employers who may be affected. Further, he has to consider the effects of his work on members of the public. Finally, the requirements apply not only to employers, but also to the self employed and persons who have, to any extent, control of premises.

3. It is important to remember that not only must the Health and Safety at Work etc Act 1974, the Factories Act 1961 and associated requirements be checked, but also the effects of any other relevant legislation.

BIBLIOGRAPHY

1. Health and Safety at Work etc Act 1974 - ELIZ II 1974 Chapter 37

2. Factories Act 1961 - 9 and 10 ELIZ 2 Chapter 34

3. Construction (General Provisions) Regulations 1961 - SI 1961 No.1580

4. Construction (Lifting Operations) Regulations 1961 - SI 1961 No.1581

5. Construction (Health and Welfare) Regulations 1966 - SI 1966 No.95

6. Asbestos Regulations 1969 - SI 1969 No.690

7. Control of Pollution Act 1974 - ELIZ II 1974 Chapter 40 and SI 1974 No. 816

8. Public Health Act 1936 - 26 GEO 5 and 1 EDW 8 CH49

9. Deposit of Poisonous Waste Act 1972 - SI 1972 No.1017

10. Control of Pollution (Licensing Waste Disposal) Regulations 1976 SI 1976 No.732

11. Electricity (Factories Act) Special Regulations 1908-1944 - SR & O 1908 No.1312 and SI 1944 No. 739

12. Diving Operations Special Regulations 1960 - SI 1960 No.688

13. Work in Compressed Air Special Regulations 1958 - SI 1958 No.61

14. Mines and Quarries Act 1954 - 2 and 3 ELIZ 2 CHAP 70

15. Alkali etc Works Regulations Act 1906 - 6 EDW 7 CHAP 14

16. The Safety Representatives and Safety Committees Regulations 1977 - SI 1977 No.500

GOVERNMENT POLICY IN RELATION TO THE CONTROL OF WASTE DISPOSAL ON LAND

R G D OSMOND

M A L ROSS

Other papers given to this Conference have concentrated largely on the technical problems of reclaiming land which has been contaminated in some way. This paper aims to show what action central government has taken and intends to take to control the disposal of waste on land and thereby to ensure that any problems in the future are minimised. The difference is that where the problem of land already contaminated is a physical one, the measures necessary to prevent further extensive contamination are primarily legislative. It is this legislative framework, formed by the Deposit of Poisonous Waste Act 1972 and Part I of the Control of Pollution Act 1974, which are firstly described in this paper.

The Control of Pollution Act is now 5 years old and there has been a good deal of experience of its workings. It has added considerably to existing planning controls, and yet industry and the public authorities concerned have cooperated excellently in implementing its provisions and making them work - and this in an economic atmosphere that has been far from favourable.

The Act is concerned with the disposal of controlled waste - that is to say with household, commercial and industrial waste. It is not concerned with agricultural waste nor with mining or quarrying spoil, nor with sewage sludge except under certain disposal situations. Under the Act, controlled waste may only be deposited on a site occupied by the holder of a licence issued by the appropriate waste disposal authority, which in England is the county council and in Wales and Scotland the district Council. It is a significant distinction between planning and waste disposal legislation that where a planning permission runs with the land, irrespective of who occupies it, a waste disposal site licence is issued to the individual operator and must be cancelled or - with the authority's agreement - transferred to someone else when he relinquishes the site. This permits the authority to take into account the suitability of the operator when considering a licence application, as well as the suitability of the site.

The Act also provides that a licence shall not be issued for a use of land for which planning permission is required unless such permission is in force. This allows questions of local amenity - the visual impact of the site, effect on traffic flows and so on - with which development control is exclusively concerned, to be considered before the more technical safety aspects which are appropriate to the site licence. It also allows the district council, which is responsible for the planning permission, to voice the local community's view before the disposal authority, which has a strategic responsibility for the whole county, considers the application.

Where a waste disposal authority proposes to issue a site licence, they are required by the Act and its associated regulations to consult the relevant collection and water authorities, the Health and Safety Executive and, in some cases, the Institute of Geological Sciences. There is thus no lack of expertise in assessing each application, and the conditions which an authority are empowered to attach to a site licence can be used to ensure that all necessary precautions to protect the environment are taken. And if a licence application is turned down, or the conditions attached to a licence are felt to be unfairly onerous, the site operator has a right of appeal to the Secretary of State.

For the potential site operator, this is a tough system but a fair one. If we are to ensure that public health and safety is not to be jeopardised by new disposal sites, and if we are to ensure that land is not to be contaminated in the way it has sometimes been in the past, then this is a system which is fair to the operator and to society at large. When the Control of Pollution Act was passed, one of the most sizeable tasks facing waste disposal authorities was the licensing of existing sites. Since the licensing provisions came into force on 14 June 1976, at least 60% of the estimated 4,200 privately-operated waste disposal sites in England have been granted licences. Possibly about 800 sites have yet to be licensed and these are believed to be mostly less important ones. This is a tribute to the cooperation which has existed between the private operators and the disposal authorities in implementing the Act.

But it is not enough simply to consider site licence applications as they arise. The problem of further land contamination can only be tackled effectively if authorities have a strategy for waste disposal in their areas. This is the purpose of section 2 of the Act. Since 1 July last year, authorities have had a duty to prepare a waste disposal plan for their areas, based on a preliminary survey of waste arisings and disposal facilities. By the end of 1978, all but two authorities had begun their surveys and 28 had started on their plans. When these are complete, each authority will have a clear idea of the type and quantity of wastes for which they will be responsible and the number and nature of sites required in the county.

Brief reference should be made to the Deposit of Poisonous Waste Act 1972. The introduction of this Act was undoubtedly catalysed by the discovery of cyanide waste at a few sites in the United Kingdom in the early 1970s. The Act provides for prosecution in the event of poisonous, noxious or polluting waste being deposited on land such that an environmental hazard is liable to arise. The Act also lays down pre- and post-notification procedures. The Control of Pollution Act 1974 has of course introduced more comprehensive controls, and at some future date it is probable that the Deposit of Poisonous Waste Act will be repealed. One likely pre-requisite is the implementation of new regulations under section 17 of the Control of Pollution Act. This section places a duty on the Secretary of State to make provision for those controlled wastes which are so difficult or dangerous to dispose of that special provision for them needs to be made. It is anticipated that the proposed regulations will do three things. Firstly, they will introduce an inclusive list of special wastes to replace the exclusive list approach of the Deposit of Poisonous Waste Act - in other words, they will say what is a special waste rather than what is not. This will create more certainty for operators and reduce the amount of paperwork for them and disposal authorities. Secondly, they will create a new consignment note system by which the producer and carrier would take responsibility for the waste at each stage of its disposal by signing the appropriate section of the prescribed form and maintaining records. This should be a deterrent against fly-tipping. Thirdly, the regulations will require site operators to keep records showing where consignments of special waste have been deposited, the disposal authority maintaining the records where the disposal licence is subsequently surrendered or revoked. These records will be available to planning authorities. With the introduction of these regulations, the United Kingdom will have fulfilled all the requirements of the EEC directive on toxic and dangerous waste.

The Government does not, however, simply provide the legislation and leave waste disposal authorities to get on with its implementation. As many of you will know, we produce a series of Waste Management Papers - we are at present up to No 18 - which give advice on waste disposal generally and on the handling and disposal of specific problem wastes. Further papers are in the pipeline, and the Department is most grateful for the continued help received from representatives with relevant experience in the public and private sectors. The Department also finances a considerable amount of research - most notably, perhaps, in the context of this conference, the research programme on the behaviour of hazardous wastes in landfill sites.

The Department's Waste Management Papers advise that wherever practicable wastes which pose long-term contact hazards should not be "stored". Waste disposal authorities are careful to ensure that the dangers of disposal sites, public and private, are minimised and they have now sufficient legal powers to ensure this. So far the idea of site operators lodging performance bonds with the disposal authority has generally not found favour in the United Kingdom. This is not only because of practical, administrative or financial difficulties but principally because it is hoped that the licensing and other legislative controls make such bonds largely superfluous. It is relevant to note that the Control of Pollution Act enables a Waste Disposal Authority to modify or even revoke a licence where they feel this to be necessary.

We are certainly not complacent about the law as it now stands. The problem with any pioneering legislation is that many problems can only be discovered as they arise. Even so, the Deposit of Poisonous Waste Act and the Control of Pollution Act provide safeguards against the uncontrolled tipping of waste, especially hazardous waste. This conference is rightly concerned with the problems created in former years and it is important to ensure that the legislation remains sufficiently comprehensive to prevent their recurrence.

of special wastes to replace the exclusive final approach of the Deposit of Poisonous Waste Act [illegible] Firstly, they will [illegible] in a [illegible] way rather than what [illegible] This will create much certainty for operators and reduce the amount of paperwork for them and disposal authorities. Secondly, they will create a new consignment note system [illegible] the producer and carrier [illegible] responsibility for the waste [illegible] by signing the appropriate section [illegible] present [illegible] preserving records [illegible] should be a [illegible] The [illegible] will require [illegible] showing where consignments of special [illegible] the disposal authority maintaining the [illegible] of the [illegible] subsequently [illegible] or [illegible] correction [illegible] to planning authorities. [illegible] These regulations [illegible] Kingdom will [illegible] requirements [illegible] directive on toxic [illegible] wastes.

The Government does not, however, simply provide the legislation and [illegible] disposal authorities to get on with its implementation. [illegible] we [illegible] produce a series of waste management [illegible] which give advice on waste [illegible] and [illegible] and [illegible] specific [illegible] the [illegible] and the Department [illegible] the [illegible] help [illegible] from [illegible] relevant experience in the public and private [illegible] a considerable amount of research [illegible] in the context of this conference, [illegible] of hazardous wastes in landfill sites.

The Department's [illegible] wherever [illegible] contact [illegible] should not be [illegible] are careful to ensure that the [illegible] and [illegible] are minimised and they have [illegible] powers [illegible] the [illegible] the disposal authority [illegible] in the United Kingdom. [illegible] financial [illegible] the licensing and [illegible] super-licence. [illegible] relevant [illegible] Control of Pollution Act [illegible] licence [illegible] revoke a licence when [illegible]

We are certainly not complacent [illegible] as it now stands. [illegible] provided with and [illegible] legislation [illegible] the Deposit of Poisonous Waste Act and the Control of Pollution Act [illegible] safeguards against [illegible] hazardous waste. This [illegible] the [illegible] waste [illegible] and [illegible] important [illegible] regulation [illegible] comprehensive [illegible] required.

DISCUSSION

Papers H1-4

Replying to questions Mr Hawkins said that the principle of 'caveat emptor' (buyer beware) applied to the purchase of land but that the vendor should disclose any knowledge which he had about waste materials' matters which could affect the use of the land. Prudent advisers (e.g. legal and valuation) to the purchaser should be able to ascertain, by asking the right questions of the vendor, and by making the proper investigations, whether a site was likely to be contaminated (the importance to the vendor of disclosing relevant information was also discussed during the Session on financial considerations).

A questioner from a regional water authority said that RWA's are statutory consultees on waste disposal licence applications. Their duty is to give advice on the measures necessary to protect water quality. This advice is based (normally) on the best evidence available at that time. There may however be cases where that advice may be incorrect and water pollution subsequently occurs as a result of waste disposal operation even though carried out in accordance with terms and conditions imposed in the licence. He asked if such circumstances occurred, in view of the Dutton vs Bognor Regis case, would the Water Authority be liable to meet costs of remedial measures necessary to remove and/or prevent further pollution?

Mr Hawkins replied that there could be some liability.

Legal precedent show the need for professional advisers, statutory undertakings and local authorities to take care in how they approach such sites and properly to document their decisions and order their reports. In order to obtain damages, should advice found to be at fault, the plaintive must normally demonstrate negligence and a reasonable defence is possible provided all reasonable professional care has been exercised. As stated in the paper all assessments should record work undertaken, documents consulted, the limitations of available information and the steps required for greater reassurance about a particular problem to be given if it was needed.

Reports should be written in the knowledge that third parties may make use of them outside of the context in which they were prepared and without supporting information from other sources.

Mr McCarthy introduced his paper by stating that he was an architect speaking on legal issues. He feels that the architects should take a greater interest in matters outside their own expertise which may influence the way in which they carry out their professional duties. There is at present no adequate legal framework to govern the use of contaminated land and one must be provided. It is incumbant upon architects and other practitioners to tell legislators and administrators what they think is required in the way of legislation and he urged all attending the conference to make use of all available channels, including their professional bodies and direct approaches to DOE to press for the necessary changes in the law.

A speaker from the floor said that if controls were to be exercised over contaminated land proper definitions would be required. It would be preferable to have a general enabling measure covering all aspects of the suitability of land requiring proper assessments to be made but this opened up all sorts of complexities and paths along which we might not wish to travel. In reply Mr McCarthy said that while separate legislation might be preferable this would be difficult and take a long time. He had sought to find ways in which the present situation could be improved as a matter of urgency by introducing new regulations under existing laws or making small amendments. There should of course be no exemptions to the provisions for contaminated land for land held by the crown or other public bodies.

Replying to a question concerning the different responsibilities of contractor and the customer to provide safety precautions Mr Watkins said he favoured a co-operative approach based upon clearly documented arrangements of where responsibilities lay. It would be unreasonable to expect duplication of all provision by both parties, but whilst not wishing to dilute in any way the contractors responsibility for his site, there are clearly some matters, such as provision of personal protective clothing and equipment, where the Clerk of Works can reasonably be expected to make his own provision. The HSE are now reviewing the responsibilities and position of all the professions involved in building and construction work.

PANEL DISCUSSION - ASSESSMENT AND ACCEPTABILITY OF HEALTH RISKS

Chairman	Dr. Shelia M. Cromwell	Department of Health & Social Security
	Professor E.D. Acheson	Dean of Medicine and Professor of Clinical Epidemiology University of Southampton
	Dr. C.F. Mills	Head: Department of Nutritional Biochemistry Rowett Research Institute
	Dr. R. Owen	Medical Adviser - TUC
	Dr. M. Sharratt	Senior Toxicologist B.P's Group Occupational Health Centre
	Mr. G.A. Thomas	Chief Officer. South Yorks County Council Environment Department

Dr. Sharratt	Assessment of Health Risks
Dr. Mills	Assessment of the Biological Aspects of Risk
Dr. Owen	Trade Union Attitudes to Risk
Mr. Thomas	Public Relations and Acceptance of Risk
Professor Acheson	Epidemiological Assessment of Risk

Assessment of Health Risks

Dr. Sharratt

Assessment of those risks to health which arise from being exposed to chemicals is a complex process. At best an assessment of risk can be imprecise and more usually it is the subject of massive uncertainties. Even when we have clear, irrefutable evidence that a set of human beings has developed some illness after being exposed to a substance, it can be difficult, even impossible, to say what the effect will be at some given exposure level or at which particular exposure level people will be safe, because there is rarely sufficient information available about the degree or length of exposure of those who have been made ill.

To predict the risks to human beings from data derived from animal studies is fraught with even greater uncertainties. I shall not expand on this except to say that just because a rat drops dead or develops a cancer when exposed to a chemical, this doesn't mean that human beings will die like flies or that the chemical will cause a cancer epidemic in the human population. Equally the fact that a substance does not appear to harm white mice does not always indicate that it is completely harmless to man.

In analysing the risk, we must nevertheless, as one component in the assessment, try to develop some idea of the intrinsic toxicological properties of the substance, i.e. its ability to cause damage to living things.

The second component of any assessment is consideration of the potency of the substance. Potency can vary enormously in substance to substance. Some chemicals can produce illness and death when only a few micrograms enter the body (botulinus toxin), while with other substances, exposure must be continuous or repeated over months or even years before adverse effects occur. It has been estimated that the potency of different substances can vary by a factor of 10^7.

The third component in the assessment is consideration of the dose or exposure levels. This again can vary enourmously in different human population. Exposure levels between different people can vary by at least a factor of 10^8.

Risks are often expressed numerically. While it is possible to do this when masses of reliable data are available on the effects of the substance on human health - e.g. risks from cigarette smoking - in the vast majority of cases it is not possible to produce meaningful figures. Attempts are made, particularly in the USA, to extrapolate animal data (particularly on cancer) to man using mathematical models. Such models have no biological basis and therefore, although they may produce interesting figures which please administrators, they have no relevance to real life.

I hope I have been able to indicate that assessment of health risks is no easy task. One is continually beset by lack of information and even toxicological and epidemiological data on which one cannot entirely rely.

In relation to the materials and situations that have been discussed at the conference, it seems likely that there will be many occasions on which there will be uncertainty about what particular chemical people are going to be exposed to, and therefore there will be even more uncertainty than in other areas as to what the intrinsic properties, or potency or exposures will be.

Assessment of the Biological Aspects of Risk

Dr. Mills

Perhaps my only qualification for being invited to participate in this discussion of the problems of assessing hazards lies in the fact that I have for some years been involved in research work on the effects of trace metal deficiency and excess and, to some extent, in work which has attempted to define the variables which influence the tolerance of potentially toxic heavy metals.

In previous discussions, emphasis was laid upon the need for definition of acceptable levels of potentially toxic materials in the environment of man. It was suggested that even arbitrary definitions of tolerance could be more acceptable than the vague basis upon which, currently, hazards have to be assessed. Basing my comments upon investigations with heavy metals, perhaps it would be useful if I gave some indication of why estimates of tolerance frequently appear to be vague.

First, it must be emphasised that relatively little is known about the effects of heavy metal toxicity or the mechanisms by which such effects arise. Evidence as to the nature of hazards accumulates from two distinct approaches. Where it is ethically acceptable, direct experimental investigation of dose/response relationships may be made; this approach is obviously precluded for the assessment of risks to man. The only alternative acceptable for all species is the epidemiological survey, work based upon large populations which include sub-populations exposed to differing degrees of risk and free from bias which may influence the interpretation of data.

Remarkably little experimental work has been carried out which is relevant to the problems of assessing the risks created by chronic low level exposure to potentially toxic heavy metals. Much more emphasis has been given to the study of the biochemical and pathological consequences of acute intoxication but recent evidence, although limited in scale, clearly indicates that extrapolation of such data is not a justifiable approach to the assessment of the long-term low level hazard. Although they are more costly, more time consuming and run a greater risk of producing complete negative results, the necessity for work on low level exposure situations has become appreciated increasingly and data are now becoming available.

Some of the more relevant findings from such studies are as follows. We are learning that, providing exposure is not too abrupt, animals can adapt to the appearance of relatively high concentrations of cadmium, zinc and copper by developing defence mechanisms. These work either by restricting the proportion of the dose which is absorbed, by restricting its transport to sensitive tissues such as the developing foetus or by hastening its excretion. An example of such a process is the capacity of the body to synthesise a novel protein, rich in metal binding groups which effectively modifies the rate of transport of the toxic metal between body tissues. One consequence of this altered distribution of metals in body tissues is that analysis of a conveniently accessible tissue such as blood does not always give a realistic assessment of total body burden of the potentially toxic metal.

We now know that the ability to adapt and to tolerate an increased exposure is not constant. It is influenced by the stage of physiological development of the subject and by the nutritional adequacy of its diet. Notable examples are the remarkably effective tolerance of a high zinc burden except during the terminal stages of pregnancy when the body appears unable to prevent the accumulation of zinc by a developing foetus and a high perinatal morality results. In marked contrast, transfer of cadmium to the foetus is effectively discriminated against and, in this instance, the period of highest risk is in the newborn gaining access to adventitious contamination of its external environment.

The nature of the diet can have a profound effect upon heavy metal tolerance. One way in which this arises is from the limited capacity of the body to distinguish some potentially toxic metals from nutritionally essential trace metals that have a similar chemical structure. Thus, for example, if the dietary supply of copper is marginal, the tolerance of cadmium is markedly reduced because cadmium competes with copper in a number of biological processes. For reasons that are not yet clear, a low iron supply also enhances cadmium toxicity; many other examples of competitive interactions between essential toxic metals are now known and recent studies suggest that it is particularly important to take such interactions into account

when attempting to assess the risks of low level heavy metal intoxication. There are also indications that the content and nature of dietary sources of protein may have a marked bearing on heavy metal toxicity. The importance of such interactions is only just becoming recognised and failure to take them into account during the planning of experimental studies has led, not surprisingly, to a relatively high frequency with which conclusions appear conflicting.

This theme is one which, obviously could be developed at some length but the two salient points that I would wish to draw from these introductory remarks are the following. First, the influence of such variables must be taken into account when attempting to interpret data from epidemiological studies and the data obtained in such surveys must be adequate to permit a realistic evaluation of the impact of variables which influence tolerance. Second, those who press for a definition of tolerable concentrations of potentially toxic materials and particularly for materials to which exposure at a low level may be prolonged, must appreciate that any such definition implies a compromise. It would be a compromise based upon a subjective appraisal of inadequate experimental and survey data and must be subject to frequent revision as better data accumulate. The comments that I have made on this topic are not an apologia for the inability of experimentalists or investigators to produce firm guidelines; rather, they are a desire to ensure that the complexity of the task that has been set should be more widely appreciated.

Some of the more relevant findings from such studies are as follows. We are learning that, providing exposure is not too abrupt, animals can adapt to the appearance of relatively high concentrations of cadmium, zinc and copper by developing defence mechanisms. These work either by restricting the proportion of the dose which is absorbed, by restricting its transport to sensitive tissues such as the developing foetus or by hastening its excretion. An example of such a process is the capacity of the body to synthesise a novel protein, rich in metal binding groups which effectively modifies the rate of transport of the toxic metal between body tissues. One consequence of this altered distribution of metals in body tissues is that analysis of a conveniently accessible tissue such as blood does not always give a realistic assessment of total body burden of the potentially toxic metal.

We now know that the ability to adapt and to tolerate an increased exposure is not constant. It is influenced by the stage of physiological development of the subject and by the nutritional adequacy of its diet. Notable examples are the remarkably effective tolerance of a high zinc burden except during the terminal stages of pregnancy when the body appears unable to prevent the accumulation of zinc by a developing foetus and a high perinatal morality results. In marked contrast, transfer of cadmium to the foetus is effectively discriminated against and, in this instance, the period of highest risk is in the newborn gaining access to adventitious contamination of its external environment.

The nature of the diet can have a profound effect upon heavy metal tolerance. One way in which this arises is from the limited capacity of the body to distinguish some potentially toxic metals from nutritionally essential trace metals that have a similar chemical structure. Thus, for example, if the dietary supply of copper is marginal, the tolerance of cadmium is markedly reduced because cadmium competes with copper in a number of biological processes. For reasons that are not yet clear, a low iron supply also enhances cadmium toxicity; many other examples of competitive interactions between essential toxic metals are now known and recent studies suggest that it is particularly important to take such interactions into account

when attempting to assess the risks of low level heavy metal intoxication. There are also indications that the content and nature of dietary sources of protein may have a marked bearing on heavy metal toxicity. The importance of such interactions is only just becoming recognised and failure to take them into account during the planning of experimental studies has led, not surprisingly, to a relatively high frequency with which conclusions appear conflicting.

This theme is one which, obviously could be developed at some length but the two salient points that I would wish to draw from these introductory remarks are the following. First, the influence of such variables must be taken into account when attempting to interpret data from epidemiological studies and the data obtained in such surveys must be adequate to permit a realistic evaluation of the impact of variables which influence tolerance. Second, those who press for a definition of tolerable concentrations of potentially toxic materials and particularly for materials to which exposure at a low level may be prolonged, must appreciate that any such definition implies a compromise. It would be a compromise based upon a subjective appraisal of inadequate experimental and survey data and must be subject to frequent revision as better data accumulate. The comments that I have made on this topic are not an apologia for the inability of experimentalists or investigators to produce firm guidelines; rather, they are a desire to ensure that the complexity of the task that has been set should be more widely appreciated.

Some of the more relevant findings from such studies are as follows. We are learning that, providing exposure is not too abrupt, animals can adapt to the appearance of relatively high concentrations of cadmium, zinc and copper by developing defence mechanisms. These work either by restricting the proportion of the dose which is absorbed, by restricting its transport to sensitive tissues such as the developing foetus or by hastening its excretion. An example of such a process is the capacity of the body to synthesise a novel protein, rich in metal binding groups which effectively modifies the rate of transport of the toxic metal between body tissues. One consequence of this altered distribution of metals in body tissues is that analysis of a conveniently accessible tissue such as blood does not always give a realistic assessment of total body burden of the potentially toxic metal.

We now know that the ability to adapt and to tolerate an increased exposure is not constant. It is influenced by the stage of physiological development of the subject and by the nutritional adequacy of its diet. Notable examples are the remarkably effective tolerance of a high zinc burden except during the terminal stages of pregnancy when the body appears unable to prevent the accumulation of zinc by a developing foetus and a high perinatal morality results. In marked contrast, transfer of cadmium to the foetus is effectively discriminated against and, in this instance, the period of highest risk is in the newborn gaining access to adventitious contamination of its external environment.

The nature of the diet can have a profound effect upon heavy metal tolerance. One way in which this arises is from the limited capacity of the body to distinguish some potentially toxic metals from nutritionally essential trace metals that have a similar chemical structure. Thus, for example, if the dietary supply of copper is marginal, the tolerance of cadmium is markedly reduced because cadmium competes with copper in a number of biological processes. For reasons that are not yet clear, a low iron supply also enhances cadmium toxicity; many other examples of competitive interactions between essential toxic metals are now known and recent studies suggest that it is particularly important to take such interactions into account

when attempting to assess the risks of low level heavy metal intoxication. There are also indications that the content and nature of dietary sources of protein may have a marked bearing on heavy metal toxicity. The importance of such interactions is only just becoming recognised and failure to take them into account during the planning of experimental studies has led, not surprisingly, to a relatively high frequency with which conclusions appear conflicting.

This theme is one which, obviously could be developed at some length but the two salient points that I would wish to draw from these introductory remarks are the following. First, the influence of such variables must be taken into account when attempting to interpret data from epidemiological studies and the data obtained in such surveys must be adequate to permit a realistic evaluation of the impact of variables which influence tolerance. Second, those who press for a definition of tolerable concentrations of potentially toxic materials and particularly for materials to which exposure at a low level may be prolonged, must appreciate that any such definition implies a compromise. It would be a compromise based upon a subjective appraisal of inadequate experimental and survey data and must be subject to frequent revision as better data accumulate. The comments that I have made on this topic are not an apologia for the inability of experimentalists or investigators to produce firm guidelines; rather, they are a desire to ensure that the complexity of the task that has been set should be more widely appreciated.

Trade Union Attitudes to Risk

Dr. Owen

This is an area of work that seldom gets the attention it merits, and I hope to approach the subject in a new way, which will be of interest to scientific, ecological and legal experts alike.

Although among the twelve million members of affiliated unions, there are in fact very large numbers of experts, scientists, and so on, there are also great numbers of craft unions who move on to sites of various kinds and in the main they are not experts on matters of contamination, although since the development of the Health and Safety Law there is a greater understanding. This understanding has been enhanced by the action of the TUC itself in their training department where we are now training some 250,000 safety representatives - about 120,000 have already been through the basic courses and these members are increasingly able to ask questions about safety. They hope to have interpretations given to them by employers' scientists, but if they are dissatisfied they have means of access to other interpretations. The Health & Safety Executive paper has already dealt with the employer's obligation. What the TUC members would expect is a real attempt to appraise a possible hazard, informing those who will undertake the work and the provision of all that is currently available in the way of protective measures.

I would like to give you a couple of examples from my days in the Medical Inspectorate. One was the demolition of a factory site where thorium compounds had been handled and the other was the construction of a tunnel under a river which entered ground below which an old gas works had been present. Here there was heavy contamination of the gravel with tars and phenols. In both cases the work was well advanced before anyone raised the question of contamination, and in the second case only when severe cases of skin damage had started to arise. It is obvious that both cases might have been predicted. A further example is the laying of telephone cables through already re-claimed land which had formerly been a refuse tip and a problem arose here.

I noticed that this conference has not only dealt with chemical and radiation matters but also the reclamation of sewage works. There is a common belief in burning as a process for removal of contaminants - this is not always justified. In burning down a building we may in fact increase any risk.

What has been mainly considered so far and certainly by my scientific colleagues is risk assessment - finding out what is there, deciding how much and what its likely properties are. The other side of the coin is risk acceptance and this is where the Trade Union members have very strong views. Quite recently I attended a meeting where they were talking about the identification of services. Clearly this is a very important matter to the construction industry during site clearance.

As Dr. Sharratt has indicated risk assessment is not always easy and therefore it is often difficult to predict a figure or to persuade people as to the relationship between the particular risk they are undertaking and other risks that are common or commonly recognised.

At the end of the scientific judgements, there has got to be some good communication through whatever machinery exists within an organisation, to transmit any findings back, and through to the actual workers on site so that these people can understand and take any necessary precautions. There is a particular difficulty on construction sites compared with the factory situation where if a risk is known specific measures to counter this can be taken, and medical surveillance of the work population can be instituted. In the rougher circumstances of site work this is not so easy so what I am really appealing for is more communication in this more difficult and perhaps transitory area.

Epidemiological Assessment of Risk

Professor E.D. Acheson

The first point I would like to make is about the use of the word "environmental" in relation to the causation of cancer. Mr. Kenny in his opening address drew attention to the fact that epidemiologists have recently shown that a very substantial proportion of cancers (perhaps as much as eighty per cent) are due to environmental factors rather than genetic factors. This term has been seriously misunderstood as referring to environmental chemicals, such as might be present, for example, in contaminated land. In fact the term was used in a much wider context to refer to all influences, other than genetic, including those of such social habits as tobacco smoking, alcohol, and dietary and sexual habits. It is probable that chemicals in the external environment independent of personal social habits contribute only a very small proprotion of the total incidence of cancer.

My second point concerns the role of the expert in setting standards. I believe that the expert's role should be to compile and present scientific evidence about the hazards related to particular levels of exposure of a substance. The final decision about the standard ought to take into account other important matters, including those related to the social and economic costs of enforcing a particular level. Such decisions should be taken not exclusively by experts but by a representative panel including informed members of the public and of labour, industry, and so on. The recent Advisory Committee on Asbestos was composed of representatives of such interests. The findings of the Committee have recently been published and the evidence on which the recommendations have been made is open to examination and discussion before final decisions are reached.

Public Relations

Mr. Thomas

Experience has shown that the public now are becoming increasingly aware of the environmental hazards that may be created by new processes in chemical manufacture. There is no doubt that certain of the observations made by the public are ill-judged as a result of wrong information but this may be created by the lack of appropriate information being given by the appropriate manufacturing authorities and other institutions. I suggested that more and more the institutions would have to leave their "ivory towers" and explain in "the market place" the results and conclusions of their work. No longer will the public accept the statements made by experts and specialists without critical analysis and assessment.

Furthermore, the influence of the press and the media should not be under-estimated. An example was given whereby the Guardian's reporting of low intensity radioactive waste was the subject of critical analysis in the United States which may lead to only one commercially operated disposal point remaining open for disposal throughout the USA. Against this, the Department of the Environment are suggesting that Waste Disposal Authorities should handle more low radio active waste on their disposal points. Due to the prior publication and the concern expressed in the USA, it would be difficult if not impossible for public officers to convince local residents of the acceptability of this proposal by the Department of the Environment.

Society is entering a situation that may require new chemicals to be guilty until found innocent. It is therefore essential that the conference realises the change in attitudes now developing in this country. In many ways this is only reflecting what has already happened in European and Scandinavian countries.

There has been a major deficiency in the conference so far - a very serious problem that must be faced very soon, in fact now, and that is the wide feeling of non-acceptance of the practices that are being adopted. I was recently in America and the public are talking more and more in the United States about environmental pollution. There was widespread publicity on Three Mile Island. As one American expert said, a minor incident everyday and a major one every three. When accused of joking he said he was not and that this sort of thing was happening all the time now. There is at this moment a very immature position on the environmental situation. This is something you are going to have to recognise. In Sweden there is now a major debate on whether nuclear power should go forward. People demand referenda. In West Germany the signs are building up and people are becoming more articulate and more strongly voiced in this country.

People will not accept what you say, they will challenge what you say. You have got to have public support of what you do. Dr. Owen developed the theme of putting it over to the operatives on site. You will have a much greater problem in putting it over to people in the field and the public. We dealt with the Ravensfield situation and our Chairman commented without exaggeration that this involved the most difficult public meeting he had ever had.

There has been, I know, a very careful assessment of the problems of low grade radioactive waste. What the people will pick up is what they read in the 'Guardian' and I quote, "There are only three commercial burial sites which accept low level radioactive wastes in the USA", and I have to consider putting more such wastes on to our local disposal sites. The minds of the public have been coloured by the accounts they have read, rightly or wrongly, and I am expected to be intermediate between the private and public sectors in putting over this kind of thing. We are under open surveillance for what we do and the public pick up, quite rightly, such question as the incidence of leukaemia. They may be immature in their assessments but their questions are intelligent and they are not immature in their acceptance that there is a problem and also in their acceptance that the specialists are not experts in divine right. I am here as a public authority because you have eventually to come to some kind of public support or public acceptance for what you want to do.

Now there are risks and risks. We have a coal mining industry in Yorkshire and we know that we are faced with the deposit of as much spoil in the next 20 years as has been dumped in the Yorkshire region in the last 200 years. That is the scale of our problem, but that is not a risk situation, that is a known situation. What is a risk situation is when you come with compounds which will eventually produce nerve gas and say dispose of these. We are the waste disposal authority charged with the licensing of sites and putting them into operation. How do we handle this and still assure our people that what we are doing is right. It is no use saying to people there is no evidence that this is causing trouble, people will say and are saying strongly that until you have evidence that proves this is safe then cease the practice. I am in the public sector that has to produce answers.

Discussion

Mr M J McCarthy

In last year's David Dimbleby lecture, Lord Rothschild severely took to task the article on risk by Lord Ashby published earlier that year in the 'New Scientist'. However, in my opinion, having identified numerous pitfalls, he promptly fell into each and every one of them. At the end of the lecture, he stated that there were greater risks from generating electricity by windmills than by nuclear power. However, he did not present any evidence to support this and, as the subsequent letters to the 'Listener' testify, the information he was drawing on was, at best, questionable and, at worst, totally erroneous.

In the final section of my paper on the Willow Tree Lane Joint Development which deals with the question of risk I state, and I quote :

"experts have no special qualifications that sanctifies their decision to declare a level of risk justifiable. It is proper to take account of public perception of a public hazard, notwithstanding that it is distorted and inconsistant with a rational assessment. All technical information leading to proposals should be freely available and open to challenge."

Does Mr Thomas consider that making all technical information freely available would help in overcoming the problems that he has described?

Mr Thomas

Yes I agree. There is one specific example in the United States. Why is the evidence still forbidden to the public in general on the action taken by the US Authorities on any disease arising from the atomic explosions in the mid 50s in the US deserts. Yes I agree with the questioner. It should be more openly available, but will the public who may say 'yes I understand that, but I still think this'. Nevertheless explanation should be brought out more openly into the Highways and byways of public discussion.

Dr Owen

The previous speakers touched on a very interesting area. There have been a long series of articles in the New Scientist on this as well as the Rothschild lecture referred to which because it was compressed had to be selective in the comparitive examination of risks. One of the problems as has been indicated, is getting over not only the assessment, but also the acceptance in the light of the way it impinges on the individual. The other day I was at a medical meeting where a discussion came up on alcoholism. One of the speakers made the point that there has been a slow but gentle trend towards regarding smoking as less acceptable with increasing public pressure against it in public places as in trains. There is very heavy public pressure in some of the other European countries in particular. There is a long term changeableness. About 7,000 people are killed every in road accidents, but there is a limit to the pressures which develop to increase the expenditure which would decrease that high fatality toll dramatically. Certainly the public attitude varies according to the nature of the risk and on the amount that each individual will accept as his contribution to its prevention.

That is in the general public. In the employment scheme things have changed much more rapidly since the Health & Safety Work Act and the extension into other areas of employment. The original Factory Act only covered seven million people and the Offices and Shops and Railway Premises Act covered an area of much lower risk of about six million people. It has now been recognised that there are large areas of risk in the other eight million people - the new entrants under the Health & Safety Legislation. There is the setting up of Research Laboratories to start really apprehending large areas of unmeasured risk, not only in relation to toxic effects but also things like stress effects which have to be set in the context of these changing judgements. I very much support this developing trend. One sees it in the EEC Committees one goes to and even international changes of attitude as to what comes over the frontier in terms of contamination from neighbouring countries.

So we are now in a much bigger scene. One tends to be somewhat myopic in focusing on the individual working situation when others have to deal with community issues.

Mr Winters (Radio chemical inspector for South Yorkshire)

In reply to Mr Thomas on the dangers of low level radiation fills, all disposal of radioactive wastes to land fill is very carefully thought about and strictly controlled. There was public concern 3 years ago and there was discussion with the Environmental Health Committee and matters were explained. Public confidence and acceptance and risk assessment are not the same.

Mr Mike Smith

As a working hypothesis I suggest that the risk to people should be no more than the risk to materials although I agree it would be more difficult to assess. How should I re-act when I am told a leachate has a positive Ames test.

Dr Sharratt

This question should never arise but if it does the reaction should be to consider all aspects of the problem rather than jump to a conclusion that someone, or everyone, is in danger. The Ames test should be considered only as one of a wide range of possible investigations. The Ames test is very sensitive to substances which can, if they enter the cell nucleus, damage the genetic material. While a high percentage of carcinogenic materials give a positive result in the test, many chemicals which do not cause cancers also give a positive test. A baked cake will give a positive result while the raw ingredients do not; how is one to react to that? I suggest only by examining all aspects of the problem rather than one; a wide range of investigations is in fact, almost always, needed to answer any question relating to the effect of chemicals on the health of man.

Dr C F Mills

Dr Sharrett's comments on the difficulty of interpreting the Ames test might also have been applied to a very wide range of other, rather arbitrary tests. Among these may be included techniques for the chemical extraction of soils in the attempt to assess the availability of adventitious heavy metals and the diffuculty of interpreting data on the heavy metal content of the blood of human subjects exposed to a chronic or acute heavy metal burden. The greatest problem lies in correlating the results of such tests with the metabolic or pathological response of the animal or human individual at risk from exposure. Very often the problem is concentrated in a target organ that is not accessible for sampling and, where we are particularly lacking data is on the validity of observed changes in the composition of tissues such as blood that might be more sensitive indicators of present or impending damage than merely the determination of freely circulating heavy metal content. This is a theme requiring much more investigational work before more useful parameters for the assessment of risk become available.

Mr Thomas

A problem may not occur until the building is there and can be well away from the source of the original pollution.

Dr Mills

Public pressure often leads to the expression and provisional advice on "scientific evidence" of a standard which would not be accepted by reputable journals but it is the best we can do at the time.

It is particularly important that the public to which Mr Thomas has referred should be given as realistic an appraisal of the state of knowledge in this field as is possible. In other words they must be given the chance to see that even experts disagree, not always because they are pugnacious but more frequently because of the scarcity of observational data. Given a relatively dispassionate appraisal of the situation they can then decide whether their anxieties are such that they wish to see more resources put into studies of this type. How much better and more helpful would television coverage of suspected cadmium problems in Somerset have been if, instead of sensationalism, the theme had been ".... what are the limitations to existing knowledge that are going to make it so difficult to determine whether the problem exists in this small community".

Mr Thomas has implied that those faced with the control of pollutants are being repeatedly let down by "experts". This may well be the case but I feel it is only fair to emphasise this is usually a reflection of a shortage of experimental or survey evidence upon biological effects rather than any lack of a conscientious approach to this topic. Turning to the role of the "media", there is very great room for improvement in the way this subject is handled as I recently became aware from personal experience. A while ago I presented a paper to the British Association for the Advancement of Science on the subject of essential and toxic trace elements in animals and man. The full text of this paper was available to the press. One "quality" newspaper gave a brief but factually accurate account of the paper. Another pounced upon one line of a 23 page typescript and produced an article which conveyed the impression that I had been talking about trace elements and sex. Two weeks later I was sent a photostat of an account in an Australian newspaper which implied that a substantial proportion of the UK male population was having diffuculty with its sex life and the Dr C F Mills of Aberdeen knew the answer ... zinc! Scientists or "experts" are not always responsible for the communications difficulties that seem to exist.

Dr Owen

How do we set standards? How can we pick out chemicals which need testing? Is it possible to feed into a computer details of people exposed and all the evidence of possible effects? Finally, could a large matrix emerge against which known evidence can be put which may "lock in" and enable prediction of risk from compounds chemically related to some with proven hazard. Epidemiological and population studies may be viewed as a human experiment. Prior evidence which alerts one to a situation is better. There is also the question of the creation of a situation.

There are risk assessments by experts and there are the people, usually workers, on the receiving end. Sometimes they confront and challenge one another. Perhaps this is due to lack of communication. Scientists must take action to persuade others of the validity of what they say and be ready to explain and justify. The scientific evidence of risk needs to be followed by agreed methods between the employer and the worker to deal with the risk. But while there are experts on the one hand and receivers on the other, action usually stems from persons in between.

Mr Rowcliffe

1 It is not unusual to find different people giving different emphases to different factors in discussion of risk assessment and acceptability of risk; and this has been reflected in the contributions from members of the panel.

2 The "experts" tend to emphasise the uncertainties and difficulties of risk assessment, and to argue against an over-cautious approach to environmental decisions, on the basis that all human activities involve some degree of risk. Those who are involved in the implementation of decisions may tend to be more concerned about public perception of risks, and the problems of persuading the public that a risk which is agreed by the "experts" to be insignificant is indeed publicly acceptable.

3 The problem for the administrator is that, however difficult the assessment of risk may be, decisions have to be taken on the basis of the best assessment that is available - even if this sometimes means over-simplification. Both central and local government are constantly involved in making decisions in which the assessment of risks is an important element.

4 Examples of such decisions have been touched on during the conference - for example, the work at Beaumont Leys has involved setting figures for heavy metal content which can be accepted as involving no significant risk for those who will live on the site. And similar problems arise in relation to all aspects of pollution; and in some cases decisions may be complicated by the need to compare widely differing types of risk (for example, in relation to energy policy how does one compare the risks inherent in nuclear energy - including disposal of radioactive waste - with the very different problems that may arise - eg through production of SO_2 - where energy is produced by the burning of fossil fuels?)

5 As the complexity of risk assessment (including assessment of relative risk) increases, the chances that public perception of the acceptability of risk will be aligned with the conclusions of "scientific" risk assessment recede. This leads to 3 questions on which the panel are invited to comment:

a. do they see any basis for a synthesis of the differing views on risk assessment/acceptability of risk that have been put forward in the disucssions so far;
b. do they see any possibility of improving public understanding of the complexities involved in this subject, so that public discussion can proceed on a more rational basis?
c. if the answer to the second question is "No", what do they see as the implications for public policy? Should we give up the attempt to get the complexities understood, and simply rely on the experts' "objective" assessment of risk?

Dr C F Mills

We are much closer to an understanding of the agricultural relevance of such situations than we are to assessing the risks arising in man. In an agricultural context an appreciable amount of work has already been done from which we are beginning to understand the effects of potentially toxic additives upon soil composition, crop yield and composition and, in some instances, effects upon animal production and health. Direct experimentation has been possible in an agricultural context which is obviously not practicable with man. One consequence of this is that we have little understanding of the extent to which, for example, tolerance of potentially toxic heavy metals may vary under different circumstances to human subjects and, for this reason, a very wide margin of safety is usually built in to any estimate of tolerable threshold concentrations. It is perhaps salutary to appreciate that if the same criteria were applied the consideration of naturally occurring toxic compounds in some common food stuffs, for example rhubarb, their consumption would hardly be recommended by any committee on food safety.

Professor Acheson

There are several lines of approach. First, there may be epidemiological evidence; second, there is animal evidence and I would refer you to the WHO series of monographs on carcinogens. For more than one hundred substances which are known to produce cancer in animals there is no evidence which enables one to come to a conclusion whether they are carcinogenic in man or not. To collect useful evidence in man one has to make use of past industrial exposure, and unfortunately it is necessary to wait at least twenty years to observe the effect. For example, in the manufacture of products like D.D.T one might investigate people who have been exposed in the distant past. If one could identify those who were exposed thirty to forty years and determine that their risk of cancer had not increased this would be an important piece of information. There is a long list substances which require this kind of work.

Dr C F Mills

While I would readily accept that studies with animal models have severe limitations when attempting to assess the likely pathological consequences of heavy metal exposure to man I sometimes wonder whether the full value of such laboratory animals is exploited. It is not uncommon to find such investigations have been carried out on as many as six different animal species. When allowance is made for the fact that metabolic and pathological consequences differ between acute and chronic toxicity, it is surprising how frequently a consistent theme of biochemical abnormalities can be detected. Such information could perhaps more frequently be used to emphasise lines of enquiry with respect to effects in man. Two examples can be quoted. The earliest consequences of chronic low level cadmium intoxication in many experimental animals are attributable to induced defects in copper and iron metabolism. How often have we looked for such effects when considering the consequences of cadmium exposure in man. Secondly, such studies with small animals are also beginning to suggest that impaired development of connective tissue may be an early consequence of these defects; the severity of the effects depends greatly upon the age of the animal at the time of exposure and other variables such as the nutritional adequacy of its diet. Although it is perhaps arrogant of me as a mere biochemist to suggest such a thing, I again wonder whether the implications of such findings are taken into account in studies of the patterns of disease incidence in man when a toxic hazard is believed to be present in the environment.

Dr Sharp

Taking Professor Acheson's point about studies of workers who have been exposed to possible carcinogens, you may like to know that this has been done in regard to certain dyestuffs, notably dichlorobenzidine by Dr Theodore Gadian who is a Medical Officer for the Clayton Aniline Co. Ltd. and reported in "Chemistry and Industry" (Issue dated 4th October 1975 page 821).

The circumstances in Clayton Aniline permitted the study of workers who had worked for a long time with suspected carcinogens and whose medical history it was possible to obtain due to the closely knit nature of the work-force in the area.

On the general matter of relations with the public, this is a difficult problem. However, if I may say so, Mr Thomas himself is a classical example of how to do it. He evidently had a very rough time over the Ravensfield tip but he won and every point won is an advantage.

Another point I would like to make is on the relationship that exists between the Civil Service, particularly the scientific Civil Service, and their opposite numbers in industry in this country. This relationship is much better than anywhere else; certainly in America there has been confrontation; in the EEC the relationship is by no means cordial. I think it would be very difficult to find elsewhere a conference of this nature, with people drawn from so many disciplines from local government and central government and industry.

Mr Jim Dixon

I am not going to ask a question but to relate an experience. One must not treat as trivial any site which one suspects to be contaminated. I am thinking of an old tar works I had to investigate on a beautiful warm day. I walked around inspecting, came home for lunch and kept the same socks on all day. At night I had bright red feet. I wonder how many little Johnnies, Willies and Tommies reached their doctors with burned feet.

Dr Naylor Firth

I wanted to put a question to the Panel that arose for me a few weeks ago (I am a scientific civil servant and not a member of the general public). I was engaged in a discussion with the general public and two things were put to me - that in Britain it is assumed that all chemicals are innocent until proved guilty and ipso facto that the general public are used as guinea pigs. How would the Panel answer that accusation?

Professor Acheson

One of the difficulties is that there is no way one can be certain about carcinogenicity in man until the substance concerned has been in use and there has been an opportunity for exposure to take place. Whilst in vitro and in vivo animal tests are important, in a few cases, for example arsenic and benzene where as far as I know there is no satisfactory evidence yet that either is carcinogenic in animals, there is no doubt that they are carcinogenic in man. Animal experiments are an important guide particularly when they are positive, but they may occasionally be misleading.

Dr Sharratt

I think the Health and Safety at Work Act did reverse very effectively the long held attitude that a substance is assumed innocent until proved guilty. Section 6 shows that you must know the toxic properties of the substance you are producing, whether it is a single substance or a formulation, so that advice can be given on how the material can be used without hazards to health. People may want to know if the substance is safe but must realise that it is virtually impossible to prove anything is safe; one cannot prove a negative. All one can do is gather information and make an informed estimate of the risk on available evidence, but with the present state of knowledge of human biology only human exposure can, with the majority of materials, give an answer which can be guaranteed absolutely.

Dr Owen

I would like to make two points. The first is in relation to making a judgement on chemicals. Quite recently I was in America where there has been pressure to ban all benzidine based dyes but not those based on orthotolidine and some other chemically related compounds. The basis of the evidence against the benzidine ones was animal experiments by the National Cancer Institute where three of these dyes did in fact produce tumours in animals. So here is a case in point, where one has animal evidence but what does one do about it?

The second compound, which I was asking about only yesterday, is very widely used - phenyl beta naphthylamine - which has been in use since the 1920s and there have been individually exposed populations and there have been epidemiological studies and no evidence has been found of excess bladder cancer in these exposed populations. Now the suspicion has been raised again about the material quoting testing of urine specimens from people who have been treated with these materials. So there you have some evidence of the possibility of a free carcinogen being released. And then you are given another piece of information that the orthohydroxyphenols which are the proximate carcinogens produced from these dye stuffs are not produced in the dog. So I am left with a rag bag of information on which somebody has got to make a judgement and then go back and explain these findings to whoever their constituents are.

On the question of constituents, in the TUC we have a very formal consultation machinery and as indicated by Professor Acheson there is a consultative document procedure which comes out through the Health and Safety Commission. The CBI has its own channels of communciation to its own membership. When I hear about representation of the public, one looks at the constitutional basis of this representation and one confesses it is very difficult to discern the mandate unless they are truly elected representatives like MPs. We have members of the general public on the Genetic Manipulation Group and they were selected on the best advice by the Minister of State for Education. When you think about it though, the only chance any of you, or any group of the public have for input, is through our ordinary constitutional Member of Parliament approach.

It is very difficult to test public reaction, except through the media or local responses and these are not always 100% representative. It is very difficult to get a representative view. The scientific community is getting more organised on this. There is running at the moment a Royal Society Working Group on risk and the public attitude to the acceptance of risk is one of the points which will come up in making value judgements.

Dr Sharratt

There is still much public distrust and this is understandable when peoples' health is involved and when, on occasions, someone in authority does not seem to understand that there is a problem. For example, a class of chemicals called nitrosamines caused cancer in all species of animals tested. It is inconceivable that if human beings are exposed to these substances they will fail to develop cancer, but even with this evidence one hears senior men in industry and medicine say 'but of course it has never been proved by epidemiological methods that these compounds cause cancer in man'. There is really only one way to answer any problem and that is to have good, sound scientific and epidemiological information. A major problem is that this can be extremely costly to obtain. In addition, it is often diffucult to say how much any investigation will cost at the outset; it may cost thousands of pounds to perform the necessary analyses to define what the chemicals are to which people could be exposed, and toxicological investigations will also be needed if the toxic properties of the substances found are not known. Such investigations will certainly cost thousands of pounds and may cost several million in a difficult case. It is essential to educate the public to understand both that investigations can be costly in time and money and that in making decisions such cost has to be weighed against the possible degree of reduction in the risk to health they achieve. We are lucky in the United Kingdom that government advisors are generally trusted; I hope we will not reach the condition in the USA where, so it appears, no one trusts anyone.

Mr Souch

In March 1979 the Royal Institute of Chemistry instituted the study entitled RECAP asking members to list the chemicals which they have used in their working life as chemists.

Through this study it is hoped to identify adverse effects of chemicals on man and in the long term to lead to better control of health risks.

Details are available from the RIC at 30 Russell Square, London WC1B 5DT.

Question

How would the Panel assess and compare a very low probability risk of high danger and a very high probability risk of low danger?

Dr Owen

I take it we are saying how does the public compare one in a thousand risk of death from a substance to a one in ten risk of, say, toxic effect short of death and I don't know the answer. I do think we have to be very careful on this question of acceptable risks. One has to be sure that the people subjected to the risk are aware of it. On many occasions still I think the knowledge of the actual workers in industry that their risk of death or injury is higher than in another industry is lacking. They accept because they don't know about it. I am not convinced they know. We must be very careful about concepts of acceptability.

Dr Sharratt

One cannot compare a slight chance of occurrence of a serious hazard with a high probability of some minor disability occurring. How can one compare a slight chance of occurrence of death or brain injury with the certainty of occurrence of a few hundred cases of dermititis? All hazard assessments must be made in relation to a single situation and all factors have to be considered in relation to it. You cannot draw solid guide lines on how to react to a theoretical situation.

Mr Thomas

Society is not going to demand any materials or any pressures that are not going to make their lives more convenient or more acceptable. The public will resist making life increasingly difficult or deprived.

Mr Khan

Risk is a subjective matter and deals in an inexact science. The risk is always related to reward. Nothing which gives rewards is free of risk. I am in toxic waste management. One is a rope walker doing balancing tricks all the time when making decisions. The question is always one of deciding options against the background of other factors, legislation, economic, social impact, public reaction, health and so on.

Dr Owen

This question of relative risk is a big problem. Many organisations are wrestling with it now. I know Dr Sharratt has been saying risks are very difficult to compare with one another, especially different degrees of injury but there is a yes/no situation and that is death and at some of the meetings I have been attending an attempt has been made to compare, say, 300 people dying one a day for a year or 300 people dying together as in a Jumbo jet crash. Now people tend to ignore the first situation but would be horrified if the second one happened every year in this country.

Take the construction industry as an example. I am interested in this particular industry as there is a Research Committee in the EEC on risks in the construction industry, which has a relatively small number of people in it but has a high number of fatal accidents.

I think one of the criteria which comes out in public acceptability is whether you can do anything about it. When we get in a jet at an airport we always hope enough has been done about it to reduce the chances enormously so that this year will not be the year for our Jumbo jet. Yet if we were members of a working population that knew that 200 to 300 chaps will be killed at work every year we wouldn't find it acceptable, yet in some industries this is the current situation. People in authority have to make judgements about priorities.

Dr C F Mills

I find it difficult to rationalise my preference for the second alternative; the low frequency high risk situation! Perhaps I am willing to accept it only when I have had the chance to hear some evidence of the nature of the risk and of its origins, and even if the opinions of experts conflicts, to have had the opportunity to make my own decision as to its acceptability.

Chairman

That is a very good note on which to end and I would like to thank the members of this Panel for contributing to this conference in the way that they have, they have been very expert, very entertaining, and very enlightening.

RECLAMATION OF CONTAMINATED LAND

SOME CONSIDERATION OF LAND VALUES

David E. Goldsmith FRICS - Chartered Surveyor

and Land Agent - Partner - Messrs. John D. Wood

INTRODUCTION

The valuation of land and property is not always the science that many consider it should be but more an art built upon the experience of the valuer and based on the interpretation of such facts and forecasts as may be available to him as the practitioner at the time of the valuation. It is an expression of opinion and often given for a particular circumstance or requirement and therefore need not always have regard to all the potentialities of the particular property. Thus in taking a note of a valuation one must have regard to the purpose for which it was asked.

In taking this subject I have assumed that the valuer is expected to consider the worth of the land in its widest sphere, both during its use for tipping and its after use. At the outset I can say that with contaminated land there is no 'rule to thumb' method that can be adopted in assessing the value and that every case must be assessed on its merits or otherwise.

Many people consider that land once filled or contamined cannot be used for any other normal use and the restrictions placed upon it are so substantial that it has virtually no value. However, this is not necessarily so and I consider it unimaginative thinking. However, we have to live with the apprehension generated by the thoughts of contamination, and this apprehension must be taken into account on most occasions.

The valuer must look beyond the immediate use to which the land can be used and apply some long term thinking to the problem and hope that in appraising the various alternatives and their respective values he can find some practical use of the land for which it can be advantageously occupied. I hope that in making this point it will stimulate some thought in the minds of the users as to whether it is economically viable to consider some greater restoration arrangements in order to promote a much better after use of the property. I sometimes, as a valuer, feel that insufficient thought has been given to this aspect - this is certainly true in the past, but may be today in these more enlightened times and as a result of what has been said at this Conference, much greater thought is being given to the subject.

BASIC INFORMATION REQUIRED

In making any valuation all the relevant facts about the property must be investigated and established. This Conference must have highlighted the advances made in the last decade on monitoring and controlling the deposition of waste materials, and the use of land for this purpose.

This should in future make a valuers life less fraught with uncertainties as it is an absolutely essential requirement in reaching an opinion to know the full background to the property one is valuing. It is therefore the hope that valuers are always given the fullest information and can be told the full facts of life however unpleasant they may be!

In the following comments I am assuming that the persons requiring a valuation take the valuer into their fullest confidence and provide him with all the facts, or will make the effort and meet the expense of establishing the facts on which a reasoned opinion can then be based. In order to formulate an opinion for a valuation the following are the main matters for consideration:-

1. Original User

   The previous use and the geophysical features of the property should be established. Was it agricultural land and if so what of its original quality and potential and what constituted the sub stratas; drainage; outfalls etc? If it were in some non agricultural use then what was this, and what services, planning potential, restrictions etc. might have applied?

2. Physical Changes

   What has physically happened to the land? Has it been excavated and has it been restored to the original levels? Is some investigation of the adjoining property necessary?

3. Purpose of Tipping

   What are the reasons behind choosing the site for the particular type and method of tipping that has occurred? The original criteria for choosing a site often gives an indication of the 'security' of the waste materials or, if as often happens, no such criteria was applied, then the warning bells can begin to sound!

4. Analysis of Tipping

   What exactly has been tipped, the state of decomposition, the presence of contamination and the likely effects this is to have. This needs a complete appraisal and here also one should investigate the standards adopted in the tipping to highlight if any future problems are likely to occur.

5. Present or Future Physical Problems

   What are the problems applying to the site such as drainage, leaching, subsidence and the likely effect of these on any future use. If there is a lack of such information then further investigation must be carried out.

6. Standard of Restoration

   If the land is already restored and planted evidence is required to establish the effectiveness of the restoration and whether there will be any complications in the full establishment of crops and the effects of contamination in the future on plant growth or the taking up of toxic elements harmful to health.

7. Planning Conditions

Are any planning conditions imposed on the property and if so are they being complied with? A straight forward question but often producing complex answers!

After this Conference it may seem easy for all this information to be available and it is reassuring to know there is so much expertise now established to provide the information, but I fear it will be many years before we can expect every land filled site to have all the requisite information about it readily available.

The background requirements I have summarised above may seem obvious, but regrettably my experience has shown that the extent of the information available from these simple questions is not always forthcoming and sometimes memories are either very short or they are the excuse for not having kept adequate records.

From this summary I hope I have made it clear that one needs to assess both the actual use to which the land has been put, the manner in which that use was conducted and the effects of that use in order that one can now look ahead to its future.

SITE LOCALITY AND TOWN AND COUNTRY PLANNING

Having investigated the basic problems of the site and before deciding if its present use is adequate or if some alternative use would be beneficial in the terms of value, one must consider the location of the site.

A land filled site in the heart of Gloucestershire countryside would have a very different valuation approach to a similarly filled site on Thameside. The former due to planning restrictions and its relative inaccessibility might only be considered for future agricultural or recreational uses, whilst the location of the latter may lead to other non agricultural uses. The planners might on Thameside recognise that the land could not be restored to its full agricultural use, but might reasonably assume some other role and take the pressure of development off other hard pressed agricultural land which would otherwise be lost for development. In other words I am suggesting that the Planning Authority should be flexible in considering these sites and realise that they can have some greater use, and thus in bringing this factor into valuation context I suppose I am almost suggesting that one has to appraise the sensibility of the Local Planning Authority!

We find that the Local Planning Authority attitudes vary throughout the country, and some are definitely reluctant to 'grasp the nettle' in contaminated land possibly owing to a lack of experience or over emotive public response. However, one should remember that planners views change as their objectives change, and as pressure for development on an area increases, so sites on which a blanket 'no' is given this year, may attract a different response in 5-10 years time.

Because one cannot always assess the time scale of an alternative use or change in planning attitudes it is sometimes difficult for a valuer to appraise this future and therefore makes the art of valuing somewhat speculative.

FUTURE USES OF THE LAND

Physically disturbed land normally takes a very long time to return to anything near its original use unless considerable care is used in the initial stages of excavation, and adequate prior thought is given to the reclamation and preservation of the topsoil overburden etc. and to ensure that adequate provision is made for the reinstatement of drainage. Thus for agriculture it can never be expected to return entirely to its original capabilities or to have a value comparative to its original undisturbed state. Thus we must look to the alternatives, and these can be very wide ranging.

If land cannot be returned to its original capabilities then some other open space used should be considered complimentary to the surrounding countryside and linked with it. Such uses are often recreational, such as a Golf Course, Public Open Space, Sports Ground, Caravan Site and Garden Centres. Such uses rely more on the surface restoration of the site rather than a reliance on the re-structuring of the lower stratas.

In urban areas other uses can prevail where land can be used complimentary to the industrial area that is likely to exist and the use for Car and Lorry Parks, Vehicle Storage and Light Warehousing can be thought of which are uses gauged to the new structure of the land, and accepting that any greater use could create enormous problems in development. To illustrate these relatively simple alternative uses I can indicate with examples.

a) Petrol Filling Station

b) Golf Course

c) Open Storage and Warehousing

TIME SCALE

In suggesting alternative future uses one must get the time scale into perspective. On occasions it is impossible to bring all these uses into being at once. I have found that some land filled sites have been allowed to lie dormant for 10-15 years after which there has been consolidation and it has been easier to accept such alternative utilisation. It would be unwise to consider the construction of rafted foundations on sites unless one is certain of the proper compaction on reclamation, but after progress loading with say surface storage, and the subsequent consolidation then the risks of rafted foundations should become less.

ANALYSIS OF DEVELOPMENT COSTS

It is all very well to consider the various alternative uses and to produce some grandiose scheme for reclaiming an unattractive land filled site, but the crux of the matter invariably comes down to the economic viability of such a scheme. Thus in assessing any such development one must be able to establish the cost of the alternative form of development. To do so is again dependent on the degree of knowledge and information available about the site.

Because of the high costs of overcoming the inherent problems some forms of development will not be viable and will have to be discounted. Thus the valuer must be aware of the cost effectiveness of alternative development and must not be carried away with grand and attractive schemes as ultimately it must be assumed that a rational buyer has to be found to create the resultant development who will need to finance the project from being able to offer the finished article at a realistic market value.

Besides the development costs a knowledge of the possibility of government sponsored grants being available, either for restoration or other purposes, and the conditions on which these grants are payable is also important, but this is the subject of another paper which we can look forward to.

EFFECTS OF CONTAMINATION ON VALUES

As I said at the outset there is no 'rule of thumb' for determining the value of contaminated land and there is a great deal more investigation and information required for these sites than almost any others. To arrive at any value there must be an appraisal at the given time of the willingness of the seller to dispose of his present interest and of the interest of the buyer with the full knowledge of the contamination to take the future risks in the matter.

As a simple example of the effect in an 'open market situation' I can recall that a block of agricultural land purchased in the 1960's by a Public Company, where part of their operations were to deposit sewage sludge which was spread on to part of the land, and then cultivated. At first sight some years later this land appeared no different from any other, and in fact the crops appeared perfectly satisfactory. The property came into the market without the full history being disclosed and a number of prospective buyers were available. A purchaser was selected who proceeded to contract but being diligent had the land surveyed for mineral deficiencies to decide on his future cropping and fertilisations. The Ministry of Agriculture Fisheries and Food who carried out this investigation highlighted that toxic elements were present, and pointed to the use for which part of the land had been put. Whilst it was borderline as to whether the quantities present in the soil would cause a great hazard to health in certain crops, the selected purchaser was apprehensive and took fright - Thus the sale went off and it was a long time before the Vendor was able to attract a purchaser, now the facts were known, who was prepared to take the risk (it was now right for this risk to be disclosed). In the end a neighbouring farmer did purchase it, but at a much reduced value.

I put this example forward as it indicates the thinking in the minds of people and we must wonder whether if the disclosure had been made at the outset, would the resultant sale have been at such a discounted figure. If the risks had been properly explained to show that there were only certain crops that could take up these elements, and the effect was concentrated over a small part of the land on which the sludge had been dumped, rather than the farm as a whole, then the effect on the overall farming policy would have been minimal. If that had been the case, then, maybe, the reaction of purchasers would not have been so great.

Another example is where there was comparatively poor reinstatement of old gravel workings which had been filled with mainly household refuse at a time when there were very limited planning conditions imposed on the consent. It was in an area close to urban development where the pressures on taking good agricultural land for development were considerable. The reinstatement of the land by the contractors was poor and a local farmer attempted for several years to re-establish the property to agricultural use but it proved not to be economically viable from that point of view. The land needed greater work on drainage and it would only be possible to treat the surface and thus could grow very limited crops.

The Local Authority decided they needed a Country Park but avoided this site and acquired adjoining property which they quite readily adapted for recreational purposes. This site was then isolated from the adjoining agricultural land, but came into the open market. There was no interest for some time, but some indiscriminate caravan parking occurred. At this point the land suddenly showed potential for other uses and in the end it was sold at a figure attractive to the Vendors, but much below its original use value. It raises the question as to whether it would not have been sensible for the Local Authority to have acquired this land and amalgamated it with their Country Park rather than taking prime agricultural land, and the difference in value of their acquisition would more than have compensated for the carrying out of site working to bring the land into the limited use they required.

CONCLUSIONS

I have tried to point out that the valuation of contaminated land is a relatively complex matter. The effectiveness of the valuer in this difficult sphere is dependent upon viewing the problems with an open and enquiring mind. It is essential in the interests of all the parties to the matter that the valuer be given the greatest amount of background information available. This information can then be interpreted by the valuer who can bring in to play his experience of other opportunities and constraints to arrive at a realistic appraisal at a particular point in time, and for the purposes of the persons requiring the valuation. Thus this is a challenging subject and one which offers a little spice to what otherwise might be a monotonous valuation life!

RECLAMATION OF CONTAMINATED LAND - OUTLINE OF RELEVANT INSURANCE

MR.B.R.WEST

COMMERCIAL UNION ASSURANCE CO.LTD.

Insurance implications are complex. For purpose of non technical insurance review consider list of interests and events in a hypothetical rehabilitation of contaminated land for building purposes.

Interest	Event
1. Owner of Contaminated land	Remedial work to rehabilitate
2. Professional adviser on rehabilitation ?	
3. Local Authority - approval required ?	
4. Contractor undertaking rehabilitation work	
5. Developer acquires land	Land to be developed
6. Soil engineer	
7. Architect	Plans prepared
8. Local (planning) Authority	Plans approved
9. Contractor undertaking building work	Buildings erected
10. Local (buildings in course of erection inspection) Authority	
11. Buildings sold to ultimate purchaser	
12. Purchasers surveyor/Building Society surveyor	

(This list is not necessarily exhaustive).

Relevant Insurance

What insurance protection is feasible related to this chain of interests and events ?

Two basic categories of insurance :-

Category (A) Insurance for specified loss or damage from specified risks.

Category (B) Insurance to indemnify legal liabilities incurred in respect of work or services performed incorrectly/negligently.

The principle standard forms of relevant insurance in these two categories are :-

Category (A)

Contract Works - will insure loss/damage of the contract works whilst in course of construction

Fire and Perils (commercial property) - will insure loss/damage of property for Fire and Special Perils e.g. lightning, explosion, aircraft, water damage, impact (NOT subsidence damage)

Fire and Perils (private property) - similar to commercial property insurance but will usually include subsidence damage

Consequential Loss (commercial property) - will insure financial loss due to interruption of business following fire and specified perils as listed above for commercial property.

Category (B)

Public Liability - will indemnify for legal liability incurred in respect of accidental bodily injury (not to employees) or loss/damage to property (not in policyholder's custody or control). Liability attaching by contract is normally included but Insurers will require notification of any material extension of Liability in this respect.

Product Liability - is an extension of Public Liability to indemnify for similar liability arising after "product" handed over.

Professional Indemnity - will indemnify for legal liability arising from professional error/omission/breach of duty.

Employers Liability - will indemnify for legal liability incurred in respect of injury to employees arising out of their employment.

(NOTE. These are "headline" descriptions only of the broad intention of the various policies. There are of course additional limitations and conditions attached to the covers)

Utilising these check lists (and related solely to the aspect of Loss/damage/liability from action to reclaim land) the following insurances may be involved.

Interest	Insurance
1. Original Owner of Land	Public Liability - Liability for injury or damage may be incurred arising from ownership of the land. This could be direct or indirect (devolving from actions of contractor) and could continue after disposal of the land in the form of undiscovered injury or damage.

2. Professional Adviser	Professional Indemnity - liability may arise from wrong advice given.
3. Local Authority	Professional Indemnity - Responsibility to inspect/approve may not have been fulfilled properly.
4. Contractor	Public Liability - Rehabilitation work may have been carried out in a defective manner resulting in Third Party Injury or Damage whilst the work is in progress or subsequently.
5. Developer	Public Liability - Repeat of liabilities outlined for original owner.
6. Soil Engineer) 7. Architect) 8. Local Authority)	Professional Indemnity - Liability may be incurred if design defective and/or responsibility for approval not properly discharged.
9. Contractor	Contract Works - Damage to contract works related to nature of Land. Public Liability/Product Liability - Work related to nature of land carried out in defective manner resulting in Third Party Injury or Damage. Employers Liability of work results in Employers injury.
10. Local Authority	Professional Indemnity - Responsibility to inspect work may not have been discharged properly.
11. Ultimate Purchaser	Fire and Perils and Consequential Loss - Damage to property from insured peril caused by defect in land.

Some Previous Insurance Claims

(a) Subsidence of made up ground used for construction of private dwellings.

(b) Industrial premises damaged when adjacent mineshaft blew its concrete plug due to uprising of industrial waste deposited in it.

(c) Spontaneous combustion in yard which (unknown to user) was previously used as rubbish tip.

As an example of practical problems the following factors arose under (a) above :-

(i) Original Builder had gone bankrupt

(ii) Local Authority responsible for Planning/Inspection absorbed by another

(iii) Houses insured for subsidence but for less than current value

(iv) Adjacent house reduced substantially in value despite no physical damage and insurance therefore inapplicable

(v) Contradiction between experts as to cause, likelihood of further subsidence, and remedy

(vi) Emotional upset of householders.

(vii) Extensive (mis-informed) media publicity.

Comments

Self evident complexity. May be reduced somewhat by "consortium" approach but various interests usually too diverse to permit blanket approach.

Loss potential and insurance protection available can be reasonably well identified in areas discussed under Category (A) (Specific insurance of specific loss/damage). However potential cost and uncertainty greater in area discussed under Category (B) Legal Liability Insurance, the fincncial potential cannot be known - it could in theory be the total cost of the project plus consequential losses. In addition the exact nature and extent of the liability incurred cannot be foreseen and is not static.

Liabilities imposed/accepted by contract must be recognised and may have a material effect on insurance needs and costs.

Total insurance protection for all potential costs and liability from something going wrong is not feasible. Substantial insurance protection can be arranged provided nature and scope of potential risk recognised by all concerned and insurance protection related to it carefully.

GOVERNMENT FINANCE

A REPRESENTATIVE OF THE DEPARTMENT OF THE ENVIRONMENT

1. A country as densely populated as Great Britain must make the best possible use of all its land. However, the costs of bringing back into use land which has been damaged as a result of industrial or other development is often so high that it cannot be borne as a normal cost of development.

2. The Government has recognised this situation and using powers under the Local Employment Act 1972 and the Local Government Act 1966 makes grants available to local authorities in England for reclaiming this damaged land. In Scotland and Wales similar arrangements are administered by the respective Development Agencies. In areas designated by the Secretary of State for Industry as special development, development, intermediate or derelict land clearance areas, and in Scotland and Wales, the rate of grant is 100% of approved costs. Elsewhere it is 50%.

3. The grants have been devised to help local authorities to meet the costs of reclaiming derelict land so that subsequent use or development is no more expensive than if it has taken place on a green field site. As a further encouragement, the initial grant covers land acquisition costs as well as the cost of the reclamation works. The amount of grant is reassessed when the value of the reclaimed site accrues to the local authority. The final grant covers only the loss incurred in reclaiming the land. When the new use of the land is public open space there is no subsequent reassessment so the full cost of the land acquisition and reclamation work is eligible for grant. However, if it can be clearly seen from the outset that the cost of the works can be covered by the increase in the value of the site following reclamation, it is not eligible for grant.

4. There is no statutory definition of derelict land. The rules for the payment of grant provide for the following definition:

> "Land so damaged by industrial or other development that it is incapable of beneficial use without treatment."

Subject to certain exclusions, and the availability of funds, grant is payable for the acquisition and reclamation of any such land which is likely to remain permanently derelict unless a grant is provided. Land that is unused because of contamination arising from its previous use clearly comes within the definition. Several of the sites mentioned in earlier papers have attracted grant on this basis. The grants are also available for abandoned industrial sites former railway land and docks, disused canals, sewage works and even burial grounds for which a different use is proposed.

5. The grants are not available for the restoration of mineral workings subject to enforceable conditions for after-treatment. Nor are they available for normal clearance and reclamation work which is part of a redevelopment scheme. Also excluded is natural derelicition, such as mudflats and land which can be described as waste land on which no development has ever taken place.

6. During 1978/79 local authority expenditure on derelict land reclamation in England amounted to about £20m of which about £16m would be provided in the form of Government grant. This excluded expenditure on reclamation which was met as part of the costs of development. In the 15 years since 1964 more than 21,000 hectares of derelict land have been reclaimed in England alone.

Acknowledgement

The text is reproduced by permission of the Controller of HMSO.

DISCUSSION

Papers J1-3

Introducing his paper Mr West said all insurance related to the possibility of something going wrong and it is self-evident that the reclamation of contaminated land offers many possibilities for financial loss, due to failure, to owners and the individuals involved. There is no all embracing insurance cover and a programme has to be drawn up for each situation, and this requires expert advice if it is to be cost effective. Replying to questions concerning sulphate attack on foundations, floor heave and the possibility of structural failure due to the use of high-alumina cement, he emphasised the difference between domestic and commercial development-cover being more likely to be obtainable for domestic dwellings. The ultimate liability for the high alumina cement problems (manufacturer/professional advisers/contractors) is still being pursued through the courts. The costs are very high and there is every incentive to try to pass on liability in such situations. Cover in respect of the structural integrity of a building should be obtainable in the light of satisfactory engineering reports at a particular point in time. But this is not specific to high-alumina cement although its presence is one factor that would have to be taken into account. Heave is probably not covered by most insurance programmes at present but this may change, at least for householders, in the future. The financial implications of such a change are being considered by insurance companies. Cover for subsidence is not usually provided for commercially owned properties - most approaches for cover come when there is some cause for concern, such as the building being on made-up ground but cover should be available after proper investigation.

The paper on government finance was presented by Mr B Fletcher and having outlined the current provisions for government assistance in dealing with derelict land, some of which may be 'contaminated land', he outlined some of the suggestions which have been made for introducing more flexibility into grant provisions, although no additional financial allocation will be made. The proposals are at an early drafting stage but possible changes include: greater flexibility in the proportion of costs which can be grant aided (at present only two rates 50 and 100%), land would no longer have to be in local authority ownership, inclusion of schemes to deal with subsidence arising from underground workings (other than that from coal mining for which the National Coal Board is responsible), survey costs would be payable even if no scheme is eventually carried out (at present abortive fees are not covered); development infrastructure and exceptional construction costs due to difficult or contaminated ground to be covered. One questioner referred to the paper by Miss Rosborough on the coal industry which showed the amount of dereliction and loss of good agricultural land which will result in the next 20 years if the NCB carries through its plans for the development of new coalfields, in NE Leicestershire, and in the longer-term possibly in Oxfordshire and Warwickshire. He asked if it would not be better to provide government money to avoid dereliction, e.g. by subsidising more expensive disposal methods, than in cleaning up subsequent dereliction. Another speaker pointed out that it is wrong to talk about derelict land-grants as Central Government Finance as though this came from a bottomless pit - it is taxpayers money being distributed for use for one purpose rather than another and the amount which can be provided is obviously limited. Mr Fletcher said that the costs of avoiding dereliction from coal mining could be tremendous and there is the danger of creating a precedent for government subsidy that other industries would seek to follow.

Replying to another question he confirmed that, although consultations are still in progress, it is the intention that any provision for remedying subsidence due to underground mining will apply to all such workings apart from coal mining operations.

Mr Ray made the contribution below which in particular draws attention to the application of Development Land Tax. As indicated in the contribution there has been some contention about the valuation of sites, particularly where residential development has been planned. Some purchasers would probably question some of the statements made in the text concerning how diligent and careful British Gas has been when making sales. The implications of possible contamination has not always been appreciated by either vendor or purchaser. You cannot make an allowance in valuation for a potential problem which no one recognises at the time. Past precedent is not therefore always helpful. I have passed a copy of the contribution to the rapporteur for the session on gas works to seek comments from the speakers in that Session as I think some of the comments by Mr Ray are more appropriate to that session.

Mr Ray, an estates surveyor for British Gas expressed the following personal views. I have listened to Mr Goldsmith's Paper with great interest, dealing as it does with both land suitable for land-fill tipping and its restoration when filled. Such farm land can, of course, be extremely valuable, the owner being able to charge a royalty for tipping rights, particularly to Local Authorities granting a license for waste disposal. One aspect not covered by Mr Goldsmith is the incidence of development land tax. Where planning consent has been granted for tipping, since the Act became law, liability for what is "material development" commences at the start of work, therefore the tax has to be paid for the whole site in advance of completion of restoration. The effect must be to increase the cost of tipping by the extra money that has to be raised by the owner. Restoration, if a separate operation, will also be potentially liable for tax, thus inhibiting the economics of most desirable schemes, as it does many others.

I would like to supplement Mr Goldsmith's Paper by looking for the effect of chemical pollution of site redevelopment of which I have some experience. Gas works sites have been re-used for the complete range of alternative developments since disposals started about 1952. The most natural re-use is as industrial or warehouse estates, many sites have been in town centres and become part of a shopping or commercial scheme or multi-storey car park, small town or village works on the other hand, have been sold for residential, agricultural use of sports grounds.

Even a 30 acre Tar Works at Greenwich was cleared, surfaced with rubble and tarmaced as a lorry and container park and let on lease. The question of ground clearance for redevelopment has, in the main, caused a number of only slight problems and this has been reflected in the price obtained for these lands. However, there has seldom, if ever, been a specific deduction for pollution: the reason for this is twofold, first the Corporation have cleared the sites to ground level, exposed known under-ground tanks and backfilled them, or in the case of gasholder tanks left them exposed, making available all known information; secondly, as has been stated before, chemical pollution is found in patches around previous plant areas and any new foundations should allow for both these hazards.

The problems that have arisen usually related to undiscovered tar wells or tanks where Works were demolished before nationalisation in 1948 and old works employees no longer exist who can recall their presence. These tanks can, however, be pumped out and residuals mixed with flyash and covered over and left in situ.

I have examined the site values received by each of the twelve Regions of the Gas Corporation over recent years and posed the question as to whether any reduction in value had been specifically made for pollution problems. The price paid for land, as my audience may know, and Mr Goldsmith has explained, is made up of several ingredients, mainly market demand, planning use and restrictions and ground conditions. It is not possible to analyse these ingredients into their component parts but allowances in the acreage figure have been made in some cases for abnormal below ground site works, such as the removal of foundations, the need to pile or lay new sewers and, of course, vice versa where existing roads, foundation and services can be re-used. There is, however, no evidence of a deduction specifically for a pollution problem as the area affected is either circumscribed, the cost of works in comparison to the overall construction cost being of little consequence or, the area avoided and perhaps landscaped.

Contention in some rare instances has arisen after sale in the case of residential redevelopment by Local Authorities where Council Members were hoping to provide low density development with open gardens, but I will refer to that later.

The site layout upon redevelopment will require, in any event, trial bore holes on a grid pattern and buildings should be grouped to allow for any uneconomic clearance works (as in any other site being re-used for new development). We have re-used such sites many times for new buildings required by the Corporation.

The word "uneconomic" is, of course, a general term seen in the context of values at the time of redevelopment. Where development is not on an investment basis, I would remind my audience that site values are not directly related to construction costs and if you look back after a few years, the additional costs incurred have often paled into insignificance compared with the increase in the value of the site.

The economics of site values/development costs is most critical on residential sites, if one is so ill-advised as to use polluted land, these decisions are made by the purchaser who needs the best advice available before using such land. The hazards are settlement and concentrated pockets of chemicals. The former can be ascertained by consultant engineers, who can advise on good loading capacity and the type of foundations required. The latter will be known from trial bore holes and analysis and if necessary, removed from site: this should be investigated prior to purchase.

Now here lies the problem caused by the Control of Pollution Act 1974, in that such materials can only be dumped at licensed tips which are now few and far between.

Parliament has therefore made the transport cost prohibitive until such time as each County has provided suitable repositories for such material. This is therefore not a problem of site ownership value, but one of disposal and what could be better than another polluted site used for such purposes and then landscaped over?

I suggest therefore that one of the points to come out of this Conference is that adequate provision should be made by Local Authorities for the receipt of polluted material within reasonable transport range of inner city and other areas re-used for residential and other development.

There are, however, many sites which have been sold by the Gas and other industries, used without any problems for residential development; particularly flats with paved play areas leaving deposits in situ.

It is a matter for the purchaser to decide on future use using all the skills of architects and engineers to achieve the optimum layout and to use suitable materials.

SUMMING UP

D H Sharp
General Secretary, Society of Chemical Industry

Our Conference has been concerned with the Reclamation of contaminated land. Contaminated land was defined for us by Mr M A Smith in an early paper on Monday as:

Land which is polluted with a sufficient quantity of toxic substances to present a threat to the health and safety of users or occupiers of land, or of workers engaged in its development or reclamation. It also includes land contaminated with an "unacceptable" amount of substances which are harmful to animals or plant life, and land from which food and fodder crops may draw excessive quantities of substances harmful to animals or man.

But it was not until Wednesday that we really began to define what we meant by reclamation. Mr Longlands said that by reclamation we really mean effecting the necessary treatment to bring land to an acceptable standard for reuse. As was pointed out by Professor Bradshaw all we really need in the case, for example, of mine tailings, is to cultivate plants, which though rapid evolution; have learned to tolerate high metal content. He gave a fascinating exposition of this at our after dinner session on Tuesday evening. We had more botanical points raised on Wednesday afternoon when considering particularly the rehabilitation of Gas Works sites.

How, whether or not you believe, in the words of the second Commandment that the Lord thy God is a jealous God visiting the sins of the fathers upon the children unto the third and fourth generation, as many speakers have said, we are dealing with the legacy of the past. For example, in a praiseworthy attempt to deal with industrial pollution the drainage of Trade Premises Act of 1937, as pointed out by Mr Kenny, required local authorities to accept trade effluent, including of course, plating waste. It was this that caused the relatively high metal content of sewage sludge which has made its use suspect or difficult in a number of applications where it would otherwise be extremely useful. Professor Bradshaw pleaded eloquently on behalf of sewage sludge and indeed, sewage sludge has received a lot of attention at the Conference. We were reassured to learn that much tighter controls will lead to a progressive diminution of metal in sludge but surprised to learn that domestic sewage sludge had quite a high metal content. This is one of the areas, uncovered by the Conference, where further work is needed.

We spent the whole of Tuesday morning considering the problems of metal contamination. A good deal of work has been done on this, by many disciplines: medicine; chemistry; botany; biology: and much more needs to be done. Chemical analysis expresses metal content as parts per million; sometimes subsidiary information is available on solubility in acetic acid or EDTA, which no one has defined but I believe stands for ethylene diamine tetraacetate. Mr Harold Egan, the government Chemist remarked pointedly fron the chair that we need to know the form in which metals are present.* Mr Davis went so far, in his paper on indicator crops, as to say that plant analysis offers a direct method of measuring available metals in soil and that no soil analysis can so far predict this - clearly another area for further work. Mr Farquhar was loudly applauded when he appealed for realism in assessing dangers from metal contamination, pointing out that someone existing entirely on a diet of lettuce in soil heavily contaminated with lead might begin to approach the maximum toleration suggested by the World Health Organisation which already, he suspected, had a safety factor of five or ten built into it. The reply was that the WHO never increases its tolerance limits but reduces them. It is also apparent that very little is known about the effect of microorganisms on metal uptake by plants.

We had a wide ranging discussion on risk by our panel this morning. One or two points struck me. Dr Owen (TUC) made the point that craftsmen frequently face transitory risk on site when presumably good ablution facilities are difficult.

Emphasis was mainly on inforamtion - and a warning from Mr Thomas that we have to take the body of public opinion very seriously.

The suggested metal limits put forward by the Department of Environment as explained by Mr Smith, in which a differentiation is made between small and large gardens, were much criticised, a number of speakers making the point that it was always unwise to put down any figures even as the declared "tentative guidelines" without the risk of their bouncing back as EEC directives. I am reminded of the first minister of the then newly formed Department of the Environment remarking the DOE was a good name to give the new department, as that was where the buck stopped!

Before anything is done on site it must be examined and sampled. A number of speakers stressed the importance of finding out as much as possible from archives and maps. We were shown how much could be gained from successive ordinance surveys and old photographs but it became apparent that records were often incomplete and that it was by no means uncommon, when working on old gas works sites, to unearth equipment and pipework that no one knew was there. Opinion is divided on whether sampling should be directed by visual inspection of vegetation and the like or whether a site should be systematically covered. In two cases where this had been done we were shown slides of three dimensional diagrams that were most effective. There was also some disagreement on whether trenches should be dug or not and at least one speaker doubted some analytical results because of sampling errors due to contamination from other levels. This is an area in which we could well do with an accepted standard or code of practice. I understand that the BSI is already considering what action it might take in the general subject matter of our Conference. Here, I suggest, is one.

*Dr Mills stated that this is very important; whether Pb as S or Ar

Apart from metals, the other hazard we have heard a good deal about is underground anaerobic fermentation leading to methane and H_2S production and, in cases where there was minimal oxygen, underground combustion which is imperfectly understood and where more work is needed.

It was also Mr Longlands who said that when the theorists have done their theorising and the pure scientists have had their say, it is left to the engineers actually to make things work. We have had some fascinating expositions of actual cases of reclamation of contaminated land and, after a little probing, have been able to establish some costs. The preferred method of dealing with gross contamination is physical - to remove a given layer of soil and replace it with virgin top soil. This is expensive - I shall come to actual costs in a minute - but the contaminated soil has to be disposed of somewhere. If by landfill we have only redistributed contaminated land and that is perhaps fair enough - to revivify a vital city site in exchange for the contamination of a less desirable site but Mr Goldsmith and Dr Bell in his after dinner speech, have commented on the pressures on good agricultural land.

Therefore the development of chemical treatment is of much importance.

It is very interesting that in two successive papers on Wednesday morning we heard the virtues of lime extolled. Mr Khan gave a very breezy presentation of cleaning up the ghastly Revenfield tip and showed dramatically how some horrible phenolic tarry waste had been cleaned up with lime. In the very next paper, Professor Bolsing described its use in cleaning up oily waste, including that from wrecked tankers. Lime is cheap and, so far as I know, there is plenty of it. Not so with selective cation exchange resins as described by Dr Heller of Bayer AG. The dramatic improvement in plant growth by the use of such resins was clearly shown in his slides. The materials are not yet commercially available and so the cost of such treatment is a matter of surmise. But, surmising, the cost of treating one hectare to a depth of one metre is given by £2,500 $\frac{pc}{x}$ where p is ppm of metal to be removed, c is the cost of resin in £ per kg, x is mg metal exchanged per gram of resin, or £1,000 $\frac{pc}{x}$ per acre. If p is 200 ppm, x = 2 and c = £5, we arrive at £500,000 per acre. This compares with £240,000 per acre to remove 3 metres of top soil at £80,000 for one metre (paper by Figg, Crunch and Tublin of Harry Stanger Ltd). But this was for the worst parts only, the average being £38 - £40,000 per acre. This compares with £20,000 per acre for clearing a Scottish Steel Works tip (Mr Gilchrist's paper) and £100,000 per acre for treating oil polluted land (Mr Bolsing).

For good measure we can throw in £7m for laying 4 km of surface water sewers at the Beckton Site (Longlands & Townsend) and so we obtain a rough cost comparison:-

		cost/acre	author
Physical treatment	(3m)	£240,000	Harry Stanger
	(1m)	£ 80,000	
average		£38 - £40,000	
Chemical treatment:-			
Lime		£100,000	Bölsing
Ion exchange resin say		£500,000	Heller
Deep sewer pipe $\rightarrow$ £2m per kilometer			

Therefore at present ion exchange treatment is probably an order of magnitude greater in cost than physical methods.

Who pays? Government, apart from development areas, pays 50%.
In 78/79 local government expenditure was £20m;
at £40,000 p acre this is only 500 acres or 200 hectares;
at £20,000 per acre this is 1000 acres.

Yet since 1964 more than 21,000 hectares (800 sq miles) has been cleared in England alone. In 15 years an average of 1,400 hectares per annum has been cleared.

Either £40,000 per acre is not the average or costs have risen or more money has been available. It is reasonable to suppose the easier sites have been tackled first.

As Mr Kenny said: Since the cost of an extensive survey may not be insignificant and since its results may affect the possible development of the site, it should preferably be done before the sale is negotiated. To complain late may earn sympathy but is unlikely to bring a cash refund.

We also had a dreadful precautionary story from Richard Hawkins in his breezy talk this morning. The Bogor Regis case seems to be a case of a never ending gravy train. The moral seems to be:

Be very careful and never commit yourself! Indeed, contrary to expectations, the legal session was one of the liveliest of the conference.

Dean and Goalby on Redevelopment of Gas works sites say that the region of British Gas responsible for the sale of a site would have included a disclaimer clause declining responsibility in the event of a problem, attributable to contamination, arising during or after development work. We are indebted to Mr McCarthy of the London Borough of Hillingdon for two excellent papers. Describing the Willow Tree Lane Joint Development, he commented on the difficulties arising from change in policy and the selling of land for private development and , in his paper on legislation, suggeste strengthening of local authority powers. Certainly there seems to be a desire to pass the buck and perhaps this particular buck should also stop at the D.o.E.

Surely tax should be remitted to bona fide development.

I cannot conclude without referring to the very interesting presentation by the landscape architects and others.

Apart from anything else we were treated to some superb photographs, particularly from Chris Baines. Clearly there is some disagreement between the requirements of macro and micro reclamation. As was well put, however, this particular session was concerned with the WHY rather than the HOW of reclamation.

I now have three short tasks. The first is to say what now? We shall have a meeting of our conference committee shortly to consider future action.

The second is the very pleasant one of thanking the committee, thanking the chairman and above all, thanking the authors, without whom there would have been no conference.

Third is to declare the Conference over.

AUTHOR INDEX